Lecture Notes in Control and Information Sciences

Volume 464

About this Series

This series aims to report new developments in the fields of control and information sciences—quickly, informally and at a high level. The type of material considered for publication includes:

1. Preliminary drafts of monographs and advanced textbooks
2. Lectures on a new field, or presenting a new angle on a classical field
3. Research reports
4. Reports of meetings, provided they are
    (a) of exceptional interest and
    (b) devoted to a specific topic. The timeliness of subject material is very important.

More information about this series at http://www.springer.com/series/642

Sorin Olaru · Alexandra Grancharova
Fernando Lobo Pereira
Editors

Developments in Model-Based Optimization and Control

Distributed Control and Industrial Applications

Springer

Editors
Sorin Olaru
Laboratory of Signals and Systems
CentraleSupélec-CNRS-Université Paris-Sud, Université Paris-Saclay
Gif-sur-Yvette
France

Fernando Lobo Pereira
Faculty of Engineering
Porto University
Porto
Portugal

Alexandra Grancharova
Department of Industrial Automation
University of Chemical Technology and Metallurgy
Sofia
Bulgaria

ISSN 0170-8643 ISSN 1610-7411 (electronic)
Lecture Notes in Control and Information Sciences
ISBN 978-3-319-26685-5 ISBN 978-3-319-26687-9 (eBook)
DOI 10.1007/978-3-319-26687-9

Library of Congress Control Number: 2015955373

Mathematics Subject Classification: 49-XX, 52-XX, 68-XX, 93-XX

Springer Cham Heidelberg New York Dordrecht London

Printed on acid-free paper

Springer International Publishing AG Switzerland is part of Springer Science+Business Media (www.springer.com)

Preface

This edited volume emerged from the two workshops dedicated to *Optimisation-based Control and Estimation* held in France at CentraleSupélec in November 2013 and November 2014, with participation of academic partners from Bulgaria, France, Italy, Norway, Portugal, Romania, Spain, and Slovakia. The aim of these workshops was to bring together specialists in control theory, applied mathematics, and from selected application domains, notably bio-reactors/industrial bioprocesses, robotic vehicle systems, and power systems, to discuss topics related to the design of advanced model-based strategies relying on optimization for identification, estimation, and control. The research teams invited for these events have been involved in several collaborative projects from whose results most of the contributions presented in this volume were extracted. The authors have been given the freedom to improve and further complement the results presented at the workshop, in order to enrich the book and highlight the relevance of the optimization-based control and estimation. The submitted chapters underwent a two-stage evaluation process involving detailed reviews and updates of the contributions which converged to the collection of chapters composing the present volume.

The support of CAMPUS France (the French national agency for the promotion of higher education, international student services, and international mobility) via bilateral projects is acknowledged here together with the partner institutions in:

- University of Porto—Pessoa project "Advanced control of a fleet of heterogeneous autonomous vehicles" coordinated in Portugal by Prof. Fernando Lobo Pereira
- Bulgarian Academy of Sciences—RILA project "Robust Distributed Model Predictive Control of Medium- and Large-Scale Systems" coordinated in Bulgaria by Assoc. Prof. Alexandra Grancharova
- Norwegian University of Science and Technology, Trondheim—Aurora project "Connections between constrained control design and the theory of positive dynamical systems" coordinated in Norway by Prof. Morten Hovd

- University of Udine—Galileo project "Set theoretic analysis of switched and time delay systems with application to fault tolerant control systems" coordinated in Italy by Prof. Stefano Miani
- University of Craiova—Brâncusi project "Predictive and adaptive control of bioprocesses (modeling, identification and control of interconnected bio-processes)" coordinated in Romania by Prof. Dan Selisteanu
- Slovak University of Technology in Bratislava—Stefanik project "Complexity, Sensitivity and Robustness of explicit predictive control laws" coordinated in Slovakia by Assoc. Prof. Michal Kvasnica
- GEPEA Laboratory Saint-Nazaire, France—with Assoc. Prof. Mariana Titica as principal investigator in the Brâncusi project
- University of Galati, Romania—with Prof. Sergiu Caraman as principal investigator in the Brâncusi project
- Polytechnic University of Catalonia—with Assoc. Prof. Carlos Ocampo Martinez as main collaborator in Spain

It is a great pleasure to thank all of the participants of the workshop for their contributions that have made these events a success. We would like to express our gratitude to our colleagues Cristina Stoica Maniu, Sihem Tebbani, and Pedro Rodriguez Ayerbe, who coordinated the research projects in France and participated skillfully and enthusiastically in the organization of the Workshops. With respect to the local arrangements, the financial support of *Direction de Recherche et Relations Internationales* of CentraleSupélec was an important asset and it is acknowledged here.

Gif-sur-Yvette
Sofia
Porto
September 2015

Sorin Olaru
Alexandra Grancharova
Fernando Lobo Pereira

Contents

Contributors

António Pedro Aguiar LSTS—Laboratory of Underwater Systems and Technologies, SYSTEC—Research Center for Systems and Technologies, ISR—Institute for Systems and Robotics, Faculty of Engineering, Porto University (FEUP), Porto, Portugal

Marian Barbu "Dunărea de Jos" University of Galati, Galaţi, Romania

Julián Barreiro-Gomez Automatic Control Department, Institut de Robòtica i Informàtica Industrial (CSIC-UPC), Universitat Politècnica de Catalunya, Barcelona, Spain; Departamento de Ingeniería Eléctrica y Electrónica, Universidad de Los Andes, Bogotá, Colombia

Gildas Besançon Grenoble-INP, Gipsa-Lab, Grenoble, France

Franco Blanchini DIMI, University of Udine, Udine, Italy

João Borges de Sousa LSTS—Laboratory of Underwater Systems and Technologies, Faculty of Engineering, Porto University (FEUP), Porto, Portugal

Sergiu Caraman "Dunărea de Jos" University of Galati, Galaţi, Romania

Daniele Casagrande DIEGM, University of Udine, Udine, Italy

Fernando A.C.C. Fontes SYSTEC—Research Center for Systems and Technologies, Institute for Systems and Robotic (ISR), Faculty of Engineering, Porto University (FEUP), Porto, Portugal

Giulia Giordano DIMI, University of Udine, Udine, Italy

Alexandra Grancharova Department of Industrial Automation, University of Chemical Technology and Metallurgy, Sofia, Bulgaria

Martin Gulan Institute of Automation, Measurement and Applied Informatics, Faculty of Mechanical Engineering, Slovak University of Technology, Bratislava, Slovakia

Morten Hovd Department of Engineering Cybernetics, Norwegian University of Science and Technology, Trondheim, Norway; NTNU, Trondheim, Norway

George Ifrim "Dunărea de Jos" University of Galati, Galaţi, Romania

Fernando Lobo Pereira LSTS—Laboratory of Underwater Systems and Technologies and SYSTEC—Research Center for Systems and Technologies, Institute for Systems and Robotic (ISR), Faculty of Engineering, Porto University (FEUP), Porto, Portugal

Joh Jairo Martínez Grenoble-INP, Gipsa-Lab, Grenoble, France

Stefano Miani DIEGM, University of Udine, Udine, Italy

Ion Necoara Automatic Control and Systems Engineering Department, University Politehnica Bucharest, Bucharest, Romania

Angelia Nedić Industrial and Enterprise Systems Engineering Department, University of Illinois at Urbana-Champaign, Urbana, IL, USA

Minh Tri Nguyen Laboratory of Signals and Systems (L2S, UMR CNRS 8506), CentraleSupélec-CNRS-Université Paris-Sud, Université Paris-Saclay, Gif-sur-Yvette cedex, France

Ngoc Anh Nguyen Laboratory of Signals and Systems (L2S, UMR CNRS 8506), CentraleSupélec-CNRS-Université Paris-Sud, Université Paris-Saclay, Gif-sur-Yvette, France

Silviu-Iulian Niculescu Laboratory of Signals and Systems (L2S, UMR CNRS 8506), CentraleSupélec-CNRS-Université Paris-Sud, Université Paris-Saclay, Gif-sur-Yvette, France

Cristian Oară Department of Automation Control and Systems Engineering, Politehnica University of Bucharest, Bucharest, Romania

Carlos Ocampo-Martinez Automatic Control Department, Institut de Robòtica i Informàtica Industrial (CSIC-UPC), Universitat Politècnica de Catalunya, Barcelona, Spain

Sorin Olaru Laboratory of Signals and Systems (L2S, UMR CNRS 8506), CentraleSupélec-CNRS-Université Paris-Sud, Université Paris-Saclay, Gif-sur-Yvette cedex, France

Andrei Patrascu Automatic Control and Systems Engineering Department, University Politehnica Bucharest, Bucharest, Romania

Ionela Prodan Laboratory of Conception and Integration of Systems (LCIS EA 3747), Université Grenoble Alpes, Valence, France

Nicanor Quijano Departamento de Ingeniería Eléctrica y Electrónica, Universidad de Los Andes, Bogotá, Colombia

Vasso Reppa Laboratory of Signals and Systems (UMR CNRS 8506), CentraleSupélec-CNRS-Univ. Paris-Sud, Université Paris-Saclay, Gif-sur-Yvette, France

Pedro Rodriguez-Ayerbe Laboratory of Signals and Systems (L2S, UMR CNRS 8506), CentraleSupélec-CNRS-Université Paris-Sud, Université Paris-Saclay, Gif-sur-Yvette, France

Boris Rohal'-Ilkiv Institute of Automation, Measurement and Applied Informatics, Faculty of Mechanical Engineering, Slovak University of Technology, Bratislava, Slovakia

Albert Rosich Flanders Mechatronics Technology Centre, Leuven, Belgium

Alessandro Rucco SYSTEC—Research Center for Systems and Technologies, Institute for Systems and Robotic (ISR), Faculty of Engineering of the University of Porto (FEUP), Porto, Portugal

John Sandoval-Moreno Grenoble-INP, G2Elab, Grenoble, France

Dan Selişteanu Department of Automation and Electronics, University of Craiova, Craiova, Romania

Dorin Şendrescu Department of Automation and Electronics, University of Craiova, Craiova, Romania

Cristina Stoica Maniu Laboratory of Signals and Systems (L2S, UMR CNRS 8506), CentraleSupélec-CNRS-Université Paris-Sud, Université Paris-Saclay, Gif-sur-Yvette, France

Florin Stoican Department of Automation Control and Systems Engineering, Politehnica University of Bucharest, Bucharest, Romania

Sihem Tebbani Laboratory of Signals and Systems, CentraleSupélec-CNRS Univ Paris Sud, Université Paris-Saclay, Gif Sur Yvette, France

Mariana Titica GEPEA, CNRS, Université de Nantes, Saint-nazaire Cedex, France

Mohsen Vatani Department of Engineering Cybernetics, Norwegian University of Science and Technology, Trondheim, Norway

Introduction

Motivation

This book concerns optimization methods as tools for decision making and control of dynamic systems in the presence of uncertainties and disturbances. By pooling broad areas of Applied Mathematics and Computation Sciences, this subject has been building a huge impact on an increasing range of diverse application areas which encompass fields such as economics, agriculture, environment and other natural resources management, social sciences, bio-systems, industrial processes, aeronautics, robotic vehicle systems, transportation power systems, electronics, as well as other engineering areas. The present volume targets the enhancement of the specific use of these tools in engineering and, more specifically, in automatic control design, with emphasis on its key components: analysis of dynamical systems, estimation, and feedback control design.

The Book Flavor and Main Contributions

This book contains a selected set of recent contributions ranging from novel formulations of model predictive control design, to the set-theoretic characterization of dynamical systems and including the recent decentralized and cooperative formulations of control laws. Together with these contributions we find eight chapters dedicated to optimization-based tools for robustness analysis, to decision making in relationship with feedback mechanisms (including fault detection and recovery, as a notable example), and to applications to biotechnology or multi-agent systems or impulsive dynamical systems.

Book Organization

This book mirrors the spiralling bidirectional interplay between conceptual optimization-based control frameworks and challenging requirements emerging from state-of-the-art applications that so fruitfully has been shaping the fast paced research in this wide area. Given the vast Research & Development network and the program of the workshops which paved the way for the present volume, we felt natural to organize the book into five parts, each reflecting a current research trend:

1. *Complexity and Structural Properties of Linear Model Predictive Control.* Structural technical issues targeting novel formulations fulfilling more sophisticated real-time specifications and performance requirements related to the predictive control design for linear systems. In Chap. 1 I. Necoara, A. Patrascu, and A. Nedić discuss the computational certifications for convex programming with a direct impact on real-time MPC. In Chaps. 2 and 3, the inverse optimality of continuous piecewise affine functions is investigated by N.A. Nguyen, M. Gulan, S. Olaru, P. Rodriguez-Ayerbe, M. Hovd, I. Necoara, and B. Rohal-Ilkiv with a straightforward interpretation of the explicit linear model predictive control design and implementation via equivalent real-time optimization schemes.
2. *Distributed-coordinated and Multi-objective Features of Model Predictive Control.* Novel formulations which highlight the interconnections of control and data sharing among subsystems are presented in this part of the volume with the ultimate goal of enabling the optimization-based coordination of decentralized systems. Chapter 4 is dedicated to distributed predictive control of interconnected polytopic systems where A. Grancharova and S. Olaru extend the recent results on this topic in order to achieve robustness. Chapter 5 by J. Sandoval Moreno, J.J. Martinez, and G. Besancon discuss a distributed-coordinated optimal control base on a price-driven approach complying with requirements arising in power generation systems. Chapter 6 is dedicated to dynamical tuning of multi-objective control where J. Barreiro-Gomez, C. Ocampo-Martinez, and N. Quijano point out the advantages of an evolutionary game-based selection of prioritization weights of predictive control objective function (and the positioning on the Pareto front of solutions).
3. *Collaborative Model Predictive Control.* This part of the book consists of a set of three chapters discussing issues arising in cooperative path following for, a possibly reconfigurable, formation of robotic vehicles in an optimization-based control design perspective. In Chap. 7 by A. Rucco, A. Pedro Aguiar, F.A.C.C. Fontes, F. Lobo Pereira, and J. Borges de Sousa, a model predictive control-based architecture that enables the decentralized cooperative path-following of multiple Unmanned Aerial Vehicles is presented. The stability of the generated trajectories with a prescribed rate of convergence while satisfying both state and control constraints are shown. I. Prodan, S. Olaru, C. Stoica Maniu, F.A.C.C. Fontes, F. Lobo Pereira, and S. Niculescu discuss, in Chap. 8, a computationally efficient predictive control-based framework for path

following. Issues concerning the optimal dynamical task assignment formulation of multi-agent systems, coupled with fault detection and isolation capabilities, are discussed by M.T. Nguyen, C. Stoica Maniu, S. Olaru, and A. Grancharova in Chap. 9 addressing MPC-based formation reconfiguration techniques.

4. *Applications of Optimization-Based Control and Identification.* Selected consolidated cases showing the benefits of customized optimization-based control that encompass applications in bio-processes are presented in this part. While, S. Tebbani, M. Titica, G. Ifrim, M. Barbu, and S. Caraman present results on the optimal operation of a lumostatic microalgae cultivation process in Chap. 10, in the ensuing chapter, heuristic optimization techniques are discussed by D. Sendrescu, S. Tebbani, and D. Selisteanu to estimate bioprocess parameters. An implementation of a real-time predictive control for both supervisor and regulatory levels of a pasteurization plant is provided by A. Rosich, and C. Ocampo-Martinez in Chap. 12.
5. *Optimization-Based Analysis and Design for Particular Classes of Dynamical Systems.* This part concerns contributions that show how optimization-based control techniques are amenable to take advantage of particular features exhibited by specific classes of dynamic models. This last part of the volume collects five chapters promoting pioneering inroads in dynamic optimization. The reader can find optimization-based control design formulation for impulsive control systems by F. Lobo Pereira, F.A.C.C. Fontes, A. Pedro Aguiar, and J. Borges de Sousa in which invariance and stability are the key issues in Chap. 13, and bilinear systems by M. Vatani, M. Hovd, and S.Olaru in which linear parameter varying control design is discussed in Chap. 14. A class of linear parameter varying systems by F. Blanchini, D. Casagrande, G. Giordano, and S. Miani in which the importance of the parameterization of the stabilizing control laws is underlined in Chap. 15, and linear systems with state-dependent bounds on disturbances by S. Olaru and V. Reppa in which robust invariance sets and ultimate bounds are derived in a set-theoretic context are discussed in Chap. 16. Finally, Chap. 17 concerns the minimal invariant set characterization by F. Stoican, C. Oara, and M. Hovd where the zonotopic disturbances are interpreted in dual terms of optimal control sequences.

It is clear from the flow of ideas throughout the various chapters that cases of the virtuous cycle of invention-discovery are demonstrated in this book. In the research areas pertinent to model-based optimization and control, it is illustrated how the persistent interaction between the dynamic "real-world challenges"-driven conceptual research (leading to new discoveries), and the consistent field trial of advanced scientific tools (leading to the invention of new scientific tools) play a fruitful role in scientific and technological progress with a huge potential societal impact.

Intended Audience

The book is intended to offer postgraduate students and researchers a perspective on new control problems involving optimization-based methods. It presents in a comprehensive manner the structural properties of the manipulated techniques, unveils the challenges, positions the research trends, and points to application from emerging fields. In this respect, it can be useful for both academic and industrial researchers.

Part I
Complexity and Structural Properties of Linear Model Predictive Control

Chapter 1
Complexity Certifications of First-Order Inexact Lagrangian Methods for General Convex Programming: Application to Real-Time MPC

Ion Necoara, Andrei Patrascu and Angelia Nedić

Abstract In this chapter, we derive the computational complexity certifications of first-order inexact dual methods for solving general smooth constrained convex problems which can arise in real-time applications, such as model predictive control. When it is difficult to project on the primal constraint set described by a collection of general convex functions, we use the Lagrangian relaxation to handle the complicated constraints and then, we apply dual (fast) gradient algorithms based on inexact dual gradient information for solving the corresponding dual problem. The iteration complexity analysis is based on two types of approximate primal solutions: the primal last iterate and an average of primal iterates. We provide sublinear computational complexity estimates on the primal suboptimality and constraint (feasibility) violation of the generated approximate primal solutions. In the final part of the chapter, we present an open-source quadratic optimization solver, referred to as DuQuad, for convex quadratic programs and for evaluation of its behavior. The solver contains the C-language implementations of the analyzed algorithms.

First author's work has been founded by UEFISCDI, project MoCOBiDS (PN II–RU–TE 2014), no. 176/01.10.2015. Second author's work has been funded by the Sectoral Operational Programme Human Resources Development 2007–2013 of the Ministry of European Funds through the Financial Agreement POSDRU/159/1.5/S/134398. Third author's work has been funded by the Office of Naval Research under grant no. N00014-12-1-0998.

I. Necoara (✉) · A. Patrascu
Automatic Control and Systems Engineering Department, University Politehnica Bucharest, Bucharest, Romania
e-mail: ion.necoara@acse.pub.ro

A. Patrascu
e-mail: andrei.patrascu@acse.pub.ro

A. Nedić
Industrial and Enterprise Systems Engineering Department, University of Illinois at Urbana-Champaign, Urbana, IL, USA
e-mail: angelia@illinois.edu

S. Olaru et al. (eds.), *Developments in Model-Based Optimization and Control*, Lecture Notes in Control and Information Sciences 464,
DOI 10.1007/978-3-319-26687-9_1

Keywords Convex problems · Lagrange duality · Inexact dual first-order methods · Iteration complexity · Model predictive control

1.1 Introduction

Nowadays, many engineering applications can be posed as general smooth constrained convex problems. Several important applications that can be modeled in this framework have attracted great attention lately, such as model predictive control for dynamical linear systems and its dual (often referred to as moving horizon estimation) [8, 11, 17, 18, 20], DC optimal power flow problem for power systems [22], and network utility maximization problems [23]. Notably, the recent advances in hardware and numerical optimization made it possible to solve linear model predictive control problems of nontrivial sizes within microseconds even on hardware platforms with limited computational power and memory.

In this chapter, we are particularly interested in real-time linear model predictive control (MPC) problems. For MPC, the corresponding optimal control problem can be recast as a smooth constrained convex optimization problem. There are numerous ways in which this problem can be solved. For example, an interior point method has been proposed in [19] and an active set method was described in [4]. Also, explicit MPC has been proposed in [2], where the optimization problem is solved offline for all possible states. In real-time (or online) applications, these methods can sometimes fail due to their overly complex iterations in the case of interior point and active set methods, or due to the large dimensions of the problem in the case of explicit MPCs. Additionally, when embedded systems are employed, computational complexity need to be kept to a minimum. As a result, second order algorithms (e.g., interior point), which most often require matrix inversions, are usually left out. In such applications, first order algorithms are more suitable [8, 10, 11, 17, 20] especially for instances when computation power and memory is limited. For many optimization problems arising in engineering applications, such as real-time MPCs, the constraints are overly complex, making projections on these sets computationally prohibitive. This is most often the main impediment of applying first-order methods on the primal optimization problem. To circumvent this, the dual approach is considered by forming the dual problem, whereby the complex constraints are moved into the objective function, thus rendering much simpler constraints for the dual variables, often being only the non-negative orthant. Therefore, we consider dual first order methods for solving the dual problem. The computational complexity certification of gradient-based methods for solving the (augmented) Lagrangian dual of a primal convex problem is studied e.g., in [1, 3, 5, 7–10, 16, 17]. However, these papers either threat quadratic problems [17] or linearly constrained smooth convex problems with simple objective function [1, 7], or the approximate primal solution is generated through averaging [8–10, 16]. On the other hand, in practice usually the last primal iterate is employed. There are few attempts to derive the computational complexity of dual gradient based methods using as an approximate primal solution the last iterate of the algorithm for particular cases of convex problems [1, 7, 9]. Moreover, from our practical experience we have observed that usually these methods converge faster in the primal last iterate than in a primal average sequence. These issues motivate our work here.

Contribution. In this chapter, we analyze the computational complexity of dual first-order methods for solving general smooth constrained convex problems. Contrary to most of the results from the literature [1, 7, 9, 16, 17], our approach allows us to use inexact dual gradient information. Another important feature of our approach is that we also provide complexity results for the primal latest iterate, while in much of the previous literature convergence rates in an average of primal iterates are given. This feature is of practical importance since usually the primal last iterate is employed in applications. More precisely, the main contributions in this chapter are:

(i) We derive the computational complexity of the dual gradient method in terms of primal suboptimality and feasibility violation using inexact dual gradients and two types of approximate primal solutions: $\mathcal{O}\left(\frac{1}{\epsilon^2}\log\frac{1}{\epsilon}\right)$ in the primal last iterate and $\mathcal{O}\left(\frac{1}{\epsilon}\log\frac{1}{\epsilon}\right)$ in an average of primal iterates, where ϵ is some desired accuracy.

(ii) We also derive the computational complexity of the dual fast gradient method in terms of primal suboptimality and feasibility violation using inexact dual gradients and two types of approximate primal solutions: $\mathcal{O}\left(\frac{1}{\epsilon}\log\frac{1}{\epsilon}\right)$ in the primal last iterate and $\mathcal{O}\left(\frac{1}{\sqrt{\epsilon}}\log\frac{1}{\epsilon}\right)$ in a primal average sequence.

(iii) Finally, we present an open-source optimization solver, termed DuQuad, consisting of the C-language implementations of the above inexact dual first-order algorithms for solving convex quadratic problems, and we study its numerical behavior.

Content. The chapter is organized as follows. In Sect. 1.2 we formulate our problem of interest and its dual, and we analyze its smoothness property. In Sect. 1.3 we introduce a general inexact dual first-order method, covering the inexact dual gradient and fast gradient algorithms, and we derive computational complexity certificates for these schemes. Finally, in Sect. 1.4 we describe briefly the DuQuad toolbox that implements the above inexact algorithms for solving convex quadratic programs in C-language, while in Sect. 1.5 we provide detailed numerical experiments.

Notation. We consider the space $\mathbb{R}^n$ composed of column vectors. For $\mathbf{x}, \mathbf{y} \in \mathbb{R}^n$, we denote the scalar product by $\langle \mathbf{x}, \mathbf{y} \rangle = \mathbf{x}^T\mathbf{y}$ and the Euclidean norm by $\|\mathbf{x}\| = \sqrt{\mathbf{x}^T\mathbf{x}}$. We denote the nonnegative orthant by $\mathbb{R}^n_+$ and we use $[\mathbf{u}]_+$ for the projection of $\mathbf{u}$ onto $\mathbb{R}^n_+$. The minimal eigenvalue of a symmetric matrix $\mathbf{Q} \in \mathbb{R}^{n\times n}$ is denoted by $\lambda_{\min}(\mathbf{Q})$ and $\|\mathbf{Q}\|_F$ denotes its Frobenius norm.

1.2 Problem Formulation

In this section, we consider the following general constrained convex optimization problem:

$$f^* = \min_{\mathbf{u}\in U} \; f(\mathbf{u}) \qquad \text{s.t.: } \mathbf{g}(\mathbf{u}) \le \mathbf{0}, \tag{1.1}$$

where $U \subseteq \mathbb{R}^n$ is a closed simple convex set (e.g., a box set), $\mathbf{0} \in \mathbb{R}^p$ is a vector of zeros, and the constraint mapping $\mathbf{g}(\cdot)$ is given by $\mathbf{g}(\cdot) = [g_1(\cdot), \dots, g_p(\cdot)]^T$. (The vector inequality $\mathbf{g}(\mathbf{u}) \le \mathbf{0}$ is to be understood coordinate-wise.) The objective function $f(\cdot)$ and the constraint functions $g_1(\cdot), \dots, g_p(\cdot)$ are convex and differentiable over their domains. Many engineering applications can be posed as the general convex problem (1.1). For example for linear model predictive control problem in condensed form [8, 11, 17, 18, 20]: f is convex (quadratic) function, U is box set describing the input constraints and $\mathbf{g}$ is given by convex functions describing the state constraints; for network utility maximization problem [1]: f is log function, $U = \mathbb{R}^n_+$ and $\mathbf{g}$ is linear function describing the link capacities; for DC optimal power flow problem [22]: f is convex function, U is box set and $\mathbf{g}$ describes the DC nodal power balance constraints.

We are interested in deriving computational complexity estimates of dual first order methods for solving the optimization problem (1.1). We make the following assumptions on the objective function and the feasible set of the problem (1.1).

Assumption 1.1 Let $U \subseteq \text{dom } f \cap \left\{\cap_{i=1}^{p} \text{dom } g_i\right\}$, and assume that:

(a) The Slater condition holds for the feasible set of problem (1.1), i.e., there exists $\bar{\mathbf{u}} \in \text{relint}(U)$ such that $\mathbf{g}(\bar{\mathbf{u}}) < \mathbf{0}$.

(b) The function f is strongly convex with constant $\sigma_f > 0$ and has Lipschitz continuous gradients with constant $L_f > 0$, i.e.:

$$\frac{\sigma_f}{2}\|\mathbf{u}-\mathbf{v}\|^2 \le f(\mathbf{u}) - (f(\mathbf{v}) + \langle \nabla f(\mathbf{v}), \mathbf{u}-\mathbf{v}\rangle) \le \frac{L_f}{2}\|\mathbf{u}-\mathbf{v}\|^2 \quad \forall \mathbf{u}, \mathbf{v} \in U.$$

(c) The function $\mathbf{g}$ has bounded Jacobians on the set U, i.e., there exists $c_g > 0$ such that $\|\nabla \mathbf{g}(\mathbf{u})\|_F \le c_g$ for all $\mathbf{u} \in U$.

Moreover, we introduce the following definition:

Definition 1.1 Given $\epsilon > 0$, a primal point $\mathbf{u}_\epsilon \in U$ is called ϵ-optimal if it satisfies:

$$|f(\mathbf{u}_\epsilon) - f^*| \le \epsilon \quad \text{and} \quad \left\|\left[\mathbf{g}(\mathbf{u}_\epsilon)\right]_+\right\| \le \epsilon.$$

Since U is assumed to be a simple set, i.e., the projection on this set is easy (e.g., (rotated) box, ellipsoid, etc.), we denote the associated dual problem of (1.1) as:

$$\max_{\mathbf{x}\geq 0} \ d(\mathbf{x}) \quad \left(= \min_{\mathbf{u}\in U} \mathcal{L}(\mathbf{u}, \mathbf{x})\right), \tag{1.2}$$

where the Lagrangian function is given by:

$$\mathcal{L}(\mathbf{u}, \mathbf{x}) = f(\mathbf{u}) + \langle \mathbf{x}, \mathbf{g}(\mathbf{u})\rangle.$$

We denote the dual optimal set with $X^* = \arg\max_{\mathbf{x}\geq \mathbf{0}} d(\mathbf{x})$. Note that Assumption 1.1(a) guarantees that strong duality holds for (1.1). Moreover, since f is a strongly convex function (see Assumption 1.1(b)), the inner subproblem $\min_{\mathbf{u}\in U} \mathcal{L}(\mathbf{u}, \mathbf{x})$ has the objective function $\mathcal{L}(\cdot, \mathbf{x})$ strongly convex for any fixed $\mathbf{x} \in \mathbb{R}^p_+$. It follows that the optimal solution $\mathbf{u}^*$ of the original problem (1.1) and $\mathbf{u}(\mathbf{x}) = \arg\min_{\mathbf{u}\in U} \mathcal{L}(\mathbf{u}, \mathbf{x})$ are unique and, thus, from Danskin's theorem [14] we get that the dual function d is differentiable on $\mathbb{R}^p_+$ and its gradient is given by:

$$\nabla d(\mathbf{x}) = \mathbf{g}(\mathbf{u}(\mathbf{x})) \qquad \text{for all } \mathbf{x} \in \mathbb{R}^p_+.$$

From Assumption 1.1(c) it follows immediately, using the mean value theorem, that the function $\mathbf{g}$ is Lipschitz continuous with constant c_g, i.e.,

$$\|\mathbf{g}(\mathbf{u}) - \mathbf{g}(\mathbf{v})\| \leq c_g \|\mathbf{u} - \mathbf{v}\| \qquad \forall \mathbf{u}, \mathbf{v} \in U. \tag{1.3}$$

In the forthcoming lemma, Assumption 1.1 (b) and (c) allow us to show that the dual function d has Lipschitz gradient. Our result is a generalization of a result in [14] given there for the case of a linear mapping $\mathbf{g}(\cdot)$ (see also [10] for a different proof):

Lemma 1.1 *Under Assumption 1.1, the dual function $d(\cdot)$ corresponding to general convex problem (1.1) has Lipschitz continuous gradient with constant $L_d = c_g^2/\sigma_f$, i.e.,*

$$\|\nabla d(\mathbf{x}) - \nabla d(\bar{\mathbf{x}})\| \leq \frac{c_g^2}{\sigma_f} \|\mathbf{x} - \bar{\mathbf{x}}\| \qquad \forall \mathbf{x}, \bar{\mathbf{x}} \in \mathbb{R}^p_+. \tag{1.4}$$

Proof Let $\mathbf{x}, \bar{\mathbf{x}} \in \mathbb{R}^p_+$. Then, by using the optimality conditions for $\mathbf{u}(\mathbf{x})$ and $\mathbf{u}(\bar{\mathbf{x}})$, we get:

$$\left\langle \nabla f(\mathbf{u}(\mathbf{x})) + \sum_{i=1}^{p} \mathbf{x}_i \nabla g_i(\mathbf{u}(\mathbf{x})), \mathbf{u}(\bar{\mathbf{x}}) - \mathbf{u}(\mathbf{x}) \right\rangle \geq 0,$$

$$\left\langle \nabla f(\mathbf{u}(\bar{\mathbf{x}})) + \sum_{i=1}^{p} \bar{\mathbf{x}}_i \nabla g_i(\mathbf{u}(\bar{\mathbf{x}})), \mathbf{u}(\mathbf{x}) - \mathbf{u}(\bar{\mathbf{x}}) \right\rangle \geq 0.$$

Adding these two inequalities and using the strong convexity of f, we further obtain

$$
\begin{aligned}
\sigma_f \|\mathbf{u}(\mathbf{x}) - \mathbf{u}(\bar{\mathbf{x}})\|^2 &\le \langle \nabla f(\mathbf{u}(\mathbf{x})) - \nabla f(\mathbf{u}(\bar{\mathbf{x}})), \mathbf{u}(\mathbf{x}) - \mathbf{u}(\bar{\mathbf{x}}) \rangle \\
&\le \left\langle \sum_i \mathbf{x}_i \nabla g_i(\mathbf{u}(\mathbf{x})) - \sum_i \bar{\mathbf{x}}_i \nabla g_i(\mathbf{u}(\bar{\mathbf{x}})), \mathbf{u}(\bar{\mathbf{x}}) - \mathbf{u}(\mathbf{x}) \right\rangle \\
&= \left\langle \sum_{i=1}^p (\mathbf{x}_i - \bar{\mathbf{x}}_i) \nabla g_i(\mathbf{u}(\mathbf{x})) - \sum_{i=1}^p \bar{\mathbf{x}}_i (\nabla g_i(\mathbf{u}(\bar{\mathbf{x}})) - \nabla g_i(\mathbf{u}(\mathbf{x}))), \mathbf{u}(\bar{\mathbf{x}}) - \mathbf{u}(\mathbf{x}) \right\rangle \\
&\le \left\langle \sum_{i=1}^p (\mathbf{x}_i - \bar{\mathbf{x}}_i) \nabla g_i(\mathbf{u}(\mathbf{x})), \mathbf{u}(\bar{\mathbf{x}}) - \mathbf{u}(\mathbf{x}) \right\rangle,
\end{aligned}
$$

where the last inequality follows from the convexity of the function g_i and $\bar{\mathbf{x}}_i \ge 0$ for all i. By the Cauchy–Schwarz inequality, we have

$$
\begin{aligned}
\sigma_f \|\mathbf{u}(\mathbf{x}) - \mathbf{u}(\bar{\mathbf{x}})\|^2 &\le \sum_{i=1}^p |\mathbf{x}_i - \bar{\mathbf{x}}_i| \|\nabla g_i(\mathbf{u}(\mathbf{x}))\| \|\mathbf{u}(\mathbf{x}) - \mathbf{u}(\bar{\mathbf{x}})\| \\
&\le \|\mathbf{x} - \bar{\mathbf{x}}\| \|\nabla \mathbf{g}(\mathbf{u}(\mathbf{x}))\|_F \|\mathbf{u}(\mathbf{x}) - \mathbf{u}(\bar{\mathbf{x}})\| \\
&\le c_g \|\mathbf{x} - \bar{\mathbf{x}}\| \|\mathbf{u}(\mathbf{x}) - \mathbf{u}(\bar{\mathbf{x}})\|,
\end{aligned}
$$

where the second inequality follows by Hölder's inequality and the last inequality follows by the bounded Jacobian assumption for $\mathbf{g}$ (see Assumption 1.1(c)). Thus, we obtain:

$$
\|\mathbf{u}(\mathbf{x}) - \mathbf{u}(\bar{\mathbf{x}})\| \le \frac{c_g}{\sigma_f} \|\mathbf{x} - \bar{\mathbf{x}}\|.
$$

Combining (1.3) with the preceding relation, we obtain that the gradient of the dual function is Lipschitz continuous with constant $L_{\mathrm{d}} = \frac{c_g^2}{\sigma_f}$, i.e.,

$$
\|\nabla d(\mathbf{x}) - \nabla d(\bar{\mathbf{x}})\| = \|\mathbf{g}(\mathbf{u}(\mathbf{x})) - \mathbf{g}(\mathbf{u}(\bar{\mathbf{x}}))\| \le c_g \|\mathbf{u}(\mathbf{x}) - \mathbf{u}(\bar{\mathbf{x}})\| \le \frac{c_g^2}{\sigma_f} \|\mathbf{x} - \bar{\mathbf{x}}\|,
$$

for all $\mathbf{x}, \bar{\mathbf{x}} \in \mathbb{R}^p_+$. □

Note that in the case of a linear mapping $\mathbf{g}$, i.e., $\mathbf{g}(\mathbf{u}) = \mathbf{G}\mathbf{u} + \mathbf{g}$, we have $c_g = \|\mathbf{G}\|_F \ge \|\mathbf{G}\|$. In conclusion, our estimate on the Lipschitz constant of the gradient of the dual function for general convex constraints $L_{\mathrm{d}} = c_g^2/\sigma_f$ can coincide with the one derived in [14] for the linear case $L_{\mathrm{d}} = \|\mathbf{G}\|^2/\sigma_f$ if one takes the linear structure of $\mathbf{g}$ into account in the proof of Lemma 1.1 (specifically, where we used Hölder's inequality). Based on relation (1.4) of Lemma 1.1, the following descent lemma holds with $L_{\mathrm{d}} = c_g^2/\sigma_f$ (see for example [14]):

$$
d(\mathbf{x}) \ge d(\mathbf{y}) + \langle \nabla d(\mathbf{y}), \mathbf{x} - \mathbf{y} \rangle - \frac{L_{\mathrm{d}}}{2} \|\mathbf{x} - \mathbf{y}\|^2 \quad \forall \mathbf{x}, \mathbf{y} \in \mathbb{R}^p_+. \tag{1.5}
$$

Using these preliminary results, in a unified manner, we analyze further the computational complexity of inexact dual first-order methods.

1.3 Inexact Dual First-Order Methods

In this section, we introduce and analyze inexact first-order dual algorithms for solving the general smooth convex problem (1.1). Since the computation of the zero-th and the first-order information of the dual problem (1.2) requires the exact solution of the inner subproblem $\min_{\mathbf{u}\in U} \mathcal{L}(\mathbf{u}, \mathbf{x})$ for some fixed $\mathbf{x} \in \mathbb{R}^p_+$, which generally cannot be computed in practice, in many practical cases, inexact dual information is available by solving the inner subproblem with a certain inner accuracy. We denote with $\tilde{\mathbf{u}}(\mathbf{x})$ the primal point satisfying the δ-optimality relations:

$$\tilde{\mathbf{u}}(\mathbf{x}) \in U, \qquad 0 \le \mathcal{L}(\tilde{\mathbf{u}}(\mathbf{x}), \mathbf{x}) - d(\mathbf{x}) \le \delta \qquad \forall \mathbf{x} \in \mathbb{R}^p_+. \tag{1.6}$$

In relation with (1.6), we introduce the following approximations for the dual function and its gradient:

$$\tilde{d}(\mathbf{x}) = \mathcal{L}(\tilde{\mathbf{u}}(\mathbf{x}), \mathbf{x}) \quad \text{and} \quad \tilde{\nabla} d(\mathbf{x}) = \mathbf{g}(\tilde{\mathbf{u}}(\mathbf{x})).$$

Then, the following bounds for the dual function $d(\mathbf{x})$ can be obtained, in terms of a linear and a quadratic model, which use only approximate information of the dual function and of its gradient (see [10, Lemma 2.5]):

$$0 \le \left(\tilde{d}(\mathbf{y}) + \langle \tilde{\nabla} d(\mathbf{y}), \mathbf{x} - \mathbf{y}\rangle\right) - d(\mathbf{x}) \le L_d \|\mathbf{x} - \mathbf{y}\|^2 + 3\delta \quad \forall \mathbf{x}, \mathbf{y} \in \mathbb{R}^p_+. \tag{1.7}$$

Note that if $\delta = 0$, then we recover the exact descent lemma (1.5). Before we introduce our algorithmic scheme, let us observe that one can efficiently solve approximately the inner subproblem if the constraint functions $g_i(\cdot)$ satisfy certain assumptions, such as either one of the following conditions:

(1) The operator $\mathbf{g}(\cdot)$ is simple, i.e., given $\mathbf{v} \in U$ and $\mathbf{x} \in \mathbb{R}^p_+$, the solution of the following optimization subproblem:

$$\min_{\mathbf{u}\in U} \left\{ \frac{1}{2}\|\mathbf{u} - \mathbf{v}\|^2 + \langle \mathbf{x}, \mathbf{g}(\mathbf{u})\rangle \right\}$$

can be obtained in linear time, i.e., $\mathcal{O}(n)$ operations. An example satisfying this assumption is the linear operator, i.e., $\mathbf{g}(\mathbf{u}) = \mathbf{G}\mathbf{u} + \mathbf{g}$, where $\mathbf{G} \in \mathbb{R}^{p\times n}$, $\mathbf{g} \in \mathbb{R}^p$ and the set U is simple.

(2) Each function $g_i(\cdot)$ has Lipschitz continuous gradients.

In such cases, based on Assumption 1.1(b) (i.e., f has Lipschitz continuous gradient), it follows that we can solve approximately the inner subproblem $\min_{\mathbf{u} \in U} \mathcal{L}(\mathbf{u}, \mathbf{x})$, for any fixed $\mathbf{x} \in \mathbb{R}^p_+$, with Nesterov's optimal method for convex problems with smooth and strongly convex objective function [15, page 143, Accelerated method]. Without loss of generality, we assume that the functions $g_i(\cdot)$ are simple. When $\mathbf{g}(\cdot)$ satisfies the above condition (2), there are minor modifications in the constants related to the convergence rate. Given $\mathbf{x} \in \mathbb{R}^p_+$, the inner approximate optimal point $\tilde{\mathbf{u}}(\mathbf{x})$ satisfying $\mathcal{L}(\tilde{\mathbf{u}}(\mathbf{x}), \mathbf{x}) - d(\mathbf{x}) \leq \delta$ is obtained with Nesterov's optimal method [15] after N_δ projections on the simple set U and evaluations of ∇f, where

$$N_\delta = \left\lfloor \sqrt{\frac{L_f}{\sigma_f}} \log\left(\frac{L_f R_p^2(\mathbf{x})}{2\delta}\right) \right\rfloor \tag{1.8}$$

with $R_p(\mathbf{x}) = \|\mathbf{v}^0 - \mathbf{u}(\mathbf{x})\|$, and $\mathbf{v}^0$ being the initial point of Nesterov's optimal method. When the simple feasible set U is compact with a diameter R_p (such as for example in MPC applications), we can bound $R_p(\mathbf{x})$ uniformly, i.e.,

$$R_p(\mathbf{x}) \leq R_p \qquad \forall \mathbf{x} \in \mathbb{R}^p_+.$$

In the sequel, *we always assume that such a bound exists*, and *we use warm-start when solving the inner subproblem*. Now, we introduce a general algorithmic scheme, called Inexact Dual First-Order Method (IDFOM), and analyze its convergence properties, computational complexity, and numerical performance.

Algorithm 1.1 IDFOM

Given $\mathbf{y}^0 \in \mathbb{R}^p_+$, $\delta > 0$, for $k \geq 0$ compute:

1. Find $\mathbf{u}^k \in U$ such that $\mathcal{L}(\mathbf{u}^k, \mathbf{y}^k) - d(\mathbf{y}^k) \leq \delta$,
2. Update $\mathbf{x}^k = \left[\mathbf{y}^k + \frac{1}{2L_\mathrm{d}} \tilde{\nabla} d(\mathbf{y}^k)\right]_+$,
3. Update $\mathbf{y}^{k+1} = (1 - \theta_k)\, \mathbf{x}^k + \theta_k \left[\mathbf{y}^0 + \frac{1}{2L_\mathrm{d}} \sum_{j=0}^{k} \frac{j+1}{2} \tilde{\nabla} d(\mathbf{y}^j)\right]_+$.

---

where $\mathbf{u}^k = \tilde{\mathbf{u}}(\mathbf{y}^k)$, $\tilde{\nabla} d(\mathbf{y}^k) = \mathbf{g}(\mathbf{u}^k)$ and the selection of the parameter θ_k is discussed next. More precisely, we distinguish two particular well-known schemes of the above framework:

- **IDGM**: by setting $\theta_k = 0$ for all $k \geq 0$, we recover the Inexact Dual Gradient Method since $\mathbf{y}^{k+1} = \mathbf{x}^k$. For this scheme, we define the dual average sequence $\hat{\mathbf{x}}^k = \frac{1}{k+1} \sum_{j=0}^{k} \mathbf{x}^j$. We redefine the dual final point (the dual last iterate $\mathbf{x}^k$ when some stopping criterion is satisfied) as $\mathbf{x}^k = \left[\hat{\mathbf{x}}^k + \frac{1}{2L_\mathrm{d}} \tilde{\nabla} d(\hat{\mathbf{x}}^k)\right]_+$. Thus, all the results concerning $\mathbf{x}^k$ generated by the algorithm **IDGM** will refer to this definition.

- **IDFGM**: by setting $\theta_k = \frac{2}{k+3}$ for all $k \geq 0$, we recover the Inexact Dual Fast Gradient Method. This variant has been analyzed in [3, 10, 14].

Note that both dual sequences are dual feasible, i.e., $\mathbf{x}^k, \mathbf{y}^k \in \mathbb{R}^p_+$ for all $k \geq 0$, and thus the inner subproblem $\min_{\mathbf{u} \in U} \mathcal{L}(\mathbf{u}, \mathbf{y}^k)$ has the objective function strongly convex. Towards estimating the computational complexity of **IDFOM**, we present an unified outer convergence rate for both schemes **IDGM** and **IDFGM** of algorithm **IDFOM** in terms of dual suboptimality. The result has been proved in [3, 10].

Theorem 1.1 [3, 10] *Given $\delta > 0$, let $\{(\mathbf{x}^k, \mathbf{y}^k)\}_{k \geq 0}$ be the dual sequences generated by algorithm **IDFOM**. Under Assumption 1.1, the following relation holds:*

$$f^* - d(\mathbf{x}^k) \leq \frac{L_d R_d^2}{k^{p(\theta)}} + 4k^{p(\theta)-1}\delta \qquad \forall k \geq 1,$$

where $p(\theta) = \begin{cases} 1, & \textit{if } \theta_k = 0 \\ 2, & \textit{if } \theta_k = \frac{2}{k+3} \end{cases}$ *and* $R_d = \min_{\mathbf{x}^* \in X^*} \|\mathbf{y}^0 - \mathbf{x}^*\|.$

Proof Firstly, consider the case $\theta_k = 0$ (which implies $p(\theta) = 1$). Note that the approximate convexity and Lipschitz continuity relations (1.7) lead to:

$$\begin{aligned} d(\mathbf{x}^k) &\geq \tilde{d}(\hat{\mathbf{x}}^k) + \langle \tilde{\nabla} d(\hat{\mathbf{x}}^k), \mathbf{x}^k - \hat{\mathbf{x}}^k \rangle - L_{\mathrm{d}} \|\mathbf{x}^k - \hat{\mathbf{x}}^k\|^2 - 3\delta \\ &\geq \tilde{d}(\hat{\mathbf{x}}^k) + L_{\mathrm{d}} \|\mathbf{x}^k - \hat{\mathbf{x}}^k\|^2 - 3\delta \\ &\overset{(6)}{\geq} d(\hat{\mathbf{x}}^k) - 3\delta, \end{aligned} \tag{1.9}$$

where in the second inequality we have used the optimality conditions of $\mathbf{x}^k = [\hat{\mathbf{x}}^k + \frac{1}{2L_{\mathrm{d}}} \tilde{\nabla} d(\hat{\mathbf{x}}^k)]_+ \in \mathbb{R}^p_+$. On the other hand, using [3, Theorem 2], the following convergence rate for the dual average point $\hat{\mathbf{x}}^k$ can be derived:

$$f^* - d(\hat{\mathbf{x}}^k) \leq \frac{L_{\mathrm{d}} R_d^2}{2k} + \delta \qquad \forall k \geq 1. \tag{1.10}$$

Combining (1.9) and (1.10) we obtain the first case of the theorem. The second case, concerning $\theta_k = \frac{2}{k+3}$, has been shown in [3, 10]. □

Our iteration complexity analysis for algorithm **IDFOM** is based on two types of approximate primal solutions: the primal last iterate sequence $\{\mathbf{v}^k\}_{k \geq 0}$ defined by $\mathbf{v}^k = \tilde{\mathbf{u}}(\mathbf{x}^k)$ or a primal average sequence $\{\hat{\mathbf{u}}^k\}_{k \geq 0}$ of the form:

$$\hat{\mathbf{u}}^k = \begin{cases} \frac{1}{k+1} \sum_{j=0}^{k} \mathbf{u}^j, & \text{if } \mathbf{IDGM} \\ \frac{2}{(k+1)(k+2)} \sum_{j=0}^{k} (j+1)\mathbf{u}^j, & \text{if } \mathbf{IDFGM}. \end{cases} \tag{1.11}$$

Note that for algorithm **IDGM** we have $\mathbf{v}^k = \mathbf{u}^k$, while for algorithm **IDFGM**, $\mathbf{v}^k \neq \mathbf{u}^k$. Without loss of generality, for the simplicity of our results, we assume:

$$\mathbf{y}^0 = 0, \qquad R_d \geq \max\left\{1, \frac{1}{c_g}, \frac{L_f}{c_g}\right\}, \qquad L_{\rm d} \geq 1. \tag{1.12}$$

If any of these conditions do not hold, then all of the results from below are valid with minor variations in the constants.

1.3.1 Computational Complexity of IDFOM in the Primal Last Iterate

In this section, we derive the computational complexity for the two main algorithms within the framework of **IDFOM**, in terms of primal feasibility violation and primal suboptimality for the last primal iterate $\mathbf{v}^k = \tilde{\mathbf{u}}(\mathbf{x}^k)$. To obtain these results, only in this section, we additionally make the following assumption:

Assumption 1.2 The primal set U is compact, i.e., $\max\limits_{\mathbf{u},\mathbf{v}\in U} \|\mathbf{u} - \mathbf{v}\| = R_p < \infty$.

Assumption 1.2 implies that the objective function f is Lipschitz continuous with constant $\bar{L}_f$, where $\bar{L}_f = \max\limits_{\mathbf{u}\in U} \|\nabla f(\mathbf{u})\|$. Now, we are ready to prove the main result of this section, given in the following theorem.

Theorem 1.2 *Let $\epsilon > 0$ be some desired accuracy and $\mathbf{v}^k = \tilde{\mathbf{u}}(\mathbf{x}^k)$ be the primal last iterate generated by algorithm **IDFOM**. Under Assumptions 1.1 and 1.2, by setting:*

$$\delta \leq \frac{L_d R_d^2}{2\alpha^{p(\theta)-1}} \left(\frac{\epsilon}{6L_d R_d^2}\right)^{4-2/p(\theta)}, \tag{1.13}$$

where $\alpha = \max\left\{1, \left(\frac{\bar{L}_f}{c_g R_d}\right)^{2/p(\theta)}\right\}$, the following assertions hold:

(i) The primal iterate $\mathbf{v}^k$ is ϵ-optimal after $\left\lfloor \alpha \left(\frac{6L_d R_d^2}{\epsilon}\right)^{2/p(\theta)} \right\rfloor$ outer iterations.

(ii) Assuming that the primal iterate $\mathbf{v}^k$ is obtained with Nesterov's optimal method [15] applied to the subproblem $\min\limits_{\mathbf{u}\in U} \mathcal{L}(\mathbf{u}, \mathbf{x}^k)$, then $\mathbf{v}^k$ is ϵ-optimal after

$$\left\lfloor \sqrt{\frac{L_f}{\sigma_f}} \left(\frac{6L_d R_d^2}{\epsilon}\right)^{\frac{2}{p(\theta)}} \left[\left(4 - \frac{2}{p(\theta)}\right) \log\left(\frac{6L_d R_d^2}{\epsilon}\right) + \log\left(\frac{L_f R_p^2 \alpha^{p(\theta)-1}}{L_d R_d^2}\right)\right]\right\rfloor$$

total number of projections on the primal simple set U and evaluations of ∇f.

Proof (i) From Assumption (1.1)(b), the Lagrangian $\mathcal{L}(\mathbf{u}, \mathbf{x})$ is σ_f-strongly convex in the variable $\mathbf{u}$ for any fixed $\mathbf{x} \in \mathbb{R}^p_+$, which gives the following inequality [13]:

$$\mathcal{L}(\mathbf{u}, \mathbf{x}) \geq d(\mathbf{x}) + \frac{\sigma_f}{2}\|\mathbf{u}(\mathbf{x}) - \mathbf{u}\|^2 \quad \forall \mathbf{u} \in U, \mathbf{x} \in \mathbb{R}^p_+. \tag{1.14}$$

Moreover, under the strong convexity assumption on f (cf. Assumption (1.1)(b)), the primal problem (1.1) has a unique optimal solution, denoted by $\mathbf{u}^*$. Using the fact that $\langle \mathbf{x}, \mathbf{g}(\mathbf{u}^*)\rangle \leq 0$ for any $\mathbf{x} \geq 0$, we have:

$$\mathcal{L}(\mathbf{u}^*, \mathbf{x}) - d(\mathbf{x}) = f(\mathbf{u}^*) + \langle \mathbf{x}, \mathbf{g}(\mathbf{u}^*)\rangle - d(\mathbf{x}) \leq f^* - d(\mathbf{x}) \quad \forall \mathbf{x} \in \mathbb{R}^p_+. \tag{1.15}$$

Combining (1.15) and (1.14), we obtain the following relation

$$\frac{\sigma_f}{2}\|\mathbf{u}(\mathbf{x}) - \mathbf{u}^*\|^2 \leq f^* - d(\mathbf{x}) \qquad \forall \mathbf{x} \in \mathbb{R}^p_+, \tag{1.16}$$

which provides the distance from $\mathbf{u}(\mathbf{x})$ to the unique optimal solution $\mathbf{u}^*$.

On the other hand, taking $\mathbf{u} = \tilde{\mathbf{u}}(\mathbf{x})$ in (1.14) and using (1.6), we have:

$$\|\mathbf{g}(\tilde{\mathbf{u}}(\mathbf{x})) - \mathbf{g}(\mathbf{u}(\mathbf{x}))\| \leq c_g\|\mathbf{u}(\mathbf{x}) - \tilde{\mathbf{u}}(\mathbf{x})\| \overset{(1.14)}{\leq} \sqrt{2L_{\mathrm{d}}\delta}, \tag{1.17}$$

where we used that $L_{\mathrm{d}} = c_g^2/\sigma_f$. From (1.16) and (1.17), we derive a link between the primal infeasibility violation and dual suboptimality gap. Indeed, using the Lipschitz continuity property of $\mathbf{g}$, we get:

$$\begin{aligned}\|\mathbf{g}(\tilde{\mathbf{u}}(\mathbf{x})) - \mathbf{g}(\mathbf{u}^*)\| &\leq \|\mathbf{g}(\tilde{\mathbf{u}}(\mathbf{x})) - \mathbf{g}(\mathbf{u}(\mathbf{x}))\| + \|\mathbf{g}(\mathbf{u}(\mathbf{x})) - \mathbf{g}(\mathbf{u}^*)\| \\ &\overset{(1.16)\,\&\,(1.17)}{\leq} \sqrt{2L_{\mathrm{d}}\delta} + \sqrt{2L_{\mathrm{d}}(f^* - d(\mathbf{x}))} \qquad \forall \mathbf{x} \in \mathbb{R}^p_+.\end{aligned}$$

Combining the above inequality with the property $\mathbf{g}(\mathbf{u}^*) \leq \mathbf{0}$, and the fact that for any $\mathbf{a} \in \mathbb{R}^p$ and $\mathbf{b} \in \mathbb{R}^p_+$ we have $\|\mathbf{a} + \mathbf{b}\| \geq \|[\mathbf{a}]_+\|$, we obtain:

$$\left\|\left[\mathbf{g}(\tilde{\mathbf{u}}(\mathbf{x}))\right]_+\right\| \leq \sqrt{2L_{\mathrm{d}}\delta} + \sqrt{2L_{\mathrm{d}}(f^* - d(\mathbf{x}))} \qquad \forall \mathbf{x} \in \mathbb{R}^p_+. \tag{1.18}$$

Secondly, we find a link between the primal and dual suboptimality. Indeed, using $\langle \mathbf{x}^*, \mathbf{g}(\mathbf{u}^*)\rangle = 0$, we have for all $\mathbf{x} \in \mathbb{R}^p_+$:

$$f^* = \langle \mathbf{x}^*, \mathbf{g}(\mathbf{u}^*)\rangle + f(\mathbf{u}^*) = \min_{\mathbf{u}\in U}\left\{f(\mathbf{u}) + \langle \mathbf{x}^*, \mathbf{g}(\mathbf{u})\rangle\right\} \leq f(\tilde{\mathbf{u}}(\mathbf{x})) + \langle \mathbf{x}^*, \mathbf{g}(\tilde{\mathbf{u}}(\mathbf{x}))\rangle.$$

Further, using the Cauchy–Schwarz inequality, we derive:

$$f(\tilde{\mathbf{u}}(\mathbf{x})) - f^* \geq -\|\mathbf{x}^*\| \|\mathbf{g}(\mathbf{u}^*) - \mathbf{g}(\tilde{\mathbf{u}}(\mathbf{x}))\| \\ \geq -R_d \left(\sqrt{2L_\mathrm{d}\delta} + \sqrt{2L_\mathrm{d}(f^* - d(\mathbf{x}))}\right) \qquad \forall \mathbf{x} \in \mathbb{R}^p_+. \qquad (1.19)$$

On the other hand, from Assumption 1.2, we obtain:

$$f(\tilde{\mathbf{u}}(\mathbf{x})) - f^* \leq \bar{L}_f \|\tilde{\mathbf{u}}(\mathbf{x}) - \mathbf{u}^*\| \leq \bar{L}_f \left(\|\tilde{\mathbf{u}}(\mathbf{x}) - \mathbf{u}(\mathbf{x})\| + \|\mathbf{u}(\mathbf{x}) - \mathbf{u}^*\|\right) \\ \overset{(1.16)\,\&\,(1.17)}{\leq} \bar{L}_f \left(\sqrt{\frac{2\delta}{\sigma_f}} + \sqrt{\frac{2}{\sigma_f}(f^* - d(\mathbf{x}))}\right). \qquad (1.20)$$

Taking $\mathbf{x} = \mathbf{x}^k$ in relation (1.18) and combining with the dual convergence rate from Theorem 1.1, we obtain a convergence estimate on primal infeasibility:

$$\left\|\left[\mathbf{g}(\mathbf{v}^k)\right]_+\right\| \leq \frac{2L_\mathrm{d}R_d}{k^{p(\theta)/2}} + \left(8L_\mathrm{d}k^{p(\theta)-1}\delta\right)^{1/2} + \left(2L_\mathrm{d}\delta\right)^{1/2}.$$

Letting $\mathbf{x} = \mathbf{x}^k$ in relations (1.19) and (1.20) and combining with the dual convergence rate from Theorem 1.1, we obtain convergence estimates on primal suboptimality:

$$-\frac{2L_\mathrm{d}R_d^2}{k^{p(\theta)/2}} - \left(8L_\mathrm{d}R_d^2k^{p(\theta)-1}\delta\right)^{1/2} - \left(2L_\mathrm{d}R_d^2\delta\right)^{1/2} \leq f(\mathbf{v}^k) - f^* \\ \leq \frac{2\bar{L}_f c_g R_d}{\sigma_f k^{p(\theta)/2}} + \bar{L}_f\left(\frac{8k^{p(\theta)-1}\delta}{\sigma_f}\right)^{1/2} + \bar{L}_f\left(\frac{2\delta}{\sigma_f}\right)^{1/2}.$$

Enforcing $\mathbf{v}^k$ to be primal ϵ-optimal solution in the two preceding primal convergence rate estimates, we obtain the stated result.
(ii) At each outer iteration $k \geq 0$, by combining the bound (1.13) with the inner complexity (1.8), Nesterov's optimal method [15] for computing $\mathbf{v}^k$ requires:

$$\left\lfloor \left(4 - \frac{2}{p(\theta)}\right)\sqrt{\frac{L_f}{\sigma_f}} \log\left(\frac{6L_\mathrm{d}R_d^2}{\epsilon}\right) + \sqrt{\frac{L_f}{\sigma_f}} \log\left(\frac{L_f R_p^2}{L_\mathrm{d}R_d^2}\alpha^{p(\theta)-1}\right)\right\rfloor$$

projections on the set U and evaluations of ∇f. Multiplying with the outer complexity given in part (i), we obtain the result. □

Thus, we obtained computational complexity estimates for primal infeasibility and suboptimality for the last primal iterate $\mathbf{v}^k$ of order $\mathcal{O}(\frac{1}{\epsilon^2}\log\frac{1}{\epsilon})$ for the scheme **IDGM** and of order $\mathcal{O}(\frac{1}{\epsilon}\log\frac{1}{\epsilon})$ for the scheme **IDFGM**. Furthermore, the inner subproblem needs to be solved with the inner accuracy δ of order $\mathcal{O}(\epsilon^2)$ for **IDGM** and of order $\mathcal{O}(\epsilon^3)$ for **IDFGM** in order for the last primal iterate $\mathbf{v}^k = \tilde{\mathbf{u}}(\mathbf{x}^k)$ to be an ϵ-optimal primal solution.

1.3.2 Computational Complexity of IDFOM in Primal Average Iterate

In this section, we analyze the computational complexity of algorithm **IDFOM** in the primal average sequence $\hat{\mathbf{u}}^k$ defined by (1.11). Similar derivations were given in [10]. For completeness, we also briefly review these results. Since the average sequence is different for the two particular algorithms **IDGM** and **IDFGM**, we provide separate results. First, we analyze the particular scheme **IDGM**, i.e., in **IDFOM** we choose $\theta_k = 0$ for all $k \geq 0$. Then, we have the identity $\mathbf{y}^{k+1} = \mathbf{x}^k$ and do not assume anymore the redefinition of the last point $\mathbf{x}^k = [\hat{\mathbf{x}}^k + \frac{1}{2L_\mathrm{d}}\tilde{\nabla} d(\hat{\mathbf{x}}^k)]_+$, i.e., algorithm **IDGM** generates one sequence $\{\mathbf{x}^k_{k\geq 0}\}$ using the classical gradient update.

Theorem 1.3 *Let $\epsilon > 0$ and $\mathbf{u}^k = \tilde{\mathbf{u}}(\mathbf{x}^k)$ be the primal sequence generated by the algorithm **IDGM** (i.e., $\theta_k = 0$ for all $k \geq 0$). Under Assumption 1.1, by setting:*

$$\delta \leq \frac{\epsilon}{3} \tag{1.21}$$

the following assertions hold:

(i) The primal average sequence $\hat{\mathbf{u}}^k$ given in (1.11) is ϵ-optimal after $\left\lfloor \frac{8L_d R_d^2}{\epsilon} \right\rfloor$ outer iterations.

(ii) Assuming that the primal iterate $\mathbf{u}^k = \tilde{\mathbf{u}}(\mathbf{x}^k)$ is obtained by applying Nesterov's optimal method [15] to the subproblem $\min_{\mathbf{u}\in U} \mathcal{L}(\mathbf{u}, \mathbf{x}^k)$, the primal average iterate $\hat{\mathbf{u}}^k$ is ϵ-optimal after:

$$\left\lfloor 8\left(\frac{L_f}{\sigma_f}\right)^{1/2} \frac{L_d R_d^2}{\epsilon} \log\left(\frac{L_f R_p^2}{\epsilon}\right)\right\rfloor$$

total number of projections on the primal simple set U and evaluations of ∇f.

Proof (*i*) Using the definition of $\mathbf{x}^{k+1}$, we have:

$$\mathbf{x}^{j+1} - \mathbf{x}^j = \left[\mathbf{x}^j + \frac{1}{2L_\mathrm{d}}\tilde{\nabla}\, d(\mathbf{x}^j)\right]_+ - \mathbf{x}^j \quad \forall j \geq 0.$$

Summing up the inequalities for $j = 0, \ldots, k$ and dividing by k, implies:

$$\frac{2L_\mathrm{d}}{k+1}(\mathbf{x}^{k+1} - \mathbf{x}^0) = \frac{2L_\mathrm{d}}{k+1}\left(\sum_{j=0}^{k}\left[\mathbf{x}^j + \frac{1}{2L_\mathrm{d}}\tilde{\nabla} d(\mathbf{x}^j)\right]_+ - \mathbf{x}^j\right)$$

$$= \frac{2L_\mathrm{d}}{k+1}\left[\sum_{j=0}^{k}\left[\mathbf{x}^j + \frac{1}{2L_\mathrm{d}}\tilde{\nabla} d(\mathbf{x}^j)\right]_+ - \left(\mathbf{x}^j + \frac{1}{2L_\mathrm{d}}\tilde{\nabla} d(\mathbf{x}^j)\right)\right] + \frac{1}{k+1}\sum_{j=0}^{k}\tilde{\nabla} d(\mathbf{x}^j).$$

Using the fact that $\tilde{\nabla} d(\mathbf{x}^j) = \mathbf{g}(\mathbf{u}^j)$, the convexity of $\mathbf{g}$ and denoting $\mathbf{z}^j = \left[\mathbf{x}^j + \frac{1}{2L_{\rm d}}\tilde{\nabla} d(\mathbf{x}^j)\right]_+ - \left(\mathbf{x}^j + \frac{1}{2L_{\rm d}}\tilde{\nabla} d(\mathbf{x}^j)\right) \in \mathbb{R}^p_+$, we get:

$$\mathbf{g}(\hat{\mathbf{u}}^k) + \frac{2L_{\rm d}}{k+1}\sum_{j=0}^{k} \mathbf{z}^j \leq \frac{2L_{\rm d}}{k+1}(\mathbf{x}^{k+1} - \mathbf{x}^0).$$

satisfies $\mathbf{a} \leq \mathbf{b}$, then $[\mathbf{a}]_+ \leq [\mathbf{b}]_+$ and $\|[\mathbf{a}]_+\| \leq \|[\mathbf{b}]_+\|$. Using these relations and the fact that $\mathbf{z}^j \geq 0$, we obtain the following convergence rate on the feasibility violation:

$$\begin{aligned} \left\|\left[\mathbf{g}(\hat{\mathbf{u}}^k)\right]_+\right\| &\leq \left\|\left[\mathbf{g}(\hat{u}^k) + \frac{2L_{\rm d}}{k+1}\sum_{j=0}^{k}\mathbf{z}^j\right]_+\right\| \leq \frac{2L_{\rm d}}{k+1}\left\|\left[\mathbf{x}^{k+1} - \mathbf{x}^0\right]_+\right\| \\ &\leq \frac{2L_{\rm d}\|\mathbf{x}^{k+1} - \mathbf{x}^0\|}{k+1}. \end{aligned} \tag{1.22}$$

On the other hand, from [10, Theorem 3.1], it can be derived that:

$$\begin{aligned} \|\mathbf{x}^{j+1} - \mathbf{x}\|^2 \leq \|\mathbf{x}^j - \mathbf{x}\|^2 - \frac{1}{L_{\rm d}}\langle\tilde{\nabla} d(\mathbf{x}^j), \mathbf{x} - \mathbf{x}^j\rangle \\ + \frac{1}{L_{\rm d}}\left(d(\mathbf{x}^{j+1}) - \tilde{d}(\mathbf{x}^j) + 3\delta\right), \end{aligned} \tag{1.23}$$

for all $\mathbf{x} \geq 0$ and $j \geq 0$. Using (1.7), i.e., $d(\mathbf{x}) \leq \tilde{d}(\mathbf{x}^j) + \langle\tilde{\nabla} d(\mathbf{x}^j), \mathbf{x} - \mathbf{x}^j\rangle$, taking $\mathbf{x} = \mathbf{x}^*$, using $d(\mathbf{x}^{j+1}) \leq d(\mathbf{x}^*)$ and summing over j from $j = 0$ to k, we obtain:

$$\|\mathbf{x}^{k+1} - \mathbf{x}^*\| \leq \|\mathbf{x}^0 - \mathbf{x}^*\| + \sqrt{\frac{3\delta(k+1)}{L_{\rm d}}}. \tag{1.24}$$

Combining the estimate for feasibility violation (1.22) and (1.24), we finally have:

$$\left\|\left[\mathbf{g}(\hat{\mathbf{u}}^k)\right]_+\right\| \leq \frac{2L_{\rm d}(\|\mathbf{x}^0 - \mathbf{x}^*\| + \|\mathbf{x}^{k+1} - \mathbf{x}^*\|)}{k+1} \leq \frac{4L_{\rm d}\|\mathbf{x}^0 - \mathbf{x}^*\|}{k+1} + 2\sqrt{\frac{3L_{\rm d}\delta}{k+1}}. \tag{1.25}$$

In order to obtain a sublinear estimate on the primal suboptimality, we write:

$$\begin{aligned} f^* &= \min_{\mathbf{u}\in U}\{f(\mathbf{u}) + \langle\mathbf{x}^*, \mathbf{g}(\mathbf{u})\rangle\} \leq f(\hat{\mathbf{u}}^k) + \langle\mathbf{x}^*, \mathbf{g}(\hat{\mathbf{u}}^k)\rangle \leq f(\hat{\mathbf{u}}^k) + \langle\mathbf{x}^*, \left[\mathbf{g}(\hat{\mathbf{u}}^k)\right]_+\rangle \\ &\leq f(\hat{\mathbf{u}}^k) + \|\mathbf{x}^*\| \left\|\left[\mathbf{g}(\hat{\mathbf{u}}^k)\right]_+\right\| \leq f(\hat{\mathbf{u}}^k) + (R_d + \|\mathbf{x}^0\|)\left\|\left[\mathbf{g}(\hat{\mathbf{u}}^k)\right]_+\right\|. \end{aligned} \tag{1.26}$$

On the other hand, taking $\mathbf{x} = \mathbf{0}$ in (1.23) and using the definition of $\tilde{d}(\mathbf{x}^j)$, we obtain:

$$\|\mathbf{x}^{j+1}\|^2 \leq \|\mathbf{x}^j\|^2 + \frac{1}{L_{\mathrm{d}}}\langle \tilde{\nabla} d(\mathbf{x}^j), \mathbf{x}^j \rangle + \frac{1}{L_{\mathrm{d}}}\left(d(\mathbf{x}^{j+1}) - f(\mathbf{u}^j) - \langle \mathbf{x}^j, \tilde{\nabla} d(\mathbf{x}^j)\rangle + 3\delta\right)$$
$$\leq \|\mathbf{x}^j\|^2 + \frac{1}{L_d}\left(f^* - f(\mathbf{u}^j) + 3\delta\right).$$

Using an inductive argument, the convexity of f and the definition of $\hat{\mathbf{u}}^k$, we get:

$$f(\hat{\mathbf{u}}^k) - f^* \leq \frac{L_{\mathrm{d}}\|\mathbf{x}^0\|^2}{k+1} + 3\delta. \tag{1.27}$$

Using the assumption $\mathbf{x}^0 = \mathbf{0}$, from (1.25), (1.26), and (1.27), we get:

$$-\frac{4L_{\mathrm{d}}R_d^2}{k+1} - 2R_d\sqrt{\frac{3L_{\mathrm{d}}\delta}{k+1}} \leq f(\hat{\mathbf{u}}^k) - f^* \leq 3\delta.$$

From assumptions on the constants R_d, L_{d}, and δ (see (1.12) and (1.21)), our first result follows.
(ii) Taking into account the relation (1.21) on δ, the inner number of projections on the simple set U at each outer iteration is given by:

$$\left\lfloor \left(\frac{L_f}{\sigma_f}\right)^{1/2} \log\left(\frac{L_f R_p^2}{\epsilon}\right)\right\rfloor.$$

Multiplying with the outer complexity obtained in (i), we get the second result. □

Further, we study the computational complexity of the second particular algorithm **IDFGM**, i.e., the scheme **IDFOM** with $\theta_k = \frac{2}{k+3}$. Note that in the framework **IDFOM** both sequences $\{\mathbf{x}^k\}_{k\geq 0}$ and $\{\mathbf{y}^k\}_{k\geq 0}$ are dual feasible, i.e., are in $\mathbb{R}^p_+$. Based on [14, Theorem 2] (see also [3, 12]), when $\theta_k = \frac{2}{k+3}$, we have the following inequality which will help us to establish the convergence properties of the particular algorithm **IDFGM**:

$$\frac{(k+1)(k+2)}{4} d(\mathbf{x}^k) + \frac{(k+1)(k+2)(k+3)}{4}\delta \tag{1.28}$$
$$\geq \max_{\mathbf{x}\geq \mathbf{0}}\left(-L_{\mathrm{d}}\|\mathbf{x} - \mathbf{y}^0\|^2 + \sum_{j=0}^{k} \frac{j+1}{2}\left[\tilde{d}(\mathbf{y}^j) + \langle \tilde{\nabla} d(\mathbf{y}^j), \mathbf{x} - \mathbf{y}^j\rangle\right]\right).$$

We now derive complexity estimates for primal infeasibility and suboptimality of the average primal sequence $\{\hat{\mathbf{u}}^k\}_{k\geq 0}$ as defined in (1.11) for algorithm **IDFGM**.

Theorem 1.4 *Let $\epsilon > 0$ and $\mathbf{u}^k = \tilde{\mathbf{u}}(\mathbf{y}^k)$ be the primal sequence generated by algorithm **IDFGM** (i.e., $\theta_k = \frac{2}{k+3}$ for all $k \geq 0$). Under Assumption 1.1, by setting:*

$$\delta \leq \frac{\epsilon^{3/2}}{8L_d^{1/2} R_d}, \tag{1.29}$$

the following assertions hold:

(i) The primal average iterate $\hat{\mathbf{u}}^k$ *given in* (1.11) *is* ϵ*-optimal after* $\left\lfloor \left(\frac{32L_d R_d^2}{\epsilon}\right)^{1/2} \right\rfloor$ *outer iterations.*

(ii) Assuming that the primal iterate $\mathbf{u}^k = \tilde{\mathbf{u}}(\mathbf{y}^k)$ *is obtained by applying Nesterov's optimal method [15] to the subproblem* $\min_{\mathbf{u} \in U} \mathcal{L}(\mathbf{u}, \mathbf{y}^k)$*, the average primal iterate* $\hat{\mathbf{u}}^k$ *is* ϵ*-optimal after:*

$$\left\lfloor \sqrt{\frac{L_f}{\sigma_f}} \left(\frac{32L_d R_d^2}{\epsilon}\right)^{1/2} \log\left(\frac{4L_d^{1/2} L_f R_p^2 R_d}{\epsilon^{3/2}}\right) \right\rfloor$$

total number of projections on the primal simple set U *and evaluations of* ∇f*.*

Proof (i) For primal feasibility estimate, we use (1.28) and the convexity of f and g:

$$\max_{\mathbf{x} \geq \mathbf{0}} \left(-\frac{4L_{\mathrm{d}}}{(k+1)^2} \|\mathbf{x} - \mathbf{y}^0\|^2 + \langle \mathbf{x}, \mathbf{g}(\hat{\mathbf{u}}^k) \rangle \right) \leq d(\mathbf{x}^k) - f(\hat{\mathbf{u}}^k) + (k+3)\delta. \tag{1.30}$$

For the right-hand side term, using the strong duality and $\mathbf{x}^* \geq \mathbf{0}$, we have:

$$\begin{aligned} d(\mathbf{x}^k) - f(\hat{\mathbf{u}}^k) &\leq d(\mathbf{x}^*) - f(\hat{\mathbf{u}}^k) = \min_{\mathbf{u} \in U} \{ f(u) + \langle \mathbf{x}^*, \mathbf{g}(\mathbf{u}) \rangle \} - f(\hat{\mathbf{u}}^k) \\ &\leq \langle \mathbf{x}^*, \mathbf{g}(\hat{\mathbf{u}}^k) \rangle \leq \left\langle \mathbf{x}^*, [\mathbf{g}(\hat{\mathbf{u}}^k)]_+ \right\rangle. \end{aligned} \tag{1.31}$$

By evaluating the left-hand side term in (1.30) at $\mathbf{x} = \frac{(k+1)^2}{8L_{\mathrm{d}}} [\mathbf{g}(\hat{\mathbf{u}}^k)]_+$ and observing that $\langle [\mathbf{g}(\hat{\mathbf{u}}^k)]_+, \mathbf{g}(\hat{\mathbf{u}}^k) - [\mathbf{g}(\hat{\mathbf{u}}^k)]_+ \rangle = 0$, we obtain:

$$\begin{aligned} &\max_{\mathbf{x} \geq \mathbf{0}} \left(-\frac{4L_{\mathrm{d}}}{(k+1)^2} \|\mathbf{x} - \mathbf{y}^0\|^2 + \langle \mathbf{x}, \mathbf{g}(\hat{\mathbf{u}}^k) \rangle \right) \\ &\geq \frac{(k+1)^2}{16L_{\mathrm{d}}} \|[\mathbf{g}(\hat{\mathbf{u}}^k)]_+\|^2 - \frac{4L_{\mathrm{d}}\|\mathbf{y}^0\|^2}{(k+1)^2} + \langle \mathbf{y}^0, [\mathbf{g}(\hat{\mathbf{u}}^k)]_+ \rangle. \end{aligned} \tag{1.32}$$

Combining (1.31) and (1.32) with (1.30), using the Cauchy–Schwarz inequality and notation $\gamma = \|[\mathbf{g}(\hat{\mathbf{u}}^k)]_+\|$, we obtain:

$$\frac{(k+1)^2}{16L_{\mathrm{d}}} \gamma^2 - (k+3)\delta - \|\mathbf{x}^* - \mathbf{y}^0\| \gamma - \frac{4L_{\mathrm{d}}\|\mathbf{y}^0\|^2}{(k+1)^2} \leq 0.$$

Thus, γ must be less than the largest root of the second-order equation, from which, together with the definition of R_{d} we get:

$$\|\left[\mathbf{g}(\hat{\mathbf{u}}^k)\right]_+\| \leq \frac{16L_{\mathrm{d}}R_{\mathrm{d}}}{(k+1)^2} + 4\sqrt{\frac{3L_{\mathrm{d}}\delta}{k+1}}. \tag{1.33}$$

For the left-hand side on primal suboptimality, using $\mathbf{x}^* \geq \mathbf{0}$, we have:

$$\begin{aligned} f^* &= \min_{\mathbf{u}\in U} \{f(\mathbf{u}) + \langle \mathbf{x}^*, \mathbf{g}(\mathbf{u})\rangle\} \leq f(\hat{\mathbf{u}}^k) + \langle \mathbf{x}^*, \mathbf{g}(\hat{\mathbf{u}}^k)\rangle \\ &\leq f(\hat{\mathbf{u}}^k) + \langle \mathbf{x}^*, [\mathbf{g}(\hat{\mathbf{u}}^k)]_+\rangle \leq f(\hat{\mathbf{u}}^k) + R_{\mathrm{d}}\|[\mathbf{g}(\hat{\mathbf{u}}^k)]_+\|. \end{aligned}$$

Using (1.33), we derive an estimate on the left-hand side primal suboptimality:

$$f(\hat{\mathbf{u}}^k) - f^* \leq \frac{16L_{\mathrm{d}}R_d^2}{(k+1)^2} + 4R_d\sqrt{\frac{3L_{\mathrm{d}}\delta}{k+1}}. \tag{1.34}$$

On the other hand, taking $\mathbf{x} = \mathbf{0}$ in (1.30) and recalling that $\mathbf{y}^0 = \mathbf{0}$, we get:

$$\begin{aligned} f(\hat{\mathbf{u}}^k) - d(\mathbf{x}^k) &\leq -\max_{\mathbf{x}\geq\mathbf{0}} \left(-\frac{4L_{\mathrm{d}}}{(k+1)^2}\|\mathbf{x}-\mathbf{y}^0\|^2 + \langle \mathbf{x}, \mathbf{g}(\hat{\mathbf{u}}^k)\rangle\right) + (k+3)\delta \\ &\leq (k+3)\delta. \end{aligned} \tag{1.35}$$

Moreover, taking into account that $d(\mathbf{x}^k) \leq f^*$, from (1.34) and (1.35), we obtain:

$$-\frac{16L_{\mathrm{d}}R_d^2}{(k+1)^2} - 4R_d\sqrt{\frac{3L_{\mathrm{d}}\delta}{k+1}} \leq f(\hat{\mathbf{u}}^k) - f^* \leq (k+3)\delta. \tag{1.36}$$

From the convergence rates (1.33) and (1.36) we obtain our first result.
(ii) Substitution of the bound (1.29) into the inner complexity estimate (1.8) leads to:

$$\left\lfloor \sqrt{\frac{L_f}{\sigma_f}} \log\left(\frac{4L_{\mathrm{d}}L_f R_d R_p^2}{\epsilon}\right)\right\rfloor$$

projections on U and evaluations of ∇f for each outer iteration. Multiplying with the outer complexity estimate obtained in part (i), we get our second result. □

Thus, we obtained computational complexity estimates for primal infeasibility and suboptimality for the average of primal iterates $\hat{\mathbf{u}}^k$ of order $\mathcal{O}(\frac{1}{\epsilon}\log\frac{1}{\epsilon})$ for the scheme **IDGM** and of order $\mathcal{O}(\frac{1}{\sqrt{\epsilon}}\log\frac{1}{\epsilon})$ for the scheme **IDFGM**. Moreover, the inner subproblem needs to be solved with the inner accuracy δ of order $\mathcal{O}(\epsilon)$ for **IDGM** and of order $\mathcal{O}(\epsilon\sqrt{\epsilon})$ for **IDFGM** so that to have the primal average sequence $\hat{\mathbf{u}}^k$ as an ϵ-optimal primal solution. Further, the iteration complexity estimates in the last primal iterate $\mathbf{v}^k$ are inferior to those estimates corresponding to an average of primal iterates $\hat{\mathbf{u}}^k$. However, in practical applications we have observed that algorithm **IDFOM** converges faster in the last primal iterate than in the primal average sequence. Note that this does not mean that our analysis is weak, since we can also construct problems which show the behavior predicted by the theory.

1.4 DuQuad Toolbox

In this section, we present the open-source solver DuQuad (see also [6]) based on C-language implementations of the framework **IDFOM** for solving quadratic programs (QP) that appear in many applications. For example, linear MPC problems are usually formulated as QPs that need to be solved at each time instant for a given state. Thus, in this toolbox we considered convex quadratic programs of the form:

$$\min_{\mathbf{u}\in U} f(\mathbf{u}) \quad \left(:= \frac{1}{2}\mathbf{u}^T \mathbf{Q}\mathbf{u} + \mathbf{q}^T \mathbf{u}\right) \qquad \text{s.t. :} \quad \mathbf{G}\mathbf{u} + \mathbf{g} \leq 0, \tag{1.37}$$

where $\mathbf{Q} \succ 0$, $\mathbf{G} \in \mathbb{R}^{p\times n}$ and $U \subseteq \mathbb{R}^n$ is a simple compact convex set, i.e., a box $U = [\mathbf{lb}\ \mathbf{ub}]$. Note that our formulation allows to incorporate in the QP either linear inequality constraints (arising e.g., in sparse formulation of predictive control and network utility maximization) or linear equality constraints (arising e.g., in condensed formulation of predictive control and DC optimal power flow). In fact the user can define linear constraints of the form: $\bar{\mathbf{lb}} \leq \bar{\mathbf{G}}\mathbf{u} + \bar{\mathbf{g}} \leq \bar{\mathbf{ub}}$ and depending on the values for $\bar{\mathbf{lb}}$ and $\bar{\mathbf{ub}}$ we have linear inequalities or equalities. Note that the objective function of (1.37) has Lipschitz gradient with constant $L_{\text{f}} = \lambda_{\max}(\mathbf{Q})$ and its dual has also Lipschitz gradient with constant $L_{\text{d}} = \frac{\|\mathbf{G}\|^2}{\lambda_{\min}(\mathbf{Q})}$. Based on the scheme **IDFOM**, the main iteration in DuQuad consists of two steps:
Step 1: for a given inner accuracy $\delta > 0$ and a multiplier $\mathbf{x} \in \mathbb{R}^p_+$, we solve approximately the inner problem with accuracy δ to obtain an approximate solution $\tilde{\mathbf{u}}(\mathbf{x})$ instead of the exact solution $\mathbf{u}(\mathbf{x})$, i.e.: $\mathcal{L}(\tilde{\mathbf{u}}(\mathbf{x}), \mathbf{x}) - d(\mathbf{x}) \leq \delta$. In DuQuad, we obtain an approximate solution $\tilde{\mathbf{u}}(\mathbf{x})$ using Nesterov's optimal method [15] and warm-start.
Step 2: Once a δ-solution $\tilde{\mathbf{u}}(\mathbf{x})$ for inner subproblem was found, we update at the outer stage the Lagrange multipliers using the scheme **IDFOM**, i.e., for updating the Lagrange multipliers we use instead of the true value of the dual gradient $\nabla d(\mathbf{x}) = \mathbf{G}\mathbf{u}(\mathbf{x}) + \mathbf{g}$, an approximate value $\tilde{\nabla} d(\mathbf{x}) = \mathbf{G}\tilde{\mathbf{u}}(\mathbf{x}) + \mathbf{g}$.

An overview of the workflow in DuQuad [6] is illustrated in Fig. 1.1. A QP problem is constructed using a Matlab script called *test.m*. Then, the function *duquad.m* is called with the problem data as input and it is regarded as a preprocessing stage for the online optimization. The binary MEX file is called, with the original problem data and the extra information as an input. The *main.c* file of the C-code includes the MEX framework and is able to convert the MATLAB data into C format. Furthermore, the converted data gets bundled into a C "struct" and passed as an input to the algorithm that solves the problem using the two steps as described above.

In DuQuad, a user can choose either algorithm **IDFGM** or algorithm **IDGM** for solving the dual problem. Moreover, the user can also choose the inner accuracy δ for solving the inner problem. In the toolbox, the default values for δ are taken as in Theorems 1.2, 1.3 and 1.4. From these theorems, we conclude that the inner QP has to be solved with higher accuracy in dual fast gradient algorithm **IDFGM** than in dual gradient algorithm **IDGM**. This shows that dual gradient algorithm **IDGM** is

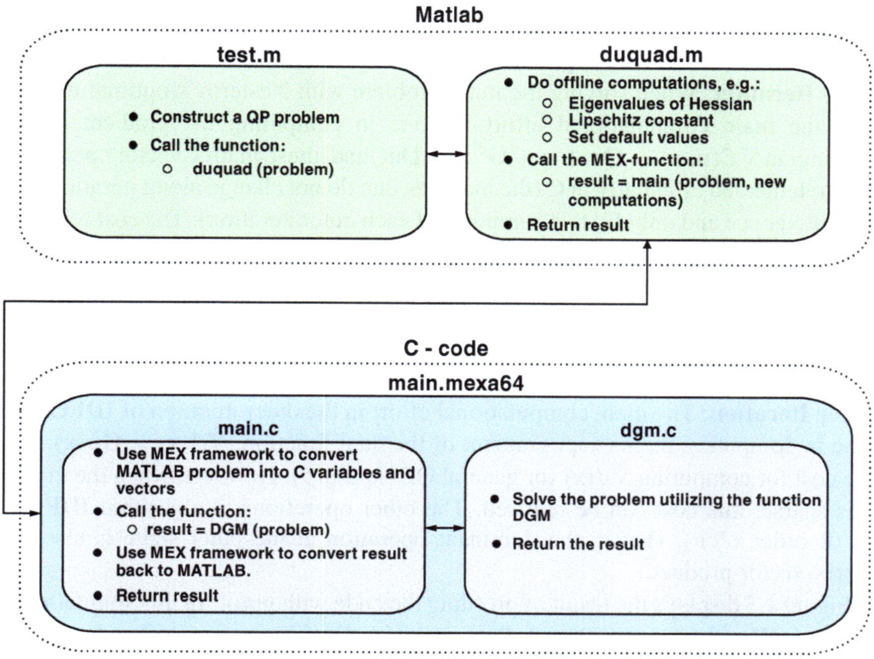

Fig. 1.1 DuQuad workflow

robust to inexact information, while dual fast gradient algorithm **IDFGM** is sensitive to inexact computations, as we can also see from plots in Fig. 1.2.

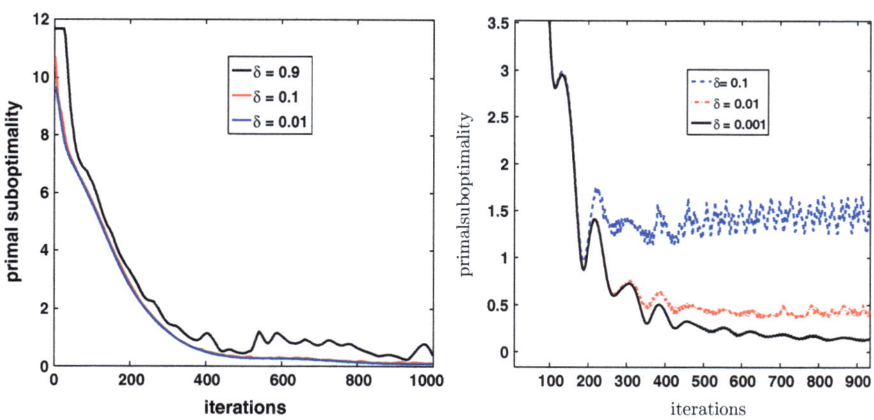

Fig. 1.2 Sensitivity of **IDGM** (*left*) and **IDFGM** (*right*) in the average of iterates in terms of primal suboptimality w.r.t. different values of the inner accuracy δ for a QP ($n = 50$ and $p = 75$) with desired accuracy $\epsilon = 0.01$

Let us analyze now the computational cost per inner and outer iteration for algorithm **IDFOM** for solving approximately the original QP problem (1.37):
Inner Iteration: When solving the inner problem with Nesterov's optimal method [15], the main computational effort is done in computing the gradient of the Lagrangian $\nabla\mathcal{L}(\mathbf{u}, \mathbf{x}) = \mathbf{Q}\mathbf{u} + \mathbf{q} + \mathbf{G}^T\mathbf{x}$. In DuQuad, these matrix-vector operations are implemented efficiently in C (the matrices that do not change along iterations are computed once and only $\mathbf{G}^T\mathbf{x}$ is computed at each outer iteration). The cost for computing $\nabla\mathcal{L}(\mathbf{u}, \mathbf{x})$ for general QPs is $\mathcal{O}(n^2)$. However, when the matrices $\mathbf{Q}$ and $\mathbf{G}$ are sparse (e.g., network utility maximization problem) the cost $\mathcal{O}(n^2)$ can be reduced substantially. The other operations in algorithm **IDFOM** are just vector operations and, hence, they are of order $\mathcal{O}(n)$. Thus, the dominant operation at the inner stage is the matrix-vector product.
Outer Iteration: The main computational effort in the outer iteration of **IDFOM** is done in computing the inexact gradient of the dual function: $\tilde{\nabla} d(\mathbf{x}) = \mathbf{G}\tilde{\mathbf{u}}(\mathbf{x}) + \mathbf{g}$. The cost for computing $\tilde{\nabla} d(\mathbf{x})$ for general QPs is $\mathcal{O}(np)$. However, when the matrix $\mathbf{G}$ is sparse, this cost can be reduced. The other operations in algorithm **IDFOM** are of order $\mathcal{O}(p)$. Hence, the dominant operation at the outer stage is also the matrix-vector product.

Figure 1.3 displays the result of profiling the code with gprof. In this simulation, a standard QP with inequality constraints, and with dimensions $n = 150$ and $p = 225$ was solved by algorithm **IDFGM**. The profiling summary is listed in the order of the time spent in each file. This figure shows that most of the execution time of the program is spent on the library module *math-functions.c*. More exactly, the dominating function is *mtx-vec-mul*, which multiplies a matrix with a vector.

In conclusion, in DuQuad the main operations are the matrix-vector products. Therefore, DuQuad is adequate for solving QP problems on hardware with limited resources and capabilities, since it does not require any solver for linear systems or

Fig. 1.3 Profiling the code with gprof

Name (location)	Sampl	Calls	Time/Call	% Time
▾ Summary	154			100.0%
▸ dfgm.c	0			0.0%
▸ fgm.c	1			0.65%
▸ general_functions.c	0			0.0%
▸ init_problem.c	0			0.0%
▸ main.c	0			0.0%
▾ math_functions.c	153			99.35%
abs_2	0	12983	0ns	0.0%
mtx_transpose	0	1	0ns	0.0%
▸ mtx_vec_mul	142	26590	53.403us	92.21%
obj	0	13234	0ns	0.0%
▸ vector_add	2	26464	755ns	1.3%
▸ vector_copy	1	27464	364ns	0.65%
vector_max	0	12860	0ns	0.0%
vector_max_with_zero	0	372	0ns	0.0%
▸ vector_min	1	12860	777ns	0.65%
▸ vector_mul	2	26718	748ns	1.3%
vector_norm_2	0	124	0ns	0.0%
▸ vector_scalar_mul	3	26215	1.144us	1.95%
▸ vector_sub	2	26960	741ns	1.3%

other complicating operations, while most of the existing solvers for QPs from the literature (such as those implementing active set or interior point methods) require the capability of solving linear systems. On the other hand, DuQuad can be also used for solving large-scale sparse QP problems since, in this case, the iterations are computationally inexpensive (only sparse matrix-vector products).

1.5 Numerical Simulations with DuQuad

For numerical experiments, using the solver DuQuad [6], we at first consider random QP problems and then a real-time MPC controller for a self balancing robot.

1.5.1 Random QPs

In this section we analyze the behavior of the dual first-order methods presented in this chapter and implemented in DuQuad for solving random QPs.

In Fig. 1.4 we plot the practical number of outer iterations on random QPs of algorithms **IDGM** and **IDFGM** for different test cases of the same dimension $n = 50$ (left) and for different test cases of variable dimension ranging from $n = 10$ to $n = 500$ (right). We have chosen the accuracy $\epsilon = 0.01$ and the stopping criteria is the requirement that both quantities

$$|f(\mathbf{u}) - f^*| \quad \text{and} \quad \|[\mathbf{Gu} + \mathbf{g}]_+\|$$

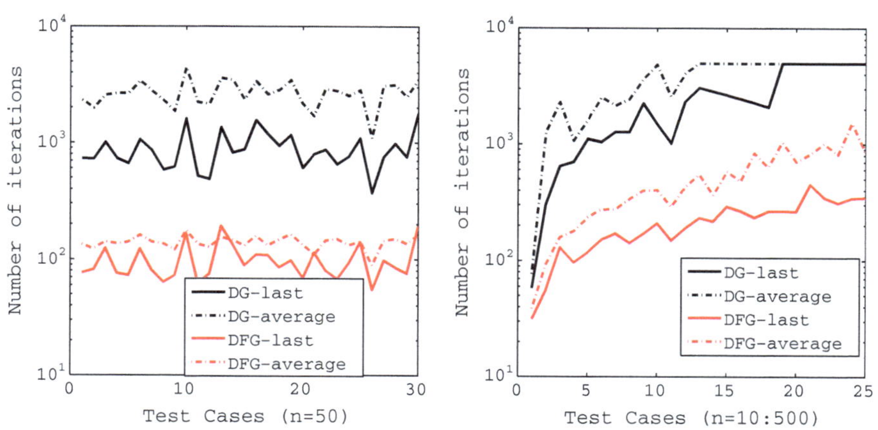

Fig. 1.4 Number of outer iterations on random QPs for **IDGM** and **IDFGM** in primal last/average of iterates for different test cases of the same dimension (*left*) and of variable dimension (*right*)

are less than the accuracy ϵ, where f^* has been computed a priori with Matlab quadprog. From this figure we observe that the number of iterations is not varying much for different test cases and, also, that the number of iterations is mildly dependent on the problem's dimension. Finally, we observe that dual first-order methods perform usually better in the primal last iterate than in the average of primal iterates.

1.5.2 Real-Time MPC for Balancing Robot

In this section we use the dual first-order methods presented in this chapter and implemented in DuQuad for solving a real-time MPC control problem.

We consider a simplified model for the self-balancing Lego mindstorm NXT extracted from [21]. The model is linear time invariant and stabilizable. The continuous linear model has the states $\mathrm{x} \in \mathbb{R}^4$ and inputs $\mathrm{u} \in \mathbb{R}$. The states for this system are the horizontal position and speed $(h, \dot{h})$, and the angle to the vertical and the angular velocity of the robot's body $(\theta, \dot{\theta})$. The input for the system represents the pulse-width modulated voltage applied to both wheel motors in percentages. We discretize the dynamical system via the zero-order hold method for a sample time of $T = 8\text{ms}$ to obtain the system matrices:

$$A = \begin{bmatrix} 1 & 0.0054 & -2 \cdot 10^{-4} & 10^{-4} \\ 0 & 0.4717 & -0.0465 & 0.0211 \\ 0 & 0.03 & 1.0049 & 0.0068 \\ 0 & 6.0742 & 1.0721 & 0.7633 \end{bmatrix}, \quad B = \begin{bmatrix} 0.0002 \\ 0.0448 \\ -0.0025 \\ -0.5147 \end{bmatrix}.$$

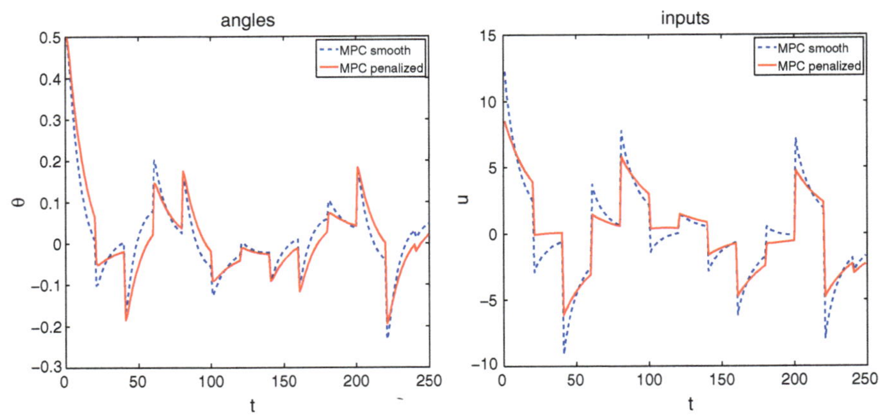

Fig. 1.5 The MPC trajectories of the state angle (*left*) and input (*right*) for $N = 10$ obtained using algorithm **IDGM** from DuQuad in the last iterate with accuracy $\epsilon = 10^{-2}$

For this linear dynamical system, we consider the duty-cycle percentage constraints for the inputs, i.e., $-12 \leq u(t) \leq 12$, and additional constraints for the position, i.e., $-0.5 \leq h \leq 0.5$, and for the body angle in degrees, i.e., $-15 \leq \theta \leq 15$. For the quadratic stage cost, we consider matrices: $\mathrm{Q} = \mathrm{diag}([1\ 1\ 6 \cdot 10^2\ 1])$ and $\mathrm{R} = 2$.

We consider two condensed MPC formulations: *MPC smooth* and *MPC penalized*, where we impose additionally a penalty term $\beta(u(t) - u(t-1))^2$, with $\beta = 0.1$, in order to get a smoother controller. Note that in both formulations, we obtain QPs [18]. Initial state is $\mathrm{x} = [0\ 0\ 0.5\ \ -0.35]^T$ and we add disturbances (of amplitude 10^{-2}) to the system at each 20 simulation steps. In Fig. 1.5, we plot the MPC trajectories of the state angle and input for a prediction horizon $N = 10$ obtained using algorithm **IDGM** in the last iterate with accuracy $\epsilon = 10^{-2}$. Similar state and input trajectories are obtained using the other versions of the scheme **IDFOM** from DuQuad. We observe a smoother behavior for MPC with the penalty term.

References

1. A. Beck, A. Nedić, A. Ozdaglar, M. Teboulle, An O(1/k) gradient method for network resource allocation problems. IEEE Trans. Control Netw. Syst. **1**(1), 64–73 (2014)
2. A. Bemporad, M. Morari, V. Dua, E.N. Pistikopoulus, The explicit linear quadtratic regulator for constrained systems. Automatica **38**(1), 3–20 (2002)
3. O. Devolder, F. Glineur, Y. Nesterov, First-order methods of smooth convex optimization with inexact oracle. Math. Program. **146**(1–2), 37–75 (2014)
4. H. Ferreau, C. Kirches, A. Potschka, H. Bock, M. Diehl, qpOASES: a parametric active-set algorithm for quadratic programming. Math. Program. Comput. **6**(4), 327–363 (2014)
5. J. Koshal, A. Nedić, U.V. Shanbhag, Multiuser optimization: distributed algorithms and error analysis. SIAM J. Optim. **21**(3), 1046–1081 (2011)
6. I. Necoara, S. Kvamme, DuQuad: a toolbox for solving convex quadratic programs using dual (augmented) first order algorithms, Conference on Decision and Control, Osaka, 2015 (DuQuad Users Manual at http://acse.pub.ro/person/ion-necoara)
7. I. Necoara, A. Patrascu, Iteration complexity analysis of dual first order methods for convex conic programming, Technical report, UPB (2014). http://arxiv.org
8. V. Nedelcu, I. Necoara, Q. Tran Dinh, Computational complexity of inexact gradient augmented Lagrangian methods: application to constrained MPC. SIAM J. Control Optim. **52**(5), 3109–3134 (2014)
9. I. Necoara, V. Nedelcu, On linear convergence of a distributed dual gradient algorithm for linearly constrained separable convex problems. Automatica **55**(5), 209–216 (2015)
10. I. Necoara, V. Nedelcu, Rate analysis of inexact dual first order methods: application to dual decomposition. IEEE Trans. Autom. Control **59**(5), 1232–1243 (2014)
11. I. Necoara, L. Ferranti, T. Keviczky, An adaptive constraint tightening approach to linear MPC based on approximation algorithms for optimization. J. Optim. Control: Appl. Methods. 1-19 (2014). doi:10.1002/oca.2121
12. I. Necoara, J. Suykens, Application of a smoothing technique to decomposition in convex optimization. IEEE Trans. Autom. Control **53**(11), 2674–2679 (2008)
13. Y. Nesterov, Introductory Lectures on Convex Optimization, Kluwer (2004)
14. Y. Nesterov, Smooth minimization of non-smooth functions. Math. Program. **103**(1), 127–152 (2005)
15. Y. Nesterov, Gradient methods for minimizing composite functions. Math. Program. **140**(1), 125–161 (2013)

16. A. Nedić, A. Ozdaglar, Approximate primal solutions and rate analysis for dual subgradient methods. SIAM J. Optim. **19**(4), 1757–1780 (2009)
17. P. Patrinos, A. Bemporad, An accelerated dual gradient-projection algorithm for embedded linear model predictive control. IEEE Trans. Autom. Control **59**(1), 18–33 (2014)
18. J. Rawlings, D. Mayne, *Model Predictive Control: Theory and Design* (Nob Hill Publishing, Madison, 2009)
19. C. Rao, S. Wright, J. Rawlings, Application of interior point methods to model predictive control. J. Optim. Theory Appl. **99**, 723–757 (1998)
20. S. Richter, C. Jones, M. Morari, Computational complexity certification for real-time MPC with input constraints based on the fast gradient method. IEEE Trans. Autom. Control **57**(6), 1391–1403 (2012)
21. Y. Yamamoto, NXTway-GS Model-Based Design–Control of self-balancing two-wheeled robot built with LEGO Mindstorms NXT (2008). www.pages.drexel.edu
22. R. Zimmerman, C. Murillo-Sanchez, R. Thomas, Matpower: steady-state operations, planning, and analysis tools for power systems research and education. IEEE Trans. Power Syst. **26**(1), 12–19 (2011)
23. E. Wei, A. Ozdaglar, A. Jadbabaie, A distributed newton method for network utility maximization-Part I and II. IEEE Trans. Autom. Control **58**(9), 2176–2188 (2013)

Chapter 2
Fully Inverse Parametric Linear/Quadratic Programming Problems via Convex Liftings

Ngoc Anh Nguyen, Sorin Olaru, Pedro Rodriguez-Ayerbe, Morten Hovd and Ion Necoara

Abstract In this chapter, we present in an unified manner the latest developments on inverse optimality problem for continuous piecewise affine (PWA) functions. A particular attention is given to convex liftings as a cornerstone for the constructive solution we advocate in this framework. Subsequently, an algorithm based on convex lifting is presented for recovering a continuous PWA function defined over a polyhedral partition of a polyhedron. We also prove that any continuous PWA function can be equivalently obtained by a parametric linear programming problem with at most one auxiliary one-dimensional variable.

Keywords Parametric convex programming problems · Convex liftings · Inverse parametric convex programming · Continuous piecewise affine functions

---

N.A. Nguyen (✉) · S. Olaru · P. Rodriguez-Ayerbe
Laboratory of Signals and Systems, CentraleSupélec-CNRS-Université Paris-Sud, Université Paris-Saclay, 3 Rue Joliot Curie, 91192 Paris, Gif-sur-Yvette, France
e-mail: Ngocanh.Nguyen@supelec.fr

S. Olaru
e-mail: Sorin.Olaru@centralesupelec.fr

P. Rodriguez-Ayerbe
e-mail: Pedro.Rodriguez@centralesupelec.fr

M. Hovd
Department of Engineering Cybernetics, Norwegian University of Science and Technology, 7491 Trondheim, Norway
e-mail: morten.hovd@itk.ntnu.no

I. Necoara
Automatic Control and Systems Engineering Department, Politehnica University of Bucharest, Splaiul Independentei, 313, Bucharest, Romania
e-mail: ion.necoara@acse.pub.ro

S. Olaru et al. (eds.), *Developments in Model-Based Optimization and Control*,
Lecture Notes in Control and Information Sciences 464,
DOI 10.1007/978-3-319-26687-9_2

2.1 Introduction

Piecewise affine (PWA) functions have been studied in Mathematics since many years. They are useful to fit nonlinear functions which are difficultly obtained via an analytic expression. In control theory, PWA functions appeared early in [40] as a new approach for nonlinear control design. Subsequently, this class of functions has been particularly of use to approximate nonlinear systems [1]. This approximation is of help to simplify control design and stability analysis for nonlinear systems. Afterward, many studies exploit PWA functions as a good candidate to approximate optimal solution of constrained optimization-based control, e.g., [7, 13, 19, 20, 35]. Subsequently, this class of functions is proved to represent an optimal solution to a minimization problem subject to linear constraints and a linear/quadratic cost function, leading to a class of *hybrid systems* called *piecewise affine* systems. This class of control laws has received significant attention in control community [8, 33, 39, 43].

However, PWA control law is shown to have two major limitations in terms of implementation, once the state-space partition includes many regions:

- the memory requirement for storing the regions and the associated control law gains, is demanding,
- the point-location problem, determining which region the current state belongs to, becomes more expensive.

Therefore, it is necessary to find other methods of implementing these control laws to overcome the above limitations. Some studies on the complexity reduction of PWA control laws are found in [21, 22, 25, 26]. In general, these studies search for complexity reduction via the simplification of state-space partition by preserving the stability property but by trading for performance degradation. An alternative direction for complexity reduction is generated by inverse optimality problem. This idea is fundamentally different by the fact that the given PWA function will be embedded into the frame of an optimization problem.

Inverse optimality problems aim at finding suitable optimization problems such that their optimal solutions are equivalent to those to the associated given functions. In particular, inverse parametric linear/quadratic programming problem focuses on recovering a continuous PWA function defined over a polyhedral partition. Some recent results are found in [6, 17, 27–29, 32]. Two different approaches are distinguished therein. The first one [17] relies on the decomposition of each component of the given continuous PWA function into the difference of two convex functions. This approach requires $2n_u$ auxiliary variables where n_u represents the dimension of the co-domain space of the given PWA function. The latter one relies on convex liftings which needs only one auxiliary variable. However, this method is restricted to continuous PWA functions defined over polytopic partitions (bounded polyhedral partition). In the same line of the works in [27, 28, 32], the result in this manuscript is also based on convex liftings. However, this chapter extends to continuous PWA functions defined over a polyhedral partition of a polyhedron (possibly unbounded).

2.2 Notation and Definitions

Apart from the common notation of the book, in this chapter we use $\mathbb{N}_{>0}$ to denote the set of positive integers. Also, for ease of presentation, we use $\mathcal{I}_N$ to denote the following index set with respect to a given $N \in \mathbb{N}_{>0}$: $\mathcal{I}_N = \{i \in \mathbb{N}_{>0} \mid i \leq N\}$.

For a given $d \in \mathbb{N}_{>0}$, we use 1_d to denote a vector in $\mathbb{R}^d$ whose elements are equal to 1.

Given two sets $P_1, P_2 \subset \mathbb{R}^d$, their Minkowski sum, denoted as $P_1 \oplus P_2$, is defined as follows:

$$P_1 \oplus P_2 := \left\{y \in \mathbb{R}^d \mid \exists x_1 \in P_1, x_2 \in P_2, \quad \text{s.t. } y = x_1 + x_2\right\}.$$

Given a set $\mathcal{S}$, we write by $\text{int}(\mathcal{S})$, $\text{conv}(\mathcal{S})$ to denote the interior, the convex hull of the set $\mathcal{S}$, respectively. Also, by $\dim(\mathcal{S})$, we denote the dimension of the affine hull of $\mathcal{S}$. With a space $\mathbb{S}$, being a subspace of $\mathbb{R}^d$, we use $\text{Proj}_{\mathbb{S}}\, \mathcal{S}$ to denote the orthogonal projection of $\mathcal{S} \subseteq \mathbb{R}^d$ onto the space $\mathbb{S}$.

A polyhedron is the intersection of finitely many halfspaces. A polytope is a bounded polyhedron. An unbounded polyhedron is known to obtain rays. An *extreme ray* is a ray which cannot be written by a convex combination of any two other rays. Given a full-dimensional polyhedron $\mathcal{S} \subset \mathbb{R}^d$, we write $\mathcal{V}(\mathcal{S})$, $\mathcal{R}(\mathcal{S})$ to denote the sets of vertices and extreme rays, of polyhedron $\mathcal{S}$, respectively. If $\mathcal{S}$ is a full-dimensional polyhedron, then its number of vertices and extreme rays are known to be finite. If $\mathcal{S}$ is a finite set of rays, i.e., $S = \{y_1, \ldots, y_n\}$ then $\text{cone}(\mathcal{S})$ represents the cone defined as follows:

$$\text{cone}(S) = \{t_1 y_1 + \cdots + t_n y_n : \ t_i \geq 0, \quad \forall\, 1 \leq i \leq n\} \tag{2.1}$$

Given two sets $\mathcal{S}_1, \mathcal{S}_2$, we write $\mathcal{S}_1 \backslash \mathcal{S}_2$ to denote the points which belong to $\mathcal{S}_1$ but do not belong to $\mathcal{S}_2$. More precisely, its mathematical description is presented as follows:

$$\mathcal{S}_1 \backslash \mathcal{S}_2 := \{x \mid x \in \mathcal{S}_1, \quad x \notin \mathcal{S}_2\}.$$

For two vectors $x, u \in \mathbb{R}^d$, $\langle x, u\rangle = x^T u$. Given a vector $u \in \mathbb{R}^d$ and a scalar $\alpha \in \mathbb{R}$, a hyperplane, denoted by $\mathcal{H}$, is defined as follows:

$$\mathcal{H} = \left\{x \in \mathbb{R}^d \mid \langle x, u\rangle = \alpha\right\}.$$

Such a hyperplane $\mathcal{H}$ is called a supporting hyperplane of a polyhedron/polytope $\mathcal{S}$ if either $\inf\{\langle x, u\rangle \mid x \in \mathcal{S}\} = \alpha$ or $\sup\{\langle x, u\rangle \mid x \in \mathcal{S}\} = \alpha$. A face of polyhedron/polytope $\mathcal{S}$ is the intersection of this set and one of its supporting hyperplanes. If $\mathcal{S} \subset \mathbb{R}^d$ denotes a full-dimensional polyhedron/polytope, then a face of dimension $k, 0 \leq k \leq d$, is briefly denoted by $k-$face. A $(d-1)-$face is called a facet, an 1-face is called an edge, a $0-$face is called a vertex. If $\mathcal{S}$ denotes a polyhedron/polytope, then by $\mathcal{F}(\mathcal{S})$, we denote the set of its facets.

For ease of presentation, given an $\epsilon \in \mathbb{R}_+$, we use $B_d(\epsilon)$ to denote a full-dimensional box in $\mathbb{R}^d$, i.e., $B_d(\epsilon) = \left\{x \in \mathbb{R}^d \mid \|x\|_\infty \leq \epsilon\right\}$.

Some necessary definitions of help for our development are presented below.

Definition 2.1 A collection of $N \in \mathbb{N}_{>0}$ full-dimensional polyhedra $\mathcal{X}_i \subset \mathbb{R}^d$, denoted by $\{\mathcal{X}_i\}_{i\in\mathcal{I}_N}$, is called a *polyhedral partition of a polyhedron* $\mathcal{X} \subseteq \mathbb{R}^d$ if

1. $\bigcup_{i\in\mathcal{I}_N} \mathcal{X}_i = \mathcal{X}$.
2. $\text{int}(\mathcal{X}_i) \bigcap \text{int}(\mathcal{X}_j) = \emptyset$ with $i \neq j,\ (i, j) \in \mathcal{I}_N^2$,

Also, $(\mathcal{X}_i, \mathcal{X}_j)$ are called neighbors if $(i, j) \in \mathcal{I}_N^2, i \neq j$ and $\dim(\mathcal{X}_i \cap \mathcal{X}_j) = d - 1$.

If $\mathcal{X}_i$ for every $i \in \mathcal{I}_N$ are polytopes, then the partition $\{\mathcal{X}_i\}_{i\in\mathcal{I}_N}$ is alternatively called *polytopic partition*. A polyhedral partition is called *cell complex*, if its facet-to-facet property [41] is fulfilled,[1] namely, any two neighboring regions share a common facet.

Definition 2.2 For a given polyhedral partition $\{\mathcal{X}_i\}_{i\in\mathcal{I}_N}$ of a polyhedron $\mathcal{X} \subseteq \mathbb{R}^d$, a *piecewise affine lifting* is described by function $z : \mathcal{X} \to \mathbb{R}$ with

$$z(x) = a_i^T x + b_i \qquad \text{for any } x \in \mathcal{X}_i, \tag{2.2}$$

and $a_i \in \mathbb{R}^d,\ b_i \in \mathbb{R},\quad \forall i \in \mathcal{I}_N$.

Definition 2.3 Given a polyhedral partition $\{\mathcal{X}_i\}_{i\in\mathcal{I}_N}$ of a polyhedron $\mathcal{X}$, a piecewise affine lifting $z(x) = a_i^T x + b_i\ \forall x \in \mathcal{X}_i$, is called *convex lifting* if the following conditions hold true:

- $z(x)$ is continuous over $\mathcal{X}$,
- for each $i \in \mathcal{I}_N$, $z(x) > a_j^T x + b_j$ for all $x \in \mathcal{X}_i \backslash \mathcal{X}_j$ and all $j \neq i,\ j \in \mathcal{I}_N$.

The *strict inequality* in Definition 2.3 implies the convexity of $z(x)$ and the fact that for any two neighboring regions $(\mathcal{X}_i, \mathcal{X}_j)$, $(a_i, b_i) \neq (a_j, b_j)$. Namely, any two neighboring regions should be lifted onto two distinct hyperplanes. The strict inequality is of help to guarantee the partition between different regions. Note that there always exits a piecewise affine lifting for any polyhedral partition. A trivial example is the one defined as in Definition 2.2 with $a_i = 0, b_i = 0$. However, it is not the case that any polyhedral partition admits a convex lifting. It is observed that a polyhedral partition with respect to the existence of a convex lifting should be a cell complex. This observation is proved via the following lemma.

Lemma 2.1 *If a given polyhedral partition* $\{\mathcal{X}_i\}_{i\in\mathcal{I}_N}$ *of a polyhedron* $\mathcal{X} \subseteq \mathbb{R}^d$ *admits a convex lifting, then it is a cell complex.*

[1] Note that a slightly more involved definition of a cell complex exists in the literature [14, 28]. However, for simplicity, we mention only the property of interest in the present context.

Proof By $z(x) = a_i^T x + b_i$, for $x \in \mathcal{X}_i$, we denote a convex lifting of the given partition $\{\mathcal{X}_i\}_{i \in \mathcal{I}_N}$. Consider a pair of neighboring regions $(\mathcal{X}_i, \mathcal{X}_j)$, $(i, j) \in \mathcal{I}_N^2$. As defined, due to the continuity of $z(x)$ at any point $x \in \mathcal{X}_i \cap \mathcal{X}_j$, the hyperplane,

$$\mathcal{H} := \left\{ x \in \mathbb{R}^d \mid a_i^T x + b_i = a_j^T x + b_j \right\},$$

separates $\mathcal{X}_i, \mathcal{X}_j$ and contains $\mathcal{X}_i \cap \mathcal{X}_j$. Also, by the second property of a convex lifting, the halfspace,

$$C_i := \left\{ x \in \mathbb{R}^d \mid a_i^T x + b_i > a_j^T x + b_j \right\},$$

contains $\mathcal{X}_i \backslash (\mathcal{X}_i \cap \mathcal{X}_j)$.

Suppose the facet-to-facet property of $\mathcal{X}_i, \mathcal{X}_j$ is not fulfilled. Then, there exists a point $x \in \mathcal{H}$, s.t. either $x \in \mathcal{X}_i, x \notin \mathcal{X}_j$, or $x \in \mathcal{X}_j, x \notin \mathcal{X}_i$. Without loss of generality, the former one happens. $x \in \mathcal{H}$ leads to $a_i^T x + b_i = a_j^T x + b_j$. Also, $x \in \mathcal{X}_i, x \notin \mathcal{X}_j$ leads to $a_i^T x + b_i > a_j^T x + b_j$. These last two inclusions are clearly contradictory. The proof is complete. □

Definition 2.4 A cell complex $\{\mathcal{X}_i\}_{i \in \mathcal{I}_N}$ of a polyhedron $\mathcal{X} \subseteq \mathbb{R}^d$ admits an *affinely equivalent polyhedron* if there exists a polyhedron $\widetilde{\mathcal{X}} \subset \mathbb{R}^{d+1}$, such that for each $i \in \mathcal{I}_N$:

1. $\exists F_i \in \mathcal{F}(\widetilde{\mathcal{X}})$ satisfying: $\text{Proj}_{\mathbb{R}^d} F_i = \mathcal{X}_i$,
2. if $\underline{z}(x) = \min\limits_{z} z$ s.t. $\left[x^T \; z\right]^T \in \widetilde{\mathcal{X}}$, then $\left[x^T \; \underline{z}(x)\right]^T \in F_i$ for $x \in \mathcal{X}_i$.

An illustration can be found in Fig. 2.1 thereby the multicolored segments along the horizontal axis represent the given polytopic partition including six regions. The pink polytope above is one of its affinely equivalent polyhedra.

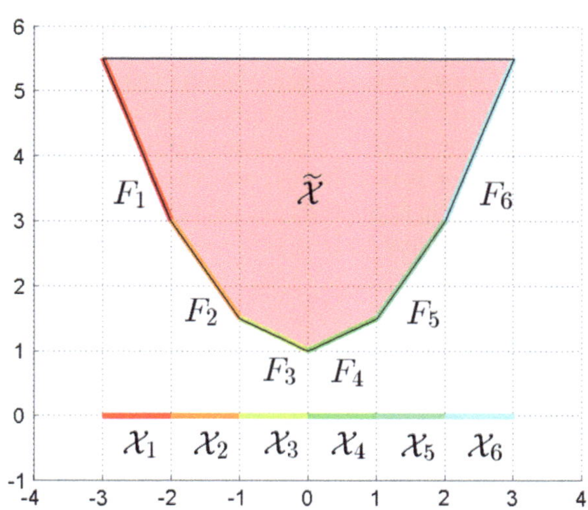

Fig. 2.1 An illustration for affinely equivalent polyhedron

Remark 2.1 Note that the lower boundary of an affinely equivalent polyhedron represents a convex lifting for a given cell complex as shown in Fig. 2.1. Therefore, starting from an affinely equivalent polyhedron $\widetilde{\mathcal{X}}$ of the given cell complex $\{\mathcal{X}_i\}_{i\in\mathcal{I}_N}$, one of its convex liftings is the optimal cost function of the following parametric linear programming problem:

$$\min_{z} z \quad \text{subject to} \quad \left[x^T \ z\right]^T \in \widetilde{\mathcal{X}},$$

where z denotes the last coordinate of $\widetilde{\mathcal{X}}$ and $x \in \mathcal{X}$. Note also that if a given cell complex is convexly liftable, the existence of convex lifting is not unique, meaning that different convex liftings can be defined over a given cell complex. However, for the practical interest in control theory, the existence is the most important property.

An algorithm for construction of convex liftings will be presented in Sect. 2.4.

2.3 Problem Statement

This section formally presents the definition of inverse optimality problem. As earlier mentioned, the main goal is to recover a continuous PWA function through an optimization problem (see also [6]). The solution relies on convex liftings.

Given a polyhedral partition $\{\mathcal{X}_i\}_{i\in\mathcal{I}_N}$ of a polyhedron $\mathcal{X} \subseteq \mathbb{R}^{n_x}$ and a continuous PWA function $f_{pwa}(\cdot) : \mathcal{X} \to \mathbb{R}^{n_u}$, the objective is to find a set of four matrices H_x, H_u, H_z, K, defining linear constraints and a linear/quadratic cost function $J(x, z, u)$ such that $f_{pwa}(x)$ can be equivalently obtained via the optimal solution to the following convex optimization problem:

$$f_{pwa}(x) = \text{Proj}_{\mathbb{R}^{n_u}} \arg \min_{\left[z^T \ u^T\right]^T} J(x, z, u) \ \text{ s.t. } \ H_x x + H_z z + H_u u \leq K. \tag{2.3}$$

It is well known that optimal solution to a parametric linear/quadratic programming problem is a PWA function defined over a polyhedral partition (see [8]). Therefore, we restrict our interest, in this manuscript, to linear constraints. Note also that a given PWA function is usually not convex/concave, the presence of an auxiliary variable z is thus of help to reinforce the convexity of the recovered optimization problem. It will be proved that scalar $z \in \mathbb{R}$ is sufficient for the recovery.

2.4 Algorithm for the Construction of a Convex Lifting

2.4.1 Existing Results on Convex Liftings

As shown in Lemma 2.1, a polyhedral partition, admitting a convex lifting, should be a cell complex. However, not every cell complex is convexly liftable. An illustration can be found in Fig. 2.2. This partition is a cell complex but not convexly liftable. Thus, being a cell complex is a necessary condition for the existence of convex liftings, but not a sufficient condition. Back to the history, we can find the trace of a large number of studies on this topic. Some prominent results for convex liftability of polyhedral partitions in $\mathbb{R}^2$ are found in [9, 10, 24, 38]. Also, some particular diagrams, e.g., Voronoi diagrams, Delaunay diagrams, and Power diagrams in the general dimensional space, are studied in [3, 5, 11, 12, 16]. Necessary and sufficient conditions for a cell complex to be convexly liftable are referred to [2, 4, 23, 28, 30, 36].

Note that these results are equivalent as proved in [36]; therefore, if a cell complex is convexly liftable, then it satisfies all these conditions. Also, due to Lemma 2.1, a polyhedral partition, whose facet-to-facet property does not hold, will not fulfill these conditions.

From the practical interest, these above conditions are only of help for recognizing a convexly liftable cell complex. They do not provide any hint for the construction of such convex liftings. As convex liftings are a tool in our constructive solution of inverse optimality, an algorithm dedicated to convex liftings will be presented in the next subsection.

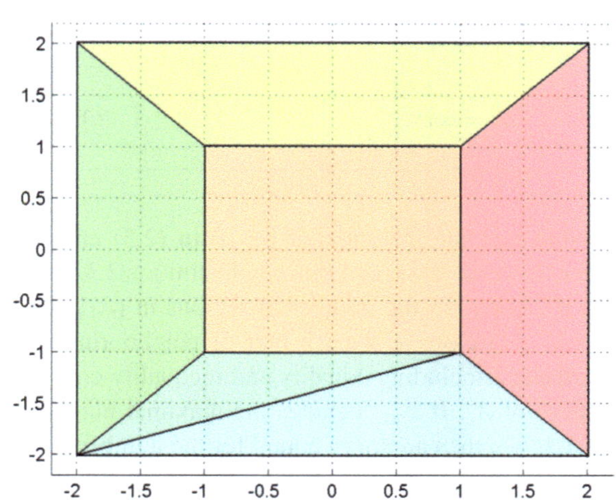

Fig. 2.2 A nonconvexly liftable cell complex

2.4.2 An Algorithm for Construction of Convex Liftings

This subsection concentrates on a construction of convex liftings for a given cell complex. The following algorithm is based on the reinforcement of continuity and convexity constraints at the vertices of the given cell complex. Clearly, the vertex representation of this cell complex is of use. Therefore, this algorithm (Algorithm 2.1) restricts the construction to a bounded cell complex $\{\mathcal{X}_i\}_{i\in\mathcal{I}_N}$, e.g., a cell complex of a polytope $\mathcal{X}$. A simple convexly liftable cell complex is presented in Fig. 2.3. One of its convex liftings is shown in Fig. 2.4.

Algorithm 2.1 Convex lifting algorithm

Input: A given cell complex $\{\mathcal{X}_i\}_{i\in\mathcal{I}_N}$ of a polytope $\mathcal{X} \subset \mathbb{R}^{n_x}$, a scalar $c > 0$.
Output: Gains a_i, b_i of a convex lifting $z(x) = a_i^T x + b_i$ for $x \in \mathcal{X}_i$.
1: Register all pairs of neighboring regions in $\{\mathcal{X}_i\}_{i\in\mathcal{I}_N}$.
2: For each pair $(i, j) \in \mathcal{I}_N^2$ such that $(\mathcal{X}_i, \mathcal{X}_j)$ are neighbors:

- Add continuity conditions:

$$a_i^T v + b_i = a_j^T v + b_j, \;\; \forall v \in \mathcal{V}(\mathcal{X}_i \cap \mathcal{X}_j). \tag{2.4}$$

- Add convexity conditions:

$$a_i^T u + b_i \geq a_j^T u + b_j + c, \;\; \forall u \in \mathcal{V}(\mathcal{X}_i), u \notin \mathcal{V}(\mathcal{X}_i \cap \mathcal{X}_j). \tag{2.5}$$

3: Solve the following convex optimization problem

$$\min_{a_i, b_i} \sum_{i\in\mathcal{I}_N} a_i^T a_i + b_i^T b_i \quad \text{subject to (2.4), (2.5)}. \tag{2.6}$$

4: Construct an affinely equivalent polyhedron

$$\widetilde{\mathcal{X}} = \text{conv}\left\{ \begin{bmatrix} v \\ z(v) \end{bmatrix} \in \mathbb{R}^{n_x+1} \mid v \in \bigcup_{i\in\mathcal{I}_N} \mathcal{V}(\mathcal{X}_i), z(v) = a_i^T v + b_i \quad \text{if} \quad v \in \mathcal{X}_i \right\}.$$

Note that the insertion of $c > 0$ in (2.5) ensures the strict inequalities called *convexity conditions*, as defined in Definition 2.3. As for the complexity of Algorithm 2.1, if N denotes the number of regions in $\{\mathcal{X}_i\}_{i\in\mathcal{I}_N}$, then step 1 considers at most $\frac{1}{2}N(N-1)$ cases. For each pair of neighboring regions, the number of imposed constraints (including equality and inequality constraints) is equal to the number of vertices of $\mathcal{X}_i$. If $v_{\max}$ denotes the maximal number of vertices among the regions in $\{\mathcal{X}_i\}_{i\in\mathcal{I}_N}$, then an upper bound for the number of constraints for (2.6) is $\frac{1}{2}N(N-1)v_{\max}$, thus scales quadratically with the number of regions. Recall that (2.6) is a quadratic programming problem and is considered to be computationally tractable with respect to the working-memory capacity of calculator.

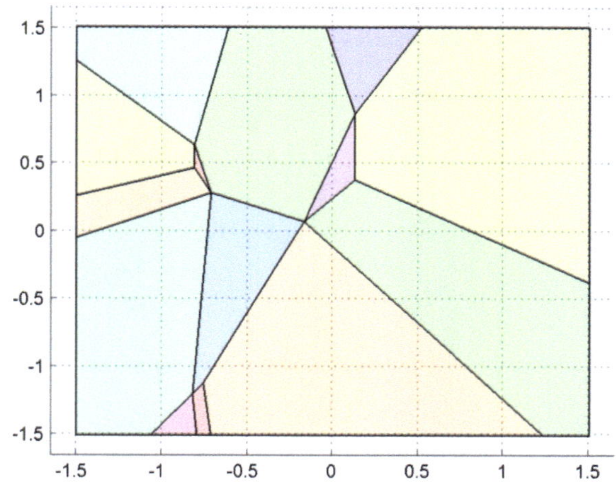

Fig. 2.3 A convexly liftable cell complex

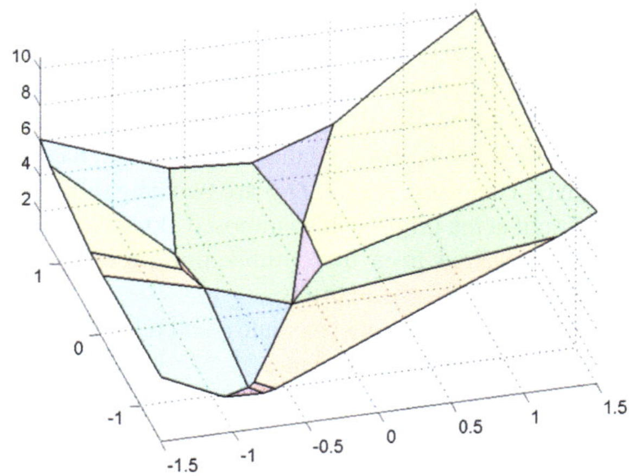

Fig. 2.4 A convex lifting of the cell complex in Fig. 2.3

Note also that the requirement that $\mathcal{X}$ is a polytope can be relaxed to compact, (not necessarily convex) polyhedral sets. As defined, a polyhedral partition is a collection of several polyhedra/polytopes. If we restrict our attention to a collection of polytopes, then their union represents a bounded set. This compact set has the boundary described by linear constraints, but it is not necessarily a polytope. Algorithm 2.1 can also construct convex liftings for such convexly liftable cell complexes if feasible. For illustration, a cell complex of a nonconvex polyhedral set is shown in Fig. 2.5. This cell complex is clearly convexly liftable.

Notice also that the feasibility of the optimization problem (2.6) is instrumental to determine whether the given partition is convexly liftable. More clearly, the given cell complex is convexly liftable if and only if problem (2.6) is feasible. Due to Lemma 2.1, for any polyhedral partition, whose facet-to-facet property does not hold, the associated optimization problem (2.6) is infeasible.

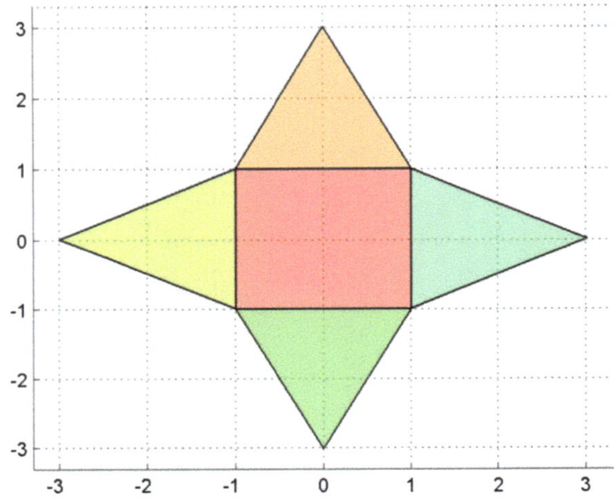

Fig. 2.5 A convexly liftable cell complex of a nonconvex polyhedral set

For a convexly liftable cell complex of a polyhedron, it can be observed that Algorithm 2.1 cannot be directly applicable. Let us take a simple example to illustrate this limitation of Algorithm 2.1. Consider a partition of four quadrants, which covers the whole $\mathbb{R}^2$. It is observed that each quadrant has only one vertex, known to be the origin 0. Therefore, if $(\mathcal{Q}_1, \mathcal{Q}_2)$ are two neighboring quadrants, only one continuity constraint at the origin will be imposed along Algorithm 2.1. It follows that $z(x) = 0$ may be resulted from the optimization problem (2.6). However, this real-valued function is not a convex lifting.

We will present an intermediate result related to the construction of convex liftings for cell complexes of polyhedra.

The following assumption is of help for our development.

Assumption 2.1 For all $x \in \bigcup_{i \in \mathcal{I}_N} \mathcal{V}(\mathcal{X}_i)$, $x \in \text{int}(B_{n_x}(\epsilon)) \subset \mathbb{R}^{n_x}$, with some suitable $\epsilon > 0$.

In view of Assumption 2.1, the following theorem is of help to construct convex liftings for cell complexes of a polyhedron.

Theorem 2.1 *Given a convexly liftable cell complex* $\{\mathcal{X}_i\}_{i \in \mathcal{I}_N}$ *of a polyhedron* $\mathcal{X} \subseteq \mathbb{R}^{n_x}$ *and a box* $B_{n_x}(\epsilon)$ *satisfying Assumption 2.1,* $f : \mathcal{X} \cap B_{n_x}(\epsilon) \to \mathbb{R}$

$$f(x) = a_i^T x + b_i \quad for \quad x \in \mathcal{X}_i \cap B_{n_x}(\epsilon),$$

is a convex lifting of the cell complex $\left\{\mathcal{X}_i \cap B_{n_x}(\epsilon)\right\}_{i \in \mathcal{I}_N}$, *if and only if the function* $g : \mathcal{X} \to \mathbb{R}$ *defined as follows:*

$$g(x) = a_i^T x + b_i \quad for \quad x \in \mathcal{X}_i,$$

is also a convex lifting of $\{\mathcal{X}_i\}_{i \in \mathcal{I}_N}$.

Proof $\longrightarrow$ First, due to Assumption 2.1, the intersection $\mathcal{X} \cap B_{n_x}(\epsilon)$ does not have any effect on the internal subdivision of $\mathcal{X}$, since any vertex of the partition $\{\mathcal{X}_i\}_{i \in \mathcal{I}_N}$ lies in the interior of $B_{n_x}(\epsilon)$.

Consider now two neighboring regions in the partition $\{\mathcal{X}_i \cap B_{n_x}(\epsilon)\}_{i \in \mathcal{I}_N}$, denoted as $\mathcal{X}_i \cap B_{n_x}(\epsilon)$, $\mathcal{X}_j \cap B_{n_x}(\epsilon)$. As assumed, $f(x)$ is a convex lifting of $\{\mathcal{X}_i \cap B_{n_x}(\epsilon)\}_{i \in \mathcal{I}_N}$, then it can be deduced from its definition that:

$$\begin{aligned} a_i^T x + b_i &= a_j^T x + b_j \quad \forall x \in (\mathcal{X}_i \cap B_{n_x}(\epsilon)) \cap (\mathcal{X}_j \cap B_{n_x}(\epsilon)) \\ a_i^T x + b_i &> a_j^T x + b_j \quad \forall x \in (\mathcal{X}_i \cap B_{n_x}(\epsilon)) \backslash (\mathcal{X}_j \cap B_{n_x}(\epsilon)). \end{aligned}$$

Note also that constraint $a_i^T x + b_i = a_j^T x + b_j$ describes the hyperplane, separating $\mathcal{X}_i \cap B_{n_x}(\epsilon)$ and $\mathcal{X}_j \cap B_{n_x}(\epsilon)$, then it separates also $\mathcal{X}_i$ and $\mathcal{X}_j$. This end leads to

$$\begin{aligned} a_i^T x + b_i &= a_j^T x + b_j \quad \forall x \in \mathcal{X}_i \cap \mathcal{X}_j, \\ a_i^T x + b_i &> a_j^T x + b_j \quad \forall x \in \mathcal{X}_i \backslash \mathcal{X}_j. \end{aligned}$$

Applying this inclusion to every pair of neighboring regions, the following inclusion can be obtained:

$$a_i^T x + b_i > a_j^T x + b_j, \quad \forall x \in \mathcal{X}_i \backslash \mathcal{X}_j, \forall j \neq i, j \in \mathcal{I}_N,$$

meaning $g(x)$ is a convex lifting of $\{\mathcal{X}_i\}_{i \in \mathcal{I}_N}$.
$\longleftarrow$ The sufficient condition can be similarly proved. □

This theorem shows that we can construct a convex lifting for a cell complex of a polyhedron from a convex lifting of an appropriate partition of a bounded set. This partition is resulted from the intersection of the given cell complex and some suitable boxes. The remaining problem is to find out one among these boxes. This task can be easily carried out from Assumption 2.1. A simple algorithm is put forward below.

Algorithm 2.2 Determining a $B_{n_x}(\epsilon)$

Input: A given cell complex $\{\mathcal{X}_i\}_{i \in \mathcal{I}_N}$ of a polyhedron $\mathcal{X} \subseteq \mathbb{R}^{n_x}$ and a scalar $c > 0$.
Output: A box $B_{n_x}(\epsilon) \subset \mathbb{R}^{n_x}$ satisfying Assumption 2.1.
1: Compute $V_x = \bigcup_{i \in \mathcal{I}_N} \mathcal{V}(\mathcal{X}_i)$.
2: Solve:

$$\min_{\epsilon} \epsilon \quad \text{s.t.} \; -(\epsilon - c)1_{n_x} \leq x \leq (\epsilon - c)1_{n_x}, \; \forall x \in V_x.$$

Note that a strictly positive scalar c needs to be inserted in Algorithm 2.2 to ensure that all $x \in \bigcup_{i \in \mathcal{I}_N} \mathcal{V}(\mathcal{X}_i)$ lie in the interior of $B_{n_x}(\epsilon)$ via constraint reinforcements. More precisely, constraints $-(\epsilon - c)1_{n_x} \leq x \leq (\epsilon - c)1_{n_x}$ imply that $\epsilon - c > 0$, leading to $\|x\|_\infty \leq \epsilon - c < \epsilon$, thus $x \in \text{int}(B_{n_x}(\epsilon))$ for all $x \in \bigcup_{i \in \mathcal{I}_N} \mathcal{V}(\mathcal{X}_i)$.

2.4.3 Nonconvexly Liftable Partitions

Taking into account the dichotomy between convexly liftable and nonconvexly liftable partitions, a natural question is how to deal with PWA functions defined over nonconvexly liftable partitions. As mentioned before, not every cell complex is convexly liftable, therefore, solving inverse parametric linear/quadratic programming problems via convex liftings needs an adaptation to deal with such particular partitions. Note that this issue has already been investigated in [28, 32]. We recall the main result here for completeness.

Theorem 2.2 *Given a nonconvexly liftable polytopic partition* $\{\mathcal{X}_i\}_{i\in\mathcal{I}_N}$, *there exists at least one subdivision, preserving the internal boundaries of this partition, such that the new cell complex is convexly liftable.*

We refer to [28, 32] for the details of proof. It is worth emphasizing that this result states only for polytopic partitions of bounded sets. However, its extension to polyhedral partitions of an unbounded set can be performed along the same arguments. This observation can formally be stated as follows.

Lemma 2.2 *For any polyhedral partition of a polyhedron, there always exists one subdivision such that the internal boundaries of this partition are preserved and the new partition is convexly liftable.*

Proof See the proof of Theorem IV.2 presented in [28].

Note that in practice a complete hyperplane arrangement is not necessary. One can find a particular case of refinement in [15].

2.5 Solution to Inverse Parametric Linear/Quadratic Programming Problems

Based on the above results, this section aims to put forward the solution to inverse optimality problem via convex liftings. This solution is viable with respect to the following standing assumption.

Assumption 2.2 The given cell complex $\{\mathcal{X}_i\}_{i\in\mathcal{I}_N}$ of a polyhedron $\mathcal{X}$ is convexly liftable.

Note that this assumption is not restrictive due to Lemma 2.2. Following this lemma, if a given polyhedral partition does not satisfy Assumption 2.2, it can be refined into a convexly liftable partition such that its internal boundaries are maintained.

We use $z(x)$ to denote a convex lifting of $\{\mathcal{X}_i\}_{i\in\mathcal{I}_N}$, $z(x) = a_i^T x + b_i$ for $x \in \mathcal{X}_i$. We want to recover a continuous PWA function, denoted by $f_{pwa}(x)$, i.e., $f_{pwa}(x) = H_i x + g_i$ for $x \in \mathcal{X}_i$. Note that these results can be extended for discontinuous PWA functions; we refer to [31] for more details.

For ease of presentation, the following sets are also defined:

$$
\begin{aligned}
V_x &= \bigcup_{i \in \mathcal{I}_N} \mathcal{V}(\mathcal{X}_i),\ R_x = \bigcup_{i \in \mathcal{I}_N} \mathcal{R}(\mathcal{X}_i),\\
V_{[x^T\, z\, u^T]^T} &= \left\{ \left[x^T\ z(x)\ f_{pwa}^T(x)\right]^T \mid x \in V_x \right\},\\
R_{[x^T\, z\, u^T]^T} &= \left\{ \left[r^T\ \widehat{z}(r)\ \widehat{f}^T(r)\right]^T \mid r \in R_x,\ \begin{array}{l} \widehat{z}(r) = a_i^T r \\ \widehat{f}(r) = H_i r \end{array} \text{ if } r \in \mathcal{R}(\mathcal{X}_i) \right\}, \qquad (2.7)\\
\Pi_v &= \operatorname{conv}(V_{[x^T\, z\, u^T]^T}),\ \Pi_r = \operatorname{cone}(R_{[x^T\, z\, u^T]^T}),\\
\Pi &= \Pi_v \oplus \Pi_r.
\end{aligned}
$$

The main result of this manuscript is presented via the following theorem which generalizes the results in [32] to general polyhedra.

Theorem 2.3 *Given a continuous PWA function $f_{pwa}(x)$, defined over a cell complex $\{\mathcal{X}_i\}_{i \in \mathcal{I}_N}$ satisfying Assumption 2.2, then $f_{pwa}(x)$ is the image via the orthogonal projection of the optimal solution to the following parametric linear programming problem:*

$$\min_{[z\ u^T]^T} z \quad \textit{subject to} \quad \left[x^T\ z\ u^T\right]^T \in \Pi. \qquad (2.8)$$

Proof Consider $x \in \mathcal{X}_i$, due to the Minkowski–Weyl theorem for polyhedra (Corollary 7.1b in [37]), x can be described as follows:

$$x = \sum_{v \in \mathcal{V}(\mathcal{X}_i)} \alpha(v) v + \sum_{r \in \mathcal{R}(\mathcal{X}_i)} \beta(r) r,$$

where $\alpha(v), \beta(r) \in \mathbb{R}_+$ and $\sum_{v \in \mathcal{V}(\mathcal{X}_i)} \alpha(v) = 1$. As a consequence, the convex lifting at x, i.e., $z(x)$ can be described by

$$
\begin{aligned}
z(x) = a_i^T x + b_i &= a_i^T \left(\sum_{v \in \mathcal{V}(\mathcal{X}_i)} \alpha(v) v + \sum_{r \in \mathcal{R}(\mathcal{X}_i)} \beta(r) r \right) + b_i,\\
&= \sum_{v \in \mathcal{V}(\mathcal{X}_i)} \alpha(v) \left(a_i^T v + b_i\right) + \sum_{r \in \mathcal{R}(\mathcal{X}_i)} \beta(r) \left(a_i^T r\right).
\end{aligned}
$$

Similarly,

$$f_{pwa}(x) = \sum_{v \in \mathcal{V}(\mathcal{X}_i)} \alpha(v)(H_i v + g_i) + \sum_{r \in \mathcal{R}(\mathcal{X}_i)} \beta(r)(H_i r).$$

It can be observed that if r is a ray of $\mathcal{X}_i$, then $\left[r^T\ a_i^T r\right]^T$ is a ray of the affinely equivalent polyhedron $\Pi_{[x^T\ z]^T}$ of $\{\mathcal{X}_i\}_{i\in\mathcal{I}_N}$, defined as follows:

$$\Pi_{[x^T\ z]^T} = \text{conv}(V_{[x^T\ z]^T}) \oplus \text{cone}(R_{[x^T\ z]^T}),$$

where

$$\begin{aligned} V_{[x^T\ z]^T} &= \left\{\left[x^T\ z(x)\right]^T \mid x \in V_x\right\}, \\ R_{[x^T\ z]^T} &= \left\{\left[r^T\ \widehat{z}(r)\right]^T \mid r \in R_x,\ \widehat{z}(r) = a_i^T r \ \text{ if } \ r \in \mathcal{R}(\mathcal{X}_i)\right\}. \end{aligned}$$

Therefore, for a region $\mathcal{X}_i$, there exists a facet of $\Pi_{[x^T\ z]^T}$, denoted by $F^{(i)}_{[x^T\ z]^T}$, such that

$$\begin{aligned} &\text{Proj}_{\mathbb{R}^{n_x}} F^{(i)}_{[x^T\ z]^T} = \mathcal{X}_i, \\ &\forall \left[x^T\ z(x)\right]^T \in F^{(i)}_{[x^T\ z]^T},\ z(x) = \min_z z \ \text{ s.t. } \left[x^T\ z\right]^T \in \Pi_{[x^T\ z]^T}. \end{aligned} \tag{2.9}$$

According to Proposition 5.1 in [27], every augmented point in $V_{[x^T\ z\ u^T]^T}$ is vertices of Π_v. Thus, lifting onto $\mathbb{R}^{n_x+n_u+1}$ leads to the existence of an n_x-face of Π, denoted by $F^{(i)}_{[x^T\ z\ u^T]^T}$ such that

$$\text{Proj}_{\mathbb{R}^{n_x+1}} F^{(i)}_{[x^T\ z\ u^T]^T} = F^{(i)}_{[x^T\ z]^T}. \tag{2.10}$$

Due to (2.9) and (2.10), the minimal value of z at a point $x \in \mathcal{X}_i$ happens when $\left[x^T\ z\ u^T\right]^T$ lies in $F^{(i)}_{[x^T\ z\ u^T]^T}$. Therefore, optimal solution to (2.8) at x can be described by

$$\begin{bmatrix} x \\ z^*(x) \\ u^*(x) \end{bmatrix} = \sum_{v\in\mathcal{V}(\mathcal{X}_i)} \alpha(v) \begin{bmatrix} v \\ a_i^T v + b_i \\ H_i v + g_i \end{bmatrix} + \sum_{r\in\mathcal{R}(\mathcal{X}_i)} \beta(r) \begin{bmatrix} r \\ a_i^T r \\ H_i r \end{bmatrix},$$

where $\alpha(v), \beta(r) \in \mathbb{R}_+$ and $\sum_{v\in\mathcal{V}(\mathcal{X}_i)} \alpha(v) = 1$. It is clear that

$$\begin{bmatrix} z^*(x) \\ u^*(x) \end{bmatrix} = \begin{bmatrix} a_i^T x + b_i \\ H_i x + g_i \end{bmatrix} = \begin{bmatrix} z(x) \\ f_{pwa}(x) \end{bmatrix}, \quad \text{for } x \in \mathcal{X}_i.$$

To complete the proof, we now need to show that the optimal solution to (2.8) is unique. In fact, at a point $x \in \mathcal{X}_i$, suppose there exist two different optimal solutions to (2.8)

$$\begin{array}{l} \left[z_1(x)\ u_1^T(x)\right]^T = \arg \min\limits_{\left[z\ u^T\right]^T} z \\ \left[z_2(x)\ u_2^T(x)\right]^T = \arg \min\limits_{\left[z\ u^T\right]^T} z \end{array} \quad \text{s.t.} \quad \left[x^T\ z\ u^T\right]^T \in \Pi.$$

It can be observed that $z_1(x) = z_2(x)$. If $u_1(x) \neq u_2(x)$, then there exist two different n_x-faces, denoted by F_1, F_2, such that $\left[x^T\ z_1(x)\ u_1^T(x)\right]^T \in F_1$ and $\left[x^T\ z_2(x)\ u_2^T(x)\right]^T \in F_2$. Therefore, $z_1(x) = z_2(x)$ leads to

$$\text{Proj}_{\mathbb{R}^{n_x+1}} F_1 = \text{Proj}_{\mathbb{R}^{n_x+1}} F_2 = F^{(i)}_{\left[x^T\ z\right]^T}.$$

Accordingly, F_1, F_2 lie in a hyperplane of dimension $n_x + 1$ which is orthogonal to the space of $\left[x^T\ z\right]^T$. An illustration can be found in Fig. 2.6. This leads to the fact that $f_{pwa}(v)$ or $\widehat{f}(r)$ in (2.7) is not uniquely defined for some $v \in \mathcal{V}(\mathcal{X}_i)$ or some $r \in \mathcal{R}(\mathcal{X}_i)$. This contradicts with the construction of Π in (2.7). Therefore, $F_1 = F_2$ meaning the optimal solution to (2.8) is unique. Further, such an n_x-face F_1 can be written in the following form:

$$\begin{aligned} F_1 &= F_1^{\#} \oplus F_2^{\#} \\ F_1^{\#} &= \text{conv}\left\{\left[v^T\ z(v)\ f_{pwa}^T(v)\right]^T \mid v \in \mathcal{V}(\mathcal{X}_i)\right\} \\ F_2^{\#} &= \text{cone}\left\{\left[r^T\ a_i^T r\ (H_i r)^T\right]^T \mid r \in \mathcal{R}(\mathcal{X}_i)\right\}. \end{aligned}$$

The proof is complete. □

Based on this result, a procedure to construct an inverse optimization problem is summarized via the following algorithm.

Algorithm 2.3 Linear equivalent optimization problem

Input: A given continuous PWA function defined over a cell complex $\{\mathcal{X}_i\}_{i\in\mathcal{I}_N}$ of a polyhedron $\mathcal{X} \subseteq \mathbb{R}^{n_x}$ satisfying Assumption 2.2.
Output: Π, $J(x, z, u)$.
1: Construct a convex lifting $z(x)$ of the cell complex $\{\mathcal{X}_i\}_{i\in\mathcal{I}_N}$.
2: Define Π as in (2.7).
3: Define $J(x, z, u) = z$.
4: Solve the following parametric linear programming problem:

$$\begin{bmatrix} z^* \\ u^* \end{bmatrix} = \arg \min_{\left[z\ u^T\right]^T} z \quad \text{subject to} \quad \left[x^T\ z\ u^T\right]^T \in \Pi.$$

5: Project the optimal solution onto $\mathbb{R}^{n_u}$ i.e., $f_{pwa}(x) = \text{Proj}_{\mathbb{R}^{n_u}} \begin{bmatrix} z^* \\ u^* \end{bmatrix}$.

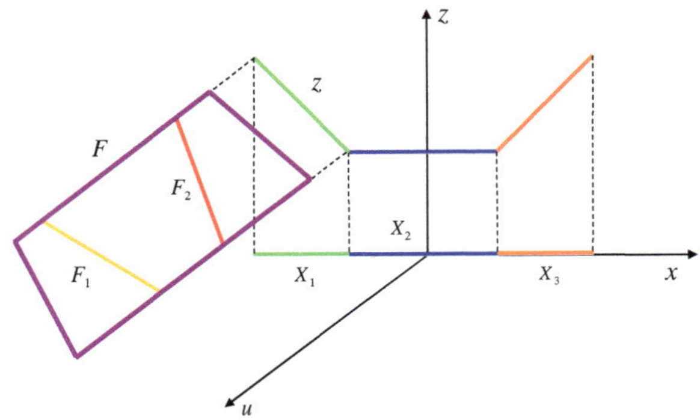

Fig. 2.6 An illustration for the uniqueness of optimal solution to (2.8)

The complexity of Algorithm 2.3 depends almost on the complexity of convex lifting computation, carried out via Algorithm 2.1. Therefore, if Algorithm 2.1 is computationally tractable, so is Algorithm 2.3.

To conclude this section, the following theorem presents an important property of the convex lifting-based method.

Theorem 2.4 *Any continuous PWA function, defined over a (not necessarily convexly liftable) polyhedral partition of a polyhedron, can be equivalently obtained via a parametric linear programming problem with at most one auxiliary* 1*-dimensional variable.*

Proof If the given polyhedral partition $\{\mathcal{X}_i\}_{i\in\mathcal{I}_N}$ is convexly liftable, following Theorem 2.3, the given continuous PWA function $f_{pwa}(x)$, defined over $\mathcal{X}$ can be obtained through a parametric linear programming problem. This optimization problem is constructed via convex lifting. This convex lifting represents an auxiliary one-dimensional variable.

If the given polyhedral partition $\{\mathcal{X}_i\}_{i\in\mathcal{I}_N}$ is not convexly liftable, Theorem 2.2 shows the existence of an equivalent cell complex $\left\{\widetilde{\mathcal{X}}_i\right\}_{i\in\mathcal{I}_{\widetilde{N}}}$ such that $\left\{\widetilde{\mathcal{X}}_i\right\}_{i\in\mathcal{I}_{\widetilde{N}}}$ is convexly liftable and the internal boundaries of $\{\mathcal{X}_i\}_{i\in\mathcal{I}_N}$ are maintained. This refinement also leads to an equivalent PWA function $\widetilde{f}_{pwa}(x)$ of $f_{pwa}(x)$ defined over a convexly liftable cell complex $\left\{\widetilde{\mathcal{X}}_i\right\}_{i\in\mathcal{I}_{\widetilde{N}}}$ of $\mathcal{X}$. Again, due to Theorem 2.3, $\widetilde{f}_{pwa}(x)$ can be obtained through a parametric linear programming problem with an auxiliary one-dimensional variable, this auxiliary variable being a convex lifting of $\left\{\widetilde{\mathcal{X}}_i\right\}_{i\in\mathcal{I}_{\widetilde{N}}}$. □

Note that we can also find a parametric quadratic programming problem which equivalently recovers the given continuous PWA function defined over a polyhedral partition of a (possibly unbounded) polyhedron, as shown in [32].

Remark 2.2 It is well known that optimal solution to a parametric linear/quadratic programming problem is a PWA function defined over a polyhedral partition. For the parametric quadratic programming case, it is shown in [8] that the optimal solution is continuous and unique. However, for the parametric linear programming case, this continuity of optimal solution may not be guaranteed. Fortunately, a continuous solution can equivalently be selected, as shown in [34, 42]. Based on the arguments presented in this chapter, a continuous optimal solution, induced from a parametric linear/quadratic programming problem, can also be equivalently obtained via an alternative parametric linear programming problem with *at most one auxiliary one-dimensional variable.* This auxiliary variable represents the convex lifting.

2.6 An Illustrative Example

This section aims to illustrate the above results via a numerical example. Suppose we need to recover the PWA function (2.11), shown in Fig. 2.7. Note that this PWA function is continuous and is defined over the whole space $\mathbb{R}$. A box $B_1(3)$ is known to satisfy Assumption 2.1. The new partition $\{\mathcal{X}_i \cap B_1(3)\}_{i\in\mathcal{I}_6}$ shown in Fig. 2.7 represents the multicolored segments along the x-axis. A convex lifting of the cell complex $\{\mathcal{X}_i\}_{i\in\mathcal{I}_6}$ is analytically presented in (2.12) and is shown in Fig. 2.8. A set of constraints for the recovered optimization problem shown in (2.13) represents the pink polyhedron in Fig. 2.9. Therein, the multicolored segments along the x-axis denote the given partition covering $\mathbb{R}$, whereas the PWA function (2.11) represents the green curve above this partition. Also, the optimal solution to (2.8) represents the solid pink curve. It can be observed that the projection of this optimal solution to the space $\begin{bmatrix}x^T & u^T\end{bmatrix}^T$ coincides with the given PWA function. It is worth recalling that the proposed approach requires only one auxiliary one-dimensional variable, denoted by z, to recover (2.11). Finally, the numerical example in this chapter is carried out in the environment of MPT 3.0 [18].

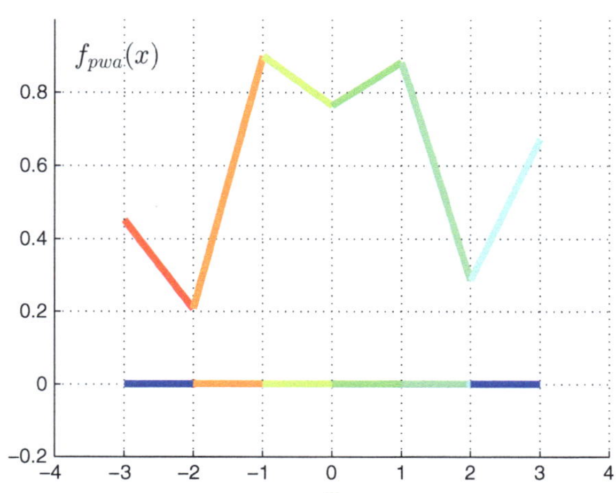

Fig. 2.7 The given PWA function (2.11) to be recovered

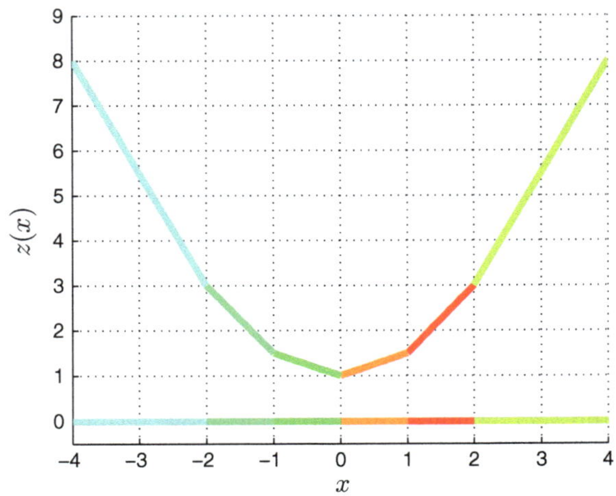

Fig. 2.8 A convex lifting of the cell complex $\{\mathcal{X}_i\}_{i\in\mathcal{I}_6}$

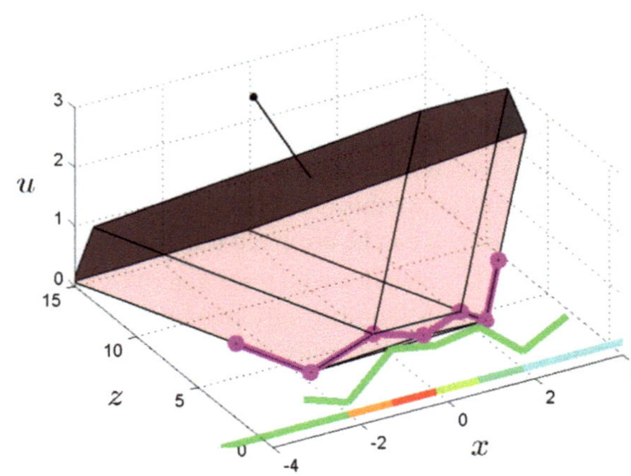

Fig. 2.9 Illustration for the optimal solution to IPL/QP via convex liftings

$$f_{pwa}(x) = \begin{cases} -0.2447x - 0.2838 & \text{for } x \leq -2 \\ 0.6940x + 1.5936 & \text{for } -2 \leq x \leq -1 \\ -0.1371x + 0.7626 & \text{for } -1 \leq x \leq 0 \\ 0.1199x + 0.7626 & \text{for } 0 \leq x \leq 1 \\ -0.5975x + 1.4800 & \text{for } 1 \leq x \leq 2 \\ 0.3883x - 0.4916 & \text{for } 2 \leq x \end{cases} \tag{2.11}$$

$$z(x) = \begin{cases} -2.5x - 2 & \text{for } x \leq -2 \\ -1.5x & \text{for } -2 \leq x \leq -1 \\ -0.5x + 1 & \text{for } -1 \leq x \leq 0 \\ 0.5x + 1 & \text{for } 0 \leq x \leq 1 \\ 1.5x & \text{for } 1 \leq x \leq 2 \\ 2.5x - 2 & \text{for } 2 \leq x \end{cases} \tag{2.12}$$

$$\Pi = \left\{ \begin{bmatrix} x \\ z \\ u \end{bmatrix} \in \mathbb{R}^3 \mid \begin{bmatrix} 1.0000 & -29.9413 & 116.5183 \\ 1.0000 & -18.4965 & 116.5183 \\ -2.2393 & -1.0000 & 1.0653 \\ -1.0000 & -1.7638 & 13.9320 \\ -1.0000 & -1.5039 & -1.8097 \\ 1.0000 & 1.7638 & -13.9320 \\ 1.0000 & 5.3390 & -50.4554 \\ 1.0000 & -13.0496 & -50.4554 \\ 2.1061 & -1.0000 & 1.0144 \\ 1.0000 & -1.4990 & -2.0893 \end{bmatrix} \begin{bmatrix} x \\ z \\ u \end{bmatrix} \leq \begin{bmatrix} 58.9139 \\ 76.0811 \\ 1.6977 \\ 10.8883 \\ -2.8839 \\ 3.3214 \\ 3.6397 \\ -51.5262 \\ 1.5013 \\ -3.0923 \end{bmatrix} \right\} \tag{2.13}$$

2.7 Conclusions

This chapter presents recent results in inverse optimality and summarizes in a concise manner a procedure to recover a continuous PWA function defined over a polyhedral partition of a polyhedron. Based on convex lifting, the study covers the general case of continuous PWA functions, presenting a full construction for the inverse optimality problem. A numerical example is finally considered to illustrate this result.

Acknowledgments The first author would like to thank Dr. Martin Gulan, Dr. Michal Kvasnica, Prof. Miroslav Fikar, Dr. Sasa Rakovic, Prof. Franz Aurenhammer, and Prof. Boris Rohal'-Ilkiv for fruitful discussions and exchanges on topics related to this work.

References

1. G.Z. Angelis, System Analysis, Modelling and Control with Polytopic Linear Models (2001)
2. F. Aurenhammer, Criterion for the affine equivalence of cell complexes in r^d and convex polyhedra in r^{d+1}. Discrete Comput. Geom. **2**, 49–64 (1987)
3. F. Aurenhammer, Power diagrams: properties, algorithms and applications. SIAM J. Comput. **16**(1), 78–96 (1987)
4. F. Aurenhammer, Recognising polytopical cell complexes and constructing projection polyhedra. J. Symb. Comput. **3**, 249–255 (1987)

5. F. Aurenhammer, Voronoi diagrams: a survey of a fundamental data structure. ACM Comput. Surv. **23**, 345–405 (1991)
6. M. Baes, M. Diehl, I. Necoara, Every continuous nonlinear control system can be obtained by parametric convex programming. IEEE Trans. Autom. Control **53**(8), 1963–1967 (2008)
7. A. Bemporad, C. Filippi, An algorithm for approximate multiparametric convex programming. Comput. Opt. Appl. **35**(1), 87–108 (2006)
8. A. Bemporad, M. Morari, V. Dua, E.N. Pistikopoulos, The explicit linear quadratic regulator for constrained systems. Automatica **38**(1), 3–20 (2002)
9. H. Crapo, W. Whiteley, Plane self stresses and projected polyhedra 1: the basic pattern. Struct. Topol. **19**, 55–73 (1993)
10. H. Crapo, W. Whiteley, Spaces of stresses, projections and parallel drawings for spherical polyhedra. Contrib. Algebra Geom. **35**(So. 2), 259–281 (1994)
11. A.J. De Loera, J. Rambau, F. Santos, Triangulations: structures for algorithms and applications, *Algorithms and Computation in Mathematics* (Springer, Berlin, 2010)
12. H. Edelsbrunner, R. Seidel, Voronoi diagrams and arrangements. Discrete Comput. Geom. **1**, 25–44 (1986)
13. A. Grancharova, T.A. Johansen, *Explicit Nonlinear Model Predictive Control* (Springer, Berlin, 2012)
14. B. Grünbaum, *Convex Polytopes* (Wiley Interscience, New York, 1967)
15. M. Gulan, N.A. Nguyen, S. Olaru, P. Rodriguez-Ayerbe, B. Rohal'-Ilkiv, Implications of inverse parametric optimization in model predictive control, in *Developments in Model-Based Optimization and Control*, ed. by S. Olaru, A. Grancharova, F.L. Pereira (Springer, Berlin, 2015)
16. D. Hartvigsen, Recognizing voronoi diagrams with linear programming. ORSA J. Comput. **4**, 369–374 (1992)
17. A.B. Hempel, P.J. Goulart, J. Lygeros, On inverse parametric optimization with an application to control of hybrid systems. IEEE Trans. Autom. Control **60**(4), 1064–1069 (2015)
18. M. Herceg, M. Kvasnica, C.N. Jones, M. Morari, Multi-Parametric Toolbox 3.0, in *Proceedings of the European Control Conference*, Zürich, Switzerland, 17–19 July 2013, pp. 502–510, http://control.ee.ethz.ch/~mpt
19. T.A. Johansen, On multi-parametric nonlinear programming and explicit nonlinear model predictive control, in *Proceedings of the 41st IEEE Conference on Decision and Control*, vol. 3 (2002), pp. 2768–2773
20. T.A. Johansen, Approximate explicit receding horizon control of constrained nonlinear systems. Automatica **40**(2), 293–300 (2004)
21. M. Kvasnica, M. Fikar, Clipping-based complexity reduction in explicit MPC. IEEE Trans. Autom. Control **57**(7), 1878–1883 (2012)
22. M. Kvasnica, J. Hledík, I. Rauová, M. Fikar, Complexity reduction of explicit model predictive control via separation. Automatica **49**(6), 1776–1781 (2013)
23. C.W. Lee, P.L.-spheres, convex polytopes, and stress. Discrete Comput. Geom. **15**, 389–421 (1996)
24. J.C. Maxwell, On reciprocal diagrams and diagrams of forces. Philos. Mag., ser. 4 **27**, 250–261 (1864)
25. H.N. Nguyen, *Constrained Control of Uncertain, Time-Varying Discrete-Time Systems. An Interpolation-Based Approach*, Lecture Notes in Control and Information Sciences (Springer, Berlin, 2014)
26. H.N. Nguyen, P.-O. Gutman, S. Olaru, M. Hovd, Explicit constraint control based on interpolation techniques for time-varying and uncertain linear discrete-time systems, in *18th IFAC World Congress*, vol. 18(1) (2011)
27. N.A. Nguyen, S. Olaru, P. Rodriguez-Ayerbe, M. Hovd, I. Necoara, Inverse parametric convex programming problems via convex liftings, in *19th IFAC World Congress* (Cape Town, South Africa, 2014)
28. N.A. Nguyen, S. Olaru, P. Rodriguez-Ayerbe, M. Hovd, I. Necoara, On the lifting problems and their connections with piecewise affine control law design, in *European Control Conference* (Strasbourg, France, 2014)

29. N.A. Nguyen, S. Olaru, P. Rodriguez-Ayerbe, On the complexity of the convex liftings-based solution to inverse parametric convex programming problems, in *European Control Conference* (Linz, Austria, 2015)
30. N.A. Nguyen, S. Olaru, P. Rodriguez-Ayerbe. Recognition of additively weighted voronoi diagrams and weighted delaunay decompositions, in *European Control Conference* (Linz, Austria, 2015)
31. N.A. Nguyen, S. Olaru, P. Rodriguez-Ayerbe, Any discontinuous PWA function is optimal solution to a parametric linear programming problem, in *54th IEEE Conference on Decision and Control* (Osaka, Japan, 2015)
32. N.A. Nguyen, S. Olaru, P. Rodriguez-Ayerbe, M. Hovd, I. Necoara, Constructive solution to inverse parametric linear/quadratic programming problems via convex liftings, https://hal-supelec.archives-ouvertes.fr/hal-01207234/file/IPCP_JOTA.pdf
33. S. Olaru, D. Dumur, A parameterized polyhedra approach for explicit constrained predictive control, in *43rd IEEE Conference on Decision and Control*, vol. 2 (2004), pp. 1580–1585
34. S. Olaru, D. Dumur, On the continuity and complexity of control laws based on multiparametric linear programs, in *45th Conference on Decision and Control* (IEEE, 2006), pp. 5465–5470
35. E.N. Pistikopoulos, M.C. Georgiadis, V. Dua, *Multi-Parametric Programming* (Wiley-VCH, Weinheim, 2007)
36. K. Rybnikov, Polyhedral partitions and stresses, Ph.D. thesis, Queen University, Kingston, Ontario, Canada (1999)
37. A. Schrijver, *Theory of Linear and Integer Programming* (Wiley, New York, 1998)
38. A. Schulz, Lifting planar graphs to realize integral 3-polytopes and topics in pseudo-triangulations, Ph.D. thesis, Fachbereich Mathematik und Informatik der Freien Universitat Berlin (2008)
39. M.M. Seron, G.C. Goodwin, J.A. Doná, Characterisation of receding horizon control for constrained linear systems. Asian J. Control **5**(2), 271–286 (2003)
40. E.D. Sontag, Nonlinear regulation: the piecewise linear approach. IEEE Trans. Autom. Control **26**(2), 346–358 (1981)
41. J. Spjøtvold, E.C. Kerrigan, C.N. Jones, P. Tøndel, T.A. Johansen, On the facet-to-facet property of solutions to convex parametric quadratic programs. Automatica **42**(12), 2209–2214 (2006)
42. J. Spjøtvold, P. Tøndel, T.A. Johansen, Continuous selection and unique polyhedral representation of solutions to convex parametric quadratic programs. J. Opt. Theory Appl. **134**(2), 177–189 (2007)
43. P. Tøndel, T.A. Johansen, A. Bemporad, An algorithm for multi-parametric quadratic programming and explicit MPC solutions. Automatica **39**(3), 489–497 (2003)

Chapter 3
Implications of Inverse Parametric Optimization in Model Predictive Control

Martin Gulan, Ngoc Anh Nguyen, Sorin Olaru, Pedro Rodriguez-Ayerbe and Boris Rohal'-Ilkiv

Abstract Recently, inverse parametric linear/quadratic programming problem was shown to be solvable via convex liftings approach [13]. This technique turns out to be relevant in explicit model predictive control (MPC) design in terms of reducing the prediction horizon to at most two steps. In view of practical applications, typically leading to problems that are not directly invertible, we show how to adapt the inverse optimality to specific, possibly convexly non-liftable partitions. Case study results moreover indicate that such an extension leads to controllers of lower complexity without loss of optimality. Numerical data are also presented for illustration.

3.1 Introduction

Interest in parametric optimization algorithms for linear and quadratic programs has been resurged during the last 15 years, following the observation that certain optimal control problems for constrained systems can be solved explicitly offline. In particular, a lot of attention has been attracted by the fact that solutions to the

M. Gulan (✉) · B. Rohal'-Ilkiv
Institute of Automation, Measurement and Applied Informatics,
Faculty of Mechanical Engineering, Slovak University of Technology,
Námestie slobody 17, 812 31 Bratislava, Slovakia
e-mail: martin.gulan@stuba.sk

B. Rohal'-Ilkiv
e-mail: boris.rohal-ilkiv@stuba.sk

N.A. Nguyen · S. Olaru · P. Rodriguez-Ayerbe
Laboratory of Signals and Systems, CentraleSupélec-CNRS-Université Paris-Sud, Université Paris-Saclay, 3, rue Joliot Curie, 911 90 Gif-sur-Yvette, France
e-mail: Ngocanh.Nguyen@supelec.fr

S. Olaru
e-mail: Sorin.Olaru@centralesupelec.fr

P. Rodriguez-Ayerbe
e-mail: Pedro.Rodriguez@centralesupelec.fr

S. Olaru et al. (eds.), *Developments in Model-Based Optimization and Control*,
Lecture Notes in Control and Information Sciences 464,
DOI 10.1007/978-3-319-26687-9_3

traditional model predictive control problems can be obtained in an explicit formulation by exploiting the concept of parametric convex programming (pCP) [3–5, 19, 22]. This allows to pre-compute the optimal solution to a linear/quadratic programming problem for a range of operating conditions of interest as a piecewise affine (PWA) function defined over a polyhedral partition of the parameter space. The implementation effort of these so-called explicit MPC (eMPC) methods [1] hence reduces to a simple function evaluation, since the controller itself maps into a lookup table of linear gains. This property naturally circumvents the main implementation drawback of the standard implicit MPC, which is the need for online optimization in the receding horizon fashion at each sampling instant. Clearly, within the real-time requirements, the solutions may become too expensive or simply infeasible, namely in the case of systems with fast-evolving dynamics. On the other hand, the efficacy of the offline solution is limited to small-dimensional systems. Therefore the eMPC algorithms necessarily encounter computational difficulties as the parameter space dimension or the prediction horizon increases. Moreover, the resulting PWA controller is often too complex; hence, if allowable, its structure may need to be treated adequately to satisfy the restrictions imposed by the implementation hardware.

In view of handling linear MPC problems often emerging from practical applications, the parametric convex programming has grown into a mature technique providing a variety of numerical algorithms available for effective solution. Recently, the concept of inverse optimality has been shown to provide a new perspective on the structural link between linear MPC and pCP problem formulation via the technique of inverse parametric convex programming (IpCP). As the name suggests, it is defined as an inverse optimality problem of parametric convex programming. It aims to build an alternative optimization problem characterized by an appropriate constraint set and a cost function such that its optimal solution coincides with the one of the original problem. More specifically, the goal of inverse parametric linear/quadratic programming is to construct a linear constraint set and a linear/quadratic cost function such that the optimal solution of this newly formulated problem is equivalent to a given PWA function defined over a given polyhedral partition. This topic has been recently investigated in [2, 7, 13], and turned out to be applicable in the complexity reduction of piecewise affine control law design. This chapter closely follows the results put forward in [13, 14]. Therein, a constructive procedure based upon convex liftings is presented, leading to an important implication in MPC design, that can be stated as follows: every continuous piecewise affine control law can be recovered via an MPC problem with a control horizon at most equal to two prediction steps. Central to the approach is the operation of convex lifting, and the related liftability condition, which plays a crucial role in the existence of an inverse optimal solution.

The main aim of this study is to address the difficulty arising in applications of IpCP in linear MPC, namely that the explicit solution of a linear MPC problem with respect to a quadratic cost function is not, in many cases, convexly liftable. With regard to this issue, in [14] it was shown that, in fact, any parameter space partition associated with a solution of a pQP can be sub-partitioned such that its convex liftability will be guaranteed. In control theory, however, such a subdivision necessarily increases the complexity of PWA control laws, which is commonly urged to be kept as low as possible for the sake of efficient eMPC implementations. Therefore, herein we rather aim at devising an alternative, yet convenient approach that would target pQP-based MPC problems which are typically not directly invertible, without generating any gain in complexity of the inherited solution. To tackle this challenge, we recall the open question of minimal complexity of the inverse optimality formulation, discussed in [12]. It is known that the IpL/QP via convex liftings replaces the constraint set of linear MPC problem based on prediction along a finite horizon with mixed constraints on the control and state variables over two prediction stages. Nevertheless, the main advantage of a significant decrease in the dimension of optimization arguments tends to come at the price of a relatively large set of constraints. Since not all the constraints practically contribute to the optimal solution, an effort is being put into reducing the constraint set towards a minimal representation necessary for the inverse optimality formulation. We exploit this fact and present an algorithmic procedure tailored for the class of MPC problems of interest. At this point the contribution of the proposed extended IpLP approach is twofold as it simultaneously treats the issue of invertibility, preconditioned by convex liftability, and eliminates the redundant constraints from the formulation. The procedure, however, exactly retrieves only a part of the solution. To establish the equivalence of the remaining portion of controller structure we adopt the concept of clipping strategy proposed in [10]. This inverse optimality-based procedure is performance lossless, i.e., the retrieved PWA function inherits all the performance, closed-loop stability and feasibility guarantees of the pQP-based/online MPC solution.

As outlined earlier, the efficient implementation of eMPC, often using embedded platforms, is closely associated with the structure of the explicit controller. Since the PWA control law is defined over a set of polytopic regions, it is clear that if the number of regions becomes large, the storage capacity of the implementation hardware may be easily exceeded. To overcome these limitations, numerous methods for complexity reduction via simplifying eMPC optimizers have been proposed. This is usually achieved by devising less complex equivalent replacement functions of the original PWA feedback with no implications on optimality, or various sophisticated, albeit suboptimal, approximations. A brief, yet recent overview can be found, e.g., in [9, 11]. We recall this topic with regard to the third feature of the approach. As it will be evidenced later, the combination of extended IpLP with the use of a simple clipping filter [10] may find its use in case a low-complexity eMPC controller is sought. Nevertheless, this ability is not an objective of this study, rather an additional gain of the approach. Hence, when an online implementation is of interest and certain assumptions are fulfilled, the less complex, yet performance-wise optimal explicit representation can offer an important advantage. In particular, real-time implementation complexity is determined not only by the memory needed to store the controller, but also by the computational time necessary to find and evaluate the corresponding control law. This is of imminent importance when targeting embedded hardware or controlling fast systems. In this light, extended IpLP with clipping shows its potential use in practical MPC applications; in comparison with the generic convex liftings-based IpCP, which as such does not yield any computational benefits within online eMPC. On the contrary, the obtained explicit solution may in case of directly non-invertible problems become too complex and thus intractable. Additionally, the proposed inverse optimization formulation may be as well cast and solved as an online MPC problem which would shed more light on the merits of its cost/constraints rearrangement.

3.2 Notation and Definitions

Throughout this study, $\mathbb{R}$, $\mathbb{N}_{>0}$ are used to denote the set of real numbers, and the field of positive integers, respectively. For a compact notation, let us define the index set $\mathcal{I}_N := \{i \in \mathbb{N}_{>0} \mid i \leq N\}$, for a given $N \in \mathbb{N}_{>0}$. For a point $x \in \mathbb{R}^d$, by $\mathbb{R}^{n_x}$ we denote the vector space containing the point x, hence in this case $\mathbb{R}^{n_x} = \mathbb{R}^d$. Next, given a finite set $\mathcal{S} = \{x_1, x_2, \ldots, x_n\}$ of n points, Card$(\mathcal{S})$ denotes the cardinal number of the set $\mathcal{S}$. By conv$(\mathcal{S})$ we denote its convex hull. Also, if $\mathcal{S}$ is a finite set of rays, i.e., $\mathcal{S} = \{y_1, \ldots, y_n\}$ then cone(S) represents the cone defined as follows:

$$\text{cone}(\mathcal{S}) = \{t_1 y_1 + \cdots + t_n y_n : \; t_i \geq 0, \; \forall 1 \leq i \leq n\} \tag{3.1}$$

A polyhedron is defined as a convex intersection of finitely many closed half-spaces. A polytope is a bounded polyhedron. Given a full-dimensional polyhedron $\mathcal{S}$ in $\mathbb{R}^d$, then $\mathcal{V}(\mathcal{S})$ denotes the set of its vertices; $\mathcal{R}(\mathcal{S})$ denotes the set of its rays. Also, int$(\mathcal{S})$ denotes its interior. $\text{Proj}_{\mathbb{S}}\mathcal{S}$ represents the orthogonal projection of the set $\mathcal{S}$ onto a subspace $\mathbb{S}$ of $\mathbb{R}^d$. Further, a face of $\mathcal{S}$ is the intersection of $\mathcal{S}$ and one of its supporting hyperplanes. A face of dimension $d-1$ is called a facet. The set of all facets of polyhedron $\mathcal{S}$ is denoted as $\mathcal{F}(\mathcal{S})$. Given a function $\kappa(x)$, dom$(\kappa(x))$ denotes its domain. In addition, given two sets $\mathcal{S}_1, \mathcal{S}_2 \subset \mathbb{R}^d$, by $\mathcal{S}_1 \oplus \mathcal{S}_2$ we denote their Minkowski sum defined as follows:

$$\mathcal{S}_1 \oplus \mathcal{S}_2 := \left\{ y \in \mathbb{R}^d \mid \exists x_1 \in \mathcal{S}_1, x_2 \in \mathcal{S}_2 \;\text{ s.t. }\; y = x_1 + x_2 \right\}.$$

Next, let us recall some necessary definitions.

Definition 3.1 A collection of $N \in \mathbb{N}_{>0}$ full-dimensional polyhedra $\mathcal{X}_i \subset \mathbb{R}^d$, denoted as $\mathcal{X} = \{\mathcal{X}_i\}_{i \in \mathcal{I}_N}$, is called a polyhedral partition of a polyhedron $\Omega \subseteq \mathbb{R}^d$ if:

(1) $\Omega = \bigcup_{i \in \mathcal{I}_N} \mathcal{X}_i$,
(2) $\text{int}(\mathcal{X}_i) \cap \text{int}(\mathcal{X}_j) = \emptyset$ with $i \neq j$, $(i, j) \in \mathcal{I}_N^2$.

In addition, we refer to polyhedra $(\mathcal{X}_i, \mathcal{X}_j)$ as neighbours, or adjacent, if $(i, j) \in \mathcal{I}_N^2$, $i \neq j$ and $\dim(\mathcal{X}_i \cap \mathcal{X}_j) = d - 1$. If a polyhedral partition is a collection of polytopes, then it is called a polytopic partition. $\mathcal{X}_i$ are referred to as regions of the partition $\mathcal{X}$.

In this chapter, a cell complex is understood as a polyhedral partition whose facet-to-facet property [20] is fulfilled, meaning that any two neighbouring regions share a common facet. Note that a complete, more general definition of cell complex can be found in [6]. However, for simplicity, we restrict our attention to the property of interest.

Definition 3.2 Given a polyhedral partition $\mathcal{X} = \{\mathcal{X}_i\}_{i \in \mathcal{I}_N}$ of a polyhedron $\Omega \subseteq \mathbb{R}^d$, a convex lifting is described by the function $z : \Omega \to \mathbb{R}$ with:

$$z(x) = a_i^T x + b_i \quad \text{for any } x \in \mathcal{X}_i, \tag{3.2}$$

and $a_i \in \mathbb{R}^d$, $b_i \in \mathbb{R}$, $\forall i \in \mathcal{I}_N$; such that the following conditions hold:

(1) $z(x)$ is continuous over Ω,
(2) for each $i \in \mathcal{I}_N$, $z(x) > a_j^T x + b_j$ for all $x \in \mathcal{X}_i \setminus \mathcal{X}_j$ and all $j \neq i$, $j \in \mathcal{I}_N$.

According to Lemma 1 in [15], a polyhedral partition admitting a convex lifting is required to be a cell complex. Conversely, a cell complex is not necessarily convexly liftable. The necessary and sufficient conditions for a cell complex to be convexly liftable[1] are referred to [14] and the references therein.

Moreover, a convex lifting can be in fact seen as a convex surface composed of the convex lower boundary of a polyhedron in an augmented space. This particular polyhedron can be described by the following definition:

Definition 3.3 A given cell complex $\mathcal{X} = \{\mathcal{X}_i\}_{i \in \mathcal{I}_N}$ of a polyhedron $\Omega \subseteq \mathbb{R}^d$ has an affinely equivalent polyhedron if there exists a polyhedron $\tilde{\mathcal{X}} \subset \mathbb{R}^{d+1}$ such that for each $i \in \mathcal{I}_N$:

(1) $\exists F_i \in \mathcal{F}(\tilde{\mathcal{X}})$ satisfying: $\text{Proj}_{\mathbb{R}^d} F_i = \mathcal{X}_i$,
(2) if $\underline{z}(x) = \min\limits_{z} z \;$ s.t. $[x^T \; z]^T \in \tilde{\mathcal{X}}$, then $[x^T \; \underline{z}(x)]^T \in F_i$ for $x \in \mathcal{X}_i$.

The second condition in Definition 3.3 implies that the set of facets of $\tilde{\mathcal{X}}$ at the lower values of z is exclusively considered. These lower facets F_i build a convex surface in $\mathbb{R}^{d+1}$ known as the previously defined convex lifting, and their image via the orthogonal projection onto $\mathbb{R}^d$ recovers the cell complex $\mathcal{X}$.

We remark that an additional set of definitions is provided within Sect. 3.5 as it closely relates to the problems treated therein.

3.3 Inverse Parametric Optimization via Convex Liftings

This section aims to recall the constructive solution to the inverse parametric linear/quadratic programming problem via convex liftings. For ease of presentation, let us start from a larger perspective—with the definition of a parametric convex[2] programming problem, which can be cast as follows:

---

[1]Note that the liftability condition does not restrict $\mathcal{X}$ to be a partition of a polytope/polyhedron. This property, however, becomes important in the context of convexity of inverse parametric linear/quadratic programming problems.

[2]Convexity of the optimization problem (3.3) is ensured by convexity of f, g_i, and h_j in θ and x.

$$\begin{aligned} \min_{\theta} \quad & f(\theta, x) \\ \text{s.t.} \quad & g_i(\theta, x) \leq 0, \ \forall i \in \mathcal{I}_p, \\ & h_j(\theta, x) = 0, \ \forall j \in \mathcal{I}_q, \end{aligned} \tag{3.3}$$

where θ, x denote the decision variable and the parameter, respectively. The terms $g_i(\theta, x)$, $h_j(\theta, x)$ represent the left-hand sides of respective inequality and equality constraints for the optimization problem (3.3).

In view of parametric linear/quadratic programming, the equality constraints are omitted for simplicity. Moreover, $g_i(\theta, x)$, $\forall i \in \mathcal{I}_p$ define linear constraints on θ and/or x, from the geometrical point of view representing a polyhedron in the augmented space of $\left[\theta^T \ x^T\right]^T$. Still, $f(\theta, x)$ stands for a linear or quadratic cost function, which has the following description:

$$f(\theta, x) = \theta^T H \theta + (F^T x + Y)^T \theta, \tag{3.4}$$

with a positive semidefinite matrix $H = H^T \succeq 0$ and appropriately dimensional matrices F, Y.

It is well known from the works of [4, 19, 22] that the optimal solution to a parametric linear/quadratic programming problem is a piecewise affine function defined over a polyhedral partition. It is shown therein that the optimal solution to a pQP problem with respect to $H \succ 0$ is continuous and unique. Otherwise, the continuity and the uniqueness properties may not be preserved for the case of a pLP. Fortunately, it is shown e.g., in [17] that an equivalently continuous optimal solution can always be selected. Therefore, the IpL/QP can focus on continuous PWA functions.

Given a continuous PWA function defined over a polyhedral partition, IpL/QP aims to find an appropriate optimization problem, characterized by a linear/quadratic cost function and a set of linear constraints such that the optimal solution to this optimization problem is equivalent to the given PWA function. Mathematically, let $f_{\text{pwa}}(x) : \Omega \subseteq \mathbb{R}^{n_x} \to \mathbb{R}^{n_u}$ denote a continuous PWA function to be recovered, defined as follows:

$$f_{\text{pwa}}(x) = F_i x + G_i \ \text{ for } x \in \mathcal{X}_i. \tag{3.5}$$

IpL/QP aims to find a cost function $J(x, z, u)$ and a set of four matrices H_x, H_u, H_z, K such that:

$$f_{\text{pwa}}(x) = \text{Proj}_{\mathbb{R}^{n_u}} \arg \min_{\left[z^T u^T\right]^T} J(x, z, u) \ \text{ s.t. } \ H_x x + H_z z + H_u u \leq K, \tag{3.6}$$

where z represents an auxiliary variable for the recovered optimization problem and will be shown to be a 1-dimensional variable.

Assumption 3.1 As reported in [13, 15], this inverse optimality problem is valid under the following standing assumptions:

A1.1 The polyhedral partition $\mathcal{X}$ is the partition of a polyhedron Ω.
A1.2 The polyhedral partition $\mathcal{X}$ is convexly liftable.

Note that Assumption A1.1 is of help in order to guarantee the convexity of the recovered optimization problem. On the other hand, Assumption A1.2 is not restrictive due to Theorem 2 and Lemma 2, also reported in [15]. Accordingly, if the given polyhedral partition does not satisfy Assumption A1.2, one can refine it such that the internal boundaries are maintained and the new partition is convexly liftable. Such subdivision will be discussed in more detail for the cases often encountered in MPC in Sect. 3.5.

As outlined in the introductory section, central to the constructive procedure of [13] is the operation of convex lifting for the given partition, recalled in Definition 3.2. An algorithm for the construction of convex liftings is referred to [14] and extended to a cell complex of a polyhedron via Theorem 1 in [15]. The following sets are also defined:

$$
\begin{aligned}
V_x &= \bigcup_{i\in\mathcal{I}_N} \mathcal{V}(\mathcal{X}_i),\ R_x = \bigcup_{i\in\mathcal{I}_N} \mathcal{R}(\mathcal{X}_i),\\
V_{[x^T\,z\,u^T]^T} &= \left\{\left[x^T\ z(x)\ f_{\text{pwa}}^T(x)\right]^T \mid x\in V_x\right\},\\
R_{[x^T\,z\,u^T]^T} &= \left\{\left[r^T\ \hat{z}(r)\ \hat{f}^T(r)\right]^T \mid r\in R_x,\ \begin{matrix}\hat{z}(r)=a_i^T r\\ \hat{f}(r)=F_i r\end{matrix}\quad \text{if } r\in\mathcal{R}(\mathcal{X}_i)\right\},\\
\Pi_v &= \operatorname{conv}(V_{[x^T\,z\,u^T]^T}),\ \Pi_r=\operatorname{cone}(R_{[x^T\,z\,u^T]^T}),\\
\Pi &= \Pi_v\oplus\Pi_r.
\end{aligned}
\tag{3.7}
$$

Following (3.7), recall that V_x, R_x denote the sets of vertices and rays of the cell complex $\mathcal{X}$, respectively. Also, $V_{[x^T\,z\,u^T]^T}$, $R_{[x^T\,z\,u^T]^T}$ stand for the sets of augmented vertices and rays of $\mathcal{X}$, into $\mathbb{R}^{n_x+n_u+1}$. Note, however, that under convex liftings, the augmented terms for a vertex are different from the one for a ray as shown in (3.7). More clearly, consider a vertex v and a ray r of region $\mathcal{X}_i$ in the cell complex $\mathcal{X}$. After embedding to the space of convex lifting, the augmented term corresponding to v is $a_i^T v + b_i$. However, under this embedding, the augmented term of r becomes $a_i^T r$. The resulting constraint set Π is computed via the Minkowski sum of Π_v and Π_r. This is basically due to Minkowski-Weyl theorem (see Corollary 7.1b in [18]).

Based on the above notation, we recall here the main result for IpL/QP problem via the following theorem.

Theorem 3.1 *Given a continuous PWA function $f_{\text{pwa}}(x)$, defined over a cell complex $\mathcal{X}=\{\mathcal{X}_i\}_{i\in\mathcal{I}_N}$ satisfying Assumptions A1.1, A1.2, then $f_{pwa}(x)$ is the image via the orthogonal projection of the optimal solution to the following parametric linear programming problem:*

$$
\min_{[z\,u^T]^T} z \quad \textit{subject to} \quad \left[x^T\,z\,u^T\right]^T \in \Pi. \tag{3.8}
$$

For the proof of Theorem 3.1 we refer to Sect. 5 of [15]. Note, however, that as discussed in [12], not all constraints of Π are meaningful in (3.8). Many of them contribute to the construction of feasible region instead of building its optimal solution. Therefore, these constraints may need to be removed to simplify the optimization problem. Algorithms put forward in [12] can be effectively utilized to carry out this procedure.

We remark that this theorem is meaningful in the context of implicit MPC where it can help to overcome its computational limitations by implementing this alternative optimization approach and solving the LP[3] (3.8) at each sampling time. In this study we rather exploit this concept for efficient explicit MPC implementations. This point is discussed in Sect. 3.5 and clarified via a case study in Sect. 3.6.

3.4 Application to Linear MPC Problems

This section aims to recall an important result of IpL/QP related to the linear model predictive control. Consider a discrete-time, linear invariant system:

$$x(k+1) = Ax(k) + Bu(k), \tag{3.9}$$

where (A, B) is stabilizable. In MPC problems, a cost function is typically defined over a finite prediction horizon $N \in \mathbb{N}_{>0}$ as follows:

$$J(U, x(k)) = \sum_{i=0}^{N-1} \ell_i(x_{k+i|k}, u_{k+i|k}) + V_N(x_{k+N|k}) \tag{3.10}$$

where $x_{k+i|k} \in \mathbb{R}^{n_x}$ is the $(k+i)$-th prediction of the system state at time k, $u_{k+i|k}$ denotes the control action at time instant $k+i$, and $U = [u_{k|k}^T, \ldots, u_{k+N-1|k}^T]^T$.

The $\ell_i(x_{k+i|k}, u_{k+i|k})$ term of the objective represents a stage cost $\forall i \in \mathcal{I}_{N-1} \cup \{0\}$ and $V_N(x_{k+N|k})$ denotes the terminal penalty. Moreover, the state and control variables are typically required to satisfy constraints:

$$x_{k+i|k} \in \mathbb{X},\ u_{k+i|k} \in \mathbb{U},\ \forall i \in \mathcal{I}_{N-1} \cup \{0\},\ x_{k+N|k} \in \mathbb{X}_f \tag{3.11}$$

with polyhedra $\mathbb{X}$, $\mathbb{U}$ containing the origin in their interior. A suitable terminal constraint set $\mathbb{X}_f \subset \mathbb{X}$ is considered to guarantee closed-loop stability.

A linear MPC problem thus aims to solve the following optimization problem:

$$\min_U J(U, x(k)) \quad \text{subject to} \quad (3.11) \tag{3.12}$$

where the stage costs and the terminal cost tend to take one of the following forms:

[3] The inverse optimality problem can be as well cast and solved implicitly/explicitly as a QP/pQP.

(1) 2-norm: $\ell_i(x_{k+i|k}, u_{k+i|k}) = x_{k+i|k}^T Q x_{k+i|k} + u_{k+i|k}^T R u_{k+i|k}$, $V_N(x_{k+N|k}) = x_{k+N|k}^T P x_{k+N|k}$, where $Q = Q^T$ is a positive semidefinite matrix and R, P are symmetric, positive definite matrices,

(2) $1/\infty$-norm: $\ell_i(x_{k+i|k}, u_{k+i|k}) = \|Qx_{k+i|k}\|_p + \|Ru_{k+i|k}\|_p$, $V_N(x_{k+N|k}) = \|Px_{k+N|k}\|_p$, where $p = 1/\infty$ and Q, R, P are of appropriate dimension.

As shown in [3, 4], the optimal solutions of such linear MPC problems of modest size can be found explicitly by parametric linear/quadratic programming, as follows:

$$U^* = \arg\min_U J(U, x(k)) \text{ s.t. } GU \leq W + Ex(k), \tag{3.13}$$

where the control input sequence U is regarded as the decision variable, the current state $x(k)$ represents the parameter, and G, W, E are appropriate matrices describing the constraints (3.11). In the implementation, the receding horizon MPC feedback becomes $u_k = \text{Proj}_{\mathbb{R}^{n_u}} U^*$. This explicit solution to a linear MPC problem has a piecewise affine structure, and therefore it inherits also the properties of an inverse optimality problem recalled earlier. Based on this property, the main result in this section is summarized via the following theorem.

Theorem 3.2 *The continuous explicit solution of a generic linear MPC problem with respect to a linear/quadratic cost function is equivalently obtained through a linear MPC problem with a linear cost function and the control horizon at most equal to two prediction steps.*

Interested reader is further referred to [13] for the proof. Therein, numerical examples can also be found, illustrating the application of the generic inverse optimality solution to problems that are directly invertible. This allows one to recover the optimal explicit continuous feedback inherited from the linear MPC problem via an inverse pL/QP problem, while the structure of the PWA control law and the underlying parameter partition are maintained. The following section is going further and focuses on more realistic MPC scenarios where the convex liftability of the state-space partition is not a priori guaranteed, hence requiring to appropriately revisit the IpCP procedure in order to find an inverse problem formulation and its solution.

3.5 A Novel Approach for a Class of Primarily Non-invertible Control Laws Obtained from MPC Problems

As outlined in the introduction, we aim at recovering linear MPC problems whose explicit solution obtained by parametric convex programming leads to a convexly non-liftable parameter partition, which therefore may not be used within the generic IpCP scheme proposed in [13] where the authors build upon the assumption that the state-space partition to be recovered is a convexly liftable cell complex (cf. Definition 3.2). Here, we are, however, interested in situations when the partition inherited from the solution of an MPC problem is found to be a convexly non-liftable polyhedral partition or cell complex, eventually. As mentioned recently in [14], the parameter space partition associated with the optimal solution to a parametric quadratic programming problem, is in many cases not convexly liftable. We remark that this tends to be the case for most of practical set-ups involving input constraints, usually yielding explicit controllers with many saturated regions.[4] One may observe that the larger their number is, the smaller the likelihood of constructing a convex lifting gets. In such a case, a straightforward scheme to approach this problem would be to rearrange the possibly disconnected and non-convex union of polyhedra sharing the same saturated control law, in a way to allow for convex liftability of the entire partition. Despite the existence of several region merging techniques, none of them can in fact guarantee this property. Even if we manage to remove the saturated regions completely, as per separation-based complexity reduction of [11], the collection of unsaturated polyhedra left to be lifted may very often be non-convex, which prevents us from constructing its affinely equivalent polyhedron while recovering the explicit controller without giving rise to some additional regions.

This problem may be as well treated by an appropriate sub-partitioning of a given convexly non-liftable polyhedral partition, as outlined recently in [14]. Therein, it is shown that any parameter space polyhedral partition associated with a solution of a pQP problem can be sub-partitioned such that its internal boundaries are preserved and the new cell complex is convexly liftable. Despite the new hyperplane arrangement allows the convex liftability to be retrieved, it naturally leads to a refining that can easily increase the complexity of a PWA feedback to a large extent. Implementation of such an eMPC controller may thus become a costly, if not intractable, alternative to the original one.

In view of the aforementioned practical aspects, let us now briefly formulate an extended inverse optimality problem to be tackled henceforth.

Problem 3.1 Given a, possibly, convexly non-liftable polyhedral partition/cell complex[5] $\mathcal{R} = \{\mathcal{R}_i\}_{i \in \mathcal{I}_R}$ of a polyhedron $\Omega \subseteq \mathbb{R}^{n_x}$, associated with a continuous piecewise affine function $\kappa(x) = f_{\text{pwa}}(x) : \Omega \to \mathbb{R}^{n_u}$, find a linear cost function $J(x, z, u)$, and matrices H_x, H_u, H_z, K such that the inverse optimal solution

$$\begin{cases} \tilde{\kappa}(x) = \text{Proj}_{\mathbb{R}^{n_u}} \arg \min\limits_{[z\, u^T]^T} J(x, z, u), \\ \text{s.t. } \; H_x x + H_z z + H_u u \leq K. \end{cases} \tag{3.14}$$

returns an equivalent replacement of $\kappa(x)$ when passed through a suitable filter $\phi(\cdot)$.

Assumption 3.2 At this point, some assumptions need to be stated to make the present approach reasonable from both the construction and practical viewpoint:

[4]The definition of a un/saturated region is given in Sect. 3.5.1.

[5]Within this section and henceforth we slightly abuse the notation $\mathcal{R}$ to refer to state-space partitions inherited from MPC problems.

A2.1 The parametric linear programming (pLP) problems are exclusively considered as possible candidates for the inverse optimality solutions.
A2.2 $\mathcal{R}$ is the partition of a polytope Ω.
A2.3 $\kappa(x)$ is the explicit optimal solution to a pQP-based linear MPC problem.
A2.4 $\kappa(x)$ is partially saturated with the corresponding polytopic regions causing $\mathcal{R}$ to be convexly non-liftable.

Assumption A2.1 provides a manageable framework for the constructive inverse optimality procedures by imposing linearity of the candidate pCP problems. Although a quadratic cost function may be used as well, in the scope of this study an IpLP is of interest for brevity, yet with a straightforward extension towards IpQP. Meanwhile, Assumption A2.2 merely requires the feasible set to be a polytope, which is usual in the context of MPC. However, it is not restrictive, and the procedure can be easily extended to the case where the feasible set is a polyhedron. The other two assumptions stem from a more practical reasoning outlined at the beginning of this section. By Assumption A2.3 we focus our attention rather on explicit solutions obtained by parametric quadratic programming. Considering pQP in this context not only relates to the nature of most linear MPC problems, but it is mainly due to the typically inherent curse of direct non-liftability of the associated solutions, defined over convexly non-liftable state-space partitions.[6] This is in contrast to pLP problems which are directly invertible without any refinement required [14], and also for this fact are not of interest here. The last assumption stems from numerous observations encountered in practical MPC problems where the explicit receding horizon feedbacks usually contain many regions for which the associated optimal control action is either constantly on the upper limit or constantly on the lower limit. This reasonable idea was also central to the works of [10, 11] aiming for complexity reduction of explicit MPC feedback laws. Recall that, in general, $\mathcal{R}$ inherited from a pQP may be convexly non-liftable since it is mostly a strict polyhedral partition, i.e., facet-to-facet property does not hold [20]; however, it may as well be a cell complex that is not convexly liftable simply due to the nature of its hyperplane arrangement (see e.g., the example in [15]). The second part of Assumption A2.4 moreover hints that the liftability issue is typically induced by the existence and structural properties of the collections of aforementioned saturated polytopic regions. This becomes more and more likely as the cardinality of such collections increases, at the price of the remaining collection of unsaturated regions, which is itself thus in a vast majority of cases found to be convexly liftable. In fact, as it is evidenced, e.g., in [10, 11], a typical explicit MPC feedback law contains a significantly smaller number of unsaturated regions as compared to the number of saturated ones. In addition, if the convex liftability of the collection of unsaturated regions, as invoked via Assumption A2.4, practically happened not to hold, it is still possible to use sub-partitioning to preserve it. With the usual structural proportions mentioned above, it would however become still very cheap compared to a complete partition refining—actually, as a matter of fact, such a possible slight increase in controller complexity would be most likely

[6] This structural property is related to the piecewise quadratic cost function. This is to be compared with the piecewise linear and convex cost function of a pLP-based MPC problem.

eliminated by a notable complexity reduction indicated by the approach proposed in the following subsection. Needless to say, no such an issue has been encountered within tested practical MPC scenarios, neither is it present in the case study put forward in Sect. 3.6.

To review the objective, by exploiting the procedure of inverse parametric optimization via convex liftings [13], we aim to reconstruct an appropriate linear programming problem (3.14) with respect to a given piecewise affine function $\kappa(x)$ defined over a given, possibly convexly non-liftable polyhedral partition/cell complex of the parameter space, $\mathcal{R}$, such that the optimal solution of this reconstructed problem is equivalent to the given PWA function. The equivalence requires that the boundary between two different regions of the parameter space partition corresponding to two different affine functions is preserved, whereas a proper rearrangement of the subset of the domain of definition, over which the PWA function is saturated, is allowable. This property is effectively exploited within this study. An example is the new convexly liftable cell complex $\tilde{\mathcal{R}}$ corresponding to a structurally modified, yet optimal PWA solution function $\phi(\tilde{\kappa}(x))$, as shown in Sect. 3.5.

The following subsections present the main contribution of this study—the extension to convex liftings-based inverse parametric convex programming procedure, namely IpLP, with the main focus on a wide and challenging class of primarily non-invertible linear MPC problems having the explicit optimal solution in the form as per Assumption 3.2.

3.5.1 Additional Notation and Definitions

Let us now append some necessary notation, related closely to the topic treated in this section. We first state a set of definitions found to be useful and hence adopted from [10] to keep the notation compact. To make the further explanation more clear, we restrict ourselves to single-input systems (cf. Remark 3.2).

Definition 3.4 Let $\overline{\kappa}$ and $\underline{\kappa}$ denote, respectively, the maximum and minimum values which the PWA function $\kappa(x) := f_i^T x + g_i$ attains over $\mathrm{dom}(\kappa(x))$. Denote by $\mathcal{I}_{\max}$ ($\mathcal{I}_{\min}$) the index set of regions where $\kappa(x)$ is saturated at the maximum (minimum), and let $\mathcal{I}_{\text{sat}} = \mathcal{I}_{\max} \cup \mathcal{I}_{\min}$. We call a region $\mathcal{R}_i$ the saturated region if it is either saturated at the minimum or at the maximum, i.e., if $i \in \mathcal{I}_{\text{sat}}$. Otherwise the region is called unsaturated. The index set of unsaturated regions is denoted by $\mathcal{I}_{\text{unsat}}$.

Definition 3.5 Given a continuous PWA function $\kappa(x)$, defined over the parameter space partition $\mathcal{R} = \{\mathcal{R}_i\}_{i=1}^{R}$, we call the PWA function $\tilde{\kappa}(x) := \tilde{f}_j^T x + \tilde{g}_j$ a suitable augmentation of $\kappa(x)$ if the following properties hold:

P1: $\tilde{\kappa}(x)$ is defined over partition $\tilde{\mathcal{R}} = \{\tilde{\mathcal{R}}_j\}_{j=1}^{\tilde{R}}$ such that $\bigcup_{j \in \mathcal{I}_{\tilde{R}}} \tilde{\mathcal{R}}_j = \bigcup_{i \in \mathcal{I}_R} \mathcal{R}_i$, i.e., $\mathrm{dom}(\tilde{\kappa}(x)) = \mathrm{dom}(\kappa(x))$.

P2: $\tilde{\kappa}(x) = \kappa(x), \ \forall x \in \mathcal{R}_{\mathcal{I}_{\text{unsat}}}$.

P3: $\tilde{\kappa}(x) \geq \overline{\kappa},\ \forall x \in \mathcal{R}_{\mathcal{I}_{\max}}$.
P4: $\tilde{\kappa}(x) \leq \underline{\kappa},\ \forall x \in \mathcal{R}_{\mathcal{I}_{\min}}$.

In addition, we introduce several polyhedral operations, which were shown to be convenient in the context of IpCP complexity analysis [12], and are utilized here to clarify the algorithmic notation.

Given a full row rank matrix $M \in \mathbb{R}^{r \times (d+1)}$, by $\mathcal{P}(M)$ we denote the polyhedron

$$\mathcal{P}(M) = \{x \in \mathbb{R}^d \mid M(\cdot, 1:d)x \leq M(\cdot, d+1)\},$$

where $M(\cdot, i)$ denotes the ith column of the matrix M and $M(\cdot, i:j)$ represents the matrix composed by the ith–jth columns of M. Conversely, given a polyhedron $\mathcal{P}$, by $\mathcal{P}^{-1}(P)$ we denote the minimal representation (in terms of dimension) of a matrix M satisfying $P = \{x \in \mathbb{R}^d \mid M(\cdot, 1:d)x \leq M(\cdot, d+1)\}$. Note that $\mathcal{P}^{-1}(P)$ is not unique due to the following observation: $\mathcal{P}(M) = \mathcal{P}(\alpha M), \alpha > 0$.

For ease of presentation let us also define an operator for removal of redundant constraints. Given two sets of constraints corresponding to two polyhedra P_M, P_N, $M \in \mathcal{P}^{-1}(P_M) \subset \mathbb{R}^{r_M \times (d+1)}$, $N \in \mathcal{P}^{-1}(P_N) \subset \mathbb{R}^{r_N \times (d+1)}$, by $\text{RmRdd}(M, N)$ we denote the set of constraints characterizing P_M which are not redundant in the representation of P_N. We remark that there exist numerous algorithms for removing redundant constraints. Herein, we recall the one presented in [16] through a mathematical description as follows: $\text{RmRdd}(M, N) = K \in \mathbb{R}^{r_K \times (d+1)}$ s.t.

- K is a sub-block of M,
- for any $i \in \mathcal{I}_{r_K}$, $\max\limits_{x | x \in P_N} K(i, 1:d)x > K(i, d+1)$.

In addition, for $P_M \subseteq P_N$, $M \in \mathcal{P}^{-1}(P_M)$, $N \in \mathcal{P}^{-1}(P_N)$, $\text{RmRdd}(N, M) = \emptyset$.

3.5.2 *Extended Convex Liftings-Based Inverse pLP with Clipping*

In this subsection we show how to recover solutions to a given class of MPC problems via another reformulated IpLP-based MPC problem with the prediction horizon equal to two prediction steps while preserving optimality.

Given a convexly non-liftable polyhedral partition/cell complex $\mathcal{R} = \{\mathcal{R}_i\}_{i \in \mathcal{I}_R}$, the proposed procedure exploits the possibly non-convex collection of unsaturated regions, $\mathcal{R}_{\mathcal{I}_{\text{unsat}}}$, which is assumed to be convexly liftable (recall Assumption A2.4 and related comments). The first step is hence to find the gains (a_i, b_i), $\forall i \in \mathcal{I}_{\text{unsat}}$ of its associated convex lifting. This can be carried out efficiently by using the constructive procedure (Algorithm 3.1 in [15]) first proposed in [14], yielding $z(x) = a_i^T x + b_i$, with $x \in \bigcup_{i \in \mathcal{I}_{\text{unsat}}} \mathcal{R}_i$. Further, instead of constructing an affinely equivalent polyhedron, we build the vertex set $V^{\text{unsat}}_{[x^T z\, u^T]^T}$ and compute its dual, half-space representation $\Pi^{\text{unsat}}_{[x^T z\, u^T]^T} = \text{conv}(V^{\text{unsat}}_{[x^T z\, u^T]^T})$ as per Theorem 5.3 in [13], slightly modified for our case. This polytope is to be used to construct a constraint set for Problem 3.1 by extending the given convex lifting and control law over the saturated regions.

As outlined in the introduction, for the further course of this study we exploit the question of minimal complexity of the inverse optimality formulation, discussed in [12]. In particular, not all the constraints—defining half-spaces of the set $\Pi_{[x^T z\, u^T]^T}$ within the generic IpCP problem formulation are meaningful from the optimization point of view, in the sense that they are not active at the optimum. Therefore, it is sufficient to exclusively consider only the active constraints, which are defined by corresponding supporting hyperplanes containing certain facets of $\Pi_{[x^T z\, u^T]^T}$. We adopt this idea to preserve the active constraints which are thus not redundant in the half-space representation of $\Pi^{\text{unsat}}_{[x^T z\, u^T]^T}$. Such a relaxation naturally modifies this set in both parameter and argument space; hence it needs to be associated with an appropriate restriction in the parameter space, which bounds the domain of validity for the optimal solution. This procedure is summarized in Algorithm 3.1.

Algorithm 3.1 Construction of constraint set for the extended IpCP

INPUT: Convexly non-liftable polyhedral partition/cell complex $\mathcal{R} = \{\mathcal{R}_i\}_{i=1}^{R}$ due to $\mathcal{R}_{\mathcal{I}_{\text{sat}}}$,

$$M^{\text{unsat}}_{[x^T z\, u^T]^T} = \mathcal{P}^{-1}(\Pi^{\text{unsat}}_{[x^T z\, u^T]^T}),\ \begin{bmatrix} \underline{z} \\ \underline{u} \end{bmatrix}_{\text{pwa}} = \arg\min_{[z\, u^T]^T} z \quad \text{s.t. } [x^T z\, u^T]^T \in \Pi^{\text{unsat}}_{[x^T z\, u^T]^T}$$

OUTPUT: Π_{min}, constraint set $\tilde{\Pi}$ associated with $\tilde{\mathcal{R}}$ (which is to be obtained from (3.15))

1: $M_{\text{min}} = [\,]$.
2: **for** $i = 1 : \text{Card}(\mathcal{I}_{\text{unsat}})$
3: Find the polyhedron P_0 described by the active constraints at $[x^T \underline{z}\, \underline{u}^T]^T$ for $x \in \mathcal{R}_{\mathcal{I}_{\text{unsat}},i}$.
4: $M_0 = \mathcal{P}^{-1}(P_0)$, $M_{\text{min}} = \begin{bmatrix} M_0 \\ M_{\text{min}} \end{bmatrix}$.
5: **end for**
6: $\Pi_{\text{min}} = \mathcal{P}(M_{\text{min}})$, $\Pi_x = \text{Proj}_{\mathbb{R}^{n_x}} \Pi_{\text{min}}$.
7: $M_x = \mathcal{P}^{-1}(\Pi_x)$, $M_{\text{f}} = \mathcal{P}^{-1}(\text{conv}(\mathcal{R}))$.
8: $\overline{M}_{\text{f}} = \text{RmRdd}(M_{\text{f}}, M_x)$.
9: $\overline{M} = \left[\overline{M}_{\text{f}}(\cdot, 1 : n_x)\ 0_{m\times(n_u+1)}\ \overline{M}_{\text{f}}(\cdot, n_x + 1)\right]$. Note: $\overline{M}_{\text{f}} \in \mathbb{R}^{m\times(n_x+1)}$
10: $M_{\text{min}} = \begin{bmatrix} M_{\text{min}} \\ \overline{M} \end{bmatrix}$, $\tilde{\Pi} = \mathcal{P}(M_{\text{min}})$.

Remark 3.1 Note that the explicit solution $[\underline{z}\, \underline{u}^T]^T$ required as input for Algorithm 3.1 is known a priori with respect to a given continuous PWA function and constructed convex lifting associated with $\mathcal{R}_{\mathcal{I}_{\text{unsat}}}$, without the need to solve the minimization problem. Step 8 is present to avoid the redundancy phenomena while adding supplementary constraints given by $\overline{M}_{\text{f}}$. In addition, we remark that for the MPC control laws designed with the feasible set $\Omega = \text{conv}(\mathcal{R})$ to be positively invariant, these constraints (Steps 6-9) can be further omitted as long as the initial state $x_0 \in \Omega$. This may reduce the complexity of the formulation in implicit MPC implementations.

This algorithmic procedure first retrieves the minimal half-space representation of the set $\Pi^{\text{unsat}}_{[x^T z\, u^T]^T}$, potentially useful for the IpCP problem formulation, and given by an unbounded polyhedron Π_{min} (Steps 1–6). Feasibility of this set is further provided through a restriction imposed on its parameter subspace (Step 7). Additional redundant constraints that may emerge at this point are effectively removed via Step 8. The resulting constraint set is thus given by $\tilde{\Pi}$ (Step 10). Now, recall that, geometrically, each active constraint corresponds to a hyperplane of dimension $n_x + n_u$. Therefore, we are interested in the constraints corresponding to the supporting hyperplanes of $\tilde{\Pi}$ which contain the n_x-faces whose orthogonal projection onto $\mathbb{R}^{n_x}$ retrieves the parameter space partition $\mathcal{R}$ associated with the optimal PWA function $\kappa(x)$. This is indeed not quite the case here, since $\mathrm{Proj}_{\mathbb{R}^{n_x+n_u}} \tilde{\Pi}$ (or $\mathrm{Proj}_{\mathbb{R}^{n_x+1}} \tilde{\Pi}$—affinely equivalent polyhedron to $\tilde{\mathcal{R}}$) is formed as a convex extension of $\mathrm{Proj}_{\mathbb{R}^{n_x+n_u}} \Pi^{\mathrm{unsat}}_{[x^T \underline{z}\, \underline{u}^T]^T}$ ($\mathrm{Proj}_{\mathbb{R}^{n_x+1}} \Pi^{\mathrm{unsat}}_{[x^T \underline{z}\, \underline{u}^T]^T}$, respectively) over $\bigcup_{i \in \mathcal{I}_{\mathrm{sat}}} \mathcal{R}_i$, bounded by the restriction in the parameter space $\mathbb{R}^{n_x}$ imposed by the feasible set $\Omega = \mathrm{dom}(\kappa(x))$. It follows that the explicit optimal solution $(\tilde{\kappa}(x), \tilde{\mathcal{R}})$ to thusly cast "inverse" problem is not equivalent anymore, in the sense that $\tilde{\kappa}(x)$ does not coincide with the original PWA feedback $\kappa(x)$, namely in its saturated portions defined over $\mathcal{R}_{\mathcal{I}_{\mathrm{sat}}}$. The cell complex $\tilde{\mathcal{R}}$ in this way inherits a collection of regions with similar (or equivalent, if $\bigcup_{i \in \mathcal{I}_{\mathrm{unsat}}} \mathcal{R}_i$ is convex) partitioning as $\mathcal{R}_{\mathcal{I}_{\mathrm{unsat}}}$.

The following theorem summarizes the minimal formulation of an extended IpLP problem resulting directly from Algorithm 3.1.

Theorem 3.3 *The image via orthogonal projection onto $\mathbb{R}^{n_u}$ of the optimal solution to the following optimization problem*

$$\min_{[z\, u^T]^T} \quad z \quad \text{s.t.} \quad [x^T z\, u^T]^T \in \tilde{\Pi} \tag{3.15}$$

where $\tilde{\Pi}$ is obtained from Algorithm 3.1, is a suitable augmentation of the given continuous PWA function $\kappa(x) : \Omega \to \mathbb{R}^{n_u}$, denoted as $\tilde{\kappa}(x)$ and associated with a convexly liftable cell complex $\tilde{\mathcal{R}} = \{\tilde{\mathcal{R}}_j\}_{j \in \mathcal{I}_{\tilde{\mathcal{R}}}}$.

Proof Follows directly from Algorithm 3.1 and from Definition 3.5 of $\tilde{\kappa}(x)$. □

As implied by Theorem 3.3, solution of LP (3.15) renders a suitable augmentation $\tilde{\kappa}(x)$ of the original piecewise affine function $\kappa(x)$. The augmented function, however, cannot be readily applied as an explicit receding horizon MPC feedback since, in general, $\tilde{\kappa}(x) \neq \kappa(x)$ for $x \in \mathcal{R}_{\mathcal{I}_{\mathrm{sat}}}$. Therefore, to achieve the equivalence of the two, we adopt the concept of a simple clipping filter proposed in [10], and recalled here as follows:

Theorem 3.4 ([10]) *Consider a saturated continuous PWA function $\kappa(x)$ and its suitable augmentation $\tilde{\kappa}(x)$ obtained per Theorem 3.3. Let*

$$\phi(\tilde{\kappa}(x)) := \max\left\{\underline{\kappa}, \min\{\tilde{\kappa}(x), \overline{\kappa}\}\right\} \tag{3.16}$$

Then the equivalence $\phi(\tilde{\kappa}(x)) = \kappa(x)$ is established for all $x \in \mathrm{dom}(\kappa(x))$.

Proof Straightforward with respect to Definition 3.5; can be found in [10]. □

Remark 3.2 As outlined in Sect. 3.5.1, the above derived procedure is clearly valid for single-input systems. However, the dimension of $\kappa(x)$ in Problem 3.1 is not necessarily equal to 1, in fact, it can be extended for the case when $\kappa(x)$ is a vector-valued function with $n_u > 1$. In that case, Definitions 3.4, 3.5, and Theorem 3.4 need to be slightly modified in an adequate manner to account for this fact; for more details see [10].

Finally, in Algorithm 3.2 we summarize the proposed constructive IpLP-based procedure towards recovering a suitable augmentation of a given continuous piecewise affine function, which when passed through a clipping filter represents the solution to Problem 3.1.

Algorithm 3.2 Extended IpLP with clipping

INPUT: Saturated continuous PWA function $\kappa(x)$ defined over a, possibly, convexly non-liftable polyhedral partition/cell complex $\mathcal{R} = \{\mathcal{R}_i\}_{i \in \mathcal{I}_R}$ of a polytope $\Omega \subset \mathbb{R}^{n_x}$.

OUTPUT: Suitable augmentation $\tilde{\kappa}(x)$ and equivalent replacement $\phi(\tilde{\kappa}(x))$, defined over a convexly liftable cell complex $\tilde{\mathcal{R}} = \{\tilde{\mathcal{R}}_j\}_{j \in \mathcal{I}_{\tilde{R}}}$.

1: Obtain the index set $\mathcal{I}_{\text{unsat}}$ containing indices of unsaturated regions.
2: Compute the gains of convex lifting for the possibly non-convex, yet convexly liftable, collection of unsaturated regions $\mathcal{R}_{\mathcal{I}_{\text{unsat}}}$ via Algorithm 3.1 (Steps 1 to 3) in [15].
3: Construct the set $\Pi^{\text{unsat}}_{[x^T z u^T]^T}$ defined over $\mathcal{R}_{\mathcal{I}_{\text{unsat}}}$, as stated in Theorem 5.3 in [13].
4: Form a new constraint set $\tilde{\Pi}$ associated with $\tilde{\mathcal{R}}$, as described in Algorithm 3.1.
5: Formulate and solve an extended IpLP problem with constraints on x, z, u given by the polytope $\tilde{\Pi}$ to obtain:

$$\begin{bmatrix} \tilde{z}^* \\ \tilde{u}^* \end{bmatrix} = \arg\min_{[z\, u^T]^T} z \quad \text{s.t. } [x^T z\, u^T]^T \in \tilde{\Pi}.$$

6: Extract the appropriate sub-component of the above optimal vector:

$$\tilde{u}^* = \text{Proj}_{\mathbb{R}^{n_u}} \begin{bmatrix} \tilde{z}^* \\ \tilde{u}^* \end{bmatrix} = \tilde{\kappa}(x)$$

to obtain a suitable augmentation $\tilde{\kappa}(x)$ of the given PWA function $\kappa(x)$, defined over a rearranged, convexly liftable cell complex $\tilde{\mathcal{R}}$.
7: Design a clipping filter $\phi(\cdot)$ as per Theorem 3.4 to maintain equivalence between $\tilde{\kappa}(x)$ and $\kappa(x)$.

3.6 A Case Study with Implication in Low-Complexity eMPC

The main aim of this section is to illustrate the proposed constructive procedure via a case study MPC problem. It is shown that by executing all the steps of Algorithm 3.2, i.e., formulating and solving an extended IpLP problem with clipping, we obtain a performance-lossless replacement of the receding horizon MPC feedback associated with the original control problem. In addition, we perform an analysis to assess the computational complexity of the scheme as well as the structural complexity of the recovered eMPC controller.

The example is taken from an active vibration control task [21], where the objective is to attenuate vibrations of a flexible cantilever beam, described by the following experimentally identified state-space model:

$$\begin{aligned} x(k+1) &= \begin{bmatrix} 0.867 & 1.119 \\ -0.214 & 0.870 \end{bmatrix} x(k) + \begin{bmatrix} 9.336\text{E}{-4} \\ 5.309\text{E}{-4} \end{bmatrix} u(k) \\ y(k) &= \begin{bmatrix} -0.553 & -0.705 \end{bmatrix} x(k) \end{aligned} \tag{3.17}$$

where the output of the model is the beam tip deflection in mm, and the input is the direct actuator voltage in V. Control inputs required to steer the system (3.17) into equilibrium can be computed by solving the following optimization problem:

$$\begin{aligned} \min_{u_0,\ldots,u_{N-1}} \quad & x_N^T P x_N + \sum_{k=0}^{N-1} \left[x_k^T Q x_k + R u_k^2 \right] \\ \text{s.t.} \quad & x_0 = x(k) \\ & x_{k+1} = A x_k + B u_k, \;\; k = 0, \ldots, N-1 \\ & u_k \in \mathbb{U}, \;\; k = 0, \ldots, N-1 \\ & x_N \in \mathbb{X}_f \end{aligned} \tag{3.18}$$

subject to constraints $|u_k| \leq 120\,\text{V}$. The MPC problem (3.18) was formulated with $Q = C^T C$, $R = 1\text{E}{-4}$, and P and $\mathbb{X}_f$ designed such that closed-loop stability is guaranteed, i.e., by setting P to the solution of DARE and using a positive control invariant terminal set $\mathbb{X}_f$. Next, it was recast and solved as a pQP. The explicit MPC feedback $\kappa(x)$ for different horizon lengths N was obtained using MPT Toolbox [8]. Each resulting PWA solution $u_0^*(x) = \kappa(x)$ was subsequently post-processed using the extended IpLP procedure described by Algorithm 3.2.

Table 3.1 reports the results obtained by the proposed extended IpLP procedure, marked by (·b), compared to the standard explicit solution via parametric quadratic programming, denoted as (·a), used for the solution of (3.18) for different lengths of prediction horizon $N \in \{10, 20, 30, 40, 50\}$. The complexity of the IpLP-based scheme is expressed in terms of the number of regions of the resulting eMPC controller, and the number of constraints (half-spaces defining the constraint set $\tilde{\Pi}$) in LP (3.15). As an example, the standard MPC problem needs 90 constraints with the horizon of length 40 to obtain the PWA control law shown in Fig. 3.1, while the formulation of IpLP with clipping requires 641 constraints (551 non-redundant "general" constraints and additional 90 constraints to conserve the solution's structure) and exactly two prediction steps to yield the same result, see Fig. 3.2. Moreover, one may also notice the complexity reduction in the number of polytopic regions of the underlying state-space partition. In this particular case, the recovered partition consists of 814 regions instead of 3397 regions of the original partition, which corresponds to the reduction by a factor of more than four.

The total memory footprint of the original function κ for the investigated scenario with 3397 regions is 408 kB. The recovered function $\tilde{\kappa}$, on the other hand, requires

Table 3.1 Selected data comparing equivalent (in terms of the resulting feedback law) formulations of the case study MPC problem for different lengths of prediction horizon N

Formulation	# of regions				# of constraints†		
	N	R	R_{unsat}	$\tilde{R}$		$\aleph(\Pi)$	$\aleph(\mathcal{A})$
(1a) Standard MPC problem	10	257	75	–	30	–	–
(1b) Extended IpLP w. clipping	2	–	–	127	–	118	30
(2a)	20	913	159	–	50	–	–
(2b)	2	–	–	320	–	232	50
(3a)	30	1955	259	–	70	–	–
(3b)	2	–	–	557	–	384	70
(4a)	40	3397	367	–	90	–	–
(4b)	2	–	–	811	–	551	90
(5a)	50	5231	367	–	110	–	–
(5b)	2	–	–	1063	–	774	110

†$\aleph(\cdot)$ denotes the number of half-spaces defining a polyhedral set. For the case of formulation (·b), $\Pi = \Pi_{\min}$ and $\mathcal{A} = \mathcal{P}(\overline{M}_{\text{f}})$ (see Algorithm 3.1)

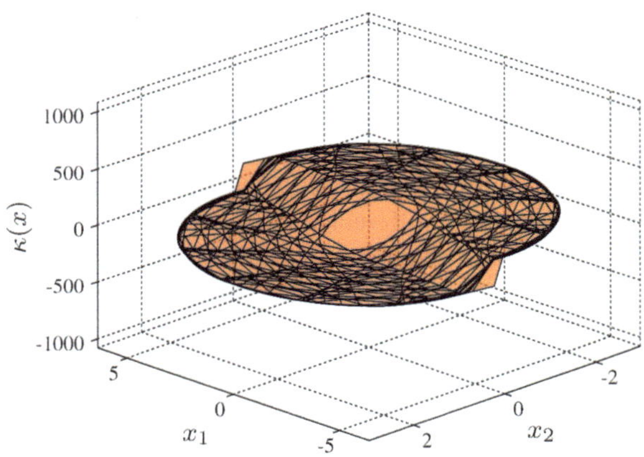

Fig. 3.1 Surface plot of the explicit MPC control law we aim to recover (3397 regions, $N = 40$)

105 kB, with additional 16 bytes to store the clipping filter $\phi(\cdot)$. This indicates reduction of memory consumption by a factor of 3.8. The worst-case computational effort needed to evaluate κ is 54332 FLOPs, which can be by employing $\tilde{\kappa}$ reduced, adequately, to 14278 FLOPs (14272 FLOPs to perform point location by sequential search and 6 FLOPs to evaluate $\phi(\tilde{\kappa}(x))$).

To review the complexity also from another practical point of view, Table 3.2 reports the CPU times spent on main algorithmic routines of the proposed extended IpCP approach and its generic pCP counterpart. The total time of the former may be split among four consecutive parts, in fact, represented by Steps 2 to 5 of Algorithm 3.2. Clearly, the last two steps represent the computationally most demanding procedures, which also scale more significantly as N increases. However, when

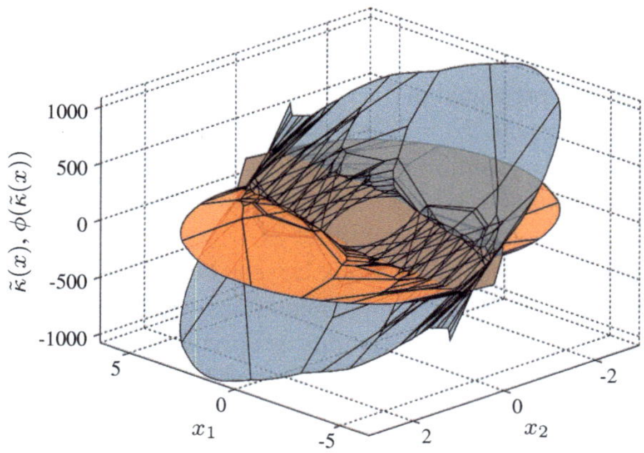

Fig. 3.2 Surface plot of the recovered eMPC controller (811 regions, $N = 2$). Result of the clipping procedure is shown in *orange colour* (illustration only)

Table 3.2 Remark on computation times of the main algorithmic features for the respective formulation of the case study MPC example

Formulation	Task execution time (s)[†]				
	$N = 10$	$N = 20$	$N = 30$	$N = 40$	$N = 50$
Standard MPC problem *pQP*	6.27	24.21	91.62	171.87	356.19
Extended IpLP with clipping *convex lifting* (Algorithm 3.2, step 2)	0.11	0.28	0.50	0.62	1.75
Facet enumeration (Algorithm 3.2, step 3)	0.05	0.07	0.12	0.17	0.23
Constraint set (Algorithm 3.2, step 4)	2.83	14.98	38.31	74.29	135.31
pLP (Algorithm 3.2, step 5)	1.67	7.33	24.46	90.81	150.48

[†]All the computations were performed on a 2.2 GHz Core i5 CPU with 4GB of RAM

summed up, the algorithmic features of extended IpCP, including the core pLP itself, are still shown to require less offline time than the original explicit solution (pQP).

Remark 3.3 Note that the two approaches may only be compared in terms of memory consumed by storage of the polyhedral partition along with the associated PWA function, or eventually the time required for online evaluation of the controller—both reduced here w.r.t. the data in Table 3.1. However, this is not the case for the offline pre-processing time spent on solving the respective pCPs since the entire IpCP procedure assumes the existence of the original explicit controller. Yet, such a study can shed more light on computational effort of the proposed approach with respect to the generic one, and in this way allow to analyse its viability.

Finally, we present a brief illustration of the aforementioned results. In particular, Fig. 3.1 shows the explicit optimal solution to the receding horizon MPC problem (3.18) for $N = 40$. The optimal feedback takes form of a continuous PWA function $\kappa(x)$ defined over a state-space partition $\mathcal{R}$. This cell complex is not convexly liftable due to the hyperplane arrangement of the two large collections of polytopes representing the controller regions over which $\kappa(x)$ is saturated at $\overline{\kappa}$ or $\underline{\kappa}$. This solution thus presents a candidate for the extended IpLP scheme. Following the algorithmic procedure presented in Sect. 3.5.1 we obtain the solution of an extended inverse MPC problem with $N = 2$, which is the PWA function $\tilde{\kappa}(x)$ depicted in Fig. 3.2. It is defined over a similar recovered partition $\tilde{\mathcal{R}}$, which is a convexly liftable cell complex. The optimality of $\tilde{\kappa}(x)$ over its entire domain of validity is achieved by passing it through a simple clipping filter $\phi(\cdot)$.

The resulting low-complexity feedback law can be readily and conveniently used within online/offline explicit MPC implementations while preserving all the performance, closed-loop stability and feasibility properties. We note that the proposed approach may be analogously exploited in implicit MPC schemes, at each sampling instant solving the equivalent "horizon 2" problem with the constraint set $\Pi_{\min}$ (cf. Remark 3.1); hence revealing the full potential efficiency of the convex liftings-based inverse optimality approach in practical model predictive control applications.

3.7 Conclusion

The chapter presents an approach exploiting the inverse parametric convex programming based upon the concept of convex lifting. As shown recently, for any continuous PWA function defined over a polyhedral partition, one may find an appropriate equivalent function by solving another parametric linear/quadratic programming problem with a supplementary variable of dimension equal to 1. Applying this idea to the model predictive control framework allowed to further state that any linear MPC formulation has an equivalent inverse reformulation with two steps of the prediction horizon. This study, in particular, deals with the invertibility of the PWA functions resulting from pQP optimization problems, which is in fact seldom guaranteed due to its structural properties. Despite being already proven that this issue can always be treated by a suitable sub-partitioning of the corresponding parameter space partition, herein the emphasis has been put mainly on devising a technique which would not increase the controller's explicit representation in terms of the polytopic regions. On the contrary, an algorithmic extension to the inverse parametric linear programming is presented, which in addition to the primary objective—to allow to target a class of common practical problems yielding convexly non-liftable cell complexes containing saturated regions, emerged to be capable of producing a lesser number of regions. This implies an expected applicability of the proposed scheme within implementations of explicit MPC, as compared to the generic convex liftings-based IpCP. It is evidenced by the presented numerical analysis which suggests that the inverse optimality concept applied to a complex but computable eMPC may lead to a lower complexity explicit control law. Subject of ongoing research remains the investigation of complexity bounds for the inverse optimality formulation. Apart from a valuable theoretic insight, the clarification of this topic shall yet improve computational tools necessary to fully exploit the efficacy of this control design approach.

Acknowledgments M. Gulan and B. Rohal̆-Ilkiv gratefully acknowledge the contribution of the Slovak Research and Development Agency under the project APVV-14-0399.

References

1. A. Alessio, A. Bemporad, A survey on explicit model predictive control, in *Nonlinear Model Predictive Control—Towards New Challenging Applications*, Lecture Notes in Control and Information Sciences, ed. by L. Magni, D.M. Raimondo, F. Allgöwer (Springer, Berlin, 2009), pp. 345–369
2. M. Baes, M. Diehl, I. Necoara, Every continuous nonlinear control system can be obtained by parametric convex programming. IEEE Trans. Autom. Control **53**(8), 1963–1967 (2008)
3. A. Bemporad, F. Borelli, M. Morari, Model predictive control based on linear programming—the explicit solution. IEEE Trans. Autom. Control **47**(12), 1974–1985 (2002)
4. A. Bemporad, M. Morari, V. Dua, E.N. Pistikopoulos, The explicit linear quadratic regulator for constrained systems. Automatica **38**(1), 3–20 (2002)
5. F. Borelli, A. Bemporad, M. Morari, The explicit linear quadratic regulator for constrained systems. J. Optim. Theory Appl. **118**(3), 515–540 (2003)
6. B. Grünbaum, *Convex Polytopes* (Wiley, New York, 1967)
7. A.B. Hempel, P.J. Goulart, J. Lygeros, Every continuous piecewise affine function can be obtained by solving a parametric linear program, in *13th European Control Conference* (Zürich, Switzerland, 2013), pp. 2657–2662
8. M. Herceg, M. Kvasnica, C.N. Jones, M. Morari, Multi-parametric toolbox 3.0, in *13th European Control Conference* (Zürich, Switzerland, 2013), pp. 502–510
9. J. Holaza, B. Takács, M. Kvasnica, S. Di Cairano, Nearly optimal simple explicit MPC controllers with stability and feasibility guarantees, in *Optimal Control Applications and Methods*. Available online (2014)
10. M. Kvasnica, M. Fikar, Clipping-based complexity reduction in explicit MPC. IEEE Trans. Autom. Control **57**(7), 1878–1883 (2012)
11. M. Kvasnica, J. Hledík, I. Rauová, M. Fikar, Complexity reduction of explicit model predictive control via separation. Automatica **49**(6), 1776–1781 (2013)
12. N.A. Nguyen, S. Olaru, P. Rodriguez-Ayerbe, On the complexity of the convex liftings-based solution to convex programming problems, in *15th European Control Conference* (Linz, Austria, 2015), pp. 3433–3438
13. N.A. Nguyen, S. Olaru, P. Rodriguez-Ayerbe, M. Hovd, I. Necoara, Inverse parametric convex programming problems via convex liftings, in *19th IFAC World Congress* (Cape Town, South Africa, 2014), pp. 2490–2494
14. N.A. Nguyen, S. Olaru, P. Rodriguez-Ayerbe, M. Hovd, I. Necoara, On the lifting problems and their connections with piecewise affine control design, in *14th European Control Conference* (Strasbourg, France, 2014), pp. 2164–2169
15. N.A. Nguyen, S. Olaru, P. Rodriguez-Ayerbe, M. Hovd, I. Necoara, Constructive solution of inverse parametric linear/quadratic programming problems, *Developments in Model-Based Optimization and Control* (Springer, 2016)
16. S. Olaru, D. Dumur, Avoiding constraints redundancy in predictive control optimization routines. IEEE Trans. Autom. Control **50**(9), 1459–1465 (2005)
17. S. Olaru, D. Dumur, On the continuity and complexity of control laws based on multiparametric linear programs, in *45th Conference on Decision and Control* (San Diego, USA, 2006), pp. 5465–5470
18. A. Schrijver, *Theory of Linear and Integer Programming* (Wiley, New York, 1998)
19. M.M. Seron, G.C. Goodwin, J.A. De Doná, Characterisation of receding horizon control for constrained linear systems. Asian J. Control **5**(2), 271–286 (2003)

20. J. Spjøtvold, E.C. Kerrigan, C.N. Jones, P. Tøndel, T.A. Johansen, On the facet-to-facet property of solutions to convex parametric quadratic programs. Automatica **42**(12), 2209–2214 (2006)
21. G. Takács, B. Rohaľ-Ilkiv, *Model Predictive Vibration Control: Efficient Constrained MPC Vibration Control for Lightly Damped Mechanical Structures* (Springer, London, 2012)
22. P. Tøndel, T.A. Johansen, A. Bemporad, An algorithm for multi-parametric quadratic programming and explicit MPC solutions. Automatica **39**(3), 489–497 (2003)

Part II
Distributed-coordinated and Multi-objective Features of Model Predictive Control

Chapter 4
Distributed Robust Model Predictive Control of Interconnected Polytopic Systems

Alexandra Grancharova and Sorin Olaru

Abstract A suboptimal approach to distributed robust MPC for uncertain systems consisting of polytopic subsystems with coupled dynamics subject to both state and input constraints is proposed. The robustness is defined in terms of the optimization of a cost function accumulated over the uncertainty and satisfying state constraints for a finite subset of uncertainties. The approach reformulates the original centralized robust MPC problem into a quadratic programming problem, which is solved by distributed iterations of the dual accelerated gradient method. A stopping condition is used that allows the iterations to stop when the desired performance, stability, and feasibility can be guaranteed. This allows for the approach to be used in an embedded robust MPC implementation. The developed method is illustrated on a simulation example of an uncertain system consisting of two interconnected polytopic subsystems.

Keywords Predictive control · Distributed control · Robust control · Distributed optimization · Accelerated gradient method · Interconnected systems · Polytopic systems · Constraints

4.1 Introduction

Model predictive control (MPC) involves the resolution at each sampling instant of a finite horizon optimal control problem subject to the system dynamics, and state and input constraints. Implementing in a centralized way MPC problem for large-scale

A. Grancharova (✉)
Department of Industrial Automation, University of Chemical Technology and Metallurgy, Bul. Kl. Ohridski 8, 1756 Sofia, Bulgaria
e-mail: alexandra.grancharova@abv.bg

S. Olaru
Laboratory of Signals and Systems, CentraleSupélec-CNRS-Université Paris-Sud, Université Paris-Saclay, 3 Rue Joliot-Curie, 91192 Gif-sur-Yvette Cedex, France
e-mail: Sorin.Olaru@centralesupelec.fr

S. Olaru et al. (eds.), *Developments in Model-Based Optimization and Control*, Lecture Notes in Control and Information Sciences 464,
DOI 10.1007/978-3-319-26687-9_4

systems may be impractical due to the topology of the plant, the need for extensive data communication and the large number of decision variables involved in the optimization. Recently, several approaches for decentralized and parallel implementation of MPC algorithms have been proposed, [3, 5, 16, 18].

In [1, 8, 10, 19, 20], approaches for distributed/decentralized MPC for systems consisting of *linear* interconnected subsystems have been developed. The approach in [8] is based on the dual decomposition methods [4, 6], where large-scale optimization problems are handled by using Lagrange multipliers to relax the couplings between the subproblems. In [10], a distributed optimization algorithm based on accelerated gradient methods using dual decomposition is proposed and its performance is evaluated on optimization problems arising in distributed MPC. Also, approaches for distributed MPC for systems composed of several *nonlinear* subsystems have been proposed (e.g., [7, 11, 14, 17]).

There are only a few papers considering the problem of robust distributed MPC of *polytopic* uncertain systems. Thus, in [22], a distributed MPC algorithm for polytopic systems subject to actuator saturation is discussed, where the distributed MPC controller is designed by solving a linear matrix inequality optimization problem. In [2], an online distributed MPC algorithm that deals explicitly with model errors is proposed. The algorithm requires decomposing the entire system into subsystems, which are coupled through their inputs. Only constraints on the inputs are considered and the upper bound on the robust performance objective is minimized by using a time-varying state-feedback controller for each subsystem. In [12], an approach to distributed robust MPC for uncertain systems consisting of *polytopic* subsystems is proposed, which applies the dynamic dual decomposition method [8] to reformulate the centralized robust MPC problem into a distributed robust MPC problem. The approach is suboptimal and is based on distributed online optimization.

In this chapter, a new approach to distributed robust MPC for uncertain systems consisting of interconnected polytopic subsystems is proposed that differs from the one in [12] in two aspects. First, it does not involve an exact online solution of quadratic programming (QP) subproblems. Instead, a suboptimal solution of the QP problem resulting from the centralized robust MPC formulation is obtained by performing distributed iterations of the dual accelerated gradient method developed for linear MPC without uncertainty [10]. This would lead to reduced complexity of the online computations and simple software and hardware implementation. Second, a stopping condition, similar to that in [9], is used that allows the iterations to stop when the desired performance, stability, and feasibility can be guaranteed.

4.2 Formulation of Robust Model Predictive Control Problem for Interconnected Polytopic Systems

Consider, a system composed by the interconnection of M linear uncertain subsystems described by the following polytopic discrete-time models:

$$\begin{aligned}
&\mathbf{x}_i(t+1) = \mathbf{A}_i(t)\mathbf{x}_i(t) + \mathbf{B}_i(t)\mathbf{u}_i(t) + \sum_{j=1, j\neq i}^{M} \mathbf{A}_{ij}\mathbf{x}_j(t) + \sum_{j=1 j\neq i}^{M} \mathbf{B}_{ij}\mathbf{u}_j(t) \\
&[\mathbf{A}_i(t),\ \mathbf{B}_i(t)] \in \Omega_i \\
&i = 1, 2, \ldots, M
\end{aligned} \tag{4.1}$$

where $\mathbf{x}_i(t) \in \mathbb{R}^{n_i}$ and $\mathbf{u}_i(t) \in \mathbb{R}^{m_i}$ are the state and control input vectors, related to the ith subsystem, $\mathbf{A}_i(t) \in \mathbb{R}^{n_i \times n_i}$ and $\mathbf{B}_i(t) \in \mathbb{R}^{n_i \times m_i}$ are uncertain time-varying matrices, $\mathbf{A}_{ij} \in \mathbb{R}^{n_i \times n_j}$, $\mathbf{B}_{ij} \in \mathbb{R}^{n_i \times m_j}$, $j = 1, 2, \ldots, M, j \neq i$ are known constant matrices, and $t \in \mathbb{Z}_{\geq 0}$ is the discrete time. For a polytopic uncertainty description, Ω_i, $i = 1, 2, \ldots, M$ are polytopes:

$$\Omega_i = Co\left\{\left[\mathbf{A}_i^1,\ \mathbf{B}_i^1\right],\ [\mathbf{A}_i^2,\ \mathbf{B}_i^2], \ldots, [\mathbf{A}_i^{L_i},\ \mathbf{B}_i^{L_i}\right]\right\}, \quad i = 1, 2, \ldots, M \tag{4.2}$$

where $Co\{\cdot\}$ denotes convex hull and $[\mathbf{A}_i^r, \mathbf{B}_i^r]$, $r = 1, 2, \ldots, L_i$ are its vertices.

The following constraints are imposed on the subsystems:

$$\mathbf{u}_i(t) \in \mathscr{U}_i, \mathbf{x}_i(t) \in \mathscr{X}_i, \quad i = 1, 2, \ldots, M \tag{4.3}$$

where $\mathscr{U}_i$ and $\mathscr{X}_i$ are the admissible sets.

The following assumption is made:

Assumption 4.1 The admissible sets $\mathscr{X}_i$ and $\mathscr{U}_i$ are bounded polyhedral sets, i.e., they are defined by:

$$\mathscr{X}_i = \left\{\mathbf{x}_i \in \mathbb{R}^{n_i} \mid \mathbf{C}_i^x \mathbf{x}_i \leq \mathbf{d}_i^x\right\} \tag{4.4}$$

$$\mathscr{U}_i = \left\{\mathbf{u}_i \in \mathbb{R}^{m_i} \mid \mathbf{C}_i^u \mathbf{u}_i \leq \mathbf{d}_i^u\right\} \tag{4.5}$$

and they include the origin in their interior. Here, $\mathbf{C}_i^x \in \mathbb{R}^{n_{c,x_i} \times n_i}$, $\mathbf{C}_i^u \in \mathbb{R}^{n_{c,u_i} \times m_i}$, $\mathbf{d}_i^x \in \mathbb{R}^{n_{c,x_i}}$, $\mathbf{d}_i^u \in \mathbb{R}^{n_{c,u_i}}$ (n_{c,x_i} and n_{c,u_i} being the number of constraints imposed on x_i and u_i, respectively).

The approach in [9] for distributed solution of MPC problems will be followed. This requires the introduction of the following tightened constraint sets:

$$(1-\delta)\mathscr{X}_i = \left\{\mathbf{x}_i \in \mathbb{R}^{n_i} \mid \mathbf{C}_i^x \mathbf{x}_i \leq (1-\delta)\mathbf{d}_i^x\right\} \tag{4.6}$$

$$(1-\delta)\mathscr{U}_i = \left\{\mathbf{u}_i \in \mathbb{R}^{m_i} \mid \mathbf{C}_i^u \mathbf{u}_i \leq (1-\delta)\mathbf{d}_i^u\right\} \tag{4.7}$$

where $\delta \in (0, 1)$ is the amount of relative constraint tightening.

Let $\mathbf{x}(t)$ and $\mathbf{u}(t)$ denote the overall state and the overall control input, i.e.:

$$\mathbf{x}(t) = [\mathbf{x}_1(t),\ \mathbf{x}_2(t), \dots, \mathbf{x}_M(t)] \in \mathbb{R}^n,\ n = \sum_{i=1}^{M} n_i \tag{4.8}$$

$$\mathbf{u}(t) = [\mathbf{u}_1(t),\ \mathbf{u}_2(t), \dots, \mathbf{u}_M(t)] \in \mathbb{R}^m,\ m = \sum_{i=1}^{M} m_i \tag{4.9}$$

From (4.3) it follows that the admissible sets for the overall state and the overall control input are $\mathscr{X} = \mathscr{X}_1 \times \mathscr{X}_2 \times \cdots \times \mathscr{X}_M$ and $\mathscr{U} = \mathscr{U}_1 \times \mathscr{U}_2 \times \cdots \times \mathscr{U}_M$. The description of the overall system dynamics, corresponding to (4.1) is:

$$\mathbf{x}(t+1) = (\mathbf{A}(t) + \tilde{\mathbf{A}})\mathbf{x}(t) + (\mathbf{B}(t) + \tilde{\mathbf{B}})\mathbf{u}(t),\ [\mathbf{A}(t),\ \mathbf{B}(t)] \in \Omega \tag{4.10}$$

where $\mathbf{A}(t)$, $\mathbf{B}(t)$, $\tilde{\mathbf{A}}$, $\tilde{\mathbf{B}}$ are block-matrices:

$$\begin{aligned} &\mathbf{A}(t) = \text{diag}\{\mathbf{A}_1(t), \dots, \mathbf{A}_M(t)\},\ \mathbf{B}(t) = \text{diag}\{\mathbf{B}_1(t), \dots, \mathbf{B}_M(t)\} \\ &\tilde{\mathbf{A}} = \begin{Bmatrix} \mathbf{A}_{ij}, & i \neq j \\ 0, & i = j \end{Bmatrix},\ \tilde{\mathbf{B}} = \begin{Bmatrix} \mathbf{B}_{ij}, & i \neq j \\ 0, & i = j \end{Bmatrix},\ i, j = 1, \dots, M \end{aligned} \tag{4.11}$$

and $\Omega = \Omega_1 \times \Omega_2 \times \cdots \times \Omega_M$. Before formulating the robust MPC problem, a set $\tilde{\Omega}$ is introduced, which is a finite subset of Ω. First, let $\Omega^{\text{vert}} = \{[\mathbf{A}^r, \mathbf{B}^r],\ r = 1, 2, \dots, L\}$ be the set of vertices of Ω and $\Omega^{\text{int}} = \{[\mathbf{A}^{L+j}, \mathbf{B}^{L+j}] \in \text{int}(\Omega),\ j = 1, 2, \dots, K\}$ be a finite set which includes interior points of the set Ω. Then, the finite set $\tilde{\Omega} \subset \Omega$ is defined as $\tilde{\Omega} = \Omega^{\text{vert}} \bigcup \Omega^{\text{int}}$ (the reason for imposing a dichotomy with respect to the elements of the set $\tilde{\Omega}$ will become clear in Remark 4.1). The set $\tilde{\Omega}$ is compactly represented as:

$$\begin{aligned} \tilde{\Omega} &= \{[\mathbf{A}^s,\ \mathbf{B}^s],\ s = 1, 2, \dots, S\} \\ &= \{[\text{diag}\{\mathbf{A}_1^s(t), \dots, \mathbf{A}_M^s(t)\},\ \text{diag}\{\mathbf{B}_1^s(t), \dots, \mathbf{B}_M^s(t)\}],\ s = 1, 2, \dots, S\} \end{aligned} \tag{4.12}$$

where $S = K + L$.

Another assumption is made about the rate of variation of parameters, mainly with respect to the prediction horizon as it will be shown next in the MPC design.

Assumption 4.2 The uncertain pairs $[\mathbf{A}_i(t), \mathbf{B}_i(t)] \in \Omega_i$, $i = 1, \dots, M$ have infrequent changes in the sense that $[\mathbf{A}_i(t), \mathbf{B}_i(t)] = \text{const}$, $i = 1, \dots, M$ for periods of time, which are not less than $\tilde{N}$ ($\tilde{N} \in \mathbb{N}$ is supposed to be sufficiently large and will be related to the prediction horizon).

Further, let $[\mathbf{A}_i(t), \mathbf{B}_i(t)] = [\mathbf{A}_i^{\text{n}}, \mathbf{B}_i^{\text{n}}] \in \Omega_i$, $i = 1, \dots, M$ be the matrices corresponding to the nominal operation of the system (4.1). It is supposed that a full measurement $\bar{\mathbf{x}} = [\bar{\mathbf{x}}_1, \bar{\mathbf{x}}_2, \dots, \bar{\mathbf{x}}_M]$ of the overall state is available at the current time

t. The robust regulation problem is considered where the goal is to steer the overall state of the system (4.1) to the origin by minimizing an accumulated cost function over the uncertainty and guaranteeing the robust feasibility of the control action. Let N be a finite horizon such that $N < \tilde{N}$. Under Assumption 4.2 it follows that $[\mathbf{A}_i(t+k), \mathbf{B}_i(t+k)] = \text{const} = [\mathbf{A}_i, \mathbf{B}_i], k = 0, 1, \ldots, N, i = 1, \ldots, M$. Then, for the current $\bar{\mathbf{x}}$, the robust regulation MPC solves the optimization problem:

Problem P1 (Centralized robust MPC):

$$V^*(\bar{\mathbf{x}}) = \min_{\mathbf{X}^{\mathrm{n}}, \mathbf{X}^1, \ldots, \mathbf{X}^S, \mathbf{U}} J(\mathbf{X}^{\mathrm{n}}, \mathbf{X}^1, \ldots, \mathbf{X}^S, \mathbf{U}, \bar{\mathbf{x}}) \tag{4.13}$$

subject to $\mathbf{x}^{\mathrm{n}}_{t|t} = \bar{\mathbf{x}}, \mathbf{x}^s_{t|t} = \bar{\mathbf{x}}, s = 1, \ldots, S$ and:

$$\mathbf{u}_{i,t+k} \in (1-\delta)\mathscr{U}_i, i = 1, \ldots, M, k = 0, 1, \ldots, N-1 \tag{4.14}$$

$$\mathbf{x}^{\mathrm{n}}_{i,t+k|t} \in (1-\delta)\mathscr{X}_i, i = 1, \ldots, M, k = 1, \ldots, N \tag{4.15}$$

$$\mathbf{x}^s_{i,t+k|t} \in (1-\delta)\mathscr{X}_i, i = 1, \ldots, M, k = 1, \ldots, N, s = 1, \ldots, S \tag{4.16}$$

$$\mathbf{x}^{\mathrm{n}}_{i,t+k+1|t} = \mathbf{A}^{\mathrm{n}}_i \mathbf{x}^{\mathrm{n}}_{i,t+k|t} + \mathbf{B}^{\mathrm{n}}_i \mathbf{u}_{i,t+k} + \sum_{j=1, j\neq i}^{M} \mathbf{A}_{ij} \mathbf{x}^{\mathrm{n}}_{j,t+k|t} + \sum_{j=1, j\neq i}^{M} \mathbf{B}_{ij} \mathbf{u}_{j,t+k}$$
$$i = 1, \ldots, M, k = 0, 1, \ldots, N-1 \tag{4.17}$$

$$\mathbf{x}^s_{i,t+k+1|t} = \mathbf{A}^s_i \mathbf{x}^s_{i,t+k|t} + \mathbf{B}^s_i \mathbf{u}_{i,t+k} + \sum_{j=1, j\neq i}^{M} \mathbf{A}_{ij} \mathbf{x}^s_{j,t+k|t} + \sum_{j=1, j\neq i}^{M} \mathbf{B}_{ij} \mathbf{u}_{j,t+k}$$
$$i = 1, \ldots, M, k = 0, 1, \ldots, N-1, s = 1, \ldots, S \tag{4.18}$$

$$\mathbf{x}^{\mathrm{n}}_{t+k|t} = [\mathbf{x}^{\mathrm{n}}_{1,t+k|t}, \mathbf{x}^{\mathrm{n}}_{2,t+k|t}, \ldots, \mathbf{x}^{\mathrm{n}}_{M,t+k|t}], k = 0, 1, \ldots, N \tag{4.19}$$

$$\mathbf{x}^s_{t+k|t} = [\mathbf{x}^s_{1,t+k|t}, \mathbf{x}^s_{2,t+k|t}, \ldots, \mathbf{x}^s_{M,t+k|t}], k = 0, 1, \ldots, N \tag{4.20}$$

$$\mathbf{u}_{t+k} = [\mathbf{u}_{1,t+k}, \mathbf{u}_{2,t+k}, \ldots, \mathbf{u}_{M,t+k}], k = 0, 1, \ldots, N-1 \tag{4.21}$$

with $\mathbf{U} = [\mathbf{u}_t, \mathbf{u}_{t+1}, \ldots, \mathbf{u}_{t+N}]$, $\mathbf{X}^{\mathrm{n}} = [\mathbf{x}^{\mathrm{n}}_{t+1|t}, \mathbf{x}^{\mathrm{n}}_{t+2|t}, \ldots, \mathbf{x}^{\mathrm{n}}_{t+N|t}]$, $\mathbf{X}^s = [\mathbf{x}^s_{t+1|t}, \mathbf{x}^s_{t+2|t}, \ldots, \mathbf{x}^s_{t+N|t}]$, $s = 1, \ldots, S$, and the cost function given by:

$$J(\mathbf{X}^{\mathrm{n}}, \mathbf{X}^1, \ldots, \mathbf{X}^S, \mathbf{U}, \bar{\mathbf{x}}) = \frac{1}{2} \sum_{k=0}^{N} l\left(\mathbf{x}^{\mathrm{n}}_{t+k|t}, \mathbf{x}^1_{t+k|t}, \ldots, \mathbf{x}^S_{t+k|t}, \mathbf{u}_{t+k}\right) \tag{4.22}$$

Here, the predicted states $\mathbf{x}^{\mathrm{n}}_{t+k|t}$ and $\mathbf{x}^s_{t+k|t}$ are determined according to (4.17)–(4.18), the accumulated cost $l(\cdot)$ over the uncertainty at stage $t+k$ is defined as:

$$l\left(\mathbf{x}^{\mathrm{n}}_{t+k|t}, \mathbf{x}^1_{t+k|t}, \ldots, \mathbf{x}^S_{t+k|t}, \mathbf{u}_{t+k}\right)$$
$$= \sum_{i=1}^{M} \left[\|\mathbf{x}^{\mathrm{n}}_{i,t+k|t}\|^2_{\mathbf{Q}_i} + \|\mathbf{u}_{i,t+k}\|^2_{\mathbf{R}_i}\right] + \sum_{i=1}^{M} \sum_{s=1}^{S} \|\mathbf{x}^s_{i,t+k|t}\|^2_{\mathbf{Q}_i} \tag{4.23}$$

and $\mathbf{Q}_i, \mathbf{R}_i \succ 0$ are weighting matrices. It should be noted that the satisfaction of the state constraints (4.15)–(4.16) guarantees a robustness of the solution in sense that the state constraints in (4.3) will be satisfied for the nominal dynamics of the system as well as for the worst-case uncertainty realizations in $\tilde{\Omega}$.

The robust MPC problem P1 can be considered as an approximation of the following robust infinite horizon optimal control problem:

Problem P2 (Robust infinite horizon optimal control):

$$V^{\infty}(\bar{\mathbf{x}}) = \min_{\mathbf{u}} \left[\sum_{i=1}^{M} \sum_{t=0}^{\infty} \left[\|\mathbf{x}_i(t)^{\mathrm{n}}\|^2_{\mathbf{Q}_i} + \|\mathbf{u}_i(t)\|^2_{\mathbf{R}_i} \right] + \sum_{i=1}^{M} \sum_{s=1}^{S} \sum_{t=0}^{\infty} \|\mathbf{x}_i^s(t)\|^2_{\mathbf{Q}_i} \right] \tag{4.24}$$

with initial state $\mathbf{x}(0) = \bar{\mathbf{x}}$ and:

$$\mathbf{u}_i(t) \in (1-\delta)\mathscr{U}_i, \;\; i = 1, 2, \ldots, M \;\text{ for all } t \tag{4.25}$$

$$\mathbf{x}_i(t)^{\mathrm{n}} \in (1-\delta)\mathscr{X}_i, \;\; i = 1, 2, \ldots, M \;\text{ for all } t \tag{4.26}$$

$$\mathbf{x}_i(t)^s \in (1-\delta)\mathscr{X}_i, \;\; i = 1, 2, \ldots, M, s = 1, 2, \ldots, S \;\text{ for all } t \tag{4.27}$$

In (4.24), (4.26), (4.27), $\mathbf{x}_i(t)^{\mathrm{n}}$ is the state trajectory of the system (4.1) for $[\mathbf{A}_i(t), \mathbf{B}_i(t)] = [\mathbf{A}_i^{\mathrm{n}}, \mathbf{B}_i^{\mathrm{n}}]$, and $\mathbf{x}_i^s(t)$ is the state trajectory obtained with $[\mathbf{A}_i(t), \mathbf{B}_i(t)] = [\mathbf{A}_i^s, \mathbf{B}_i^s]$, $s = 1, 2, \ldots, S$.

In Sect. 4.3.2, the infinite horizon performance of the distributed robust MPC is compared to the optimal infinite horizon cost $V^{\infty}(\bar{\mathbf{x}})$. It is supposed that $V^{\infty}(\bar{\mathbf{x}})$ is finite which implies the stability of the centralized robust MPC.

Remark 4.1 By assuming that $[\mathbf{A}(t+k), \mathbf{B}(t+k)] = \text{const} = [\mathbf{A}, \mathbf{B}]$, $k = 0, 1, \ldots, N$, the predicted overall state is:

$$\mathbf{x}_{t+k|t} = (\mathbf{A} + \tilde{\mathbf{A}})^k \mathbf{x} + \sum_{j=0}^{k-1} (\mathbf{A} + \tilde{\mathbf{A}})^j (\mathbf{B} + \tilde{\mathbf{B}}) \mathbf{u}_{t+k-1-j} \tag{4.28}$$

Then, the state constraints $\mathbf{x}_{t+k|t} \in \mathscr{X}, k = 1, 2, \ldots, N$ will be in general nonconvex with respect to the uncertain matrices **A** and **B**, because the predicted state is nonlinear with respect to them. Therefore, considering only the vertices of the sets Ω when requiring robust feasibility may not be sufficient. For this reason, the finite uncertainty set $\tilde{\Omega}$ should include some interior elements in addition to the vertices. The richer the set of interior points, the more reduced the level of suboptimality with respect to the optimization problem corresponding to $[\mathbf{A}, \mathbf{B}] \in \Omega$.

Remark 4.2 In the present work, the terminal constraints are dropped with respect to a classical robust MPC formulation. The main issue we concentrate on is the reformulation of (4.13)–(4.21) in the form of a distributed optimization. As such, in the next sections, we will consider that (4.13)–(4.21) represent an acceptable centralized control in terms of stability, performance, and feasibility.

4.3 Distributed Optimization Approach to Robust MPC

4.3.1 *Representation of the Robust MPC Problem as a Distributed Optimization Problem*

By stacking all decision variables into one vector $\mathbf{Y} \in \mathbb{R}^{n_\mathbf{Y}}$ (with the dimension given by $n_\mathbf{Y} = \sum_{i=1}^{M} [N(n_i + m_i) + SNn_i]$):

$$\mathbf{Y} = \left[\begin{array}{ll} \begin{array}{c}\text{1st}\\ \text{subsystem}\end{array} & \left\{ \begin{array}{l} \mathbf{x}^{\mathrm{n}}_{1,t+1|t}, \mathbf{u}_{1,t}, \mathbf{x}^{\mathrm{n}}_{1,t+2|t}, \mathbf{u}_{1,t+1}, \ldots, \mathbf{x}^{\mathrm{n}}_{1,t+N|t}, \mathbf{u}_{1,t+N-1}, \\ \mathbf{x}^{1}_{1,t+1|t}, \ldots, \mathbf{x}^{1}_{1,t+N|t}, \ldots, \mathbf{x}^{S}_{1,t+1|t}, \ldots, \mathbf{x}^{S}_{1,t+N|t}, \end{array} \right. \\ \vdots & \\ \begin{array}{c}M\text{th}\\ \text{subsystem}\end{array} & \left\{ \begin{array}{l} \mathbf{x}^{\mathrm{n}}_{M,t+1|t}, \mathbf{u}_{M,t}, \mathbf{x}^{\mathrm{n}}_{M,t+2|t}, \mathbf{u}_{M,t+1}, \ldots, \mathbf{x}^{\mathrm{n}}_{M,t+N|t}, \mathbf{u}_{M,t+N-1}, \\ \mathbf{x}^{1}_{M,t+1|t}, \ldots, \mathbf{x}^{1}_{M,t+N|t}, \ldots, \mathbf{x}^{S}_{M,t+1|t}, \ldots, \mathbf{x}^{S}_{M,t+N|t} \end{array} \right. \end{array} \right] \tag{4.29}$$

the optimization problem (4.13)–(4.21) can be written as the following quadratic programming (QP) problem:

Problem P3 (QP problem):

$$V^*(\bar{\mathbf{x}}) = \min_{\mathbf{Y}} \frac{1}{2} \mathbf{Y}^{\mathrm{T}} \bar{\mathbf{H}} \mathbf{Y} \tag{4.30}$$

$$\text{subject to} \quad \bar{\mathbf{A}}\mathbf{Y} = \bar{\mathbf{B}}\bar{\mathbf{x}} \tag{4.31}$$

$$\bar{\mathbf{C}}\mathbf{Y} \leq (1 - \delta)\bar{\mathbf{d}} \tag{4.32}$$

Here, $\bar{\mathbf{H}} \in \mathbb{R}^{n_\mathbf{Y} \times n_\mathbf{Y}}$, $\bar{\mathbf{A}} \in \mathbb{R}^{n_{\bar{\mathbf{A}}} \times n_\mathbf{Y}}$, $\bar{\mathbf{B}} \in \mathbb{R}^{n_{\bar{\mathbf{A}}} \times n_{\bar{\mathbf{x}}}}$, $\bar{\mathbf{C}} \in \mathbb{R}^{n_{\bar{\mathbf{C}}} \times n_\mathbf{Y}}$, $\bar{\mathbf{d}} \in \mathbb{R}^{n_{\bar{\mathbf{C}}}}$ $(n_{\bar{\mathbf{A}}} = \sum_{i=1}^{M} (S+1)(N-1)n_i, n_{\bar{\mathbf{x}}} = \sum_{i=1}^{M} n_i, n_{\bar{\mathbf{C}}} = \sum_{i=1}^{M} (S+1)N(n_{c,x_i} + n_{c,u_i}))$. In (4.30)–(4.32) we have:

$$\begin{aligned} \bar{\mathbf{H}} &= \mathrm{diag}\{\bar{\mathbf{H}}_1, \bar{\mathbf{H}}_2, \ldots, \bar{\mathbf{H}}_M\} \\ \bar{\mathbf{A}} &= \left[\bar{\mathbf{A}}_1 \mid \bar{\mathbf{A}}_2 \mid \ldots \mid \bar{\mathbf{A}}_M\right]^{\mathrm{T}}, \bar{\mathbf{B}} = \left[\bar{\mathbf{B}}_1 \mid \bar{\mathbf{B}}_2 \mid \ldots \mid \bar{\mathbf{B}}_M\right]^{\mathrm{T}} \\ \bar{\mathbf{C}} &= \left[\bar{\mathbf{C}}_1 \mid \bar{\mathbf{C}}_2 \mid \ldots \mid \bar{\mathbf{C}}_M\right]^{\mathrm{T}}, \bar{\mathbf{d}} = \left[\bar{\mathbf{d}}_1 \mid \bar{\mathbf{d}}_2 \mid \ldots \mid \bar{\mathbf{d}}_M\right]^{\mathrm{T}} \end{aligned} \tag{4.33}$$

For the ith subsystem the matrix $\bar{\mathbf{H}}_i \in \mathbb{R}^{n_{\mathbf{Y}_i} \times n_{\mathbf{Y}_i}}$ $(n_{\mathbf{Y}_i} = N(n_i + m_i) + SNn_i)$ is defined as follows:

$$\bar{\mathbf{H}}_i = \mathrm{diag}\{\underbrace{\mathbf{W}_i, \mathbf{W}_i, \ldots, \mathbf{W}_i}_{N \text{ elements}}, \overbrace{\underbrace{\mathbf{Q}_i, \mathbf{Q}_i, \ldots, \mathbf{Q}_i}_{N \text{ elements}}, \ldots, \underbrace{\mathbf{Q}_i, \mathbf{Q}_i, \ldots, \mathbf{Q}_i}_{N \text{ elements}}}^{S \text{ elements}}\} \tag{4.34}$$

where $\mathbf{W}_i = \left[\begin{smallmatrix} \mathbf{Q}_i & 0 \\ 0 & \mathbf{R}_i \end{smallmatrix}\right]$, and the matrices $\bar{\mathbf{A}}_i \in \mathbb{R}^{n_{\bar{\mathbf{A}}_i} \times n_{\mathbf{Y}}}$, $\bar{\mathbf{B}}_i \in \mathbb{R}^{n_{\bar{\mathbf{A}}_i} \times n_{\bar{\mathbf{x}}}}$, $\bar{\mathbf{C}}_i \in \mathbb{R}^{n_{\bar{\mathbf{C}}_i} \times n_{\mathbf{Y}}}$ and the vector $\bar{\mathbf{d}}_i \in \mathbb{R}^{n_{\bar{\mathbf{C}}_i}}$, ($n_{\bar{\mathbf{A}}_i} = (S+1)(N-1)n_i$, $n_{\bar{\mathbf{C}}_i} = (S+1)N(n_{c,x_i} + n_{c,u_i})$), $i = 1, 2, \ldots, M$ are:

$$\bar{\mathbf{A}}_i = \left[\bar{\mathbf{A}}_i^{\mathrm{n}} \,|\, \bar{\mathbf{A}}_i^{1} \,|\, \ldots \,|\, \bar{\mathbf{A}}_i^{S}\right]^{\mathrm{T}}, \bar{\mathbf{B}}_i = \left[\bar{\mathbf{B}}_i^{\mathrm{n}} \,|\, \bar{\mathbf{B}}_i^{1} \,|\, \ldots \,|\, \bar{\mathbf{B}}_i^{S}\right]^{\mathrm{T}}, \bar{\mathbf{C}}_i = \mathrm{diag}\{\bar{\mathbf{C}}_i^{\mathrm{n}}, \bar{\mathbf{C}}_i^{1}, \ldots, \bar{\mathbf{C}}_i^{S}\}$$

$$\bar{\mathbf{d}}_i = [\underbrace{\mathbf{d}_i^{x,u}, \mathbf{d}_i^{x,u}, \ldots, \mathbf{d}_i^{x,u}}_{N \text{ elements}}, \overbrace{\underbrace{\mathbf{d}_i^{x,u}, \mathbf{d}_i^{x,u}, \ldots, \mathbf{d}_i^{x,u}}_{N \text{ elements}}, \ldots, \underbrace{\mathbf{d}_i^{x,u}, \mathbf{d}_i^{x,u}, \ldots, \mathbf{d}_i^{x,u}}_{N \text{ elements}}}^{S \text{ elements}}] \quad (4.35)$$

where $\mathbf{d}_i^{x,u} = \left[\mathbf{d}_i^{x}, \mathbf{d}_i^{u}\right]$. In (4.35) we have:

$$\bar{\mathbf{A}}_i^{\mathrm{n}} = \begin{bmatrix} \bar{\mathbf{A}}_{i,1,1}^{\mathrm{n}} & \bar{\mathbf{A}}_{i,1,2}^{\mathrm{n}} & \cdots & \bar{\mathbf{A}}_{i,1,M}^{\mathrm{n}} \\ \bar{\mathbf{A}}_{i,2,1}^{\mathrm{n}} & \bar{\mathbf{A}}_{i,2,2}^{\mathrm{n}} & \cdots & \bar{\mathbf{A}}_{i,2,M}^{\mathrm{n}} \\ \vdots & \vdots & \ddots & \vdots \\ \bar{\mathbf{A}}_{i,N-1,1}^{\mathrm{n}} & \bar{\mathbf{A}}_{i,N-1,2}^{\mathrm{n}} & \cdots & \bar{\mathbf{A}}_{i,N-1,M}^{\mathrm{n}} \end{bmatrix}, \bar{\mathbf{B}}_i^{\mathrm{n}} = \begin{bmatrix} -\mathbf{A}_{i1} & -\mathbf{A}_{i2} & \ldots & -\mathbf{A}_i^{\mathrm{n}} & \ldots & -\mathbf{A}_{iM} \\ 0 & 0 & \ldots & 0 & \ldots & 0 \\ \vdots & \vdots & \ddots & \vdots & \ddots & \vdots \\ 0 & 0 & \ldots & 0 & \ldots & 0 \end{bmatrix} \quad (4.36)$$

$$\bar{\mathbf{A}}_i^{s} = \begin{bmatrix} \bar{\mathbf{A}}_{i,1,1}^{s} & \bar{\mathbf{A}}_{i,1,2}^{s} & \cdots & \bar{\mathbf{A}}_{i,1,M}^{s} \\ \bar{\mathbf{A}}_{i,2,1}^{s} & \bar{\mathbf{A}}_{i,2,2}^{s} & \cdots & \bar{\mathbf{A}}_{i,2,M}^{s} \\ \vdots & \vdots & \ddots & \vdots \\ \bar{\mathbf{A}}_{i,N-1,1}^{s} & \bar{\mathbf{A}}_{i,N-1,2}^{s} & \cdots & \bar{\mathbf{A}}_{i,N-1,M}^{s} \end{bmatrix}, \bar{\mathbf{B}}_i^{s} = \begin{bmatrix} -\mathbf{A}_{i1} & -\mathbf{A}_{i2} & \ldots & -\mathbf{A}_i^{s} & \ldots & -\mathbf{A}_{iM} \\ 0 & 0 & \ldots & 0 & \ldots & 0 \\ \vdots & \vdots & \ddots & \vdots & \ddots & \vdots \\ 0 & 0 & \ldots & 0 & \ldots & 0 \end{bmatrix}$$
$$i = 1, 2, \ldots, M, s = 1, 2, \ldots, S \quad (4.37)$$

$$\bar{\mathbf{C}}_i^{\mathrm{n}} = \bar{\mathbf{C}}_i^{1} = \cdots = \bar{\mathbf{C}}_i^{S} = \mathrm{diag}\{\underbrace{\mathbf{C}_i^{x,u}, \mathbf{C}_i^{x,u}, \ldots, \mathbf{C}_i^{x,u}}_{N \text{ elements}}\} \quad (4.38)$$

where $\mathbf{C}_i^{x,u} = \begin{bmatrix} \mathbf{C}_i^{x} & 0 \\ 0 & \mathbf{C}_i^{u} \end{bmatrix}$. In (4.36) $\bar{\mathbf{A}}_{i,k,j}^{\mathrm{n}}$, $k = 1, \ldots, N-1, j = 1, \ldots, M$ depend on the matrices $\mathbf{A}_i^{\mathrm{n}}$, $\mathbf{B}_i^{\mathrm{n}}$, $\mathbf{A}_{ij}$, $\mathbf{B}_{ij}$, $j = 1, \ldots, M$ of the system (4.1) and in (4.37) $\bar{\mathbf{A}}_{i,k,j}^{s}$, $k = 1, \ldots, N-1, j = 1, \ldots, M$, $s = 1, \ldots, S$ depend on the matrices $\mathbf{A}_i^{s}$, $\mathbf{B}_i^{s}$, $\mathbf{A}_{ij}$, $\mathbf{B}_{ij}$, $j = 1, \ldots, M$, $s = 1, \ldots, S$ of the system (4.1).

The robust MPC problem (4.13)–(4.21) is solved distributedly by applying the dual accelerated gradient algorithm in [10]. The distribution is enabled by solving the dual problem to (4.30)–(4.32), which is created by introducing dual variables $\boldsymbol{\lambda} \in \mathbb{R}^{n_{\bar{\mathbf{A}}}}$ for the equality constraints (4.31) and dual variables $\boldsymbol{\mu} \in \mathbb{R}^{n_{\bar{\mathbf{C}}}}$ for the inequality constraints (4.32). It is shown in [10] that the dual problem can be written as:

$$\max_{\boldsymbol{\lambda}, \boldsymbol{\mu} \geq 0} D(\bar{\mathbf{x}}, \boldsymbol{\lambda}, \boldsymbol{\mu}) \quad (4.39)$$

where $D(\bar{\mathbf{x}}, \boldsymbol{\lambda}, \boldsymbol{\mu})$ is the dual cost function:

$$D(\bar{\mathbf{x}}, \boldsymbol{\lambda}, \boldsymbol{\mu}) = -\frac{1}{2}(\bar{\mathbf{A}}^{\mathrm{T}}\boldsymbol{\lambda} + \bar{\mathbf{C}}^{\mathrm{T}}\boldsymbol{\mu})^{\mathrm{T}}\bar{\mathbf{H}}^{-1}(\bar{\mathbf{A}}^{\mathrm{T}}\boldsymbol{\lambda} + \bar{\mathbf{C}}^{\mathrm{T}}\boldsymbol{\mu}) - \boldsymbol{\lambda}^{\mathrm{T}}\bar{\mathbf{B}}\bar{\mathbf{x}} - \boldsymbol{\mu}^{\mathrm{T}}\bar{\mathbf{d}}(1-\delta) \tag{4.40}$$

The distributed algorithm in [10] is applied to solve the QP problem (4.30)–(4.32). The algorithm is a dual accelerated gradient method described by the following global iterations:

$$\mathbf{Y}^r = -\bar{\mathbf{H}}^{-1}(\bar{\mathbf{A}}^{\mathrm{T}}\boldsymbol{\lambda}^r + \bar{\mathbf{C}}^{\mathrm{T}}\boldsymbol{\mu}^r) \tag{4.41}$$

$$\bar{\mathbf{Y}}^r = \mathbf{Y}^r + \frac{r-1}{r+2}(\mathbf{Y}^r - \mathbf{Y}^{r-1}) \tag{4.42}$$

$$\boldsymbol{\lambda}^{r+1} = \boldsymbol{\lambda}^r + \frac{r-1}{r+2}(\boldsymbol{\lambda}^r - \boldsymbol{\lambda}^{r-1}) + \frac{1}{L}(\bar{\mathbf{A}}\bar{\mathbf{Y}}^r - \bar{\mathbf{B}}\bar{\mathbf{x}}) \tag{4.43}$$

$$\boldsymbol{\mu}^{r+1} = \max\left(0,\, \boldsymbol{\mu}^r + \frac{r-1}{r+2}(\boldsymbol{\mu}^r - \boldsymbol{\mu}^{r-1}) + \frac{1}{L}(\bar{\mathbf{C}}\bar{\mathbf{Y}}^r - \bar{\mathbf{d}}(1-\delta))\right) \tag{4.44}$$

Here, r is the iteration number and $L = \|[\bar{\mathbf{A}}^{\mathrm{T}}, \bar{\mathbf{C}}^{\mathrm{T}}]^{\mathrm{T}}\bar{\mathbf{H}}^{-1}[\bar{\mathbf{A}}^{\mathrm{T}}, \bar{\mathbf{C}}^{\mathrm{T}}]\|$ is the Lipschitz constant to the gradient of the dual function (4.40). In order to distribute the dual accelerated gradient iterations (4.41)–(4.44), the following vector $\mathbf{Y}_i \in \mathbb{R}^{n_{\mathbf{Y}_i}}$ ($n_{\mathbf{Y}_i} = N(n_i + m_i) + SNn_i$) of decision variables, associated to the ith subsystem is introduced:

$$\mathbf{Y}_i = \begin{bmatrix} \mathbf{x}^{\mathrm{n}}_{i,\,t+1|t},\ \mathbf{u}_{i,\,t},\ \mathbf{x}^{\mathrm{n}}_{i,\,t+2|t},\ \mathbf{u}_{i,\,t+1},\ \ldots,\ \mathbf{x}^{\mathrm{n}}_{i,\,t+N|t},\ \mathbf{u}_{i,\,t+N-1}, \\ \mathbf{x}^{1}_{i,\,t+1|t},\ \ldots,\ \mathbf{x}^{1}_{i,\,t+N|t},\ \ldots,\ \mathbf{x}^{S}_{i,\,t+1|t},\ \ldots,\ \mathbf{x}^{S}_{i,\,t+N|t} \end{bmatrix} \tag{4.45}$$

Also, let $\boldsymbol{\lambda}_i \in \mathbb{R}^{n_{\bar{\mathbf{A}}_i}}$ and $\boldsymbol{\mu}_i \in \mathbb{R}^{n_{\bar{\mathbf{C}}_i}}$ be the dual variables for the equality and the inequality constraints, related to the ith subsystem. Then, the distributed iterations of the dual accelerated gradient method are:

$$\mathbf{Y}_i^r = -\bar{\mathbf{H}}_i^{-1}\left(\bar{\mathbf{A}}_i^{\mathrm{T}}\boldsymbol{\lambda}^r + \bar{\mathbf{C}}_i^{\mathrm{T}}\boldsymbol{\mu}_i^r\right) \tag{4.46}$$

$$\bar{\mathbf{Y}}_i^r = \mathbf{Y}_i^r + \frac{r-1}{r+2}\left(\mathbf{Y}_i^r - \mathbf{Y}_i^{r-1}\right) \tag{4.47}$$

$$\boldsymbol{\lambda}_i^{r+1} = \boldsymbol{\lambda}_i^r + \frac{r-1}{r+2}\left(\boldsymbol{\lambda}_i^r - \boldsymbol{\lambda}_i^{r-1}\right) + \frac{1}{L}\left(\bar{\mathbf{A}}_i\bar{\mathbf{Y}}^r - \bar{\mathbf{B}}_i\bar{\mathbf{x}}_i\right) \tag{4.48}$$

$$\boldsymbol{\mu}_i^{r+1} = \max\left(0,\, \boldsymbol{\mu}_i^r + \frac{r-1}{r+2}\left(\boldsymbol{\mu}_i^r - \boldsymbol{\mu}_i^{r-1}\right) + \frac{1}{L}\left(\bar{\mathbf{C}}_i\bar{\mathbf{Y}}_i^r - \bar{\mathbf{d}}_i(1-\delta)\right)\right) \tag{4.49}$$

$$i = 1, 2, \ldots, M \tag{4.50}$$

Note that because of the couplings in the dynamics models of the subsystems, in (4.46) the computation of the decision variables $\mathbf{Y}_i^r$ for the ith subsystem requires to dispose of information about the dual variables $\boldsymbol{\lambda}$ for the whole system. For the same reason, in (4.48) the update of the dual variables $\boldsymbol{\lambda}_i$ associated for the ith subsystem uses the information about the decision variables $\bar{\mathbf{Y}}^r$ for the entire system. Since there is no couplings in the control input and state constraint of the subsystems (cf. (4.3)), in (4.49) the update of the dual variables $\boldsymbol{\mu}_i$ for the ith subsystem requires information only about the decision variables $\bar{\mathbf{Y}}_i^r$ for this subsystem.

4.3.2 Algorithm for Distributed Robust MPC

As pointed out in [9], in distributed MPC it is important to keep the number of iterations in the solution algorithm, i.e., the amount of communication between subsystems, as small as possible. At the same time, the number of iterations must be large enough to give a feasible solution to the optimization problem and to guarantee stability of the closed-loop system and the desired performance. In [9], a stopping condition to the distributed optimization algorithm that guarantees these properties is presented, based on a novel adaptive constraint tightening approach. Here, the same approach is applied to solve distributedly the optimization problem (4.30)–(4.32), which results from the centralized robust MPC problem formulation (4.13)–(4.21).

Let $\mathbf{Y}^r$ be the vector of decision variables obtained at the rth iteration of update routine resumed by the steps (4.46)–(4.49). Denote by $\mathbf{U}^r$ the control input trajectory for the overall system, corresponding to $\mathbf{Y}^r$, i.e., $\mathbf{U}^r = [\mathbf{u}_t^r, \mathbf{u}_{t+1}^r, \ldots, \mathbf{u}_{t+N-1}^r]$. Then, according to the receding horizon policy the robust MPC law applied at time t is $\mathbf{u}_{\text{MPC}}(t) = [\mathbf{u}_{1,\text{MPC}}(t), \ldots, \mathbf{u}_{M,\text{MPC}}(t)] = \mathbf{u}_t^r$ and the dynamics of the uncertain closed-loop system is described by:

$$\mathbf{x}_i(t+1) = \mathbf{A}_i(t)\mathbf{x}_i(t) + \mathbf{B}_i(t)\mathbf{u}_{i,\text{MPC}}(t) + \sum_{j=1, j\neq i}^{M} \mathbf{A}_{ij}\mathbf{x}_j(t) + \sum_{j=1, j\neq i}^{M} \mathbf{B}_{ij}\mathbf{u}_{j,\text{MPC}}(t)$$
$$[\mathbf{A}_i(t), \mathbf{B}_i(t)] \in \Omega_i, \quad i = 1, 2, \ldots, M \tag{4.51}$$

Given, the initial state $\mathbf{x}(0) = \bar{\mathbf{x}}$ of the system (4.51), the infinite horizon performance of the robust MPC law is defined as:

$$V_{\text{MPC}}^{\infty}(\bar{\mathbf{x}}) = \sum_{i=1}^{M}\sum_{t=0}^{\infty}\left[\|\mathbf{x}_i(t)^{\text{n}}\|_{\mathbf{Q}_i}^2 + \|\mathbf{u}_{i,\text{MPC}}(t)\|_{\mathbf{R}_i}^2\right] + \sum_{i=1}^{M}\sum_{s=1}^{S}\sum_{t=0}^{\infty}\|\mathbf{x}_i^s(t)\|_{\mathbf{Q}_i}^2 \tag{4.52}$$

Here, $\mathbf{x}_i(t)^{\text{n}}$ is the state trajectory of the system (4.51) corresponding to $[\mathbf{A}_i, \mathbf{B}_i] = [\mathbf{A}_i^{\text{n}}, \mathbf{B}_i^{\text{n}}]$, while $\mathbf{x}_i^s(t)$ is the state trajectory obtained with $[\mathbf{A}_i, \mathbf{B}_i] = [\mathbf{A}_i^s, \mathbf{B}_i^s]$, $s = 1, 2, \ldots, S$.

The infinite horizon performance of the robust MPC law will be compared to the optimal infinite horizon cost $V^{\infty}(\bar{\mathbf{x}})$ obtained by solving the robust infinite horizon optimal control problem P2 (cf. Sect. 4.2). The relation between $V_{\text{MPC}}^{\infty}(\bar{\mathbf{x}})$ and $V^{\infty}(\bar{\mathbf{x}})$ is obtained by using the results of the relaxed dynamic programming [13]. Based on such a construction, it can be claimed that for a given performance parameter

$\alpha \in (0, 1]$ a decrease in the optimal cost function (as defined in (4.13)) along the evolution of the state of system (4.51):

$$V^*(\mathbf{x}(t)) \geq V^*(\mathbf{x}(t+1)) + \alpha l(\mathbf{x}(t)^{\mathrm{n}}, \mathbf{x}(t)^1, \ldots, \mathbf{x}(t)^S, \mathbf{u}_{\mathrm{MPC}}(t))$$
$$\forall[\mathbf{A}(t), \mathbf{B}(t)] \in \Omega \tag{4.53}$$

for every $t \geq 0$, ensure the robust asymptotical stability of the closed-loop system (4.51) with performances that satisfy:

$$\alpha V^{\infty}_{\mathrm{MPC}}(\bar{\mathbf{x}}) \leq V^{\infty}(\bar{\mathbf{x}}) \tag{4.54}$$

In order to apply the stopping condition from [9], let us define the minimum of the stage cost l for given initial state $\mathbf{x}(0) = \bar{\mathbf{x}}$ of the system (4.1) as:

$$l^*(\mathbf{x}(t)^{\mathrm{n}}, \mathbf{x}(t)^1, \ldots, \mathbf{x}(t)^S) = \min_{\mathbf{u} \in \mathscr{U}} l(\mathbf{x}(t)^{\mathrm{n}}, \mathbf{x}(t)^1, \ldots, \mathbf{x}(t)^S, \mathbf{u}(t)) \tag{4.55}$$

where $\mathbf{x}(t)^{\mathrm{n}} = \mathbf{x}(t)^1 = \cdots = \mathbf{x}(t)^S = \bar{\mathbf{x}}$.

Let $\mathbf{U}^r = [\mathbf{u}^r_t, \mathbf{u}^r_{t+1}, \ldots, \mathbf{u}^r_{t+N-1}]$ (extracted from $\mathbf{Y}^r$), $\boldsymbol{\lambda}^r$ and $\boldsymbol{\mu}^r$ be the control input trajectory and the dual variables obtained at the rth iteration of performing the steps (4.46)–(4.49). Denote with $\mathbf{U}^r_q = [\mathbf{u}^r_{t+1}, \mathbf{u}^r_{t+2}, \ldots, \mathbf{u}^r_{t+N-1}, 0]$ the shifted control input sequence. Further, define the primal cost as follows:

$$P(\bar{\mathbf{x}}, \mathbf{U}) = \begin{cases} \sum_{k=0}^{N} l(\mathbf{x}^{\mathrm{n}}_{t+k|t}, \mathbf{x}^1_{t+k|t}, \ldots, \mathbf{x}^S_{t+k|t}, \mathbf{u}_{t+k}), \text{ if } \mathbf{u}_{t+k} \in \mathscr{U},\ k = 0, 1, \ldots, N-1 \\ \quad \text{and } \mathbf{x}^{\mathrm{n}}_{t+k|t} \in \mathscr{X}, \mathbf{x}^s_{t+k|t} \in \mathscr{X},\ k = 1, 2, \ldots, N,\ s = 1, 2, \ldots, S \\ \infty \quad \text{otherwise} \end{cases} \tag{4.56}$$

In (4.56), $\mathbf{x}^{\mathrm{n}}_{t+k|t}$ and $\mathbf{x}^s_{t+k|t}$, $s = 1, \ldots, S$ are the predicted states obtained according to (4.17)–(4.20). Let $\varepsilon > 0$, $\varepsilon < \alpha$ be the specified relative optimality tolerance. The stopping condition in [9] is modified and it consists in adapting the amount of constraint tightening δ to satisfy:

$$\delta \bar{\mathbf{d}}^{\mathrm{T}} \boldsymbol{\mu}^r \leq \varepsilon l^*(\mathbf{x}(t)^{\mathrm{n}}, \mathbf{x}(t)^1, \ldots, \mathbf{x}(t)^S) \tag{4.57}$$

and

$$D(\bar{\mathbf{x}}, \boldsymbol{\lambda}^r, \boldsymbol{\mu}^r) \geq P\left(\mathbf{A}^{\mathrm{n}}\bar{\mathbf{x}} + \mathbf{B}^{\mathrm{n}}\mathbf{u}^r_t, \mathbf{U}^r_q\right) + \alpha l\left(\mathbf{x}(t)^{\mathrm{n}}, \mathbf{x}(t)^1, \ldots, \mathbf{x}(t)^S, \mathbf{u}^r_t\right) \tag{4.58}$$

$$D(\bar{\mathbf{x}}, \boldsymbol{\lambda}^r, \boldsymbol{\mu}^r) \geq P\left(\mathbf{A}^{s}\bar{\mathbf{x}} + \mathbf{B}^{s}\mathbf{u}^r_t, \mathbf{U}^r_q\right) + \alpha l\left(\mathbf{x}(t)^{\mathrm{n}}, \mathbf{x}(t)^1, \ldots, \mathbf{x}(t)^S, \mathbf{u}^r_t\right)$$
$$s = 1, 2, \ldots, S \tag{4.59}$$

Then it will hold that:

$$V^*(\bar{\mathbf{x}}) \geq V^*\left(\mathbf{A}^{\mathrm{n}}\bar{\mathbf{x}} + \mathbf{B}^{\mathrm{n}}\mathbf{u}^r_t\right) + (\alpha - \varepsilon) l\left(\mathbf{x}(t)^{\mathrm{n}}, \mathbf{x}(t)^1, \ldots, \mathbf{x}(t)^S, \mathbf{u}^r_t\right) \tag{4.60}$$

$$V^*(\bar{\mathbf{x}}) \geq V^*\left(\mathbf{A}^{s}\bar{\mathbf{x}} + \mathbf{B}^{s}\mathbf{u}^r_t\right) + (\alpha - \varepsilon) l\left(\mathbf{x}(t)^{\mathrm{n}}, \mathbf{x}(t)^1, \ldots, \mathbf{x}(t)^S, \mathbf{u}^r_t\right)$$
$$s = 1, 2, \ldots, S \tag{4.61}$$

Then it follows that:

$$(\alpha - \varepsilon) V_{\mathrm{MPC}}^{\infty}(\bar{\mathbf{x}}) \leq V^{\infty}(\bar{\mathbf{x}})$$
$$\text{for } [\mathbf{A},\ \mathbf{B}] = [\mathbf{A}^{\mathrm{n}},\ \mathbf{B}^{\mathrm{n}}] \ \text{ or } [\mathbf{A},\ \mathbf{B}] = [\mathbf{A}^{s},\ \mathbf{B}^{s}],\ s \in \{1, 2 \ldots, S\} \tag{4.62}$$

The following algorithm is a slight modification of the algorithm in [9].

Algorithm 4.1 Robust MPC by distributed iterations

1. Given α, δ_{init} and ε. Let $t = 0$ and $\mathbf{Y}(0) = 0$, $\boldsymbol{\lambda}(0) = 0$, $\boldsymbol{\mu}(0) = 0$.
2. Let the state at time t be $\mathbf{x}(t) = \bar{\mathbf{x}} = [\bar{\mathbf{x}}_1, \ldots, \bar{\mathbf{x}}_M]$.
3. Set $r = 0$, $l = 0$, and $\delta = \delta_{\mathrm{init}}$.
4. Initialize algorithm (4.46)–(4.50) with $\mathbf{Y}^0 = \mathbf{Y}^{-1} = \mathbf{Y}(t)$, $\boldsymbol{\lambda}^0 = \boldsymbol{\lambda}^{-1} = \boldsymbol{\lambda}(t)$, $\boldsymbol{\mu}^0 = \boldsymbol{\mu}^{-1} = \boldsymbol{\mu}(t)$.
5. **Do**
6. **If** $D(\bar{\mathbf{x}}, \boldsymbol{\lambda}^r, \boldsymbol{\mu}^r) \geq P(\bar{\mathbf{x}}, \mathbf{U}^r) - \frac{\varepsilon}{l+1} l^*(\mathbf{x}(t)^{\mathrm{n}}, \mathbf{x}(t)^1, \ldots, \mathbf{x}(t)^S)$
7. **or** $\delta \bar{\mathbf{d}}^{\mathrm{T}} \boldsymbol{\mu}^r > \varepsilon l^*(\mathbf{x}(t)^{\mathrm{n}}, \mathbf{x}(t)^1, \ldots, \mathbf{x}(t)^S)$
8. Let $\delta := \delta/2$, $l := l + 1$, $r = 0$.
9. **end if**
10. Run Δr iterations of (4.46)–(4.50).
11. Let $r := r + \Delta r$.
12. **while** $D(\bar{\mathbf{x}}, \boldsymbol{\lambda}^r, \boldsymbol{\mu}^r) \geq P(\mathbf{A}^{\mathrm{n}}\bar{\mathbf{x}} + \mathbf{B}^{\mathrm{n}}\mathbf{u}_t^r, \mathbf{U}_q^r) + \alpha l(\mathbf{x}(t)^{\mathrm{n}}, \mathbf{x}(t)^1, \ldots, \mathbf{x}(t)^S, \mathbf{u}_t^r)$,
$D(\bar{\mathbf{x}}, \boldsymbol{\lambda}^r, \boldsymbol{\mu}^r) \geq P(\mathbf{A}^{s}\bar{\mathbf{x}} + \mathbf{B}^{s}\mathbf{u}_t^r, \mathbf{U}_q^r) + \alpha l(\mathbf{x}(t)^{\mathrm{n}}, \mathbf{x}(t)^1, \ldots, \mathbf{x}(t)^S, \mathbf{u}_t^r)$,
$s = 1, 2, \ldots, S$
and $\delta \bar{\mathbf{d}}^{\mathrm{T}} \boldsymbol{\mu}^r \leq \varepsilon l^*(\mathbf{x}(t)^{\mathrm{n}}, \mathbf{x}(t)^1, \ldots, \mathbf{x}(t)^S)$.
13. Apply to the overall system the control input $\mathbf{u}_{\mathrm{MPC}}(t) = \mathbf{u}_t^r$.
14. Let $\mathbf{Y}(t) = \mathbf{Y}^r$, $\boldsymbol{\lambda}(t) = \boldsymbol{\lambda}^r$, $\boldsymbol{\mu}(t) = \boldsymbol{\mu}^r$.
15. Let $t = t + 1$ and go to step 2.

There are several parameters in Algorithm 4.1, which are set according to [9]. The first is the performance parameter $\alpha \in (0, 1]$ which guarantees closed-loop performance as specified by (4.54). The larger values of α are associated with better performance, but then a longer control horizon N will be needed to guarantee the specified performance. The initial constraint tightening parameter $\delta_{\mathrm{init}} \in (0, 1)$ (from which the constraint tightening parameter δ is adapted to satisfy (4.57)) is usually chosen $\delta_{\mathrm{init}} = 0.2$. The third parameter is the relative optimality tolerance $\varepsilon > 0$ where $\varepsilon < \alpha$, which must be chosen to satisfy a technical condition derived in [9].

4.4 Numerical Example

Consider, the following system composed of two interconnected polytopic subsystems S_1 and S_2:

$$\begin{aligned} S_1 &: \mathbf{x}_1(t+1) = \mathbf{A}_1(t)\mathbf{x}_1(t) + \mathbf{B}_1 u_1(t) + \mathbf{A}_{12}\mathbf{x}_2(t), \mathbf{A}_1(t) \in \Omega_1 \\ S_2 &: \mathbf{x}_2(t+1) = \mathbf{A}_2(t)\mathbf{x}_2(t) + \mathbf{B}_2 u_2(t) + \mathbf{A}_{21}\mathbf{x}_1(t), \mathbf{A}_2(t) \in \Omega_2 \end{aligned} \tag{4.63}$$

where:

$$\begin{aligned} \mathbf{A}_1(t) &= \begin{bmatrix} \beta_1(t) & -0.09 \\ 0.17 & 0.79 \end{bmatrix}, \beta_1(t) \in [0.43, 0.83] \\ \mathbf{A}_2(t) &= \begin{bmatrix} \beta_2(t) & -0.09 \\ 0.17 & 0.69 \end{bmatrix}, \beta_2(t) \in [0.53, 0.93] \\ \mathbf{B}_1 &= \begin{bmatrix} 0.06 \\ 0.01 \end{bmatrix}, \mathbf{B}_2 = \begin{bmatrix} 0.07 \\ 0.01 \end{bmatrix}, \mathbf{A}_{12} = \mathbf{A}_{21} = \begin{bmatrix} 0 & 0 \\ 0 & 0.1 \end{bmatrix} \end{aligned} \tag{4.64}$$

Here, β_1 and β_2 are uncertain parameters. The sets Ω_1 and Ω_2 have two vertices corresponding to $\beta_1 = 0.43$, $\beta_1 = 0.83$ and $\beta_2 = 0.53$, $\beta_2 = 0.93$, respectively. The finite sets $\tilde{\Omega}_1$ and $\tilde{\Omega}_2$ are defined as:

$$\begin{aligned} \tilde{\Omega}_1 &= \{[\mathbf{A}_1(\beta_1),\ \mathbf{B}_1], \beta_1 \in \{0.43,\ 0.63,\ 0.83\}\} \\ \tilde{\Omega}_2 &= \{[\mathbf{A}_2(\beta_2),\ \mathbf{B}_2], \beta_2 \in \{0.53,\ 0.73,\ 0.93\}\} \end{aligned} \tag{4.65}$$

The nominal matrices are $\mathbf{A}_1^{\mathrm{n}} = \mathbf{A}_1(0.63)$, $\mathbf{A}_2^{\mathrm{n}} = \mathbf{A}_2(0.73)$. The following state and input constraints are imposed on the system (4.63):

$$\begin{bmatrix} -0.1 \\ -0.1 \end{bmatrix} \le \mathbf{x}_i(t), -2 \le u_i(t) \le 2, \quad i = 1, 2 \tag{4.66}$$

The prediction horizon is $N = 5$ and the weighting matrices are $\mathbf{Q}_i = \mathbf{I}$, $\mathbf{R}_i = 0.01$, $i = 1, 2$. The centralized robust MPC problem (problem P1) is represented as a distributed optimization problem (problem P3) by applying the approach described in Sect. 4.3.1. Algorithm 4.1 with parameters $\alpha = 0.95$, $\delta_{\text{init}} = 0.5$, $\varepsilon = 0.05$ is used to generate the two control inputs for an initial state of the overall system $\mathbf{x}(0) = [2.5\,2.5\,2.5\,2.5]$. The simulations are performed for the variations of the uncertain parameters, shown in Fig. 4.1. The computed trajectories of the control inputs u_1, u_2 and the states x_1^1, x_1^2 and x_2^1, x_2^2, associated to the subsystems S_1 (i.e., $\mathbf{x}_1 = [x_1^1, x_1^2]$) and S_2 (i.e., $\mathbf{x}_2 = [x_2^1, x_2^2]$) are depicted in Figs. 4.2, 4.3, 4.4. They are compared with the trajectories corresponding to the centralized robust MPC and with the response of the uncertain system (4.63) in closed-loop with a nominal distributed MPC (designed for the nominal system with matrices $\mathbf{A}_1^{\mathrm{n}} = \mathbf{A}_1(0.63)$, $\mathbf{A}_2^{\mathrm{n}} = \mathbf{A}_2(0.73)$).

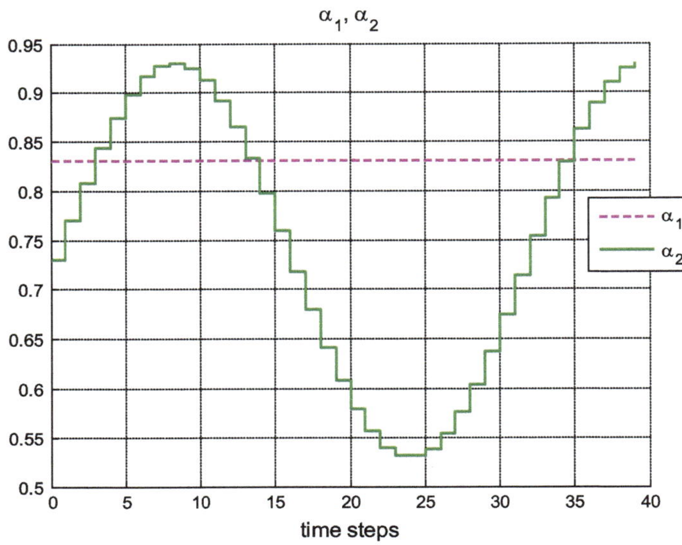

Fig. 4.1 The parameters α_1, α_2

The results show that the suboptimal trajectories obtained with the distributed robust MPC keep both the state and input constraints and do not differ significantly from the centralized robust MPC trajectories. There are two reasons why the results with the distributed robust MPC are so close to those with the centralized robust MPC. First, a low level of suboptimality is allowed by specifying a large value for the performance parameter α ($\alpha = 0.95$) and a small value for the relative optimality tolerance parameter ε ($\varepsilon = 0.05$). Thus, according to (4.62) $V^{\infty}_{\text{MPC}}(\bar{\mathbf{x}})/V^{\infty}(\bar{\mathbf{x}}) \leq (1/0.9 \approx 1.11)$ and therefore the level of suboptimality is at most 11 %. Second, in the cost function (4.22)–(4.23) the weighting coefficients for the state have relatively large values in comparison to those for the control input, which is another reason to have a very small difference between the state trajectories with the distributed and the centralized robust MPCs (see Figs. 4.3 and 4.4). Also, it can be seen from Figs. 4.3 and 4.4 that the nominal distributed MPC violates some of the state constraints. This is not related to the suboptimality of the distributed approach, but to the fact that the nominal MPC is designed by assuming nominal system matrices and ignoring the uncertainty. Therefore, the satisfaction of constraints with the nominal MPC can be guaranteed only for the nominal system.

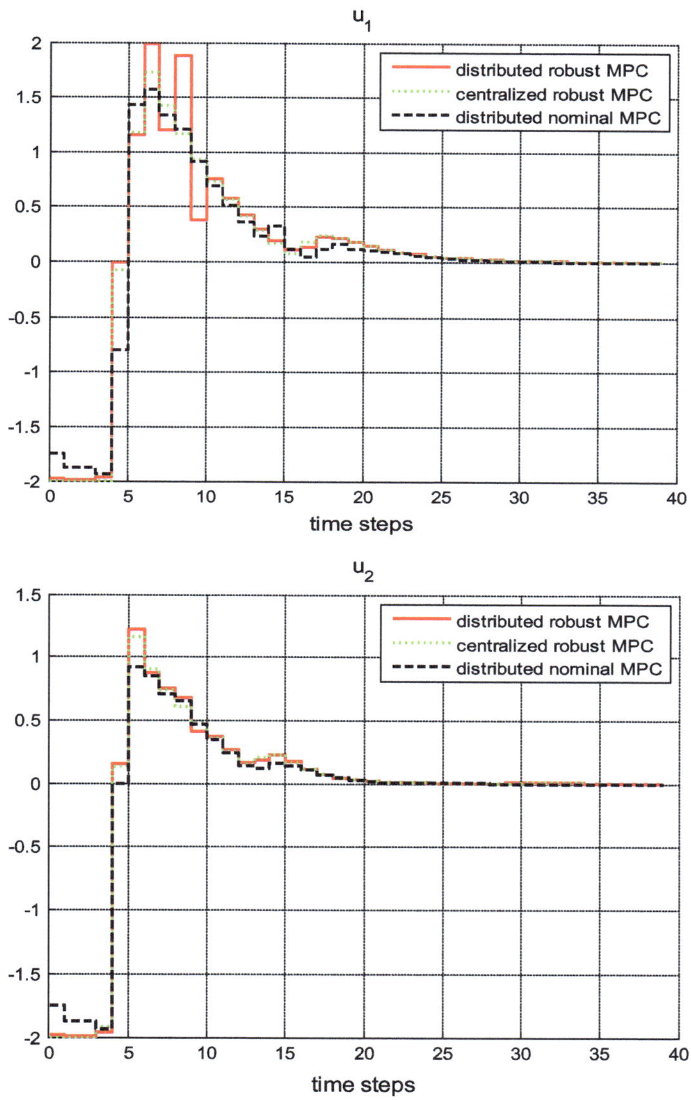

Fig. 4.2 The control inputs u_1 and u_2 for subsystems S_1 and S_2

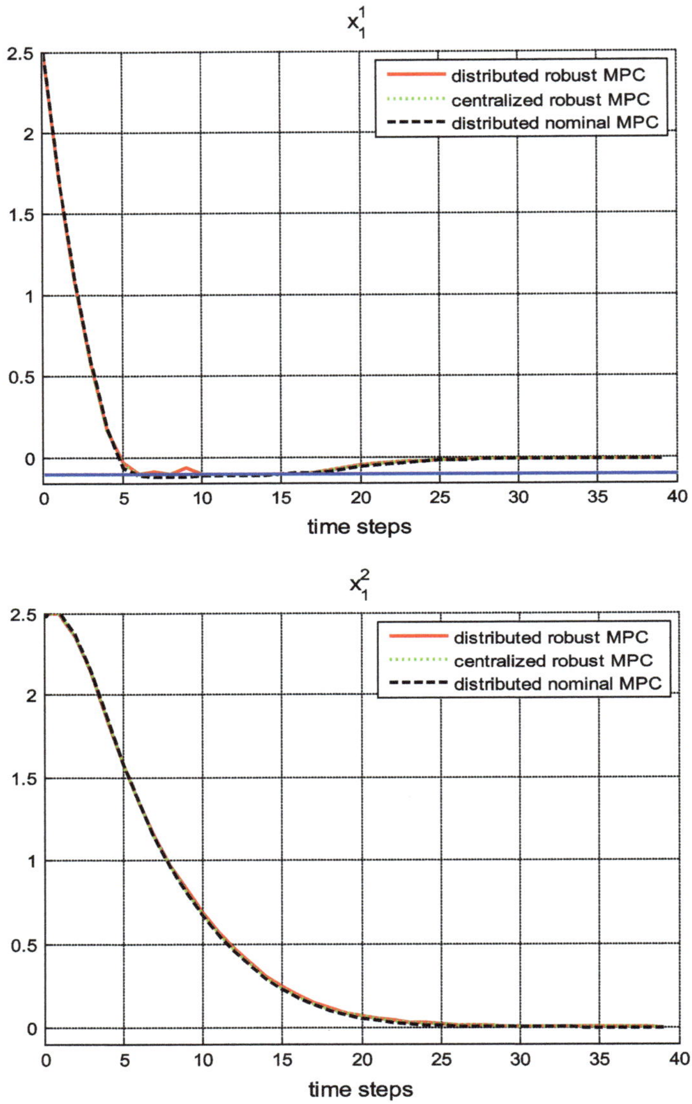

Fig. 4.3 The states x_1^1, x_1^2 of subsystem S_1. The *blue line* corresponds to the constraint

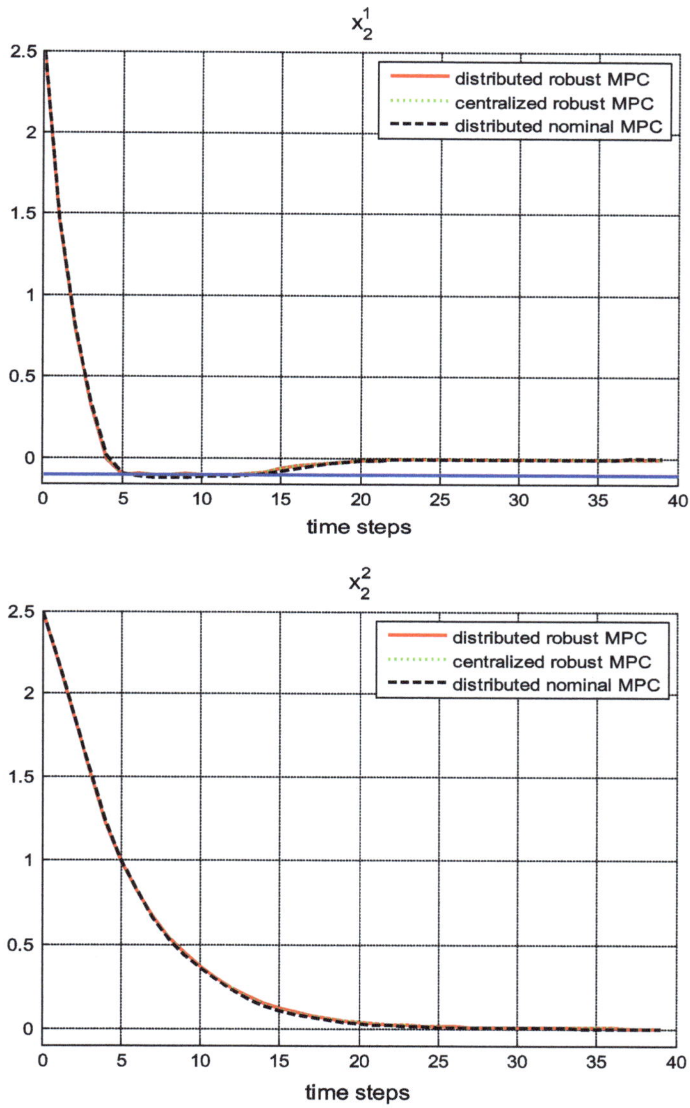

Fig. 4.4 The states x_2^1, x_2^2 of subsystem S_2. The *blue line* corresponds to the constraint

4.5 Conclusions

In this chapter, a suboptimal approach to distributed robust MPC for uncertain systems consisting of interconnected constrained polytopic subsystems is proposed. The approach reformulates the robust MPC problem into a QP problem, which is solved efficiently by distributed iterations of the dual accelerated gradient method with a stopping condition. The robust feasibility and the robust performance are defined for a finite set, which includes the vertices and some interior points of the polytopic uncertainty set. The richer the set of interior points is, higher would be the guarantee for robust feasibility over the polytopic uncertainty set and more reduced would be the level of suboptimality with respect to the robust optimization problem defined for the whole uncertainty set. However, this should be traded off against the increased computational complexity associated to the larger size of the resulting QP problem. The level of suboptimality of the distributed robust MPC with respect to the centralized robust MPC (both defined for the finite uncertainty set) depends on the values of the performance parameter and the relative tolerance parameter, as it is shown in the simulation example. A further extension of the approach would include consideration of the network-induced constraints (cf. [15, 21]), associated to a networked control system structure.

Acknowledgments This work was financed by the National Science Fund of the Ministry of Education and Science of Republic of Bulgaria, contract No. DRila 01/12 and the Partenariats Hubert Curien (PHC) Programme Rila of the French Government, contract No. 29401YJ "Robust distributed predictive control of complex systems."

References

1. A. Alessio, D. Barcelli, A. Bemporad, Decentralized model predictive control of dynamically coupled linear systems. J. Process Control **21**, 705–714 (2011)
2. W. Al-Gherwi, H. Budman, A. Elkamel, A robust distributed model predictive control algorithm. J. Process Control **21**, 1127–1137 (2011)
3. P.D. Christofides, R. Scattolini, D. Muñoz de la Peña, J. Liu, Distributed model predictive control: A tutorial review and future research directions. Comput. Chem. Eng. **51**, 21–41 (2013)
4. G. Cohen, B. Miara, Optimization with an auxiliary constraint and decomposition. SIAM J. Control Optim. **28**, 137–157 (1990)
5. G.A. Constantinides, Parallel architectures for model predictive control, in *Proceedings of the European Control Conference* (Budapest, Hungary, 2009)
6. G.B. Dantzig, P. Wolfe, The decomposition algorithm for linear programs. Econometrica **29**, 767–778 (1961)
7. W.B. Dunbar, Distributed receding horizon control of dynamically coupled nonlinear systems. IEEE Trans. Autom. Control **52**, 1249–1263 (2007)
8. P. Giselsson, A. Rantzer, Distributed model predictive control with suboptimality and stability guarantees, in *Proceedings of the Conference on Decision and Control* (Atlanta, GA, 2010)
9. P. Giselsson, A. Rantzer, On feasibility, stability and performance in distributed model predictive control. IEEE Trans. Autom. Control **59**, 1031–1036 (2014)

10. P. Giselsson, M.D. Doan, T. Keviczky, B. De Schutter, A. Rantzer, Accelerated gradient methods and dual decomposition in distributed model predictive control. Automatica **49**, 829–833 (2013)
11. A. Grancharova, T.A. Johansen, Distributed model predictive control of interconnected nonlinear systems by dynamic dual decomposition, in *Maestre JM*, ed. by R.R. Negenborn (Springer, Distributed Model Predictive Control Made Easy, 2014)
12. A. Grancharova, S. Olaru, An approach to distributed robust model predictive control of discrete-time polytopic systems, in *Proceedings of the 19th IFAC World Congress*, (Cape Town, South Africa, 2014)
13. L. Grüne, A. Rantzer, On the infinite horizon performance of receding horizon controllers. IEEE Trans. Autom. Control **53**, 2100–2111 (2008)
14. M. Heidarinejad, J. Liu, D. Muñoz de la Peña, J.F. Davis, P.D. Christofides, Multirate Lyapunov-based distributed model predictive control of nonlinear uncertain systems. J. Process Control **21**, 1231–1242 (2011)
15. J.P. Hespanha, P. Naghshtabrizi, Y. Xu, A survey of recent results in networked control systems. Proc. IEEE Spec. Issue Technol. Netw. Control Syst. **95**, 138–162 (2007)
16. J.M. Maestre, R.R. Negenborn, Distributed model predictive control made easy, in *Series: Intelligent Systems, Control and Automation: Science and Engineering*, vol. 69 (Springer, Hidelberg, 2014)
17. D.M. Raimondo, L. Magni, R. Scattolini, Decentralized MPC of nonlinear systems: An input-to-state stability approach. Int. J. Robust Nonlinear Control **17**, 1651–1667 (2007)
18. R. Scattolini, Architectures for distributed and hierarchical model predictive control—A review. J. Process Control **19**, 723–731 (2009)
19. A.N. Venkat, I.A. Hiskens, J.B. Rawlings, S.J. Wright, Distributed MPC strategies with application to power system automatic generation control. IEEE Trans. Control Syst. Technol. **16**, 1192–1206 (2008)
20. Y. Zhang, S. Li, Networked model predictive control based on neighbourhood optimization for serially connected large-scale processes. J. Process Control **17**, 37–50 (2007)
21. L. Zhang, H. Gao, O. Kaynak, Network-induced constraints in networked control system—A survey. IEEE Trans. Ind. Inf. **9**, 403–416 (2013)
22. L. Zhang, J. Wang, C. Li, Distributed model predictive control for polytopic uncertain systems subject to actuator saturation. J. Process Control **23**, 1075–1089 (2013)

Chapter 5
Optimal Distributed-Coordinated Approach for Energy Management in Multisource Electric Power Generation Systems

John Sandoval-Moreno, John Jairo Martínez and Gildas Besançon

Abstract In the context of distributed power generation systems, the energy management and coordination of generators are imperative tasks to be done. Such systems, typically considered as large-scale systems, can include different dynamical and functional characteristics in both, generators and loads. In this sense, the use of distributed-coordinated control strategies, including operational constraints, becomes an interesting alternative for these applications. This chapter proposes a novel *price-driven coordination* technique. The approach considers that a centralized optimal control problem can be splitted into several unconstrained controlled subsystems, all coordinated by an agent which is intended for accomplishing the global performance, while assuring the system constraints. The approach is applied to a microgrid that combines different generation technologies and load profiles.

Keywords Optimal coordination · Decentralized control · Model predictive control for linear systems · Power plants

John Sandoval-Moreno acknowledges the Colombian Administrative Department of Science, Technology and Innovation (COLCIENCIAS) for the "Francisco José de Caldas" scholarship that financed his doctoral studies in France.

J. Sandoval-Moreno (✉)
Grenoble-INP, G2Elab, F-38000 Grenoble, France
e-mail: john-anderson.sandoval-moreno@g2elab.grenoble-inp.fr; johnasanm@gmail.com

J.J. Martínez · G. Besançon
Grenoble-INP, Gipsa-Lab, 38000 Grenoble, France
e-mail: john.martinez@gipsa-lab.grenoble-inp.fr

J.J. Martínez · G. Besançon
Institut Universitaire de France, Paris, France
e-mail: gildas.besancon@gipsa-lab.grenoble-inp.fr

S. Olaru et al. (eds.), *Developments in Model-Based Optimization and Control*, Lecture Notes in Control and Information Sciences 464,
DOI 10.1007/978-3-319-26687-9_5

5.1 Introduction

In the last decades, the advances in renewable-based power generators technologies, such as fuel cells, photovoltaic panels, and wind turbines [12, 18, 19], have impulsed the inclusion of smaller power generators in the electric network, along the bulk power generators such as hydro or thermoelectrical ones [13]. The aim is to migrate to sustainable generation technologies that cover the rising needs of the society. Actually, developments in *smart grid* technologies [3, 6, 20] aim to implement efficient communication and control infrastructures for maintaining the power network performance and stability, considering the heterogeneity of generators, storage unities, and charges (vehicles, residences, industries).

Modern trends in power networks management are keen to obtain decentralized control structures. In this sense, the application of *distributed control* methods is useful for developing such algorithms [15], considering the *large-scale* characteristic of the power networks. The power control strategies are developed under two principles: system stability, related with voltage and frequency regulation [13] and optimal performance that aims to optimize the power generated by the different sources and is achieved by minimizing a cost function.

The algorithms are based in the decomposition of the control structures in smaller ones, typically known as *local controllers* or *agents*. In the case, where the local controllers are interacting between them for performing a mission, according to a global criteria, the global control system is known as a *distributed control* system. However, in the case where the local controllers are exchanging information with a higher hierarchical unit (known as *coordinator*), the control scheme is considered as a *distributed-coordinated* one.The model decomposition can be done by simple inspection, or using some methods like the proposed in [15, 16, 26].

The distributed-coordination can be performed by using different methods. There are two well-known coordination methods in the literature, the *interaction prediction* method and the *price-driven* method. Both methods are based on the current states values for each subsystem. The first method propose local control values to be implemented for every subsystem after the computation of the *optimal* expected interactions to achieve the global optimal control [7, 28]. In the case of the *price-driven* coordination method, the coordinator considers the subsystems interactions as a family of equality constraints to be respected. Thus, the coordinator solves a global optimization problem by introducing a Lagrange multiplier (the price) that is distributed to every subsystem to locally perform the control [1, 8, 10]. This scheme reduces the amount of information between the coordinator and local controllers, while allowing independent local control structures that could eventually work independently from the coordinator actions [9, 14, 22, 27].

Considering the requirements of global performance while assuring system stability, in the present chapter, is presented a methodology for conceiving a price-driven coordination scheme based on local *model predictive control* (MPC) algorithms [4, 10]. This chapter represents an extension of a preliminary work presented in [22]. Here, the proposed approach considers that the global objective function can be optimized while dealing with the operative constraints. The idea is to decompose the global optimization problem in several local and coordinated optimization problems. The local controllers perform a state feedback control law, obtained by the *explicit unconstrained solution* of local MPC problems. These local controllers receive, from the coordinator, an optimal coordination (tuning) signal that is obtained after solving a constrained optimization problem that considers the *global system constraints*. The coordination problem is obtained after directly integrate the local explicit solution, receiving only the subsystems current data (states, disturbances, and setpoint values). The coordination approach is applied to a multisource power generation system that combines different generation technologies and load profiles.

This chapter is organized as follows: Sect. 5.2 includes the definition of the centralized MPC problem, written in two forms: with and without explicit *interactions* handling. Section 5.3 shows the details of the optimization problem decomposition and coordinations strategy. Section 5.4 includes the description and simulation of a two-areas power system in which the proposed control approach is compared with a centralized MPC strategy. In the final section, some conclusions are added.

5.2 Centralized Optimal Control Problem with System Interactions

In this section is presented a reformulation of Model Predictive Control (MPC) for linear systems, introduced in [10]. In the referenced work, the authors present a MPC formulation that considers the system model as a centralized structure. In the current work, the same problem is reformulated by adding an explicit representation of the system *interactions*. As done in [10], the proposed formulation is written as a *quadratic programming* (QP) problem, which can be solved using available numerical methods.

5.2.1 System Model Including Subsystem Interactions

Considering the following discrete-time dynamical system:

$$\boldsymbol{x}(k+1) = \boldsymbol{A}\boldsymbol{x}(k) + \boldsymbol{B}\boldsymbol{u}(k) + \boldsymbol{E}\boldsymbol{d}(k) \tag{5.1}$$

where k denotes the current time step, $\boldsymbol{x}(k) \in \mathbb{R}^n, \boldsymbol{u}(k) \in \mathbb{R}^m, \boldsymbol{d}(k) \in \mathbb{R}^q$ stand for the state, input and disturbances vector, respectively, while $\boldsymbol{A} \in \mathbb{R}^{n\times n}, \boldsymbol{B} \in \mathbb{R}^{n\times m}$ and $\boldsymbol{E} \in \mathbb{R}^{n\times q}$ are the states, input and disturbance matrices, respectively. The expression (5.1) represents the system model from a global perspective. Using simple inspection or particular model analysis techniques [15, 16, 26], the global dynamical model can be decomposed in smaller parts, known as *subsystems*, identified each one with the subindex $i = 1 \cdots z$.

The subsystem i is represented by local states, inputs, and disturbance vectors $\boldsymbol{x}_i(k) \in \mathbb{R}^{n_i}$, $\boldsymbol{u}_i(k) \in \mathbb{R}^{m_i}$ and $\boldsymbol{d}_i(k) \in \mathbb{R}^{q_i}$, respectively, as well as an *interactions* vector $\boldsymbol{v}_i(k) \in \mathbb{R}^{r_i}$ that represents the effect of the subsystems $i \neq j$ at the system i. In this sense, (5.1) can be written in the following extended form:

$$\begin{bmatrix} \boldsymbol{x}_1(k+1) \\ \vdots \\ \boldsymbol{x}_z(k+1) \end{bmatrix} = \boldsymbol{A} \begin{bmatrix} \boldsymbol{x}_1(k) \\ \vdots \\ \boldsymbol{x}_z(k) \end{bmatrix} + \boldsymbol{B} \begin{bmatrix} \boldsymbol{u}_1(k) \\ \vdots \\ \boldsymbol{u}_z(k) \end{bmatrix} + \boldsymbol{E} \begin{bmatrix} \boldsymbol{d}_1(k) \\ \vdots \\ \boldsymbol{d}_z(k) \end{bmatrix} \tag{5.2}$$

From (5.2), it is possible to write the subsystems dynamics in the following form:

$$\boldsymbol{x}_i(k+1) = \boldsymbol{A_{ii}} \boldsymbol{x}_i(k) + \boldsymbol{B_{ii}} \boldsymbol{u}_i(k) + \boldsymbol{E_{ii}} \boldsymbol{d}_i(k) + \boldsymbol{v}_i(k) \tag{5.3}$$

where $\boldsymbol{v}_i(k)$ has the following structure:

$$\boldsymbol{v}_i(k) = \sum_{j=1; j \neq i}^{z} \boldsymbol{A_{ij}} \boldsymbol{x}_j(k) + \boldsymbol{B_{ij}} \boldsymbol{u}_j(k) + \boldsymbol{E_{ij}} \boldsymbol{d}_j(k) \tag{5.4}$$

The full system *interaction vector* $\boldsymbol{v}(k) \in \mathbb{R}^r$ (for some $r \leq n$), based on (5.4), satisfies the following expression:

$$\begin{bmatrix} \boldsymbol{v}_1(k) \\ \vdots \\ \boldsymbol{v}_z(k) \end{bmatrix} = \boldsymbol{v_A} \begin{bmatrix} \boldsymbol{x}_1(k) \\ \vdots \\ \boldsymbol{x}_z(k) \end{bmatrix} + \boldsymbol{v_B} \begin{bmatrix} \boldsymbol{u}_1(k) \\ \vdots \\ \boldsymbol{u}_z(k) \end{bmatrix} + \boldsymbol{v_E} \begin{bmatrix} \boldsymbol{d}_1(k) \\ \vdots \\ \boldsymbol{d}_z(k) \end{bmatrix} \tag{5.5}$$

equivalently,

$$\boldsymbol{v}(k) = \boldsymbol{v_A} \boldsymbol{x}(k) + \boldsymbol{v_B} \boldsymbol{u}(k) + \boldsymbol{v_E} \boldsymbol{d}(k) \tag{5.6}$$

where $\boldsymbol{v_A}$, $\boldsymbol{v_B}$, and $\boldsymbol{v_E}$ are the matrices $\boldsymbol{A}$, $\boldsymbol{B}$, and $\boldsymbol{E}$, but replacing by zero matrices their main diagonal. Now, one can define a global dynamics model that considers the subsystems interactions in a more explicit way. That is,

$$\boldsymbol{x}(k+1) = \boldsymbol{A_d} \boldsymbol{x}(k) + \boldsymbol{B_d} \boldsymbol{u}(k) + \boldsymbol{E_d} \boldsymbol{d}(k) + \boldsymbol{v}(k) \tag{5.7}$$

where $\boldsymbol{A_d} = \boldsymbol{A} - \boldsymbol{v_A}$, $\boldsymbol{B_d} = \boldsymbol{B} - \boldsymbol{v_B}$, and $\boldsymbol{E_d} = \boldsymbol{E} - \boldsymbol{v_E}$. The signals in the modified dynamical model (5.7) are defined for the following admissible sets (according to the physical restrictions of the system):

$$\begin{array}{ll} \boldsymbol{x}(k) \in \mathcal{X} & \text{where } \mathcal{X} = \{\boldsymbol{x}(k) \in \mathbb{R}^n : \boldsymbol{x_{min}} \leq \boldsymbol{x}(k) \leq \boldsymbol{x_{max}}\} \\ \boldsymbol{u}(k) \in \mathcal{U} & \text{where } \mathcal{U} = \{\boldsymbol{u}(k) \in \mathbb{R}^m : \boldsymbol{u_{min}} \leq \boldsymbol{u}(k) \leq \boldsymbol{u_{max}}\} \\ \boldsymbol{d}(k) \in \mathcal{D} & \text{where } \mathcal{D} = \{\boldsymbol{d}(k) \in \mathbb{R}^q : \boldsymbol{d_{min}} \leq \boldsymbol{d}(k) \leq \boldsymbol{d_{max}}\} \\ \boldsymbol{v}(k) \in \mathcal{V} & \text{where } \mathcal{V} = \{\boldsymbol{v}(k) \in \mathbb{R}^n : \boldsymbol{v_{min}} \leq \boldsymbol{v}(k) \leq \boldsymbol{v_{max}}\} \end{array} \tag{5.8}$$

where the values $\boldsymbol{v_{min}}$ and $\boldsymbol{v_{max}}$ are obtained by replacing the corresponding limits defined in (5.8) for the states, inputs and disturbances, in (5.6).

5.2.2 Augmented Centralized Optimization Problem

For writing the optimization problem, based on the MPC formulation, define the following vectors $\boldsymbol{x_s} \equiv \boldsymbol{x}(\infty)$, $\boldsymbol{u_s} \equiv \boldsymbol{u}(\infty)$ and $\boldsymbol{v_s} \equiv \boldsymbol{v}(\infty)$ as the desired steady-state values of the states, inputs, and interactions vectors, respectively, which would be achieved when a permanent disturbance vector $\boldsymbol{d_s}$ affects the system.

The MPC problem is devoted to *minimize* the tracking errors $\boldsymbol{x}'(k) = \boldsymbol{x}(k) - \boldsymbol{x_s}$, $\boldsymbol{u}'(k) = \boldsymbol{u}(k) - \boldsymbol{u_s}$ and $\boldsymbol{v}'(k) = \boldsymbol{v}(k) - \boldsymbol{v_s}$, under the effect of some disturbance variations around an steady-state value $\boldsymbol{d}'(k) = \boldsymbol{d}(k) - \boldsymbol{d_s}$ during the finite horizon N, while considering a final state tracking error $\boldsymbol{x}'(N) = \boldsymbol{x}(N) - \boldsymbol{x_s}$ [10]. The cost function $J_{gv}(\boldsymbol{x}, \boldsymbol{u}, \boldsymbol{v})$ (the subindex gv stands for *global with interactions*), that is used for computing the optimal input signal is represented in the following form:

$$
\begin{aligned}
J_{gv}(\boldsymbol{x}, \boldsymbol{u}, \boldsymbol{v}) = \tfrac{1}{2} \sum_{k=0}^{N-1} \left(\boldsymbol{x}'^{T}(k)\boldsymbol{Q}\boldsymbol{x}'(k) + \boldsymbol{u}'^{T}(k)\boldsymbol{R_u}\boldsymbol{u}'(k) + 2\boldsymbol{u}'^{T}(k)\boldsymbol{S_u}\boldsymbol{x}'(k)\right) \\
+ \tfrac{1}{2} \sum_{k=0}^{N-1} \left(\boldsymbol{v}'^{T}(k)\boldsymbol{R_v}\boldsymbol{v}'(k) + 2\boldsymbol{v}'^{T}(k)\boldsymbol{S_v}\boldsymbol{x}'(k)\right) + \tfrac{1}{2}\boldsymbol{x}'^{T}(N)\boldsymbol{P}\boldsymbol{x}'(N)
\end{aligned}
\tag{5.9}
$$

where $\boldsymbol{P}$, $\boldsymbol{Q}$, $\boldsymbol{R_u}$, $\boldsymbol{S_u}$, $\boldsymbol{R_v}$, and $\boldsymbol{S_v}$ represent weighting matrices with the following properties: $\boldsymbol{R_u} \geq 0$, $\boldsymbol{Q} = \boldsymbol{Q}^T \geq 0$, $\boldsymbol{S_u}$ requires that $\boldsymbol{Q} \geq \boldsymbol{S_u}\boldsymbol{R_u}^{-1}\boldsymbol{S_u}^T$, $\boldsymbol{R_v}$ should be chosen conveniently to guarantee convexity of the problem [9, 27], $\boldsymbol{S_v}$ is typically equal to zero and $\boldsymbol{P} = \boldsymbol{P}^T \geq 0$ is obtained after solving the discrete-time Ricatti algebraic equation [10, 26] for the system dynamics (5.1) while using the matrices $\boldsymbol{Q}$, $\boldsymbol{R_u}$ and $\boldsymbol{S_u}$.

Differently from the cost function formulation of [10], in the cost function (5.9) is included a penalization term for the interactions, considering that the proposed approach treats the interactions explicitly.

At this stage, the idea is to optimize (5.9), considering the system dynamics (5.7), the interactions (5.6), the constraints defined at (5.8), and a good knowledge of the disturbances vector $\boldsymbol{d}(k)$ for the prediction horizon N. The problem can be written in the following form:

$$
\begin{aligned}
\min_{\boldsymbol{u}} \quad & J_{gv}(\boldsymbol{x}, \boldsymbol{u}, \boldsymbol{v}) \\
\text{subject to } & \boldsymbol{L}\boldsymbol{u}(k) \leq \boldsymbol{W} \\
& \boldsymbol{x}(k+1) = \boldsymbol{A_d}\boldsymbol{x}(k) + \boldsymbol{B_d}\boldsymbol{u}(k) + \boldsymbol{E_d}\boldsymbol{d}(k) + \boldsymbol{v}(k),\ 0 \leq k \leq N-1 \\
& \boldsymbol{v}(k) = \boldsymbol{v_A}\boldsymbol{x}(k) + \boldsymbol{v_B}\boldsymbol{u}(k) + \boldsymbol{v_E}\boldsymbol{d}(k),\ 0 \leq k \leq N-1 \\
& \boldsymbol{x}(0) \in \mathcal{X}, \boldsymbol{x}(k) \in \mathcal{X}, \boldsymbol{u}(k) \in \mathcal{U}, \boldsymbol{v}(k) \in \mathcal{V}, \boldsymbol{d}(k) \in \mathcal{D},\ 0 \leq k \leq N-1
\end{aligned}
\tag{5.10}
$$

Here, the objective is to express the problem (5.10) as a QP problem. For doing so, define the following sequences for the signal vectors and their reference values:

$$\bar{\boldsymbol{x}} \equiv \left[\boldsymbol{x}^T(1) \cdots \boldsymbol{x}^T(N)\right]^T \quad \bar{\boldsymbol{u}} \equiv \left[\boldsymbol{u}^T(0) \cdots \boldsymbol{u}^T(N-1)\right]^T \quad \bar{\boldsymbol{d}} \equiv \left[\boldsymbol{d}^T(0) \cdots \boldsymbol{d}^T(N-1)\right]^T$$
$$\bar{\boldsymbol{x}}_s \equiv \left[\boldsymbol{x}_s^T \cdots \boldsymbol{x}_s^T\right]^T \quad \bar{\boldsymbol{u}}_s \equiv \left[\boldsymbol{u}_s^T \cdots \boldsymbol{u}_s^T\right]^T \quad \bar{\boldsymbol{d}}_s \equiv \left[\boldsymbol{d}_s^T \cdots \boldsymbol{d}_s^T\right]^T$$
$$\bar{\boldsymbol{v}} \equiv \left[\boldsymbol{v}^T(0) \cdots \boldsymbol{v}^T(N-1)\right]^T \quad \bar{\boldsymbol{v}}_s \equiv \left[\boldsymbol{v}_s^T \cdots \boldsymbol{v}_s^T\right]^T \tag{5.11}$$

Taking into account that the interaction variable $\boldsymbol{v}(k)$ is linked with the effects of $\boldsymbol{u}(k)$ as seen in (5.6), it is proposed to be used the following decision variable for the QP problem:

$$\boldsymbol{u_{ext}} \equiv \left[\boldsymbol{u}^T(0)\ \boldsymbol{u}^T(1) \cdots \boldsymbol{u}^T(N-1) \mid \boldsymbol{v}^T(0)\ \boldsymbol{v}^T(1) \cdots \boldsymbol{v}^T(N-1)\right] = [\bar{\boldsymbol{u}}^T\ \ \bar{\boldsymbol{v}}^T]^T \tag{5.12}$$

With the elements introduced in (5.11)–(5.12), one can write the *predicted states vector* $\bar{\boldsymbol{x}}$ as in (5.13), and the *predicted states error vector* $\bar{\boldsymbol{x}}'$ as in (5.14), as function of the current state value $\boldsymbol{x}(0)$, the disturbances sequences $\bar{\boldsymbol{d}}$, the extended variable $\boldsymbol{u_{ext}}$ and the steady-state values of these vectors $\boldsymbol{x_s}$, $\bar{\boldsymbol{d}}_s$, and $\boldsymbol{u_{ext-s}} \equiv [\bar{\boldsymbol{u}}_s{}^T\ \bar{\boldsymbol{v}}_s{}^T]^T$.

In fact, $\bar{\boldsymbol{x}}$ can be obtained by developing the state Eq. (5.7) for $k = 1$ to $k = N$ (further details to perform this operation are included in [10]).

$$\bar{\boldsymbol{x}} = \boldsymbol{\Omega_d}\boldsymbol{x}(0) + \boldsymbol{\Psi_d}\boldsymbol{u_{ext}} + \boldsymbol{\Theta_d}\bar{\boldsymbol{d}} \tag{5.13}$$

$$\bar{\boldsymbol{x}}' = \bar{\boldsymbol{x}} - \bar{\boldsymbol{x}}_s = \boldsymbol{\Omega_d}\boldsymbol{x}'(0) + \boldsymbol{\Psi_d}\boldsymbol{u}'_{\boldsymbol{ext}} + \boldsymbol{\Theta_d}\bar{\boldsymbol{d}}' \tag{5.14}$$

where $\boldsymbol{x}'(0) = \boldsymbol{x}(0) - \boldsymbol{x_s}$, $\boldsymbol{u}'_{\boldsymbol{ext}} = \boldsymbol{u_{ext}} - \boldsymbol{u_{ext-s}}$, $\bar{\boldsymbol{d}}' = \bar{\boldsymbol{d}} - \bar{\boldsymbol{d}}_s$, while the matrices $\boldsymbol{\Omega_d}$, $\boldsymbol{\Gamma_d}$, $\boldsymbol{\Theta_d}$, $\boldsymbol{\Lambda_d}$ are obtained from the deployment of (5.7) between $k = 1$ and $k = N$, with $\boldsymbol{\Psi_d} \equiv [\boldsymbol{\Gamma_d}\ \ \boldsymbol{\Lambda_d}]$:

$$\boldsymbol{\Omega_d} \equiv \begin{bmatrix} A_d \\ A_d^2 \\ \vdots \\ A_d^N \end{bmatrix} \quad \boldsymbol{\Gamma_d} \equiv \begin{bmatrix} B_d & 0 & \cdots & 0 \\ A_d B_d & B_d & \cdots & 0 \\ \vdots & \vdots & \ddots & \vdots \\ A_d^{N-1} B_d & A_d^{N-2} B_d & \cdots & B_d \end{bmatrix}$$
$$\boldsymbol{\Theta_d} \equiv \begin{bmatrix} E_d & 0 & \cdots & 0 \\ A_d E_d & E_d & \cdots & 0 \\ \vdots & \vdots & \ddots & \vdots \\ A_d^{N-1} E_d & A_d^{N-2} E_d & \cdots & E_d \end{bmatrix} \quad \boldsymbol{\Lambda_d} \equiv \begin{bmatrix} I & 0 & \cdots & 0 \\ A_d & I & \cdots & 0 \\ \vdots & \vdots & \ddots & \vdots \\ A_d^{N-1} & A_d^{N-2} & \cdots & I \end{bmatrix} \tag{5.15}$$

With similar considerations, the *predicted interaction vector* can be expressed in terms of the system dynamics and the interactions equations, also considering the variable $\boldsymbol{u_{ext}}$ as follows:

$$\overline{\boldsymbol{\Lambda}}\bar{\boldsymbol{v}} = \overline{\boldsymbol{\Gamma}}\bar{\boldsymbol{u}} + \overline{\boldsymbol{\Omega}}\boldsymbol{x}(0) + \overline{\boldsymbol{\Theta}}\bar{\boldsymbol{d}} \tag{5.16}$$

that is equivalent to:

$$0 = \overline{\boldsymbol{\Psi}}\boldsymbol{u_{ext}} + \overline{\boldsymbol{\Omega}}\boldsymbol{x}(0) + \overline{\boldsymbol{\Theta}}\bar{\boldsymbol{d}} \tag{5.17}$$

where matrices $\overline{\boldsymbol{\Omega}}$, $\overline{\boldsymbol{\Gamma}}$ and $\overline{\boldsymbol{\Theta}}$ are defined by developing (5.6) from $k=0$ to $k=N-1$, while $\overline{\boldsymbol{\Psi}} \equiv [\overline{\boldsymbol{\Gamma}} \;\; -\overline{\boldsymbol{\Lambda}}]$:

$$\overline{\boldsymbol{\Omega}} \equiv \begin{bmatrix} \boldsymbol{v_A} \\ \boldsymbol{v_A A_d} \\ \vdots \\ \boldsymbol{v_A A_d^{N-1}} \end{bmatrix} \overline{\boldsymbol{\Gamma}} \equiv \begin{bmatrix} \boldsymbol{v_B} & 0 & \cdots & 0 \\ \boldsymbol{v_A B_d} & \boldsymbol{v_B} & \cdots & 0 \\ \vdots & \vdots & \ddots & \vdots \\ \boldsymbol{v_A A_d^{N-2} B_d} & \boldsymbol{A_d^{N-3} B_d} & \cdots & \boldsymbol{v_B} \end{bmatrix}$$
$$\overline{\boldsymbol{\Theta}} \equiv \begin{bmatrix} \boldsymbol{v_E} & 0 & \cdots & 0 \\ \boldsymbol{v_A E_d} & \boldsymbol{v_E} & \cdots & 0 \\ \vdots & \vdots & \ddots & \vdots \\ \boldsymbol{v_A A_d^{N-2} E_d} & \boldsymbol{v_A A_d^{N-3} E_d} & \cdots & \boldsymbol{v_E} \end{bmatrix} \overline{\boldsymbol{\Lambda}} \equiv \begin{bmatrix} I & 0 & \cdots & 0 \\ -\boldsymbol{v_A} & I & \cdots & 0 \\ \vdots & \vdots & \ddots & \vdots \\ -\boldsymbol{v_A A_d^{N-2}} & -\boldsymbol{v_A A_d^{N-3}} & \cdots & I \end{bmatrix} \tag{5.18}$$

As a final stage to write the problem (5.10) in QP form, it is defined the following extended weighting matrices:

$$\begin{gathered} \overline{\boldsymbol{Q}} = diag\{\boldsymbol{Q} \cdots \boldsymbol{Q}\ \boldsymbol{P}\} \\ \overline{\boldsymbol{R}}_{\boldsymbol{u}} = diag\{\boldsymbol{R}_u \cdots \boldsymbol{R}_u\},\ \overline{\boldsymbol{R}}_{\boldsymbol{v}} = diag\{\boldsymbol{R}_v \cdots \boldsymbol{R}_v\},\ \boldsymbol{R_{ext}} = diag\{\overline{\boldsymbol{R}}_{\boldsymbol{u}}\ \overline{\boldsymbol{R}}_{\boldsymbol{v}}\} \\ \overline{\boldsymbol{S}}_{\boldsymbol{u}} = diag\{\boldsymbol{S}_u \cdots \boldsymbol{S}_u\},\ \ \overline{\boldsymbol{S}}_{\boldsymbol{v}} = diag\{\boldsymbol{S}_v \cdots \boldsymbol{S}_v\},\ \ \boldsymbol{S_{ext}} = diag\{\overline{\boldsymbol{S}}_{\boldsymbol{u}}\ \overline{\boldsymbol{S}}_{\boldsymbol{v}}\} \end{gathered} \tag{5.19}$$

Then using (5.11) and (5.19), the expression (5.9) takes the following form, as a function of $\bar{\boldsymbol{x}}'$ and $\boldsymbol{u}'_{\boldsymbol{ext}}$ (see (5.14)):

$$J_{gv}(\boldsymbol{x}, \boldsymbol{u_{ext}}) = \frac{1}{2}\left[\boldsymbol{x}'^T(0)\boldsymbol{Q}\boldsymbol{x}'(0) + \bar{\boldsymbol{x}}'^T\overline{\boldsymbol{Q}}\bar{\boldsymbol{x}}' + \boldsymbol{u}'^T_{\boldsymbol{ext}}\boldsymbol{R_{ext}}\boldsymbol{u}'_{\boldsymbol{ext}} + 2\boldsymbol{u}'^T_{\boldsymbol{ext}}\boldsymbol{S_{ext}}\left(\boldsymbol{G_x}\bar{\boldsymbol{x}}' + \boldsymbol{G_x^*}\bar{\boldsymbol{x}}'(0)\right)\right] \tag{5.20}$$

where the matrices $\boldsymbol{G_x}$ and $\boldsymbol{G_x^*}$ are defined as:

$$\boldsymbol{G_x} = \begin{bmatrix} 0 & 0 & \cdots & 0 \\ I_n & 0 & \cdots & 0 \\ \vdots & \ddots & \ddots & \vdots \\ 0 & \cdots & I_n & 0 \end{bmatrix} \quad \boldsymbol{G_x^*} = \begin{bmatrix} I_n \\ 0 \\ \vdots \\ 0 \end{bmatrix} \tag{5.21}$$

The equality constraints imposed by the system dynamics and interactions of the problem (5.10) are included in the QP problem in the following way:

- The predicted states error vector $\bar{\boldsymbol{x}}'$, represented by (5.14) is replaced in the cost function (5.20).
- The interactions equality (5.17) can be included in the cost function by using a *Lagrange multiplier* vector $\boldsymbol{p} = [\boldsymbol{p}(0) \cdots \boldsymbol{p}(N-1)]^T$. In this way, the optimization problem is relaxed and a new parameter should be computed [1]. This constraint is represented by expression (5.17).

With these considerations, the cost function $J_{gv}(\boldsymbol{x}, \boldsymbol{u}, \boldsymbol{v})$ is transformed in the following quadratic cost function $J_{ext}(\boldsymbol{u_{ext}}, \boldsymbol{p})$, with $\overline{V}_{ext}$ gathering all terms independent of variables $\boldsymbol{u_{ext}}$ and $\boldsymbol{p}$:

$$J_{ext}(\boldsymbol{u_{ext}}, \boldsymbol{p}) = \tfrac{1}{2}\boldsymbol{u^T_{ext}} \boldsymbol{H_c} \boldsymbol{u_{ext}} + \boldsymbol{u^T_{ext}}[\boldsymbol{K_{1c}}(\boldsymbol{x}(0) - \boldsymbol{x_s}) + \boldsymbol{K_{2c}}(\bar{\boldsymbol{d}} - \bar{\boldsymbol{d}}_s) - \boldsymbol{H_c}\boldsymbol{u_{ext-s}}] \\ + \boldsymbol{p}^T(\overline{\boldsymbol{\Omega}}\boldsymbol{x}(0) + \overline{\boldsymbol{\Theta}}\bar{\boldsymbol{d}} + \boldsymbol{K^T_{3c}}\boldsymbol{u_{ext}}) + \bar{V}_{ext} \tag{5.22}$$

where matrices $\boldsymbol{H_c}$, $\boldsymbol{K_{1c}}$, $\boldsymbol{K_{2c}}$ and $\boldsymbol{K_{3c}}$ are defined as:

$$\begin{aligned} \boldsymbol{H_c} &= \boldsymbol{\Psi_d}^T\overline{\boldsymbol{Q}}\boldsymbol{\Psi_d} + \boldsymbol{R_{ext}} + 2\boldsymbol{S_{ext}}\boldsymbol{G_x}\boldsymbol{\Psi_d} & \boldsymbol{K_{1c}} &= \boldsymbol{\Psi_d}^T\overline{\boldsymbol{Q}}\boldsymbol{\Omega_d} + \boldsymbol{S_{ext}}\boldsymbol{G_x}\boldsymbol{\Omega_d} \\ \boldsymbol{K_{2c}} &= \boldsymbol{\Psi_d}^T\overline{\boldsymbol{Q}}\boldsymbol{\Theta_d} + \boldsymbol{S_{ext}}\boldsymbol{G_x}\boldsymbol{\Theta_d} & \boldsymbol{K_{3c}} &= \overline{\boldsymbol{\Psi}}^T \end{aligned} \tag{5.23}$$

and the matrix $\boldsymbol{H_c}$ should be positive defined for assuring the convexity of the cost function [10]. After these manipulations, the following *augmented centralized optimization problem*, commonly refereed in the literature as the *primal problem* [1, 10] is obtained:

$$\begin{aligned} &\max_{\boldsymbol{p}} \min_{\boldsymbol{u_{ext}}} J_{ext}(\boldsymbol{u_{ext}}, \boldsymbol{p}) \\ &\text{subject to } \boldsymbol{L_{ext}}\boldsymbol{u_{ext}} \le \boldsymbol{W_{ext}} \end{aligned} \tag{5.24}$$

where the matrices $\boldsymbol{L_{ext}}$ and $\boldsymbol{W_{ext}}$ have the following structure, considering the sets (5.8), the sequences (5.13) and (5.16), the term $\boldsymbol{u_{ext-1}}$ as the last time step extended vector value $\boldsymbol{u_{ext-1}} \equiv [\bar{\boldsymbol{u}}_{-1}{}^T \; \bar{\boldsymbol{v}}_{-1}{}^T]^T$, and $\boldsymbol{\delta_{umax}}$, $\boldsymbol{\delta_{umin}}$ as the sequences of admissible input's slew rate values:

$$\boldsymbol{L_{ext}} = \begin{bmatrix} \boldsymbol{\phi_{ext}} \\ -\boldsymbol{\phi_{ext}} \end{bmatrix}; \; \boldsymbol{W_{ext}} = \begin{bmatrix} \bar{\boldsymbol{\Delta}}_{ext} \\ -\underline{\boldsymbol{\Delta}}_{ext} \end{bmatrix} + \begin{bmatrix} -\boldsymbol{\xi_{ext}} \\ \boldsymbol{\xi_{ext}} \end{bmatrix}\boldsymbol{x}(0) + \begin{bmatrix} -\boldsymbol{\xi_{u-ext}} \\ \boldsymbol{\xi_{u-ext}} \end{bmatrix}\boldsymbol{u_{ext-1}} \tag{5.25}$$

$$\begin{aligned} &\boldsymbol{\phi_{ext}} = \begin{bmatrix} \boldsymbol{\Psi_d} \\ I_{N\times m} \; 0_{N\times r} \\ 0_{N\times m} \; I_{N\times r} \\ \boldsymbol{E_{\delta,ext}} \; 0_{N\times r} \end{bmatrix} \quad \bar{\boldsymbol{\Delta}}_{ext} = \begin{bmatrix} \boldsymbol{x_{max}} - \boldsymbol{\Theta d_{max}} \\ \boldsymbol{u_{max}} \\ \boldsymbol{v_{max}} \\ \boldsymbol{\delta u_{max}} \end{bmatrix} \quad \underline{\boldsymbol{\Delta}}_{ext} = \begin{bmatrix} \boldsymbol{x_{min}} - \boldsymbol{\Theta d_{min}} \\ \boldsymbol{u_{min}} \\ \boldsymbol{v_{min}} \\ \boldsymbol{\delta u_{min}} \end{bmatrix} \\ &\boldsymbol{\xi_{ext}} = \begin{bmatrix} \boldsymbol{\Omega_d} \\ 0_{N\times m} \\ 0_{N\times r} \\ 0_{N\times m} \end{bmatrix} \quad \boldsymbol{\xi_{u-ext}} = \begin{bmatrix} 0_{N(m+r)\times(m+r)} \\ 0_{Nm\times(m+r)} \\ 0_{Nr\times(m+r)} \\ \boldsymbol{E_{-1,ext}} \end{bmatrix} \quad \boldsymbol{E_{\delta,ext}} = \begin{bmatrix} I_m & 0_m & \cdots & 0_m \\ -I_m & I_m & \cdots & 0_m \\ \vdots & \vdots & \ddots & \vdots \\ 0_m & \cdots & -I_m & I_m \end{bmatrix} \\ &\boldsymbol{E_{-1,ext}} = \begin{bmatrix} I_m \\ 0_{(N-1)\times m} \end{bmatrix} \end{aligned} \tag{5.26}$$

5.3 Decomposition and Coordination Problems

In the last section, it was obtained (5.24) that is indeed the definition for a centralized model constrained optimization problem, where the interactions are treated in an explicit way. Although, the interactions are described in the model, the optimal input signals should correspond after solving the problem described in [10], under the same operative conditions. The solution of the problem (5.24) delivers the signals $\boldsymbol{u_{ext}}$ and $\boldsymbol{p}$. Only the first m elements of $\boldsymbol{u_{ext}}$ are used as inputs vector.

However, the system structure allows to distribute this problem accordingly, by decomposing the cost function into the z identified subsystems that compose the global system. This procedure is introduced, in first place in the current section. Then, it is shown how the *price-driven coordination* is established [8, 14, 27], including an analysis related to the stability of the global system under coordination.

5.3.1 Computing Explicit Local Solutions

The extended cost function (5.9) in its original form, and (5.22) in quadratic form, can now be divided in z parts, each one associated with a subsystem. Remember that each ith subsystem is described by the dynamical model (5.3) and has its own states, inputs, disturbances, and interactions are named $\boldsymbol{x_i}(k)$, $\boldsymbol{u_i}(k)$, $\boldsymbol{d_i}(k)$ and $\boldsymbol{v_i}(k)$, respectively, and its corresponding steady-state values $\boldsymbol{x_{i,s}}$, $\boldsymbol{u_{i,s}}$, $\boldsymbol{d_{i,s}}$ and $\boldsymbol{v_{i,s}}$.

Before going further, the local extended input-interaction vector $\boldsymbol{\bar{u}_{ext,i}}(k)$, as well as the local matrices $\boldsymbol{R_{ext,i}}$, $\boldsymbol{S_{ext,i}}$ are introduced:

$$\boldsymbol{\bar{u}_{ext,i}}(k) = [\boldsymbol{u_i}^{\boldsymbol{T}}\boldsymbol{(k)}\ \ \boldsymbol{v_i}^{\boldsymbol{T}}\boldsymbol{(k)}]^T,\ \boldsymbol{R_{ext,i}} = diag\{\boldsymbol{R_{u,i}}\ \boldsymbol{R_{v,i}}\},\ \boldsymbol{S_{ext,i}} = diag\{\boldsymbol{S_{u,i}}\ \boldsymbol{S_{v,i}}\} \tag{5.27}$$

One can write then, the global cost function (5.9) in the following form (the subindexes *ext* and *eq* stand stand for "extended form" and "equalities," respectively):

$$J_{ext}(\boldsymbol{x}, \boldsymbol{u}, \boldsymbol{v}) = \sum_{i=1}^{z} J_{ext,i} = \sum_{i=1}^{z} J_{gv,i} + \sum_{i=1}^{z} J_{eq,i}, \tag{5.28}$$

where the cost functions $J_{ext,i}$ and $J_{eq,i}$ have the following form:

$$\begin{aligned} J_{gv,i} &= \tfrac{1}{2}\boldsymbol{x_i'}^T(N)\boldsymbol{P_i}\boldsymbol{x_i'}(N) + \tfrac{1}{2}\sum_{k=0}^{N-1}\left(\boldsymbol{x_i'}^T(k)\boldsymbol{Q_i}\boldsymbol{x_i'}(k) + \boldsymbol{\bar{u}'^T_{ext,i}}(k)\boldsymbol{R_{ext,i}}\boldsymbol{\bar{u}'_{ext,i}}(k)\right) \\ &\quad + \tfrac{1}{2}\sum_{k=0}^{N-1}\left(2\boldsymbol{\bar{u}'^T_{ext,i}}(k)\boldsymbol{S_{ext,i}}\boldsymbol{x'}_i(k)\right) \\ J_{eq,i} &= \boldsymbol{p}^T\sum_{k=0}^{N-1}\left[\boldsymbol{M_i}\boldsymbol{\bar{u}_{ext,i}}(k) + \eta_i(\boldsymbol{x_i}(k), \boldsymbol{d_i}(k))\right] \end{aligned} \tag{5.29}$$

where $\boldsymbol{\bar{u}'_{ext,i}}(k) = \boldsymbol{\bar{u}_{ext,i}}(k) - \boldsymbol{\bar{u}_{ext,i-s}}$ is the difference between the current local extended vector and its steady-state value and $\boldsymbol{x_i'}(k) = \boldsymbol{x_i}(k) - \boldsymbol{x_{i,s}}$ is the local states error vector $t = k$. The matrix $\boldsymbol{M_i}$ in (5.29) represents the *effect of inputs at subsystem i over the full system interactions vector*, while η_i represents the effects of local states and disturbances. This matrix is obtained after obtaining the vectors $\boldsymbol{\bar{u}_{ext,i}}(k)$ from (5.5). Here is illustrated how this matrix is obtained for a process with two subsystems:

$$\begin{aligned}
\begin{bmatrix} \boldsymbol{v}_1(k) \\ \boldsymbol{v}_2(k) \end{bmatrix} &= \begin{bmatrix} 0 & \boldsymbol{A}_{12} \\ \boldsymbol{A}_{21} & 0 \end{bmatrix} \begin{bmatrix} \boldsymbol{x}_1(k) \\ \boldsymbol{x}_2(k) \end{bmatrix} + \begin{bmatrix} 0 & \boldsymbol{E}_{12} \\ \boldsymbol{E}_{21} & 0 \end{bmatrix} \begin{bmatrix} \boldsymbol{d}_1(k) \\ \boldsymbol{d}_2(k) \end{bmatrix} + \begin{bmatrix} 0 & \boldsymbol{B}_{12} \\ \boldsymbol{B}_{21} & 0 \end{bmatrix} \begin{bmatrix} \boldsymbol{u}_1(k) \\ \boldsymbol{u}_2(k) \end{bmatrix} \\
0 &= \sum_{i=1}^{z} \eta_i(k) + \begin{bmatrix} 0 & -I \\ \boldsymbol{B}_{21} & 0 \end{bmatrix} \begin{bmatrix} \boldsymbol{u}_1(k) \\ \boldsymbol{v}_1(k) \end{bmatrix} + \begin{bmatrix} \boldsymbol{B}_{12} & 0 \\ 0 & -I \end{bmatrix} \begin{bmatrix} \boldsymbol{u}_2(k) \\ \boldsymbol{v}_2(k) \end{bmatrix} \\
0 &= \sum_{i=1}^{z} \eta_i(k) + \boldsymbol{M}_1 \bar{\boldsymbol{u}}_{\boldsymbol{ext},\boldsymbol{1}}(k) + \boldsymbol{M}_2 \bar{\boldsymbol{u}}_{\boldsymbol{ext},\boldsymbol{2}}(k)
\end{aligned} \tag{5.30}$$

One can then write the local dynamics (5.3), in terms of $\bar{\boldsymbol{u}}_{\boldsymbol{ext},\boldsymbol{i}}(k)$ as follows, with $\bar{\boldsymbol{B}}_{\boldsymbol{i}} \equiv [\boldsymbol{B}_{\boldsymbol{ii}} \;\; I_{v,i}]$:

$$\boldsymbol{x}_{\boldsymbol{i}}(k+1) = \boldsymbol{A}_{\boldsymbol{ii}} \boldsymbol{x}_{\boldsymbol{i}}(k) + \bar{\boldsymbol{B}}_{\boldsymbol{i}} \bar{\boldsymbol{u}}_{\boldsymbol{ext},\boldsymbol{i}}(k) + \boldsymbol{E}_{\boldsymbol{ii}} \boldsymbol{d}_{\boldsymbol{i}}(k) \tag{5.31}$$

The current objective is to express the local dynamics as function of the local state value $\boldsymbol{x}_{\boldsymbol{i}}(0)$ and the $N-steps$ local sequences for the inputs and disturbances. For doing this, the following sequences, within their steady-state values are defined for the subsystem i:

$$\begin{aligned}
\bar{\boldsymbol{x}}_{\boldsymbol{i}} &\equiv \left[\boldsymbol{x}_{\boldsymbol{i}}^{\boldsymbol{T}}(1) \cdots \boldsymbol{x}_{\boldsymbol{i}}^{\boldsymbol{T}}(N)\right]^T & \bar{\boldsymbol{x}}_{\boldsymbol{i},\boldsymbol{s}} &\equiv \left[\boldsymbol{x}_{\boldsymbol{i},\boldsymbol{s}}^{\boldsymbol{T}} \cdots \boldsymbol{x}_{\boldsymbol{i},\boldsymbol{s}}^{\boldsymbol{T}}\right]^T \\
\hat{\boldsymbol{u}}_{\boldsymbol{ext},\boldsymbol{i}} &\equiv \left[\bar{\boldsymbol{u}}_{\boldsymbol{ext},\boldsymbol{i}}^{\boldsymbol{T}}(0) \cdots \bar{\boldsymbol{u}}_{\boldsymbol{ext},\boldsymbol{i}}^{\boldsymbol{T}}(N-1)\right]^T & \hat{\boldsymbol{u}}_{\boldsymbol{ext},\boldsymbol{i}-\boldsymbol{s}} &\equiv \left[\bar{\boldsymbol{u}}_{\boldsymbol{ext},\boldsymbol{i}-\boldsymbol{s}} \cdots \bar{\boldsymbol{u}}_{\boldsymbol{ext},\boldsymbol{i}-\boldsymbol{s}}\right]^T \\
\hat{\boldsymbol{d}}_{\boldsymbol{i}} &\equiv \left[\boldsymbol{d}_{\boldsymbol{i}}^{\boldsymbol{T}}(0) \cdots \boldsymbol{d}_{\boldsymbol{i}}^{\boldsymbol{T}}(N-1)\right]^T & \hat{\boldsymbol{d}}_{\boldsymbol{i},\boldsymbol{s}} &\equiv \left[\boldsymbol{d}_{\boldsymbol{i},\boldsymbol{s}}^{\boldsymbol{T}} \cdots \boldsymbol{d}_{\boldsymbol{i},\boldsymbol{s}}^{\boldsymbol{T}}\right]^T
\end{aligned} \tag{5.32}$$

With this result, one can write the *predicted local states vector* $\bar{\boldsymbol{x}}_{\boldsymbol{i}}$ as in (5.33), as well as the *predicted local states error vector*, presented as in (5.34)

$$\bar{\boldsymbol{x}}_{\boldsymbol{i}} = \boldsymbol{\Omega}_{\boldsymbol{i}} \boldsymbol{x}_{\boldsymbol{i}}(0) + \boldsymbol{\Gamma}_{\boldsymbol{i}} \hat{\boldsymbol{u}}_{\boldsymbol{ext},\boldsymbol{i}} + \boldsymbol{\Theta}_{\boldsymbol{i}} \hat{\boldsymbol{d}}_{\boldsymbol{i}} \tag{5.33}$$

$$\bar{\boldsymbol{x}}'_i = \bar{\boldsymbol{x}}_{\boldsymbol{i}} - \bar{\boldsymbol{x}}_{\boldsymbol{i},\boldsymbol{s}} = \boldsymbol{\Omega}_{\boldsymbol{i}} \boldsymbol{x}'_{\boldsymbol{i}}(0) + \boldsymbol{\Gamma}_{\boldsymbol{i}} \hat{\boldsymbol{u}}'_{\boldsymbol{ext},\boldsymbol{i}} + \boldsymbol{\Theta}_{\boldsymbol{i}} \hat{\boldsymbol{d}}'_{\boldsymbol{i}} \tag{5.34}$$

where $\boldsymbol{x}'_{\boldsymbol{i}}(0) = \boldsymbol{x}_{\boldsymbol{i}}(0) - \boldsymbol{x}_{\boldsymbol{i},\boldsymbol{s}}$, $\hat{\boldsymbol{u}}'_{\boldsymbol{ext},\boldsymbol{i}} = \hat{\boldsymbol{u}}_{\boldsymbol{ext},\boldsymbol{i}} - \hat{\boldsymbol{u}}_{\boldsymbol{ext},\boldsymbol{i}-\boldsymbol{s}}$, $\hat{\boldsymbol{d}}'_{\boldsymbol{i}} = \hat{\boldsymbol{d}}_{\boldsymbol{i}} - \hat{\boldsymbol{d}}_{\boldsymbol{i},\boldsymbol{s}}$ are the local error vectors, while the matrices $\boldsymbol{\Omega}_{\boldsymbol{i}}$, $\boldsymbol{\Gamma}_{\boldsymbol{i}}$ and $\boldsymbol{\Theta}_{\boldsymbol{i}}$ are defined as follows:

$$\boldsymbol{\Omega}_i = \begin{bmatrix} \boldsymbol{A}_{ii} \\ \boldsymbol{A}_{ii}^2 \\ \vdots \\ \boldsymbol{A}_{ii}^N \end{bmatrix} \quad \boldsymbol{\Gamma}_i = \begin{bmatrix} \bar{\boldsymbol{B}}_i & 0 & \cdots & 0 \\ \boldsymbol{A}_{ii}\bar{\boldsymbol{B}}_i & \bar{\boldsymbol{B}}_i & \cdots & 0 \\ \vdots & \vdots & \ddots & \vdots \\ \boldsymbol{A}_{ii}^{N-1}\bar{\boldsymbol{B}}_i & \boldsymbol{A}_{ii}^{N-2}\bar{\boldsymbol{B}}_i & \cdots & \bar{\boldsymbol{B}}_i \end{bmatrix} \quad \boldsymbol{\Theta}_i = \begin{bmatrix} \boldsymbol{E}_{ii} & 0 & \cdots & 0 \\ \boldsymbol{A}_{ii}\boldsymbol{E}_i & \boldsymbol{E}_{ii} & \cdots & 0 \\ \vdots & \vdots & \ddots & \vdots \\ \boldsymbol{A}_{ii}^{N-1}\boldsymbol{E}_{ii} & \boldsymbol{A}_{ii}^{N-2}\boldsymbol{E}_{ii} & \cdots & \boldsymbol{E}_{ii} \end{bmatrix} \tag{5.35}$$

With the different elements introduced along this section, the idea is to write the subsystem cost function (5.29) as a QP form in $\bar{\boldsymbol{u}}_{\boldsymbol{ext(i)}}$ and $\mathbf{p}$. For obtaining such formulation, consider the following local cost matrices, whose selection is done from the global system desired performance, except matrix $\boldsymbol{P}_{\boldsymbol{i}}$ that is selected after solving a local Ricatti algebraic equation, for the local dynamics (5.31):

$$\begin{array}{c} \bar{\boldsymbol{Q}}_i = diag\{\boldsymbol{Q}_i \cdots \boldsymbol{P}_i\} \\ \overline{\boldsymbol{R}}_{u,i} = diag\{\boldsymbol{R}_{u,i} \cdots \boldsymbol{R}_{u,i}\}\ \overline{\boldsymbol{R}}_{v,i} = diag\{\boldsymbol{R}_{v,i} \cdots \boldsymbol{R}_{v,i}\}\ \boldsymbol{R}_{ext,i} = diag\{\overline{\boldsymbol{R}}_{u,i}\ \overline{\boldsymbol{R}}_{v,i}\} \\ \overline{\boldsymbol{S}}_{u,i} = diag\{\boldsymbol{S}_{u,i} \cdots \boldsymbol{S}_{u,i}\}\ \ \overline{\boldsymbol{S}}_{v,i} = diag\{\boldsymbol{S}_{v,i} \cdots \boldsymbol{S}_{v,i}\}\ \ \boldsymbol{S}_{ext,i} = diag\{\overline{\boldsymbol{S}}_{u,i}\ \overline{\boldsymbol{S}}_{v,i}\} \end{array} \tag{5.36}$$

the local cost function in quadratic form takes the following form, considering again the error vectors $\boldsymbol{x}_i'(0)$, $\hat{\boldsymbol{u}}'_{ext,i}$ and $\hat{\boldsymbol{d}}_i'$ used in (5.34), with $\bar{V}_{ext,i)}$ representing the independent terms:

$$J_{ext(i)} = \frac{1}{2}\hat{\boldsymbol{u}}_{ext,i}^T \boldsymbol{H}_i \hat{\boldsymbol{u}}_{ext,i} + \hat{\boldsymbol{u}}_{ext,i}^T[\boldsymbol{K}_{1,i}\boldsymbol{x}_i'(0) + \boldsymbol{K}_{2(i)}\hat{\boldsymbol{d}}_i' + \boldsymbol{K}_{3(i)}\boldsymbol{p} - \boldsymbol{H}_{(i)}\hat{\boldsymbol{u}}_{ext,i-s}] + \bar{V}_{ext,i} \tag{5.37}$$

where the gains $\boldsymbol{H}_i$, $\boldsymbol{K}_{1,i}$, $\boldsymbol{K}_{2,i}$, $\boldsymbol{K}_{3,i}$ are defined as follows:

$$\begin{array}{ll} \boldsymbol{H}_i = \boldsymbol{\Gamma}_i^T\overline{\boldsymbol{Q}}_i\boldsymbol{\Gamma}_i + \boldsymbol{R}_{ext,i} + 2\overline{\boldsymbol{S}}_i\boldsymbol{\Gamma}_i, & \boldsymbol{K}_{1,i} = \boldsymbol{\Gamma}_i^T\overline{\boldsymbol{Q}}_i\boldsymbol{\Omega}_i + \boldsymbol{S}_{ext,i}\boldsymbol{\Omega}_i \\ \boldsymbol{K}_{2,i} = \boldsymbol{\Gamma}_i^T\overline{\boldsymbol{Q}}_i\boldsymbol{\Theta}_i + \boldsymbol{S}_{ext,i}\boldsymbol{\Theta}_i, & \boldsymbol{K}_{3,i} = \mathrm{diag}\{\boldsymbol{M}_i \cdots \boldsymbol{M}_i\}^T \end{array} \tag{5.38}$$

From the subsystem cost function (5.37) the following *local unconstrained explicit optimal solution* is obtained:

$$\hat{\boldsymbol{u}}_{ext,i}^{opt} = -\boldsymbol{H}_i^{-1}\left(\boldsymbol{K}_{1,i}\boldsymbol{x}_i'(0) + \boldsymbol{K}_{2,i}\hat{\boldsymbol{d}}_i' + \boldsymbol{K}_{3,i}\boldsymbol{p} - \boldsymbol{H}_i\hat{\boldsymbol{u}}_{ext,i-s}\right) \tag{5.39}$$

that is equivalent to:

$$\hat{\boldsymbol{u}}_{ext,i}^{opt} = -\boldsymbol{K}_{x,i}\boldsymbol{x}_i'(0) - \boldsymbol{K}_{d,i}\hat{\boldsymbol{d}}_i' - \boldsymbol{K}_{p,i}\boldsymbol{p} - \hat{\boldsymbol{u}}_{ext,i-s} \tag{5.40}$$

The subsystem control loop is depicted in Fig. 5.1. In this structure, it is seen a *state feedback controller* that stabilizes the local dynamics, according to the properties of the local weighting matrices in the cost function (5.29). The structure also includes a *feedforward component* for compensating the local disturbance effect, and not less important, an external tuning signal represented by the Lagrange Multiplier vector $\boldsymbol{p}$.

An interesting aspect of this control configuration is that the subsystems can be stabilized by local controllers (explicit MPC local solutions), whereas $\boldsymbol{p}$, a common signal for all the local controllers, is used as external element for modifying the subsystem control signals, according to a *global performance indicator*. Evidently,

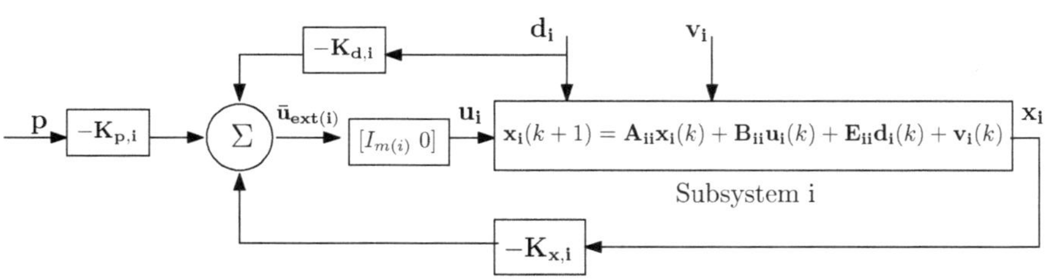

Fig. 5.1 Information flow for the Coordination algorithm

the subsystems controllability is a condition that should be verified for applying the proposed methodology.

This principle, and the possibility of obtaining explicit local solutions from MPC controllers are the main characteristics of the coordination strategy that is described in the following section.

5.3.2 Global coordination Based on Local Explicit Solutions

Taking the fact that each subsystem computes a control sequence vector influenced by vector $\boldsymbol{p}$, the following *unconstrained global explicit* solution is obtained, after taking all the local solutions (5.40):

$$\hat{\boldsymbol{u}}_{ext}^{opt} = [(\hat{\boldsymbol{u}}_{ext,1}^{opt})^T \ (\hat{\boldsymbol{u}}_{ext,2}^{opt})^T \cdots (\hat{\boldsymbol{u}}_{ext,z}^{opt})^T]^T \quad (5.41)$$

This vector takes the following form, in terms of system signals and the price vector $\boldsymbol{p}$:

$$\hat{\boldsymbol{u}}_{ext}^{opt} = -\widetilde{\boldsymbol{K}}_1 \boldsymbol{x}'(0) - \widetilde{\boldsymbol{K}}_2 \hat{\boldsymbol{d}}' - \widetilde{\boldsymbol{K}}_3 \boldsymbol{p} + \hat{\boldsymbol{u}}_{ext-s} \quad (5.42)$$

where its matrices and vectors are defined as follows:

$$\widetilde{\boldsymbol{K}}_1 = diag\{\boldsymbol{K}_{x,1} \cdots \boldsymbol{K}_{x,z}\}, \ \widetilde{\boldsymbol{K}}_2 = diag\{\boldsymbol{K}_{d,1} \cdots \boldsymbol{K}_{d,z}\} \ \widetilde{\boldsymbol{K}}_3 = [\boldsymbol{K}_{p,1} \cdots \boldsymbol{K}_{p,z}]^T \quad (5.43)$$

$$\boldsymbol{x}'(0) = \begin{bmatrix} \boldsymbol{x}_1(0) - \boldsymbol{x}_{1,s} \\ \vdots \\ \boldsymbol{x}_z(0) - \boldsymbol{x}_{z,s} \end{bmatrix} \ \hat{\boldsymbol{d}}' = \begin{bmatrix} \hat{\boldsymbol{d}}_1' - \hat{\boldsymbol{d}}_{1,s} \\ \vdots \\ \hat{\boldsymbol{d}}_z' - \hat{\boldsymbol{d}}_{z,s} \end{bmatrix} \ \hat{\boldsymbol{u}}_{ext-s} = \begin{bmatrix} \hat{\boldsymbol{u}}_{ext,1-s} \\ \vdots \\ \hat{\boldsymbol{u}}_{ext,z-s} \end{bmatrix} \quad (5.44)$$

The optimization problem (5.24) is written in terms of the global extended input vector sequence $\boldsymbol{u}_{ext}$ and the global extended disturbance vector $\bar{\boldsymbol{d}}$. For integrating the optimizing solution (5.42) to the problem (5.24), consider the transformations $\bar{\boldsymbol{d}} = Y_d \hat{\boldsymbol{d}}$ and $\boldsymbol{u}_{ext} = Y_u \hat{\boldsymbol{u}}_{ext}$, being Y_d and Y_u non singular matrices, built by inspection. The *global optimal unconstrained solution* now takes the following form:

$$\boldsymbol{u}_{ext}^{opt} = -\varphi_1(\boldsymbol{x}(0) - \boldsymbol{x}_s) - \varphi_2(\bar{\boldsymbol{d}} - \bar{\boldsymbol{d}}_s) - \varphi_3 \boldsymbol{p} + \boldsymbol{u}_{ext-s} \quad (5.45)$$

where the matrices $\varphi_1, \varphi_2, \varphi_3$ are defined as follows:

$$\varphi_1 = Y_u \widetilde{\boldsymbol{K}}_1, \varphi_2 = Y_u \widetilde{\boldsymbol{K}}_2 Y_d^{-1}, \varphi_3 = Y_u \widetilde{\boldsymbol{K}}_3 \quad (5.46)$$

Given, the time invariant matrices $\varphi_1, \varphi_2, \varphi_3$, it is possible to use the *explicit extended control expression* (5.45) as a feasible optimizer for the optimization problem (5.24) [1]. After performing this action, the following cost function is obtained:

$$J_{ext}(\boldsymbol{p}) = \bar{V}_p + \frac{1}{2}\boldsymbol{p}^T \boldsymbol{H_p p} + \boldsymbol{p}^T[\boldsymbol{K_{1p}x}'(0) + \boldsymbol{K_{2p}}\bar{\boldsymbol{d}}' + \boldsymbol{K_{3p}u_{ext_s}}] + \boldsymbol{p}^T(\bar{\boldsymbol{\Omega}}\boldsymbol{x}(0) + \bar{\boldsymbol{\Theta}}\bar{\boldsymbol{d}}) \tag{5.47}$$

where $\bar{V}_p$ are independent terms of $\boldsymbol{p}$ and the gains $\boldsymbol{H_p}, \boldsymbol{K_{1p}}, \boldsymbol{K_{2p}}, \boldsymbol{K_{3p}}$ are defined as (see (5.23) for details over $\boldsymbol{H_c}, \boldsymbol{K_{1c}}, \boldsymbol{K_{2c}}, \boldsymbol{K_{3c}}$):

$$\begin{array}{ll} \boldsymbol{K_{3p}} = \boldsymbol{K_{3c}^T} & \boldsymbol{H_p} = \varphi_3^T \boldsymbol{H_c}\varphi_3 - 2\varphi_3^T \boldsymbol{K_{3p}}, \\ \boldsymbol{K_{1p}} = \varphi_3^T(\boldsymbol{H_c}\varphi_1 + \boldsymbol{K_{1c}}) - \boldsymbol{K_{3p}}\varphi_1 & \boldsymbol{K_{2p}} = \varphi_3^T(\boldsymbol{H_c}\varphi_2 + \boldsymbol{K_{2c}}) - \boldsymbol{K_{3p}}\varphi_2 \end{array} \tag{5.48}$$

The new cost function (5.47) is written only in terms of the Lagrange Multiplier vector $\boldsymbol{p}$. In this function, the matrix $\boldsymbol{H_p}$ must be assured to be *definite positive* also to guarantee its convexity.

Taking into account that it has been only obtained a feasible optimizer from (5.45) for the primal problem, it should be assured that constraints are respected in the subsystems, while assuring the global performance defined by (5.47).

This is done by the tuning properties of the vector $\boldsymbol{p}$ in the local controllers, according to Fig. 5.1. Therefore, it is necessary to establish some constraints for the new decision variable $\boldsymbol{p}$, as:

$$\boldsymbol{L_p p} \leq \boldsymbol{W_p} \tag{5.49}$$

where the matrices $\boldsymbol{L_p}, \boldsymbol{W_p}$ are obtained after replacing (5.45) in the constraints expression of the problem (5.24), obtaining the following structures:

$$\boldsymbol{L_p} = \begin{bmatrix} \boldsymbol{\phi_p} \\ -\boldsymbol{\phi_p} \end{bmatrix} \; \boldsymbol{W_p} = \begin{bmatrix} \bar{\boldsymbol{\Delta}}_p - \boldsymbol{\delta_{sp}} \\ -\underline{\boldsymbol{\Delta}}_p + \boldsymbol{\delta_{sp}} \end{bmatrix} + \begin{bmatrix} -\boldsymbol{\xi_{xp}} \\ \boldsymbol{\xi_{xp}} \end{bmatrix} \boldsymbol{x}(0) + \begin{bmatrix} -\boldsymbol{\xi_{dp}} \\ \boldsymbol{\xi_{dp}} \end{bmatrix} \bar{\boldsymbol{d}} + \begin{bmatrix} -\boldsymbol{\xi_{u-ext}} \\ \boldsymbol{\xi_{u-ext}} \end{bmatrix} \boldsymbol{u_{ext-1}} \tag{5.50}$$

$$\begin{array}{lll} \boldsymbol{\phi_p} = -\boldsymbol{\phi_{ext}}\varphi_3 & \bar{\boldsymbol{\Delta}}_p = \bar{\boldsymbol{\Delta}}_{ext} & \underline{\boldsymbol{\Delta}}_p = \underline{\boldsymbol{\Delta}}_{ext} \\ \boldsymbol{\delta_{sp}} = \boldsymbol{\phi_{ext}}(\boldsymbol{\varphi_1 x_s} + \boldsymbol{\varphi_2}\bar{\boldsymbol{d}}_s + \boldsymbol{u_{ext s}}) & \boldsymbol{\xi_{xp}} = \boldsymbol{\xi_{ext}} - \boldsymbol{\phi_{ext}}\varphi_1 & \boldsymbol{\xi_{dp}} = -\boldsymbol{\phi_{ext}}\varphi_2 \end{array} \tag{5.51}$$

It can be noticed in (5.50) that $\boldsymbol{W_p}$ includes also elements defined by the admissible rank of the signals, as well as the current signal values, and the term $\boldsymbol{\delta_{sp}}$ that is related to the desired set point values for the signals, as seen in (5.51). For stability analysis, $\boldsymbol{\delta_{sp}}$ can be turned into 0.

After these manipulations, the following *global optimization problem*, known as the *dual problem* of (5.24) is:

$$\begin{array}{ll} \max\limits_{\boldsymbol{p}} & J_{ext}(\boldsymbol{p}) \\ \text{subject to} & \boldsymbol{L_p p} \leq \boldsymbol{W_p} \end{array} \tag{5.52}$$

Considering that the present approach is devoted to obtain a control structure in a distributed-coordinated form, this problem is solved by a superior level entity which receives the subsystems signals (states, disturbances, and set points vectors) and return to each subsystem the vector $\boldsymbol{p}$. The subsystems work with stabilizer feedback controllers, and receive from a *coordinator* the signal $\boldsymbol{p}$ that modifies the local control signals in such way that all subsystems respect their constraints, while

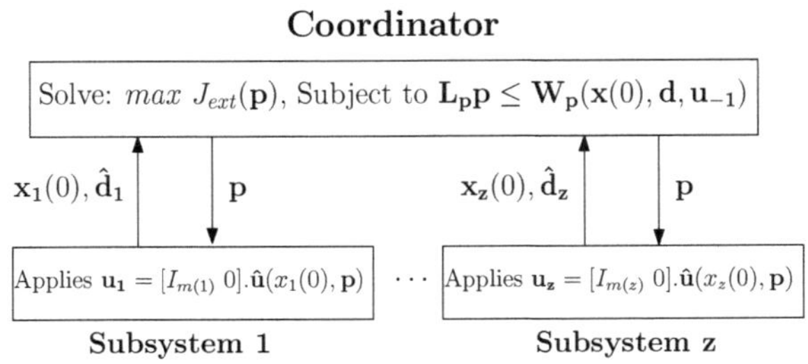

Fig. 5.2 Information flow for the coordination scheme

assuring an optimal global performance. Thus, the *price-driven* coordination scheme is performed according to the definition [15], because a price vector is deployed in the system subcontrollers.

The proposed distributed-coordination principle is shown in Fig. 5.2, and the details of the implementation of this optimization problem are summarized in the Algorithm 5.1.

Algorithm 5.1 Proposed Coordination Algorithm for local receding horizon-based controllers

Require: Offline computation of matrices from the centralized QP problem (5.24): $\boldsymbol{\Gamma_d}, \boldsymbol{\Omega_d}, \boldsymbol{\Theta_d}, \boldsymbol{\Psi_d}, \boldsymbol{H}_c, \boldsymbol{K_{1c}}, \boldsymbol{K_{2c}}, \boldsymbol{K_{3c}}, \overline{\boldsymbol{\Gamma}}, \overline{\boldsymbol{\Omega}}, \overline{\boldsymbol{\Theta}}, \overline{\boldsymbol{\Psi}}$ as indicated in (5.15), (5.18), and (5.23).

Require: Offline computation of matrices for the original coordination QP problem constraints polyhedral: $\boldsymbol{L_{ext}}, \boldsymbol{W_{ext}}$, as indicated in (5.25), (5.26).

Require: Offline computation of matrices for the local QP problems: $\boldsymbol{\Gamma_i}, \boldsymbol{\Omega_i}, \boldsymbol{\Theta_i}, \boldsymbol{H}_i, \boldsymbol{K_{1,i}}, \boldsymbol{K_{2,i}}, \boldsymbol{K_{3,i}}$, according to (5.35) and (5.38).

Require: Offline computation of matrices for the coordination problem (5.52): $\varphi_1, \varphi_2, \varphi_3, \boldsymbol{K_{1p}}, \boldsymbol{K_{2p}}, \boldsymbol{K_{3p}}$ according to (5.46) and (5.48).

Require: Off-line computation of the constraints for the coordination problem (5.52): $\boldsymbol{L_p}, \boldsymbol{W_p}$, according to (5.50), (5.51).

START

S1: At time $t = k$, the coordinator receives the state, disturbance and reference vectors from each subsystem: $\boldsymbol{x_i}(0), \boldsymbol{x_{i,s}}, \hat{\boldsymbol{d}}_i, \hat{\boldsymbol{d}}_{i,s}, \boldsymbol{u_{ext,i}}$, and $\boldsymbol{u_{ext,i-s}}$, $i = 1 \cdots z$. Then, it merges the vectors accordingly to obtain $\boldsymbol{x}(k), \boldsymbol{d}$ and $\boldsymbol{u_{ext-1}}$(if necessary).

S2: The coordinator solves the optimization problem (5.52), obtaining **p**.

S3: The coordinator sends the vector **p** to each subsystem. Use (5.42) to obtain the local extended control sequence $\hat{\boldsymbol{u}}^{\boldsymbol{opt}}_{\boldsymbol{ext,i}}$ at time $t = k$.

S4: Each subsystem applies the first m_i control elements of $\hat{\boldsymbol{u}}^{\boldsymbol{opt}}_{\boldsymbol{ext,i}}$ as the local control. Go back to **S1** at the next sampling time ($t = k + 1$).

5.3.3 *Stability Analysis of the Coordination Strategy*

Considering that all subsystems are locally stabilized by an unconstrained optimal controller that integrates a Lagrange Multiplier $\boldsymbol{p}$ in an additive form (see Fig. 5.3), the stability conditions can be established by using the small-gain theorem [26].

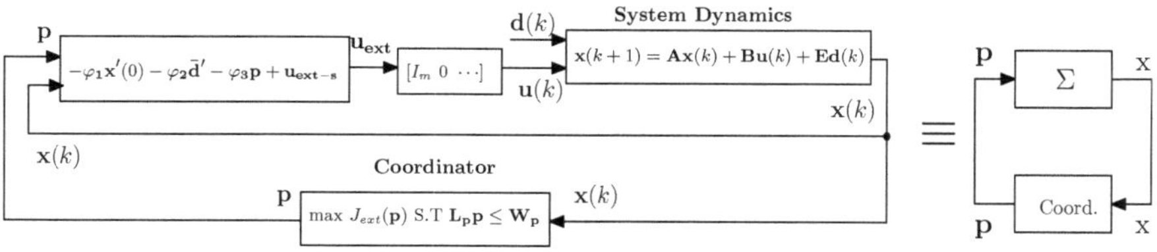

Fig. 5.3 Closed-loop coordinated system configuration

Therefore, the stability properties of the coordinated system depends on the total open-loop subsystems-coordinator gain.

Considering a stable *local-controlled* subsystems in closed loop with a coordinator, as depicted in Fig. 5.3, this system is stable if the following condition holds:

$$\gamma_c \gamma_s < 1 \tag{5.53}$$

where the gains γ_s stands for the local-controlled system gain and γ_c stands for the coordinator-agent gain. These bounds can be computed using the bounded real-lemma [2], and by considering the worst case admissible values of the Lagrange Multiplier $\boldsymbol{p}$. The latter is determined by the constraints in $\boldsymbol{p}$ that depend on the current states $\boldsymbol{x}$ (see (5.49) and (5.50)).

Due to space limitations for this chapter, the stability analysis for the proposed approach will be better exposed in further publications. However, a deeper discussion and results are presented in [23].

5.4 Coordination of a Two-Area Power Generation System

As application for the distributed-coordinated control scheme, it is proposed a system, where the load-frequency control (LFC) is required, due to the interaction of synchronized generators [5, 9, 17]. In Fig. 5.4 is depicted the configuration of the test system. The system has two generation areas, represented as *Area 1* and *Area 2*, which are interconnected by power lines. Each area has an hydroelectric generator, a non-dispatchable generation unit operated at maximum power (photovoltaic or wind generator), and an energy storage device.

Fig. 5.4 Configuration of the proposed two-area power generation system

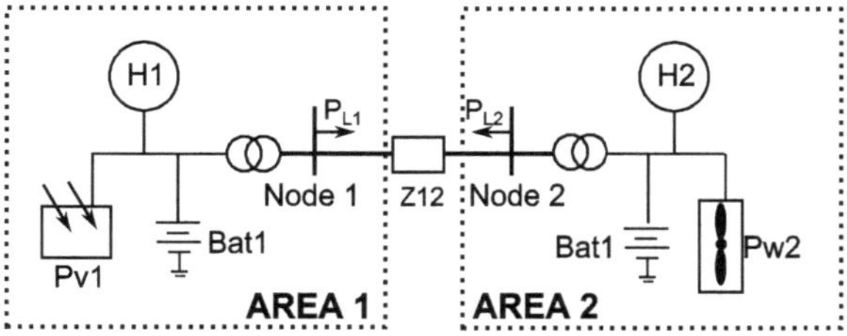

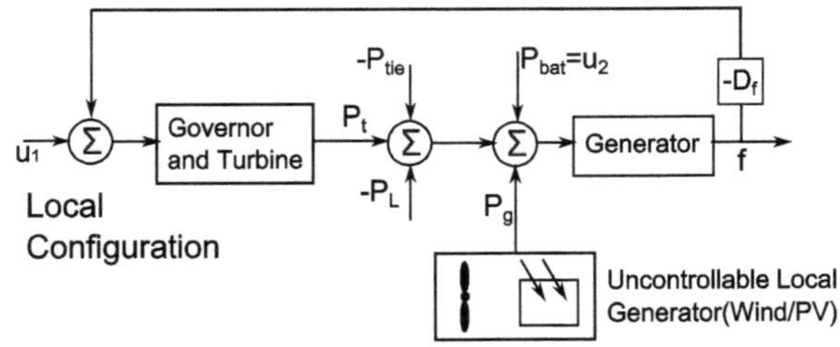

Fig. 5.5 Control structure for each local power generation system

In the LFC problem, the idea is to maintain the system frequency. This is done by acting over the mechanical power that after conversion, turns into electric power. The general scheme of each generation area is depicted in Fig. 5.5. In the diagram, it is seen how the mechanical power P_t is affected by the electrical powers P_L (the local load), the non-dispatchable power power P_g injected by the alternative-based generator and the power injected by a *controllable backup battery* $P_{bat} = u_2$. The governor input signal is represented by a control signal u_1, generated by the *high-level control* strategy that represents a *reference load value*, and a droop-frequency component, which is proportional to the frequency variation [13, 24]. The high-level controller must assure the system stability under power injections by the alternative generators and local loads (maintain the line frequency around 50 Hz), optimize the power shared between the areas and maintain, under suitable levels, the stored power in the backup batteries [3, 11, 21, 25]. In this case study only active power is considered.

5.4.1 Dynamical Model and System Parameters

For the following equations, it is assumed that variables can be defined as $x = X_0 + \tilde{x}$, where X_0 is the equilibrium (initial) operative condition and $\tilde{x}$ is the variation around this value.

Power Lines Interactions

For the proposed system, it is considered two power nodes i and j, each one with respective voltage magnitudes V_i, V_j and angles θ_i, θ_j, as well as link impedance $Z_{fg} = j\,X_{fg}$, where X_{fij} is the power link reactance (active power lossless assumption). The instantaneous active power ($P_{tie,ij}$) transmitted from node i to node j [13] is:

$$P_{tie,ij} = \frac{V_i V_j}{X_{ij}} sin(\theta_i - \theta_j) \tag{5.54}$$

From (5.54) it can be seen how voltage magnitude and angle at each node, fix the amount of active power transmitted between them. Considering small voltage variations, the following small signal model for power transmission is [13]:

$$\widetilde{p}_{tie,ij} \simeq \frac{V_{i0}V_{j0}}{X_{ij}} cos(\theta_{i0} - \theta_{j0}) \, (\widetilde{\theta}_i - \widetilde{\theta}_j) \simeq K_{tij} \, (\widetilde{\theta}_i - \widetilde{\theta}_j) \tag{5.55}$$

Microgrid Dynamical Model

Due to the similar configuration of each generator and their identical connectivity with the other ones, a similar dynamical model represents the dynamics of each area. For a particular area i and based on the local configuration scheme included in Fig. 5.5, the following expressions are obtained (recall that $\tilde{u}_1$ is the reference power for the governor and $\tilde{u}_2$ is the injected power from the battery):

$$\begin{aligned} \dot{\widetilde{\theta}}_i &= 2\pi \widetilde{f}_i \\ \dot{\widetilde{f}}_i &= \tfrac{1}{T_{Ni}}[-\widetilde{f}_i + K_{Ni}(\widetilde{p}_{t,i} + \widetilde{p}_{g,i} + \widetilde{u}_{i,2} - \widetilde{p}_{L,i} - \widetilde{p}_{tie,i})] \\ \dot{\widetilde{p}}_{t,i} &= \tfrac{1}{T_{Gi}}[-\widetilde{p}_{t,i} + G_{ti}(\widetilde{u}_{i,1} - D_{fi}\widetilde{f}_i)] \\ \dot{\widetilde{soc}}_i &= -K_{si}.\widetilde{u}_{i,2} \end{aligned} \tag{5.56}$$

where θ_i is the voltage angle, f_i is the frequency, $p_{t,i}$ is the transmitted mechanical power, soc_i is the state of charge of the battery, $p_{g,i}$ is the power injected by the uncontrollable generator unit, $p_{L,i}$ is the local load and $p_{tie,i}$ is the total power transmitted from the unit to the other ones. T_{Ni} and K_{Ni} are the generator equivalent time constant and gain, T_{Gi} and G_{ti} are the time constant and the gain of the governor-turbine system and D_{fi} is the droop-frequency coefficient and K_{si} is a gain associated with the capacity of the battery.

In typical hydroelectric generators, $T_{Gi} << T_{Ni}$, making $\dot{\widetilde{p_{t,i}}} \cong 0$. Also, due to unique connection between the areas, the term $\widetilde{p}_{tie,i}$ can be directly obtained from (5.55). The following simplified model is obtained for each generation area:

$$\begin{aligned} \dot{\widetilde{\theta}}_i &= 2\pi \widetilde{f}_i \\ \dot{\widetilde{f}}_i &= \tfrac{1}{T_{Ni}}[-a_i \widetilde{f}_i + b_i\widetilde{u}_{i,1} + K_{Ni}(\widetilde{u}_{i,2} + \widetilde{p}_{gi} - \widetilde{p}_{L1}) - c_i\widetilde{\theta}_i] + \tfrac{1}{T_{Ni}}c_i\widetilde{\theta}_j \\ \dot{\widetilde{soc}}_i &= -K_{si}.\widetilde{u}_{i,2} \end{aligned} \tag{5.57}$$

where $a_i = 1 + K_{Ni}G_{ti}D_{fi}$, $b_i = K_{Ni}G_{ti}$, and $c_i = K_{Ni}K_{tij}$. It is noticed how the frequency of the area i is affected by angle variations of area j. Precisely, the right-most term of the frequency dynamics is the proposed interactions term for each area. Defining the states vector $\boldsymbol{x_i} = [\widetilde{\theta}_i \;\; \widetilde{f}_i \;\; \widetilde{soc}_i]^T$, control vector $\boldsymbol{u_i} = [\widetilde{u}_{i,1} \;\; \widetilde{u}_{i,2}]^T$, disturbances vector $\boldsymbol{d_i} = [\widetilde{p}_{Li} \;\; \widetilde{p}_{gi}]^T$ and interactions vector $\boldsymbol{v_i}$ that only relies on $\boldsymbol{x_j}$, the following state space model in discrete time, with sampling time T_s is obtained for each area i:

Table 5.1 Parameters for the Two areas microgrid

Parameter	Value (units)
Voltages magnitudes (V_1, V_2)	1.00, 0.9525 (p.u)
Initial voltages angles (θ_1, θ_2)	0.000, – 0.3094 (rad)
Line reactances (X_{12})	0.9 (p.u)
Equivalent generator gain (K_{N1}, K_{N2})	110, 80
Equivalent governor-turbine gain (G_{t1}, G_{t2})	1.00, 1.00
Equivalent generator time constant (T_{N1}, T_{N2})	25, 15 (s)
Droop frequency gain (D_{f1}, D_{f2})	0.25, 0.40
Max. generated power ($P_{T,max}$, each unit)	0.3 (p.u)
Max. injectable power ($P_{g,max}$, each unit)	0.25 (p.u)
Initial injected power ($P_{g1,0}, P_{g2,0}$)	0.00, 0.00 (p.u)
Max. load power ($P_{L,max}$, each unit)	0.5 (p.u)
Initial local load powers ($P_{L1,0}, P_{L2,0}$)	0.20, 0.20 (p.u)

$$
\begin{aligned}
\boldsymbol{x_i}(k+1) &= \boldsymbol{A_{ii}x_i}(k) + \boldsymbol{B_{ii}u_i}(k) + \boldsymbol{E_{ii}d_i}(k) + \boldsymbol{v_i}(k) \\
\boldsymbol{x_i}(k+1) &= \begin{bmatrix} 1 & 2\pi T_s & 0 \\ -c_i T_s/T_{Ni} & 1 - a_i T_s/T_{Ni} & 0 \\ 0 & 0 & 1 \end{bmatrix} \boldsymbol{x_i}(k) + \begin{bmatrix} 0 & 0 \\ b_i T_s/T_{Ni} & K_{Ni}T_s/T_{Ni} \\ 0 & -K_{si} \end{bmatrix} \boldsymbol{u_i}(k) \\
&+ \begin{bmatrix} 0 & 0 \\ -K_{Ni}T_s/T_{Ni} & K_{Ni}T_s/T_{Ni} \\ 0 & 0 \end{bmatrix} \boldsymbol{d_i}(k) + \frac{T_s}{T_{Ni}} \begin{bmatrix} 0 & 0 & 0 \\ K_{tij} & 0 & 0 \\ 0 & 0 & 0 \end{bmatrix} \boldsymbol{x_j}(k)
\end{aligned}
\tag{5.58}
$$

Numerical Values

The parameters as well as initial values for voltage magnitudes and angles, obtained after a *load flow analysis*, are shown in Table 5.1, based in information taken from [5, 13].

5.4.2 Control Objectives

For this case study, the control strategy should minimize the frequency and state-of-charge deviations deviation, as well as minimize the voltage angle variations, penalizing the power transmitted between both areas. The selected cost function structure is similar to (5.9). Regarding the cost functions for each subsystem, the following ones were selected, where it is strongly penalized the angle and state-of-charge variations, as well as the batteries actions in each area:

$$Q_1 = Q_2 = diag\{0.1\ 10\ 2\}\ ;\ \boldsymbol{R_{u,1}} = \boldsymbol{R_{u,2}} = diag\{0.1\ 0.5\}\ ;\ \boldsymbol{S_{u,1}} = \boldsymbol{S_{u,2}} = 0$$
$$\boldsymbol{R_{v,1}} = \boldsymbol{R_{v,2}} = diag\{10\ 10\ 10\}\ ;\ \boldsymbol{S_{v,1}} = \boldsymbol{S_{v,2}} = 0$$

The matrix P in the cost function is computed from the discrete-time Riccati algebraic equation with Q, R and S. With this matrix, one can ensure that a potential centralized strategy would *a priori* stabilize the system, having in considerations the disturbance effects [10]. The admissible signals values are defined according to the values reported in Table 5.1, referred to the initial conditions. For this reason, the constraints are defined such that the origin is one admissible point in the optimization process. These ranges are:

$$\begin{array}{ll} \boldsymbol{x_{1,max}} = \boldsymbol{x_{2,max}} = [\pi/4\ \ 2\ \ 0.4]^T & \boldsymbol{x_{1,min}} = \boldsymbol{x_{2,min}} = [-\pi/4\ \ -2\ \ -0.3]^T \\ \boldsymbol{u_{1,max}} = \boldsymbol{u_{2,max}} = [0.45\ \ 0.5]^T & \boldsymbol{u_{1,min}} = \boldsymbol{u_{2,min}} = [-0.45\ \ -0.5]^T \\ \boldsymbol{d_{1,max}} = \boldsymbol{d_{2,max}} = [0.8\ \ 0.45]^T & \boldsymbol{d_{1,min}} = \boldsymbol{d_{2,min}} = [-0.5\ \ -0.05]^T \end{array}$$

5.4.3 Simulation Results

For the selected case study, the distributed-coordination approach is considered to accomplish the control objectives for the system. Considering that the proposed approach is obtained from the MPC problem definition for the centralized model according to [10], the obtained results should be compared against this centralized control methodology. The idea is to analyse the system performance of the proposed approach against the well-known MPC controller, using the same weights and constraints.

For this system, an initial load flow analysis [13] was performed to obtain the voltage magnitudes and phase angles for the proposed start-up conditions. Such initial values are included in Table 5.1, where it is seen that angles and disturbance are within the operative ranges.

The simulations were executed for both control strategies, considering the disturbances profiles (uncontrolled generation in chart *Injected Generation* and loads in the chart *Loads*) that are represented in Fig. 5.6. The simulations results are shown in Fig. 5.7. For each chart, **C** refers to the centralized MPC results, while **DC** refers to the distributed-coordinated proposed approach, in a simulation of about 450s. For both systems, the prediction horizon was chosen equal to $N = 2$ and the sampling period $T_s = 1$ s. The simulations were performed in a PC with a processor Atmel A6-3420(1.5 Ghz, 4 Gb of RAM), under Matlab R2013a. In Fig. 5.7 are included the states (angle and battery's state of charge in the charts *Angles* and *SoC*), as well as the power from generators and batteries (charts *Hydro Power* and *Bat. Power*. The frequency is not represented, but given the small angle variations and their stable

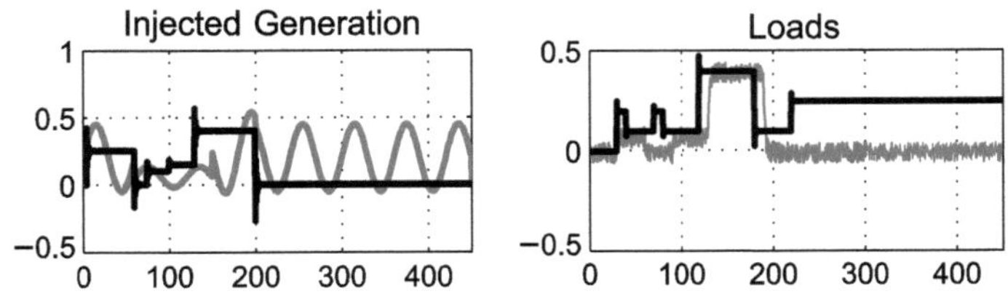

Fig. 5.6 Injected generation and load profiles for the power system simulation (*Bold* Area 1, *Light* Area 2)

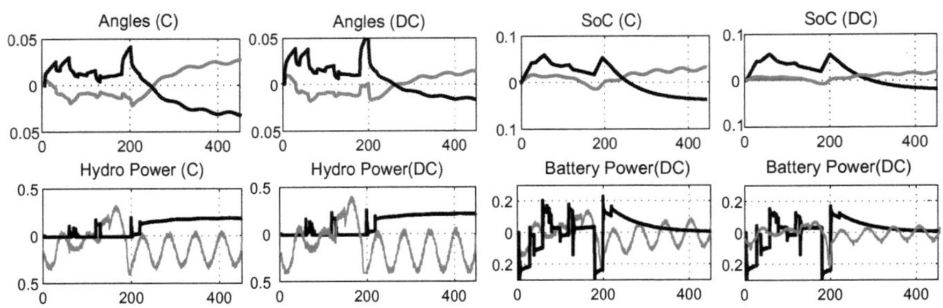

Fig. 5.7 Simulation results of Centralized MPC (C) and Distributed-Coordinated MPC (DC) approach for the multisource application (*Bold* Area 1, *Light* Area 2)

behavior, is maintained around the nominal value in both scenarios. Looking at the simulation results, the following elements are highlighted:

- For both implementations, the system constraints, defined in Sect. 5.4.2 are respected. The system performance is not degraded in their presence.
- The angle values are sufficiently small, implying null frequency variation. Also, the batteries state of charge is also regulated in an optimal fashion. In fact, for the DMPC case their variation is smaller, aspect that is related with the interactions penalization in this scheme.
- The batteries action, combined with the hydropower regulation respond in good way, against power injection and loads. It is seen how hydropower generation is reduced when the alternative source injects its power, particularly in *Area 2*.
- The proposed control algorithm performance was compared against the centralized control approach, using the mean cost ratio $\bar{R}_J = 1/L \sum_{l=1}^{L} \frac{J_{DC,l}}{J_{C,l}}$, with $J_{DC,l}$ the cost for the proposed approach while $J_{C,l}$ the cost for the centralized control, at the simulation l. For $L = 10$ simulations under the same initial and load conditions, $\bar{R}_J \approx 1.05$, showing that both approaches have an equivalent performance, even when $J_{DC,l}$ considers the weighting matrices $\boldsymbol{R_{v,i}}, \boldsymbol{S_{v,i}}$ for penalizing the interactions.

However, in this case study, the lost of coordination was not tested, due to possible large angle variations. Under high frequency variations, the total generated hydro power should increase to compensate these changes. If this power generation is sustained during, there is a possibility to expend the stored water [13].

5.5 Conclusions

In this work, a price-driven-based optimization approach was proposed for coordination of distributed large-scale systems. The approach uses explicit unconstrained local solutions, along with a coordinator agent, for assuring the global performance, while respecting the operative constraints. The proposed solution is based in the decomposition of the global cost function along the subsystems, which solve their own unconstrained problem, whose solutions are then combined to partially solve the global optimization problem. After this, the coordinator solves the new optimization problem and finds a price vector (Lagrange Multiplier) that satisfies the system constraints. One of the advantages of the proposed technique is that each subsystem can operate with a local control strategy, and uses the coordination signal as a tuning parameter for achieving the global constraints. The technique has been applied in a two-area power generation system.

References

1. S. Boyd, *Convex Optimization* (Cambridge University Press, Cambridge, 2004)
2. S. Boyd, L. Ghaoui, E. Feron, V. Balakrishnan, Linear matrix inequalities in system and control theory, *Society for Industrial and Applied Mathematics*, Studies in Applied Mathematics (1994)
3. M. Brenna, E. De Berardinis, L. Delli Carpini, F. Foiadelli, P. Paulon, P. Petroni, G. Sapienza, G. Scrosati, D. Zaninelli, Automatic distributed voltage control algorithm in smart grids applications. IEEE Trans. Smart Grids **4**(2), 877–885 (2013)
4. E. Camacho, C. Bordons, *Model Predictive Control*, Advanced Textbooks in Control and Signal Processing (Springer, Berlin, 2004)
5. E. Camponogara, D. Jia, B.H. Krogh, S. Talukdar, Distributed model predictive control. IEEE Control Syst. Mag. **22**(1), 44–52 (2002)
6. G. Carpinelli, G. Celli, S. Mocci, F. Mottola, F. Pilo, D. Proto, Optimal integration of distributed energy storage devices in smart grids. IEEE Trans. Smart Grids **4**(2), 985–995 (2013)
7. P. Chawdhry, S. Ahson, Application of interaction-prediction approach to load-frequency control (lfc) problem. IEEE Trans Syst. Man Cybern. **12**(1), 66–71 (1982)
8. G. Cohen, Optimization by decomposition and coordination: a unified approach. IEEE Trans. Autom. Control **23**(2), 222–232 (1978)
9. D. Georges, Distributed model predictive control based on decomposition-coordination and networking, in *Proceedings of the 10th European Control Conference (ECC)* (2009)
10. G. Goodwin, M. Seron, J. De Doná, *Constrained Control and Estimation: An Optimisation Approach*, Communications and Control Engineering (Springer, Berlin 2005)
11. T. Green, M. Prodanović, Control of inverter-based micro-grids. Electric Power Syst. Res. **77**(9), 1204–1213 (2007)
12. S. Heier, *Grid Integration of Wind Energy Conversion Systems* (Wiley, New York, 2006)
13. P. Kundur, N.J. Balu, M.G. Lauby, *Power System Stability and Control* (McGraw-Hill, New York, 1994)

14. G.K.H. Larsen, J. Pons, S. Achterop, J. Scherpen, Distributed MPC applied to power demand side control, in *Proceedings of the European Control Conference (ECC)* (2013)
15. M. Mesarovic, D. Macko, Y. Takahara, *Theory of Hierarchical, Multilevel Systems* (Academic Press, New York, 1970)
16. C. Ocampo-Martinez, D. Barcelli, V. Puig, A. Bemporad, Hierarchical and decentralised model predictive control of drinking water networks: application to Barcelona case study. IET Control Theory Appl. **6**(1), 62–71 (2012)
17. S. Papathanassiou, N. Hatziargyriou, K. Strunz, et al., A benchmark low voltage microgrid network, in *Proceedings of the CIGRE Symposium: Power System with Disperse Generation* (2005)
18. M.R. Patel, *Wind and Solar Power Systems: Design, Analysis, and Operation* (CRC Press, Boca Raton, 2012)
19. J. Pukrushpan, A. Stefanopoulou, H. Peng, Control of fuel cell breathing. IEEE Control Syst. Mag. **24**(2), 30–46 (2004)
20. W. Qi, J. Liu, P.D. Christofides, A distributed control framework for smart grid development: energy/water system optimal operation and electric grid integration. J. Process Control **21**(10), 1504–1516 (2011)
21. K. Rudion, A. Orths, Z. Styczynski, K. Strunz, Design of benchmark of medium voltage distribution network for investigation of DG integration, in *Proceedings of the IEEE Power Engineering Society General Meeting* (2006)
22. J. Sandoval-Moreno, G. Besançon, J.J. Martinez Molina, Lagrange multipliers based price driven coordination with constraints consideration for multisource power generation systems, in *Proceedings of the 13th European Control Conference (ECC)* (2014)
23. J. Sandoval-Moreno, Contribution to the coordination of distributed model predictive controllers, applied to the electric power production. Ph.D. Thesis, University of Grenoble (2014)
24. C. Sao, P. Lehn, Autonomous load sharing of voltage source converters. IEEE Trans. Power Deliv. **20**(2), 1009–1016 (2005)
25. J. Schiffer, T. Seel, J. Raisch, T. Sezi, A consensus-based distributed voltage control for reactive power sharing in microgrids, in *Proceedings of the 13th European Control Conference (ECC)* (2014)
26. S. Skogestad, I. Postlethwaite, *Multivariable Feedback Control* (Wiley, New York, 2006)
27. J. Zárate-Florez, G. Besancon, J.J. Martinez Molina, F. Davelaar, Explicit price method for coordinated control and hydro-power production example, in *Proceedings of the 8th IFAC Symposium on Power Plant and Power System Control* (2012)
28. J. Zárate-Florez, J. Martinez, G. Besancon, D. Faille, Decentralized-coordinated model predictive control for a hydro-power valley. Math. Comput. Simul. **91**, 108–118 (2012)

Chapter 6
Evolutionary Game-Based Dynamical Tuning for Multi-objective Model Predictive Control

Julián Barreiro-Gomez, Carlos Ocampo-Martinez and Nicanor Quijano

Abstract Model predictive control (MPC) is one of the most used optimization-based control strategies for large-scale systems, since this strategy allows to consider a large number of states and multi-objective cost functions in a straightforward way. One of the main issues in the design of multi-objective MPC controllers, which is the tuning of the weights associated to each objective in the cost function, is treated in this work. All the possible combinations of weights within the cost function affect the optimal result in a given Pareto front. Furthermore, when the system has time-varying parameters, e.g., periodic disturbances, the appropriate weight tuning might also vary over time. Moreover, taking into account the computational burden and the selected sampling time in the MPC controller design, the computation time to find a suitable tuning is limited. In this regard, the development of strategies to perform a dynamical tuning in function of the system conditions potentially improves the closed-loop performance. In order to adapt in a dynamical way the weights in the MPC multi-objective cost function, an evolutionary game approach is proposed. This approach allows to vary the prioritization weights in the proper direction taking as a reference a desired region within the Pareto front. The proper direction for the prioritization is computed by only using the current system values, i.e., the current optimal control action and the measurement of the current states, which establish the system cost function over a certain point in the Pareto front. Finally, some simulations of a multi-objective MPC for a real multivariable case study show a comparison between the system performance obtained with static and dynamical tuning.

J. Barreiro-Gomez (✉) · C. Ocampo-Martinez
Automatic Control Department, Institut de Robòtica i Informàtica Industrial (CSIC-UPC), Universitat Politècnica de Catalunya, Llorens i Artigas, 4-6, 08028 Barcelona, Spain
jbarreiro@iri.upc.edu

C. Ocampo-Martinez
cocampo@iri.upc.edu

J. Barreiro-Gomez · N. Quijano
Departamento de Ingeniería Eléctrica y Electrónica, Universidad de Los Andes, Carrera 1 No 18A-10, Bogotá, Colombia
nquijano@uniandes.edu.co; j.barreiro135@uniandes.edu.co

S. Olaru et al. (eds.), *Developments in Model-Based Optimization and Control*,
Lecture Notes in Control and Information Sciences 464,
DOI 10.1007/978-3-319-26687-9_6

Keywords Model predictive control · Evolutionary game theory · Dynamical tuning · Large-scale systems

6.1 Introduction

Model predictive control (MPC) is an optimization-based control strategy widely used in the solution of the control of large-scale systems since it can manage a large number of variables in a straightforward manner, it may consider several objectives, and it can consider a variety of physical and operational constraints. The versatility of the MPC controller is reflected on the amount of elements that can be adjusted, e.g., control-oriented model of the system, horizon for the states prediction, control horizon, or weights in the multi-objective cost function. In this regard, one of the main issues of the multi-objective MPC design is the selection of all these parameters. This work focuses particularly on the tuning issue given by the selection of the weights in the cost function. These weights assign a prioritization to each objective affecting the solution of the optimization problem that is solved at each iteration by the controller. Consequently, tuning these weights might improve considerably the control performance.

The design problem of tuning has been already treated by many researchers using different strategies. Most of the existing strategies to tune an MPC controller utilize an offline approach, and sometimes it is a trial and error procedure. In [5], a review of some tuning strategies has been made, and some approaches such as offline and online strategies have been classified. Since the conditions over the system might vary over time, it has become a relevant issue in the development of strategies that allow to tune MPC controllers permanently in a dynamical manner. Moreover, it must be taken into account that an online tuning strategy necessarily implies an extra computational burden. For instance in [1], a tuning strategy is presented based on a linear approximation between the closed-loop predicted output, and the parameters that may be tuned in the MPC controller. Also, it has highlighted its simplicity as an advantage for implementation. More approaches to solve this problem have been proposed after the review presented in [5]. In [4], a tuning methodology based on a matching to a desired reference controller has been proposed. This method allows to select the MPC weight matrices, making the MPC perform as a desired linear controller. Afterward, this methodology has been generalized in [23]. The use of a linear controller as a reference has also been studied for multiple-input-multiple-output systems in [21]. Other alternatives to perform the tuning of an MPC controller has been explored. For instance, a self-tuning terminal cost approach is applied in [13] for an economic MPC controller. In [22], a normalization procedure and a computation of the minimum distance from the Pareto front to a management point have been proposed as a tuning strategy. Other approaches use learning systems. For example, a learning approach based on artificial neural networks and fuzzy logic has been studied for performing the tuning of a predictive controller in [6]. Then, it is concluded that other learning techniques might be implemented in order to solve the problem of dynamical tuning for predictive controllers.

On the other hand, game theory has been used as a learning approach for a large variety of control systems. In [12], the use of game theory applied to distributed control design is discussed. The game theoretical approach has been used for the design of multiagent systems, and to solve optimization problems [10, 11, 25]. Other perspective is the evolutionary game approach [20, 24]. For instance in [2, 3, 14, 17–19], a population dynamics approach has been presented for control and/or optimization purposes. Motivated by all the applications of this game theoretical approach in control systems, this work uses the evolutionary game theory as a learning approach to propose a dynamical tuning strategy for multi-objective MPC controllers.

The contribution of this chapter is a novel dynamical tuning strategy based on evolutionary game theory. This approach varies the prioritization weights trying to maintain the values of the objective functions within a preestablished *management region* over the Pareto front. The *management region* is selected according to a desired performance of the system, and determines the proper direction for the evolution of the prioritization weights when disturbances in the system make objective functions take undesired values over the Pareto front. Furthermore, the population dynamics approach only requires information about the current condition of the system, which determine a current value in the Pareto front. In this regard, the present methodology demands less computational burden with respect to other tuning strategies that need to compute more than one value over the Pareto front. The proposed evolutionary game-based dynamical tuning is tested for an MPC controller with a drinking water network (DWN) as a case study.

This chapter is organized as follows. Section 6.2 introduces a brief background of MPC and population dynamics. Section 6.3 presents the proposed evolutionary game-based dynamical tuning for a multi-objective MPC. Section 6.4 describes the real case study that has been used to test the proposed dynamical tuning approach. In this section, the results are shown and the proposed tuning strategy performance is compared with the performance of an MPC tuned with the static strategy. Finally, some concluding remarks are made and further work is pointed out in Sect. 6.5.

Notation

All column vectors are denoted by bold style, e.g., $\mathbf{x}$. Matrices are denoted by bold upper case, e.g., $\mathbf{A}$. In contrast, scalars are denoted by nonbold style, e.g., N. The sets are denoted by calligraphic upper case, e.g., $\mathcal{S}$. The norm $||\mathbf{x}||$ of the vector $\mathbf{x} \in \mathbb{R}^{n_x}$ is defined as $||\mathbf{x}|| = \sqrt{\mathbf{x}^\top \mathbf{x}}$. Finally, real numbers are denoted by $\mathbb{R}$, all the nonnegative numbers are denoted by $\mathbb{R}_+$, and all the nonzero positive real numbers are denoted by $\mathbb{R}_{>0}$. Similarly, the integer numbers, and nonnegative integer numbers are denoted by $\mathbb{Z}$, and $\mathbb{Z}_+$, respectively. Throughout this document, discrete-time and continuous-time systems are treated. Therefore, the subindex $k \in \mathbb{Z}_+$ denotes that the system is described in discrete time, whereas the use of time t in the continuous-time expressions is mostly omitted in order to simplify the notation.

6.2 Background

This section introduces some preliminaries such as the problem statement regarding multi-objective MPC, and the basic concepts within the framework of population dynamics, particularly regarding the Smith dynamics. These preliminaries are used later on in the statement of the proposed novel dynamical tuning.

6.2.1 Model Predictive Control

Consider a system represented by the following discrete-time state-space model:

$$\mathbf{x}(k+1) = \mathbf{A}\mathbf{x}(k) + \mathbf{B}\mathbf{u}(k) + \mathbf{B}_l\mathbf{d}(k), \tag{6.1}$$

where $\mathbf{x} \in \mathbb{R}^{n_x}$ is the state vector, $\mathbf{u} \in \mathbb{R}^{n_u}$ is the vector of manipulated variables, $\mathbf{d} \in \mathbb{R}^{n_d}$ denotes the vector of disturbances affecting the system, $k \in \mathbb{Z}_+$ denotes the discrete time, and A, B, and B_l are the state-space system matrices with suitable dimensions. The states and control actions are subject to bounds and physical constraints, which define feasible sets given by

$$\mathcal{X} \triangleq \left\{\mathbf{x} \in \mathbb{R}^{n_x} : \mathbf{G}\mathbf{x} \leq \mathbf{g}\right\}, \tag{6.2a}$$
$$\mathcal{U} \triangleq \left\{\mathbf{u} \in \mathbb{R}^{n_u} : \mathbf{H}\mathbf{u} \leq \mathbf{h}\right\}, \tag{6.2b}$$

where $\mathbf{G}$, $\mathbf{H}$, $\mathbf{g}$, and $\mathbf{h}$ are matrices of suitable dimensions. Let $\hat{\mathbf{u}}$ be the control action sequence for a fixed-time prediction horizon H_p, $\hat{\mathbf{x}}$ be the state sequence resulting from applying the control action sequence over the system (6.1) from the initial state $\mathbf{x}(0|k) \triangleq \mathbf{x}(k)$, and $\hat{\mathbf{d}}$ be the disturbance sequence along H_p, i.e.,

$$\hat{\mathbf{u}} \triangleq \{\mathbf{u}(0|k), \mathbf{u}(1|k), \ldots, \mathbf{u}(H_p-1|k)\}, \tag{6.3a}$$
$$\hat{\mathbf{x}} \triangleq \{\mathbf{x}(1|k), \mathbf{x}(2|k), \ldots, \mathbf{x}(H_p|k)\}, \tag{6.3b}$$
$$\hat{\mathbf{d}} \triangleq \{\mathbf{d}(0|k), \mathbf{d}(1|k), \ldots, \mathbf{d}(H_p-1|k)\}. \tag{6.3c}$$

The system (6.1) is controlled by a multi-objective MPC controller whose optimization problem is composed by a cost function with N objectives, each one with an associated weight $\gamma_i,\ i = 1, \ldots, N$ that assigns a prioritization, i.e.,

$$\min_{\hat{\mathbf{u}}} J(\mathbf{x}(0), \mathbf{u}) = \sum_{j=1}^{N} \gamma_j J_j(\mathbf{x}(0), \mathbf{u}), \tag{6.4a}$$

subject to

$$\mathbf{x}(i+1|k) = \mathbf{A}\mathbf{x}(i|k) + \mathbf{B}\mathbf{u}(i|k) + \mathbf{B}_l\mathbf{d}(i|k), \quad i \in [0, H_p - 1] \subset \mathbb{Z}_+, \tag{6.4b}$$
$$\mathbf{u}(i|k) \in \mathcal{U}, \quad i \in [0, H_p - 1] \subset \mathbb{Z}_+, \tag{6.4c}$$
$$\mathbf{x}(i|k) \in \mathcal{X}, \quad i \in [0, H_p] \subset \mathbb{Z}_+. \tag{6.4d}$$

The issue treated in this work concerns the proper tuning for the optimization problem (6.4), i.e., how to find the proper values for the weight factors $\gamma_1, \ldots, \gamma_N$.

Assuming that the optimization problem (6.4) is feasible, there is an optimal input sequence given by

$$\hat{\mathbf{u}}^* \triangleq \{\mathbf{u}^*(0|k), \mathbf{u}^*(1|k), \ldots, \mathbf{u}^*(H_p - 1|k)\} \in \mathcal{U},$$

and due to the fact that only one control action of the sequence is applied to the system, then the final optimal control action is given by

$$\mathbf{u}^*(k) \triangleq \mathbf{u}^*(0|k).$$

Once the optimal control action $\mathbf{u}^*(k)$ is applied to the system, a new optimization problem of the form in (6.4) is solved to compute the next optimal control action. Then, a new optimal sequence is computed for the iteration $k+1$ repeating the mentioned procedure and using a new measurement of the system states as an initial condition in the prediction model.

6.2.2 *Population Dynamics*

Consider a population composed by a large and finite number of rational agents involved in a strategic game. During the interaction, each agent chooses a strategy from the set of the N available strategies in the population, which is denoted by $\mathcal{S} = \{1, \ldots, N\}$. Each objective in the cost function of the MPC is associated to a strategy in the population game. The fact that agents are rational implies that they make decisions in order to improve their benefits known as payoffs, which are determined by a fitness function. Let $p_i \in \mathbb{R}_+$ be the portion of agents in the population choosing the strategy $i \in \mathcal{S}$. Thus, the vector $\mathbf{p} = [p_1 \ \cdots \ p_N]^\top$ corresponds to a strategic distribution of agents among all the strategies. The set of possible strategic distributions within the population is given by a simplex denoted by

$$\Delta = \left\{ \mathbf{p} \in \mathbb{R}_+^N : \sum_{i=1}^{N} p_i = 1 \right\}, \tag{6.5}$$

where the unit in the sum of proportions is associated to the total amount of agents in the population, and the interior of the simplex is defined as

$$\text{int}\,\Delta = \left\{ \mathbf{p} \in \mathbb{R}^N_{>0} : \sum_{i=1}^{N} p_i = 1 \right\}. \tag{6.6}$$

Each fitness function is a mapping $f_i : \Delta \mapsto \mathbb{R}$ that takes a strategic distribution in the population and returns a real value corresponding to the payoff that the portion of agents p_i receives for selecting the strategy $i \in \mathcal{S}$. Similarly, the fitness function defined by the mapping $\mathbf{F} : \Delta \mapsto \mathbb{R}^N$ is the vector of all fitness functions, i.e., $\mathbf{F}(\mathbf{p}) = \left[f_1(\mathbf{p}) \;\; f_2(\mathbf{p}) \;\; \cdots \;\; f_N(\mathbf{p}) \right]^\top$.

The vector of fitness $\mathbf{F}$ determines the evolution of the game. For instance, the framework of the proposed dynamical tuning methodology is given by stable games. This condition establishes conditions over the fitness functions. The formal definition of stable games is stated next [8].

Definition 6.1 The population game $\mathbf{F} : \Delta \mapsto \mathbb{R}^N$ is a stable game if

$$(\mathbf{p} - \mathbf{q})^\top (\mathbf{F}(\mathbf{p}) - \mathbf{F}(\mathbf{q})) \leq 0, \;\; \text{for all } \mathbf{p}, \mathbf{q} \in \Delta, \tag{6.7}$$

and this condition is equivalent to the condition that $D\mathbf{F}(\mathbf{p})$ is negative semidefinite, where $\left[D\mathbf{F}(\mathbf{p}) \right]_{ij} = \frac{\partial f_i(\mathbf{p})}{\partial p_j}$. ◇

The process of selecting an agent and making decisions to change strategies in order to improve the payoffs is mathematically described by the population dynamics, e.g., replicator dynamics, projection dynamics, or Smith dynamics. In this chapter, the Smith dynamics have been selected and their features, as the previously mentioned property, are presented and explained below.

Smith Dynamics

The Smith dynamics are one of the six fundamental population dynamics [20]. Any of these six fundamental population dynamics can be used for the proposed tuning strategy. However, in this chapter, the Smith dynamics have been chosen for the following reasons: (i) they satisfy nonnegativeness of proportion of agents, and (ii) proportions do not extinct under the Smith dynamics (i.e., if a $p_i(t_1) = 0$ for some $t_1 \geq 0$ and $\mathbf{p}^* \in \text{int}\Delta$, then there exists a $t_2 > t_1$ such that $p_i(t_2) > 0$, and a $t_3 > t_2$ such that $\mathbf{p}(t_3) = \mathbf{p}^*$). The Smith dynamics are given by the following equation:

$$\dot{p}_i = \sum_{j=1}^{N} p_j \left[f_i(\mathbf{p}) - f_j(\mathbf{p}) \right]_+ - p_i \sum_{j=1}^{N} \left[f_j(\mathbf{p}) - f_i(\mathbf{p}) \right]_+, \;\; \text{for all } i \in \mathcal{S}, \tag{6.8}$$

where $[\cdot]_+ = \max(0, \cdot)$. Notice that the Smith dynamics can be rewritten as

$$\dot{p}_i = \frac{1}{2} \sum_{j=1}^{N} \left((1 - \phi_{ij}) p_i + (1 + \phi_{ij}) p_j \right) \left[f_i(\mathbf{p}) - f_j(\mathbf{p}) \right], \;\; \text{for all } i \in \mathcal{S}, \tag{6.9}$$

where $\phi_{ij} = \text{sgn}\left(f_i(\mathbf{p}) - f_j(\mathbf{p})\right)$. If the equilibrium point of the Smith dynamics is $\mathbf{p}^* \in \text{int}\Delta$, then $p_i^*, p_j^* > 0$, for all $i, j \in \mathcal{S}$, and the equilibrium point in (6.9) implies that $f_i(\mathbf{p}^*) = f_j(\mathbf{p}^*)$, for all $i, j \in \mathcal{S}$.

Proposition 6.1 *The simplex Δ is an invariant set under the Smith dynamics (6.8), i.e., if the initial condition of the population state $\mathbf{p}(0) \in \Delta$, then $\mathbf{p}(t) \in \Delta$, for all $t \geq 0$.*

Proof The simplex is determined by the set of proportions such that $\sum_{i=1}^{N} p_i = 1$, and $p_i \geq 0$, for all $i \in \mathcal{S}$. First, the proof consists of showing that $\sum_{i=1}^{N} \dot{p}_i = 0$, i.e.,

$$\begin{aligned} \sum_{i=1}^{N} \dot{p}_i &= \sum_{i=1}^{N} \left\{ \sum_{j=1}^{N} p_j \left[f_i(\mathbf{p}) - f_j(\mathbf{p})\right]_+ - \sum_{j=1}^{N} p_i \left[f_j(\mathbf{p}) - f_i(\mathbf{p})\right]_+ \right\}, \\ &= \sum_{i=1}^{N} \sum_{j=1}^{N} p_j \left[f_i(\mathbf{p}) - f_j(\mathbf{p})\right]_+ - \sum_{j=1}^{N} \sum_{i=1}^{N} p_j \left[f_i(\mathbf{p}) - f_j(\mathbf{p})\right]_+, \\ &= 0. \end{aligned}$$

This shows the invariance of the set given by condition $\sum_{i=1}^{N} p_i = 1$. Now suppose that the population states are in the limit of the simplex Δ, i.e., a proportion of agents $p_i = 0$, then the Smith equation associated to $i \in \mathcal{S}$ is given by

$$\dot{p}_i = \sum_{j=1}^{N} p_j \left[f_i(\mathbf{p}) - f_j(\mathbf{p})\right]_+,$$

and consequently, $\dot{p}_i \geq 0$, then the positiveness of proportion of agents is satisfied. This completes the proof. □

Once it has been shown that the simplex is an invariant set under the Smith dynamics, it is necessary to show the convergence to the equilibrium point $\mathbf{p}^* \in \Delta$ as it is stated in the following theorem.

Theorem 6.1 *Let $\mathbf{F}$ be a continuously differentiable stable game, then the equilibrium point $\mathbf{p}^* \in \Delta$ is asymptotically stable under the Smith dynamics (6.8).*

Proof The proof of this theorem is reported in [20]. However, a sketch of the proof is presented. Consider the Lyapunov candidate $V(\mathbf{p})$ given by

$$V(\mathbf{p}) = \frac{1}{2} \sum_{i=1}^{N} \sum_{j=1}^{N} p_i \left[f_j(\mathbf{p}) - f_i(\mathbf{p})\right]_+^2,$$

where $V(\mathbf{p}^*) = 0$, and $V(\mathbf{p}) > 0$, for all $\mathbf{p} \neq \mathbf{p}^*$. Its derivative is

$$\dot{V}(\mathbf{p}) = \dot{\mathbf{p}}^\top D\mathbf{F}(\mathbf{p})\dot{\mathbf{p}} + \frac{1}{2}\sum_{i=1}^{N}\sum_{j=1}^{N} x_j \left[f_i(\mathbf{p}) - f_j(\mathbf{p})\right]_+ \left(\sum_{k=1}^{N}\left[f_k(\mathbf{p}) - f_i(\mathbf{p})\right]_+^2 - \left[f_k(\mathbf{p}) - f_j(\mathbf{p})\right]_+^2\right).$$

Notice that the first element $\dot{\mathbf{p}}^\top D\mathbf{F}(\mathbf{p})\dot{\mathbf{p}} \leq 0$ since $\mathbf{F}$ is a stable game. In order to analyze the second term, suppose that $f_i(\mathbf{p}) > f_j(\mathbf{p})$, then $\left[f_i(\mathbf{p}) - f_j(\mathbf{p})\right]_+ > 0$. Now $f_k(\mathbf{p}) - f_i(\mathbf{p}) < f_k(\mathbf{p}) - f_j(\mathbf{p})$, and due to the fact that $[\cdot]_+$ is nondecreasing, then $\sum_{k=1}^{N}\left[f_k(\mathbf{p}) - f_i(\mathbf{p})\right]_+^2 - \left[f_k(\mathbf{p}) - f_j(\mathbf{p})\right]_+^2 \leq 0$. Finally, if $f_i(\mathbf{p}) < f_j(\mathbf{p})$, then $\left[f_i(\mathbf{p}) - f_j(\mathbf{p})\right]_+ = 0$ making zero the second term. As conclusion, $\dot{V}(\mathbf{p}) \leq 0$. Moreover using the La Salle's invariance principle, the equality $\dot{V}(\mathbf{p}) = 0$ holds for $f_i(\mathbf{p}^*) = f_j(\mathbf{p}^*)$, for all $i, j = 1, \ldots, N$ completing the proof. □

Due to the fact that the MPC controller works in discrete time, and the population dynamics evolve in continuous time, a way to sample the population dynamics is presented, i.e., a sampled Smith dynamics are established using the continuous evolution of proportions as introduced next. Consider a sampling time denoted by τ to sample the evolution of the population states $\mathbf{p}(t)$ under the Smith dynamics (6.8), i.e., every time τ, the population states $\mathbf{p}(\tau)$ evolve as a discrete evolution denoted by $\tilde{\mathbf{p}}(k)$ (Sampled Smith Dynamics). Notice that the population dynamics sampling time must be shorter than the MPC controller sampling time since the evolution of the population dynamics determine the prioritization weights for next iteration in the MPC controller, i.e., $\tau < \Delta t$. Suppose that the initial condition is given by $\tilde{\mathbf{p}}(0) = \mathbf{p}(0)$, then the evolution of the discrete population states is given by $\tilde{\mathbf{p}}(k+1) = \mathbf{p}(\tau)$, $\tilde{\mathbf{p}}(k+2) = \mathbf{p}(2\tau)$, and so on, i.e.,

$$\tilde{p}_i(k+b) = p_i(b\tau), \text{ where } b \in \mathbb{Z}, \text{ and for all } i \in \mathcal{S}.$$

Suppose an arbitrary evolution of the proportion of agents playing strategy $i \in \mathcal{S}$ in continuous time $p_i(t)$ as it is shown in Fig. 6.1. Then, it is obtained the evolution of the same proportion of agents by saving its values every time τ. Figure 6.1 also shows the discrete evolution of the proportion of agents $\tilde{p}_i(k)$ under the sampled Smith dynamics for different values of τ, i.e., $\tau = 6$ s, $\tau = 8$ s, and $\tau = 10$ s. Furthermore, the dynamical prioritization weights are given by the discrete proportion of agents $\tilde{\mathbf{p}}(k)$.

6.3 Proposed Dynamical Tuning Strategy

This section introduces the proposed evolutionary game-based dynamical tuning for multi-objective MPC. The dynamical tuning strategy is divided in two parts. First, it is necessary to make a normalization procedure, and then a strategy to assign weights dynamically. Both stages are presented next.

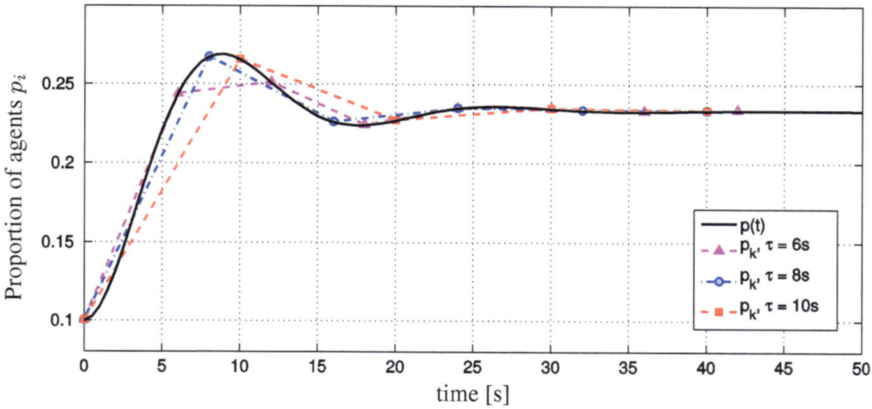

Fig. 6.1 Example of proportion of agents for different values τ

6.3.1 Normalization

The cost function in (6.4a) involves multiple control objectives, making necessary to establish appropriate weights for each one. In general, each objective has a different nature and, as consequence of this, it is not a trivial issue to find the proper set of weights to obtain the desired control performance. In order to establish a proper distribution of weights in the objective functions, it is necessary first to normalize the cost function [9].

Let $\mathbf{x}_i^*, \mathbf{u}_i^*$ be the optimal solution for the single objective optimization of the ith objective function J_i. The solution $\mathbf{x}_i^*, \mathbf{u}_i^*$ is obtained by solving the optimization problem of the MPC controller (6.4) with $\gamma_i = 1$ and $\gamma_j = 0$, for all $j \neq i$. Then, the utopia point, denoted by $\mathbf{J}^{\text{utopia}}$, is found as follows:

$$\mathbf{J}^{\text{utopia}} = \left[J_1(\mathbf{x}_1^*, \mathbf{u}_1^*) \quad J_2(\mathbf{x}_2^*, \mathbf{u}_2^*) \quad \cdots \quad J_N(\mathbf{x}_N^*, \mathbf{u}_N^*) \right]. \tag{6.10}$$

The ith nadir value is denoted by

$$J_i^{\text{nadir}} = \max \left(J_i(\mathbf{x}_1^*, \mathbf{u}_1^*) \quad J_i(\mathbf{x}_2^*, \mathbf{u}_2^*) \quad \cdots \quad J_i(\mathbf{x}_N^*, \mathbf{u}_N^*) \right), \tag{6.11}$$

and the nadir point $\mathbf{J}^{\text{nadir}}$ is given by

$$\mathbf{J}^{\text{nadir}} = \left[J_1^{\text{nadir}} \quad J_2^{\text{nadir}} \quad \cdots \quad J_N^{\text{nadir}} \right]. \tag{6.12}$$

Finally, the normalized multi-objective cost function has the form

$$\tilde{J}(\mathbf{x}, \mathbf{u}) = \sum_{i=1}^{N} \tilde{J}_i(\mathbf{x}, \mathbf{u}),$$

where each normalized objective is

$$\tilde{J}_i(\mathbf{x}, \mathbf{u}) = \frac{J_i(\mathbf{x}, \mathbf{u}) - J_i^{\text{utopia}}}{J_i^{\text{nadir}} - J_i^{\text{utopia}}}.$$

After the normalization, weights in the cost function determine a prioritization without being affected by the order of magnitude of each objective.

6.3.2 Dynamical Weighting Procedure

Once the objective function has been normalized, it is adequate to establish weights $\tilde{\mathbf{p}}(k)$ to each one of the objectives at each discrete-time instant, i.e., the weight for the ith objective at $k \in \mathbb{Z}_+$ is given by $\tilde{p}_i(k)$. Then, the optimization problem for the normalized MPC controller is stated as follows:

$$\min_{\hat{\mathbf{u}}} \sum_{j=1}^{N} \tilde{p}_j(k)\tilde{J}_j(\mathbf{x}(0), \mathbf{u}), \tag{6.13a}$$

subject to

$$\mathbf{x}(i+1|k) = \mathbf{A}\mathbf{x}(i|k) + \mathbf{B}\mathbf{u}(i|k) + \mathbf{B}_l\mathbf{d}(i|k), \quad i \in [0, H_p - 1] \subset \mathbb{Z}_+, \tag{6.13b}$$

$$\mathbf{u}(i|k) \in \mathcal{U}, \quad i \in [0, H_p - 1] \subset \mathbb{Z}_+, \tag{6.13c}$$

$$\mathbf{x}(i|k) \in \mathcal{X}, \quad i \in [0, H_p] \subset \mathbb{Z}_+, \tag{6.13d}$$

where $\tilde{\mathbf{p}}(k) = \begin{bmatrix}\tilde{p}_1(k) & \tilde{p}_2(k) & \cdots & \tilde{p}_N(k)\end{bmatrix}^\top$ and $\sum_{i=1}^{N} \tilde{p}_i(k) = 1$. The unitary value in the equality constraint represents the total mass population according to (6.5). The proper prioritization of these objectives might vary over time as exogenous disturbances affecting the system also vary. In order to overcome this issue, it is proposed a dynamical tuning using a population dynamics approach. Then, the fitness functions $f_i(p_i) \triangleq f_i(\tilde{p}_i(k))$ are selected to be function of each objective evaluated at the current optimal control action, i.e., $\tilde{J}_i(\hat{\mathbf{x}}^*(k), \hat{\mathbf{u}}^*(k))$. Note that this selection of fitness is appropriate since more priority tends to be assigned to those objectives with greater values.

Furthermore, it is desired to assign a prioritization over a region in the Pareto front known as management region (MR). The importance assigned over the MR is determined by a weight w_i in the ith fitness function of the Smith dynamics, i.e.,

$$f_i(\tilde{p}_i(k)) = w_i \tilde{J}_i(\hat{\mathbf{x}}^*(k), \hat{\mathbf{u}}^*(k)). \tag{6.14}$$

A region is selected over the Pareto front instead of a point as reported in [22]. The selection of a management point as in [22] implies to have to compute several different prioritization weights at each iteration in order to find the proper combination of weights. This procedure must be made at every iteration since conditions in the system vary over time as disturbances in the system also vary. Moreover, the disturbances behave in a stochastic manner, for which it is not possible to determine a strategy that uses a limited number of close values to the last one over the Pareto front. The selection of an MR helps to determine the proper direction for each weight by only disposing of a single value over the Pareto front at each iteration. Notice that this proper direction can be computed despite the stochastic behavior of disturbances in the system since only the current condition is required. Furthermore, this relaxation of the point for a region allows to reduce the computational burden.

Remark 6.1 Note that the prioritization in (6.14) assigns an importance to a region in the Pareto front (i.e., at MR) for the population dynamics evolution, and the terms w_i, for $i = 1, \ldots, N$, do not appear in the optimization problem of the MPC, and should not be confused with the weights of the cost function in the MPC controller. ◇

The differences between the MR and the static weights in the multi-objective optimization problem are discussed. To do so, consider a simple and general optimization problem given by

$$\min_{\mathbf{z}} J(\mathbf{z}) = p_1 J_1(\mathbf{z}) + p_2 J_2(\mathbf{z}), \tag{6.15a}$$

subject to

$$\mathbf{V}\mathbf{z} \leq \mathbf{v} + \mathbf{c}, \tag{6.15b}$$

where $\mathbf{z} \in \mathbb{R}^m$ is the decision variable, and $\mathbf{V} \in \mathbb{R}^{l \times m}$ is a constant matrix with suitable dimension. The values $p_1, p_2 \in \mathbb{R}$ establish a static prioritization for the objectives $J_1(\mathbf{z})$ and $J_2(\mathbf{z})$, respectively. The vector $\mathbf{v} \in \mathbb{R}^l$ is a constant component in the constraint, whereas the vector $\mathbf{c} \in \mathbb{R}^l$ is a time-varying component. For instance, the time-variant value of the vector $\mathbf{c} \in \mathbb{R}^l$ may be associated to a disturbance $\mathbf{d} \in \mathbb{R}^l$ involved in a constraint in the optimization problem of an MPC controller.

First, suppose that $\mathbf{c} = \mathbf{c}_1$ in (6.15b), being $\mathbf{c}_1 \in \mathbb{R}^l$ a vector of arbitrary entries. For this case, suppose that the obtained Pareto front is the one presented in Fig. 6.2a, and its normalized Pareto front is the one presented in Fig. 6.2b. This figure shows an example in which the management region is given by $w_1 = w_2 = 0.5$, and shows the solution for the optimization problem when static weights in the multi-objective functions are assigned as $p_1 = p_2 = 0.5$ to objectives $J_1(\mathbf{z})$ and $J_2(\mathbf{z})$, respectively. Notice the difference between the selection of the MR and the assignment of the weights in the cost function.

Now, suppose that $\mathbf{c}$ in (6.15b) varies, e.g., $\mathbf{c} = \mathbf{c}_2$, where the entries of $\mathbf{c}_1$ and $\mathbf{c}_2$ are near values, i.e., $\mathbf{c}_1 - \mathbf{c}_2 \approx \mathbf{0}$. In this case, the Pareto front varies. Suppose that the new Pareto front is the one obtained in Fig. 6.2c, with its corresponding normalized front presented in Fig. 6.2d. When making this modification over $\mathbf{c}$, the solution of the optimization problem for the weights $p_1 = p_2 = 0.5$ changes dramatically over the Pareto front (this fact illustrates the effect when the disturbances, denoted by $\mathbf{d}$, vary in the optimization problem (6.13)). However, notice that the MR is still defined

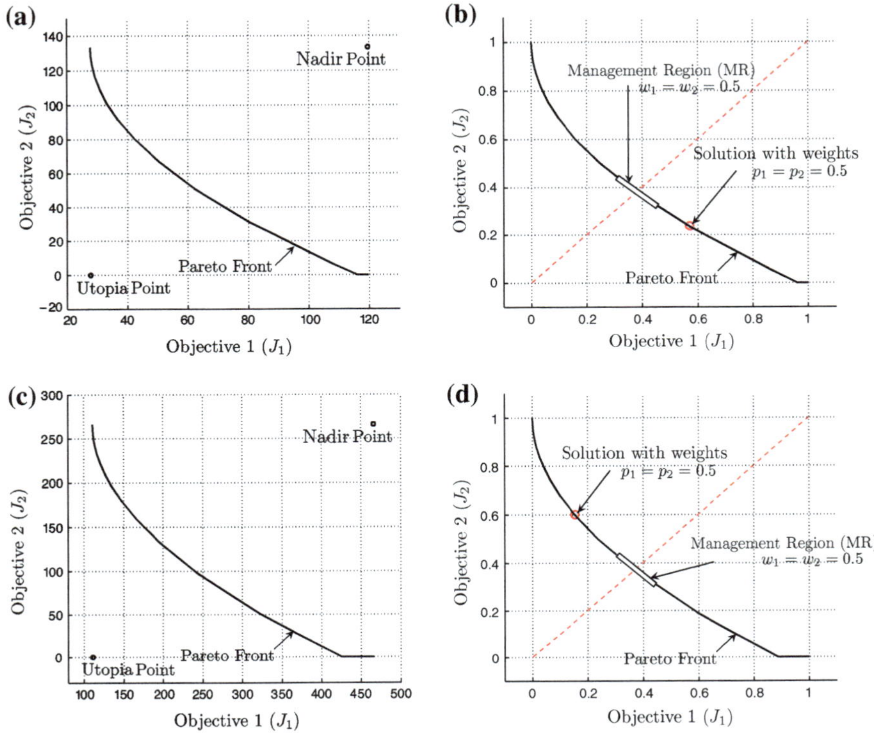

Fig. 6.2 Comparison between the MR and the optimization prioritizing weights

as a region where the objective functions have a equitable value for the particular case $w_1 = w_2 = 0.5$.

When the MR is defined, the dynamical tuning strategy is in charge of finding the proper weights $\tilde{p}_1$ and $\tilde{p}_2$ in the normalized cost function, such that the solution lies inside the MR. This philosophy is different from the static tuning strategy where the weights are determined previously. The process to assign dynamically the tuning weights is performed using the population dynamics, and then in order to guarantee that the Smith dynamics have a stable behavior, it is necessary the following assumption according to the definition of a stable game.

Assumption 6.1 The fitness function $f_i(p_i)$ is a decreasing function with respect to p_i. Then the game $\mathbf{F}$ is a stable game, and stability of the population game is ensured according to Theorem 1. Note that it is expected that the value of the objective $\tilde{J}_i(\hat{\mathbf{x}}^*(k), \hat{\mathbf{u}}^*(k))$ decreases as bigger weight $\tilde{p}_i(k)$ is assigned to it when solving the corresponding optimization problem. ◇

A detailed procedure to implement the evolutionary game-based dynamical tuning for multi-objective model predictive control is presented in Algorithm 6.1.

Algorithm 6.1 Evolutionary game-based dynamical tuning for multi-objective MPC

procedure INITIALIZATION
  $H_s \leftarrow$ simulation length
  $H_p \leftarrow$ prediction horizon
  $N \leftarrow$ number of objectives
  $\mathbf{x}(k) \leftarrow \mathbf{x}(0) \in \mathbb{R}^{n_x}$ states initial condition
  $\mathbf{p}(0) \leftarrow \mathbf{p} \in \mathbb{R}_+^N$ proportion initial condition for continuous Smith dynamics
  $\tilde{\mathbf{p}}(k) \leftarrow \mathbf{p}(0) \in \mathbb{R}_+^N$ discrete proportion initial condition
  $\tau \leftarrow$ time for population dynamics
end procedure
while $k \leq H_s$ **do**
  procedure NORMALIZATION
    $i \leftarrow 1$ initialization index for local objectives
    while $i \leq N$ **do**
      $\mathbf{u}_i^* \leftarrow \arg\min_{\hat{\mathbf{u}}} J_i(\mathbf{x}, \mathbf{u})$ with constraints
      $J_i^{\text{utopia}} \leftarrow J_i(\mathbf{x}_i^*, \mathbf{u}_i^*)$
      $i \leftarrow i + 1$
    end while
    $j \leftarrow 1$ initialization index for nadir points
    while $j \leq N$ **do**
      $J_j^{\text{nadir}} \leftarrow \max\begin{pmatrix} J_j(\mathbf{x}_1^*, \mathbf{u}_1^*) & J_j(\mathbf{x}_2^*, \mathbf{u}_2^*) & \cdots & J_j(\mathbf{x}_N^*, \mathbf{u}_N^*)\end{pmatrix}$
      $j \leftarrow j + 1$
    end while
  end procedure
  procedure NORMALIZED MPC
    $\hat{\mathbf{x}}^*(k), \hat{\mathbf{u}}^*(k) \leftarrow \arg\min \sum_{i=1}^{N} \tilde{p}_i(k)\tilde{J}_i(\mathbf{x}, \mathbf{u})$ with constraints
    $\mathbf{u}^*(k) \leftarrow \mathbf{u}^*(0|k) \in \mathbb{R}^{n_u}$ optimal control action
  end procedure
  procedure COMPUTATION OF FITNESS FUNCTIONS
    $i \leftarrow 1$ initialization index for local objectives
    while $i \leq N$ **do**
      $f_i(p_i) \triangleq f_i(\tilde{p}_i(k)) \leftarrow \tilde{J}_i(\hat{\mathbf{x}}^*(k), \hat{\mathbf{u}}^*(k))$
      $i \leftarrow i + 1$
    end while
  end procedure
  procedure CONTINUOUS-TIME SMITH DYNAMICS $\forall\, i \in \mathcal{S}$
    $\dot{p}_i = \sum_{j=1}^{N} p_j \left[f_i(p_i) - f_j(p_j)\right]_+ - p_i \sum_{j=1}^{N} \left[f_j(p_j) - f_i(p_i)\right]_+$, for $0 \leq t \leq \tau$
    $\tilde{\mathbf{p}}(k) \leftarrow \mathbf{p}(\tau)$ update of discrete agent proportions
    $\mathbf{p}(0) \leftarrow \tilde{\mathbf{p}}(k)$ new initial condition for Smith dynamics
  end procedure
  procedure OPTIMAL CONTROL ACTION APPLIED TO THE SYSTEM
    $\mathbf{x}(k) \leftarrow \mathbf{A}\mathbf{x}(k) + \mathbf{B}\hat{\mathbf{u}}^*(k) + \mathbf{B}_l\mathbf{d}(k)$
    $k \leftarrow k + 1$
  end procedure
end while

6.4 Case Study

The Barcelona Drinking Water Network (DWN) is a large-scale system composed by tanks, valves, pumps, drinking water sources, and water demands as reported in [16]. The volumes in tanks compose the state vector $\mathbf{x} \in \mathbb{R}^{n_x}$, the flows through the valves and pumps compose the vector of manipulated control actions $\mathbf{u} \in \mathbb{R}^{n_u}$, and the water-demanded flows are collected in vector $\mathbf{d} \in \mathbb{R}^{n_d}$. The corresponding discrete-time model is given by

$$\mathbf{x}(k+1) = \mathbf{A}\mathbf{x}(k) + \mathbf{B}\mathbf{u}(k) + \mathbf{B}_l\mathbf{d}(k), \tag{6.16a}$$

$$\mathbf{0} = \mathbf{E}_u\mathbf{u}(k) + \mathbf{E}_d\mathbf{d}(k), \tag{6.16b}$$

where the difference equation in (6.16a) describes the dynamics of the storage tanks in the system, and Eq. (6.16b) describes the static relations given by the mass balance at junction nodes within the network. Moreover, $\mathbf{0}$ is a column vector whose entries are null, and $\mathbf{A}$, $\mathbf{B}$, $\mathbf{B}_l$, $\mathbf{E}_u$, and $\mathbf{E}_d$ are constant matrices with suitable dimensions determined by the DWN topology [7].

6.4.1 System Management Criteria

The cost function for the MPC controller is determined by operational objectives, which are established by the company in charge of the DWN. These objectives are usually determined by the following three aspects: (i) economic operation, (ii) smoothness operation, and (iii) safety operation. For the economical aspect, there are two costs associated to the DWN operation. The first cost is related to water depending on the selected source to get water during the day, it is given by $\boldsymbol{\alpha}_1 \in \mathbb{R}^{n_u}$ and whose units are economic units per flow unit ([e.u.]/[m^3/s]). The second cost is time variant during the day, associated to the energy required to operate the active elements in the DWN (i.e., valves and pumps), and it is given by $\boldsymbol{\alpha}_2 \in \mathbb{R}^{n_u}$ in economic units per flow unit ([e.u.]/[m^3/s]). In general, the economic operation objective consists in minimizing the water production and transport costs given in economic units (e.u.), i.e.,

$$J_1(\mathbf{u}(k)) \triangleq \left|(\boldsymbol{\alpha}_1 + \boldsymbol{\alpha}_2(k))^\top \mathbf{u}(k)\right|. \tag{6.17}$$

Regarding the smoothness operation, it is related to the variations of the control actions along the time, i.e., $\Delta\mathbf{u}(k) = \mathbf{u}(k) - \mathbf{u}(k-1)$. This objective consists in minimizing

$$J_2(\mathbf{u}(k)) \triangleq \|\Delta\mathbf{u}(k)\|^2. \tag{6.18}$$

Finally, the safety operation consists in guaranteeing that there is enough stored water to satisfy the demands during certain period of time. Due to the fact that demand is supposed to be obtained from a forecasting procedure, this operation objective is managed by the following soft constraint:

$$\mathbf{x}(k) \geq \mathbf{x}_s(k) - \boldsymbol{\xi}(k), \text{ for all } k, \tag{6.19}$$

where $\mathbf{x}_s \in \mathbb{R}^{n_x}$ is the vector of safety volumes for all the tanks. The variable $\boldsymbol{\xi} \in \mathbb{R}^{n_x}$ does not have a direct relationship with any element of the system, and it is introduced as a decision variable in the optimization problem to manage the safety volumes. Furthermore, it is desired that $\boldsymbol{\xi}$ tends to zero in order to avoid violations of the constraint (which implies the depletion of such safety volumes). Then, the third objective related to the system states is given by the minimization of

$$J_3(\boldsymbol{\xi}(k)) \triangleq \|\boldsymbol{\xi}(k)\|^2 . \tag{6.20}$$

It is worth to highlight that it is already known the importance in the prioritization of objectives, which is determined by the company in charge of the management of the network. This known importance among the objective functions is commonly used to determine a static tuning for the system. The most important objective is the economical aspect, and the second is to guarantee the safety volumes. This fact is used below with the case study to determine different and possible cases to test the performance of the MPC controllers.

6.4.2 *Optimization Problem of the Predictive Controller*

Once the system management criteria have been established with the objectives J_1, J_2, and J_3, it can be set the normalized optimization problem behind the MPC controller design, i.e.,

$$\begin{aligned}\min_{\hat{\mathbf{u}},\hat{\xi}} J(\mathbf{u}, \boldsymbol{\xi}) = &\sum_{j=0}^{H_p-1} \tilde{p}_1(k)\tilde{J}_1(\mathbf{u}(k+j)) + \sum_{j=0}^{H_p-1} \tilde{p}_2(k)\tilde{J}_2(\mathbf{u}(k+j))\\ &+ \sum_{j=0}^{H_p-1} \tilde{p}_3(k)\tilde{J}_3(\boldsymbol{\xi}(k+j)),\end{aligned}$$

subject to

$$\begin{aligned}\mathbf{x}(i+1|k) &= \mathbf{A}\mathbf{x}(i|k) + \mathbf{B}\mathbf{u}(i|k) + \mathbf{B}_l\mathbf{d}(i|k), \quad i \in [0, H_p - 1] \subset \mathbb{Z}_+,\\ \mathbf{0} &= \mathbf{E}_u\mathbf{u}(i|k) + \mathbf{E}_d\mathbf{d}(i|k), \quad i \in [0, H_p - 1] \subset \mathbb{Z}_+,\\ \mathbf{u}(i|k) &\in \mathcal{U}, \quad i \in [0, H_p - 1] \subset \mathbb{Z}_+,\\ \mathbf{x}(i|k) &\in \mathcal{X}, \quad i \in [0, H_p] \subset \mathbb{Z}_+,\\ \mathbf{x}(i|k) &\geq \mathbf{x}_s(k) - \boldsymbol{\xi}(i|k), \quad i \in [0, H_p] \subset \mathbb{Z}_+,\end{aligned}$$

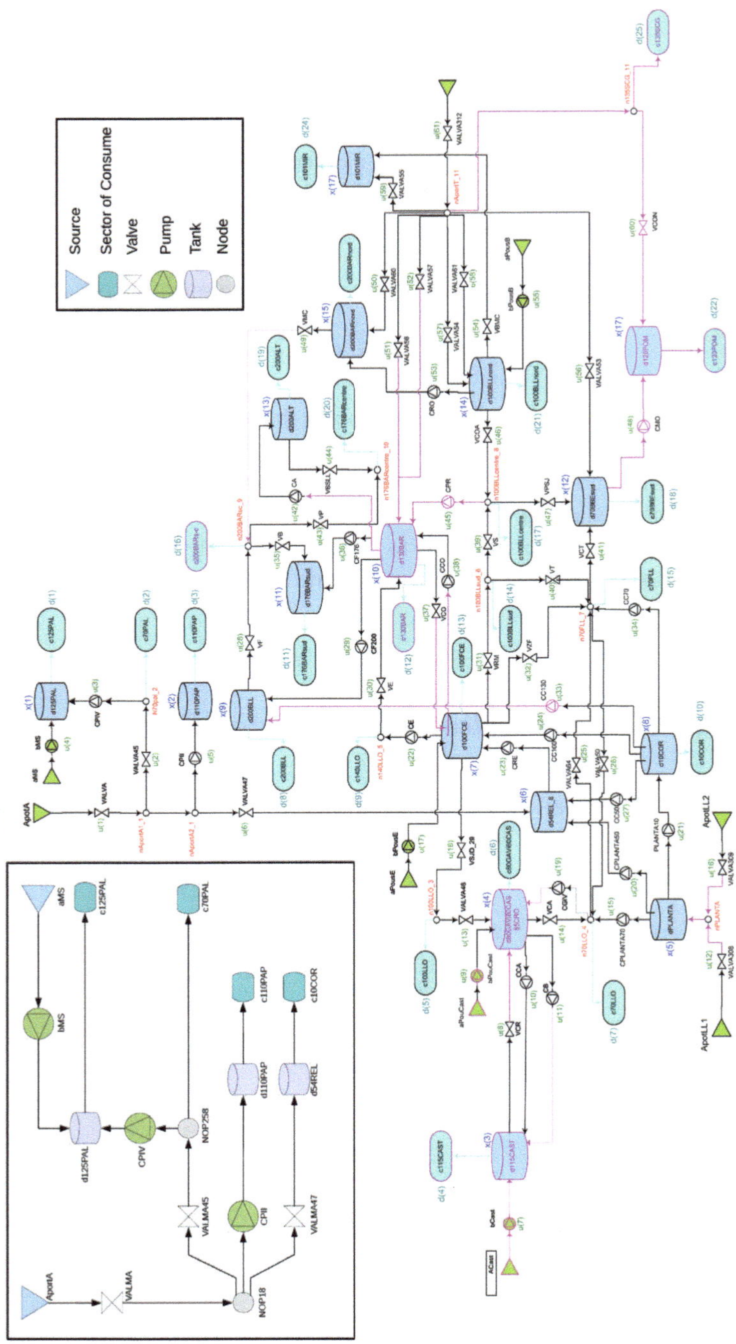

Fig. 6.3 Case study. **a** Topology of the three-tank DWN. **b** Topology of the 17-tank DWN

where the feasible sets for the control actions and the system states are given by $\mathcal{U} = \{\mathbf{u} \in \mathbb{R}^{n_u} | \mathbf{u}_{\min} \leq \mathbf{u} \leq \mathbf{u}_{\max}\}$ and $\mathcal{X} = \{\mathbf{x} \in \mathbb{R}^{n_x} | \mathbf{x}_{\min} \leq \mathbf{x} \leq \mathbf{x}_{\max}\}$, where $\mathbf{u}_{\min}$ and $\mathbf{u}_{\max}$ are the minimum and maximum limits for the control actions, respectively. Similarly, $\mathbf{x}_{\min}$ and $\mathbf{x}_{\max}$ are the minimum and maximum limits for the system states. Finally, similarly as in (6.3), $\hat{\boldsymbol{\xi}}$ is a sequence during the horizon H_p.

Figure 6.3 shows two different significative portions of the Barcelona DWN. Figure 6.3a is a portion of the DWN involving three tanks (states), three valves, and three pumps (control actions), two drinking water sources, and four water demands (disturbances). Then, the matrices and limits for its discrete model are given by

$$
\begin{aligned}
\mathbf{A} &= \begin{bmatrix} 1 & 0 & 0 \\ 0 & 1 & 0 \\ 0 & 0 & 1 \end{bmatrix}, \quad \mathbf{B} = \begin{bmatrix} 0 & 0 & 0 & 1 & 1 & 0 \\ 0 & 0 & 0 & 0 & 0 & 1 \\ 0 & 0 & 1 & 0 & 0 & 0 \end{bmatrix} \Delta t, \quad \mathbf{B}_l = \begin{bmatrix} -1 & 0 & 0 & 0 \\ 0 & 0 & -1 & 0 \\ 0 & 0 & 0 & -1 \end{bmatrix} \Delta t, \\
\mathbf{E}_u &= \begin{bmatrix} 1 & -1 & -1 & 0 & 0 & -1 \\ 0 & 1 & 0 & 0 & -1 & 0 \end{bmatrix}, \quad \mathbf{E}_d = \begin{bmatrix} 0 & 0 & 0 & 0 \\ 0 & -1 & 0 & 0 \end{bmatrix}, \quad \mathbf{x}_{\min} = [0 \quad 0 \quad 0]^\top, \\
\mathbf{x}_{\max} &= [470 \quad 960 \quad 3100]^\top, \quad \mathbf{u}_{\min} = [0 \quad 0 \quad 0 \quad 0 \quad 0 \quad 0]^\top, \\
\mathbf{u}_{\max} &= [1.2970 \quad 0.05 \quad 0.12 \quad 0.0150 \quad 0.0317 \quad 0.0220]^\top,
\end{aligned}
$$

where the sampling time $\Delta t = 3600$ s. On the other hand, Fig. 6.3b shows a portion with 17 tanks (states), 61 manipulated flows (control actions), nine water sources, and 25 water demands (disturbances). Matrices and limits for this discrete model are not presented because of lack of space.

6.4.3 Scenarios

Two different scenarios are presented in order to analyze the performance of the proposed dynamical tuning strategy. In general, demand has a periodic behavior (seasonality), maintaining the same mean value and with a regular amplitude in time. However, it is considered the case in which the periodic demand varies unexpectedly during time, i.e., the case in which the demand profile varies its mean value and its regular amplitude. The purpose of these abrupt changes is to analyze how the prioritization weights are adapted when the system conditions suffer variations. Moreover, these scenarios can certainly occur because of unexpected situations such as public events, damages in the network as leaks, move of population, growth of the system, etc.

Consequently, it is possible to analyze both the performance when the demand decreases, and when demand increases unexpectedly as shown in Fig. 6.4, i.e.,

- **Scenario 1**: decreasing in the demand profiles (see Fig. 6.4a).
- **Scenario 2**: increasing in the demand profiles (see Fig. 6.4b).

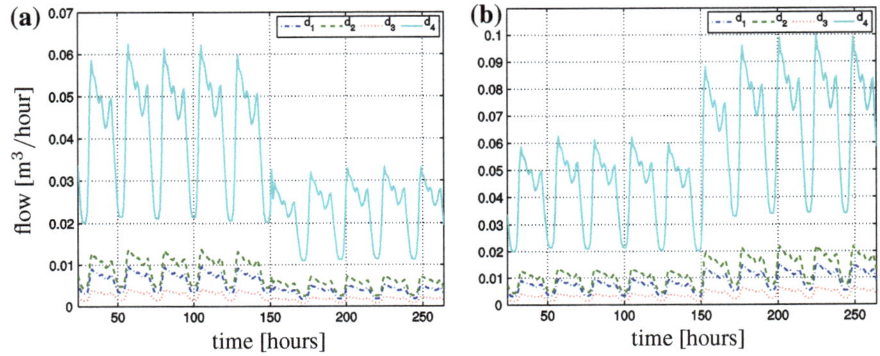

Fig. 6.4 Demand profile for: **a** Scenario 1 with a decrease in the demands, and **b** Scenario 2 with an increment in the demands

Both scenarios are analyzed to illustrate that the dynamical tuning strategy may adapt a proper combination of weights in the cost function of the optimization problem behind the MPC controller, for any change in the nominal system behavior.

In the proposed dynamical tuning methodology, the first step is to normalize the cost function by computing the nadir and the utopia points. After making this procedure, then the following step is to assign a prioritization to the MR where it is desired that different objectives evolve around. For this case study, the prioritization is given by w_1 for economic objective, w_2 for smoothness objective, and w_3 for the safety objective, where $\sum_{i=1}^{N} w_i = 1$.

In order to make a fair comparison between the performance of a multi-objective MPC with conventional static tuning and the performance of an multi-objective MPC with the proposed dynamical tuning strategy, there must be established a relationship between the weights for objective functions. The weights $\gamma_1, \ldots, \gamma_N$ for the cost function in Problem (6.4), and the weights for the MR $w_1, \ldots, w_N$ in (6.14), are selected to be the same as $w_i = \gamma_i$, for all $i \in \mathcal{S}$. This criterion establishes a fair comparison for a prioritization over the cost function without any tuning strategy, and a prioritization using a dynamical tuning strategy.

6.4.4 Results and Discussion

The performance of the controllers is determined by the economical costs C given in economic units (e.u.) during the total number of simulation days (in this case 11 days), i.e.,

$$C = \sum_{k=0}^{264} (\boldsymbol{\alpha}_1 + \boldsymbol{\alpha}_2(k))^\top \mathbf{u}(k), \tag{6.22}$$

Table 6.1 Economic results for Scenario 1 and Scenario 2 in the case study of three states in Fig. 6.3a

	Tuning case	Dynamical tuning costs C_D 11 days (e.u.)	Static tuning costs C_S 11 days (e.u.)	Reduction of costs $C_S - C_D$ (e.u.)
Scenario 1	1	5649.45	5660.61	11.16
	2	5656.89	5666.13	9.24
	3	5657.53	5677.65	20.12
	4	5657.01	5738.12	81.11
Scenario 2	1	8983.86	8984.57	0.71
	2	8986.35	8993.64	7.29
	3	8985.69	9011.49	25.80
	4	8986.33	9075.49	89.16

Notice that for the comparison of data the management region corresponds to the prioritization of the MPC controller with static tuning, i.e., $[w_1 \;\; w_2 \;\; w_3]^\top = [\gamma_1 \;\; \gamma_2 \;\; \gamma_3]^\top$

where the costs are denoted by C_D for the dynamical tuning case, and by C_S for the static tuning case. For each scenario, four different cases corresponding to four MRs are tested:

- Tuning case 1: $[\gamma_1 \;\; \gamma_2 \;\; \gamma_3]^\top = [0.7 \;\; 0.1 \;\; 0.2]^\top$,
- Tuning case 2: $[\gamma_1 \;\; \gamma_2 \;\; \gamma_3]^\top = [0.6 \;\; 0.15 \;\; 0.25]^\top$,
- Tuning case 3: $[\gamma_1 \;\; \gamma_2 \;\; \gamma_3]^\top = [0.5 \;\; 0.2 \;\; 0.3]^\top$,
- Tuning case 4: $[\gamma_1 \;\; \gamma_2 \;\; \gamma_3]^\top = [0.4 \;\; 0.25 \;\; 0.35]^\top$.

Notice that these different cases for tuning satisfy the prioritization explained in Sect. 6.4.1, i.e., $w_1 > w_3 > w_2$.

Table 6.1 shows the comparison between the costs of the MPC with the proposed dynamical tuning and with a static prioritization for the four cases, and for Scenarios 1 and 2, for the case study of three states presented in Fig. 6.3a. Table 6.1 also shows the difference of costs, i.e., a reduction of costs when changing the static tuning for the dynamical tuning strategy given by $C_S - C_D$. More specifically, results for Scenario 1 show a reduction of costs with the dynamical tuning for all the tested cases. During the 11 days, reduction of costs between 9.24 e.u. and 81.11 e.u. can be obtained.

For Scenario 2, it can be seen that for all the cases a reduction of costs is obtained if the dynamical tuning strategy is adopted. These reductions during 11 days oscillate between 0.70 e.u. and 89.16 e.u. depending on the management region case.

Regarding the dynamical behavior of the proposed strategy, and the evolution of weights, Fig. 6.5 shows the performance of the MPC controller with the dynamical tuning strategy for a management point given by $\mathbf{w} = [0.6 \;\; 0.15 \;\; 0.25]^\top$, and for the Scenario 1. Similarly, Fig. 6.6 shows the performance of the MPC controller with the dynamical tuning strategy for a management point given by $\mathbf{w} = [0.7 \;\; 0.1 \;\; 0.2]^\top$, and for the Scenario 2. In the performance of the dynamical weights, it can be seen that they oscillate with the same period as the disturbances in the system. Moreover,

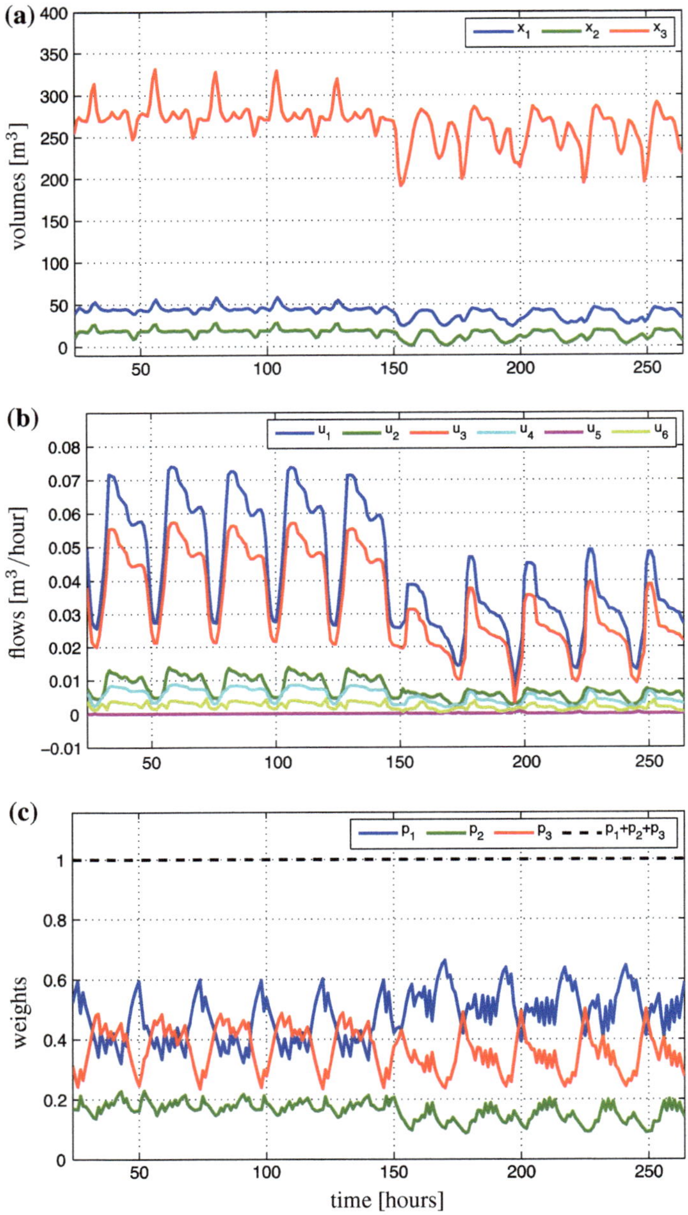

Fig. 6.5 MPC controller with evolutionary game-based dynamical tuning for the Scenario 1 with a decrease in the demands and a management region given by $\mathbf{w} = [0.6 \quad 0.15 \quad 0.25]^{\top}$. Subfigures correspond to: **a** system states $\mathbf{x}$, **b** control actions $\mathbf{u}$, and **c** dynamical tuning $\tilde{\mathbf{p}}(k)$

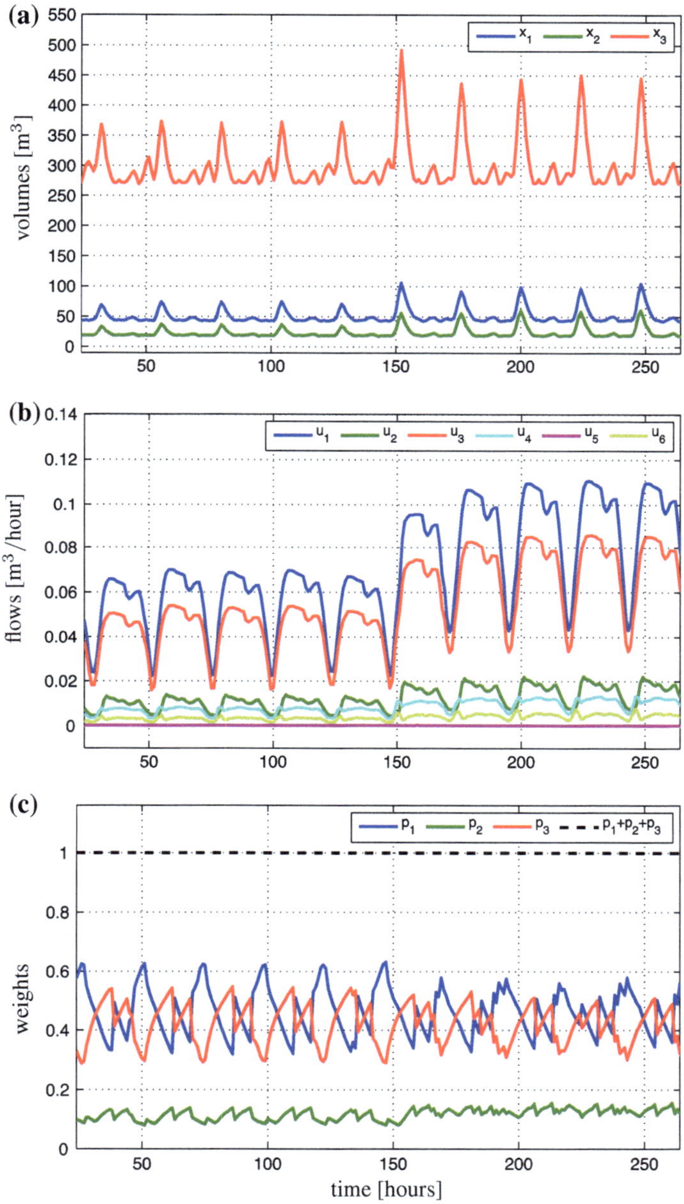

Fig. 6.6 MPC controller with evolutionary game-based dynamical tuning for the Scenario 2 with an increment in the demands and a management region given by $\mathbf{w} = [0.7 \quad 0.1 \quad 0.2]^\top$. Subfigures correspond to: **a** system states $\mathbf{x}$, **b** control actions $\mathbf{u}$, and **c** dynamical tuning $\tilde{\mathbf{p}}(k)$

Table 6.2 Economic results for Scenario 1 in the case study of three states in Fig. 6.3b

Dynamical tuning costs C_D 11 days (e.u.)	Static tuning costs C_S 11 days (e.u.)	Reduction of costs $C_S - C_D$ (e.u.)
398645.19	420894.99	22249.80

Notice that for the comparison of data the management region corresponds to the prioritization of the MPC controller with static tuning, and for the tuning case 2, i.e., $[w_1 \quad w_2 \quad w_3]^\top = [\gamma_1 \quad \gamma_2 \quad \gamma_3]^\top = [0.6 \quad 0.15 \quad 0.25]^\top$

it can be seen that the mean value of each weight varies when the behavior of the demands changes on the seventh day.

The previously presented results are a proof of concept to see how tuning is adapted dynamically as conditions over the system vary. The case study shown in Fig. 6.3a does not contain redundant paths to satisfy water demand, and involves a reduced number of states and control actions. Consequently, there is less freedom in order to adjust the proper prioritization values to potentially improve the performance.

Then, the dynamical tuning strategy is implemented in a bigger case study shown in Fig. 6.3b for Scenario 1, and the tuning case 2. This implementation is made in order to check the improvement of the performance for a large-scale system with the proposed tuning approach.

Table 6.2 shows the results for the case study shown in Fig. 6.3b. It can be seen a higher reduction of costs when adopting the dynamical tuning strategy in a larger case study. The considerable reduction is obtained since the case study in Fig. 6.3b has redundant paths to satisfy the water demand.

6.5 Conclusions and Further Work

A novel dynamical tuning strategy based on evolutionary game theory has been proposed. Similarly as other tuning strategies suggest, in the proposed tuning strategy it is necessary to normalize the cost function. In this regard, the proposed strategy does not imply to have higher computation burden with respect to other online tuning strategies. Once the problem is normalized, it is not ensured to generate several points in the Pareto front. This is an advantage of the proposed strategy in comparison to other tuning strategies that require the computation of several points in the Pareto front to establish a proper tuning. However, it should be satisfied that the Pareto front satisfy an assumption clearly defined in this work.

The results obtained in this chapter reflect an improvement in the reduction of economical costs. Moreover, a higher reduction of costs is obtained with the 17 variable states network, than with the smaller system of three variable states. For future work, it is necessary to test the dynamical tuning in a larger system, whose topology includes redundancy paths, and more actuators and constraints (e.g., the whole Barcelona network that is composed by 63 states can be found in [26]). Then, a more considerable improvement between the performance of an MPC with static tuning and the performance of an MPC with the proposed dynamical tuning might be obtained. Finally, the prediction horizon is considered within the MPC parameters that compose the issue of tuning, and it can be included in the dynamical tuning strategy.

Acknowledgments Authors would like to thank COLCIENCIAS (grant 6172) and Agència de Gestió d'Ajust Universitaris i de Recerca AGAUR for supporting J. Barreiro-Gómez. This work has been partially supported by the projects "Drenaje urbano y cambio climático: Hacia los sistemas de alcantarillados del futuro, Fase II. COLCIENCIAS", and ECOCIS (Ref. DPI2013-48243-C2-1).

References

1. A. Al-Ghazzawi, E. Ali, A. Nouh, E. Zafiriou, On-line tuning strategy for model predictive controllers. J. Process Control **11**, 265–284 (2001)
2. J. Barreiro-Gomez, N. Quijano, C. Ocampo-Martinez, Distributed constrained optimization based on population dynamics, in *Proceedings of the IEEE Conference on Decision and Control (CDC)* (Los Angeles, USA, 2014), pp. 4260–4265
3. J. Barreiro-Gomez, N. Quijano, C. Ocampo-Martinez, Distributed control of drinking water networks using population dynamics: Barcelona case study, in *Proceedings of the IEEE Conference on Decision and Control (CDC)* (Los Angeles, USA, 2014), pp. 3216–3221
4. S. Di Cairano, A. Bemporad, Model predictive control tuning by controller matching. IEEE Trans. Autom. Control **55**, 185–190 (2010)
5. J.L. Garriga, M. Soroush, Model predictive control tuning methods: a review. Ind. Eng. Chem. Res. (I&EC) **49**, 3505–3515 (2010)
6. J.M. Grosso, C. Ocampo-Martinez, V. Puig, Learning-based tuning of supervisory model predictive control for drinking water networks. Eng. Appl. Artif. Intell. **26**, 1741–1750 (2013)
7. J.M. Grosso, C. Ocampo-Martinez, V. Puig, B. Joseph, Chance-constrained model predictive control for drinking water networks. J. Process Control **24**, 504–516 (2014)
8. J. Hofbauer, W.H. Sandholm, Stable games and their dynamics. J. Econ. Theory **144**(4), 1665–1693 (2009)
9. I.Y. Kim, O.L. de Weck, Adaptive weighted sum method for multiobjective optimization: a new method for pareto front generation. Struct. Multidiscipl. Optim. **31**, 105–116 (2006)
10. N. Li, J. Marden, Designing games for distributed optimization with a time varying communication graph, in *Proceedings of the IEEE Conference on Decision and Control (CDC)* (Maui, Hawaii, 2012), pp. 7764–7769
11. N. Li, J. Marden, Decoupling coupled constraints through utility design. IEEE Trans. Autom. Control **59**, 2289–2294 (2014)
12. J. Marden, J. Shamma, Game theory and distributed control, in *Handbook of Game Theory*, vol. 4, ed. by H.P. Young, S. Zamir (Elsevier Science, Amsterdam, 2013)
13. M.A. Müller, D. Angeli, F. Allgöwer, On the performance of economic model predictive control with self-tuning terminal cost. J. Process Control **24**, 1179–1186 (2014)
14. G. Obando, A. Pantoja, N. Quijano, Building temperature control based on population dynamics. IEEE Trans. Control Syst. Technol. **22**(1), 404–412 (2014)
15. C. Ocampo-Martinez, D. Barcelli, V. Puig, A. Bemporad, Hierarchical and decentralised model predictive control of drinking water networks: application to barcelona case study. IET Control Theory Appl. **6**(1), 62–71 (2012)
16. C. Ocampo-Martinez, V. Puig, G. Cembrano, J. Quevedo, Application of predictive control strategies to the management of complex networks in the urban water cycle. IEEE Control Syst. Mag. **33**(1), 15–41 (2013)
17. A. Pantoja, N. Quijano, A population dynamics approach for the dispatch of distributed generators. IEEE Trans. Ind. Electron. **58**(10), 4559–4567 (2011)

18. A. Pantoja, N. Quijano, Distributed optimization using population dynamics with a local replicator equation, in *Proceedings of the IEEE Conference on Decision and Control (CDC)* (Maui, Hawaii, 2012), pp. 3790–3795
19. E. Ramirez-Llanos, N. Quijano, A population dynamics approach for the water distribution problem. Int. J. Control **83**(9), 1947–1964 (2010)
20. W.H. Sandholm, *Population Games and Evolutionary Dynamics* (MIT Press, Cambridge, 2010)
21. G. Shah, S. Engell, Tuning MPC for desired closed-loop performance for MIMO systems, in *Proceedings of the American Control Conference (ACC)* (San Francisco, USA, 2011), pp. 4404–4409
22. R. Toro, C. Ocampo-Martinez, F. Logist, J.V. Impe, V. Puig, Tuning of predictive controllers for drinking water networked systems, in *Proceedings of the 18th IFAC World Congress* (Milan, Italy, 2011), pp. 14507–14512
23. Q.N. Tran, R. Octaviano, L. Özkan, A.C.P.M. Backx, Generalized predictive control tuning by controller matching, in *Proceedings of the American Control Conference (ACC)* (Portland, USA, 2014), pp. 4889–4894
24. J.W. Weibull, *Evolutionary Game Theory* (The MIT Press, London, 1997)
25. J. Zhang, D. Qi, G. Zhao, A game theoretical formulation for distributed optimization problems, in *Proceeding of the IEEE International Conference on Control and Automation (ICCA)* (2013), pp. 1939–1944
26. Y. Wang, C. Ocampo-Martinez, V. Puig, J. Quevedo, Gaussian-process-based demand forecasting for predictive control of drinking water networks, in *9th International Conference on Critical Information Infrastructures Security* (Limassol, Cyprus, 2014), pp. 13–15

Part III
Collaborative Model Predictive Control

Chapter 7
A Model Predictive Control-Based Architecture for Cooperative Path-Following of Multiple Unmanned Aerial Vehicles

Alessandro Rucco, António Pedro Aguiar, Fernando A.C.C. Fontes, Fernando Lobo Pereira and João Borges de Sousa

Abstract This chapter proposes a sampled-data model predictive control (MPC) architecture to solve the decentralized cooperative path-following (CPF) problem of multiple unmanned aerial vehicles (UAVs). In the cooperative path-following proposed scenario, which builds on previous work on CPF, multiple vehicles are required to follow pre-specified paths at nominal speed profiles (that may be path dependent) while keeping a desired, possibly time-varying, geometric formation pattern. In the proposed framework, we exploit the potential of optimization-based control strategies with significant advantages on explicitly addressing input and state constraints and on the ability to allow the minimization of meaningful cost functions. An example consisting of three fixed wing UAVs that are required to follow a given desired maneuver illustrates the proposed framework. We highlight and discuss some features of the UAVs trajectories.

Keywords Model predictive control · Cooperative path-following · Multiple UAVs

A. Rucco · A.P. Aguiar (✉) · F.A.C.C. Fontes
SYSTEC—Research Center for Systems and Technologies, Institute for Systems and Robotic (ISR), Faculty of Engineering, Porto University (FEUP), 4200-465 Porto, Portugal
e-mail: pedro.aguiar@fe.up.pt

A. Rucco
e-mail: alessandrorucco@fe.up.pt

F.A.C.C. Fontes
e-mail: faf@fe.up.pt

F. Lobo Pereira
LSTS—Laboratory of Underwater Systems and Technologies and SYSTEC—Research Center for Systems and Technologies, Institute for Systems and Robotic (ISR), Faculty of Engineering, Porto University (FEUP), 4200-465 Porto, Portugal
e-mail: flp@fe.up.pt

J. Borges de Sousa
LSTS—Laboratory of Underwater Systems and Technologies, Faculty of Engineering, Porto University (FEUP), 4200-465 Porto, Portugal
e-mail: jtasso@fe.up.pt

S. Olaru et al. (eds.), *Developments in Model-Based Optimization and Control*,
Lecture Notes in Control and Information Sciences 464,
DOI 10.1007/978-3-319-26687-9_7

7.1 Introduction

In the last few years, Unmanned Aerial Vehicles (UAVs) have been playing an increasing important role in several complex tasks for both civilian and military missions. A subset of representative operation scenarios includes search and rescue operations, long endurance missions, and environmental measurements tasks including the use of multiple UAVs in cooperation, see, e.g., [7, 27, 34]. These type of missions require the computation and execution of (in some cases) aggressive and complex maneuvers, especially for the case of multiple UAVs working in cooperation.

Once the desired UAV motion is defined, the next step is to make the UAV to execute it. To this end, two types of approaches are possible: trajectory tracking and path-following methods. The former is concerned with the design of control laws that force the UAV to reach and follow a time-parameterized reference (i.e., a geometric path with an associated timing law), see, e.g., [5, 29, 32], while the latter requires the vehicle to converge to and follow a geometric path that is specified without a temporal law, see, e.g., [15, 28, 31, 35, 36]. The underlying assumption in path-following method is that the vehicle's forward speed tracks a desired speed profile, while the controller acts on the vehicle's orientation to drive it to the path. We refer to [1, 3] for a comparison and discussion between the trajectory tracking and the path-following approaches. In [33] a trajectory optimization strategy for UAVs is proposed. The strategy is based on a virtual target vehicle (VTV) perspective. The strategy takes into account the velocity of the VTV, which helps to improve the convergence of the actual path to the desired one. In this chapter, we extend and adapt the trajectory optimization strategy presented in [33] for path-following of multiple UAVs, where the vehicles are required to follow pre-specified paths at nominal speed profiles.

The success of more challenging missions requires the employment of multiple vehicles working in cooperation toward the same goal. To this effect, one naive possibility is to make the vehicles to share all internal and external information to improve coordination performance. However, such approach is not in general feasible in terms of bandwidth and computational complexity. Moreover, the communication topology may vary over time due to connectivity or even vehicle failure. A suitable communication constraints representation is a methodology based on a framework that relates the concept of Graph Laplacian to represent links between vehicles as addressed in [14]. In particular, in [13] the authors show how the Graph Laplacian associated to a formation interconnection structure plays a fundamental role in assessing stability of the behavior of the components in coordination. In [20], the authors consider two types of communication topologies. The first captures the case where the communication graph is alternately connected and disconnected (called brief connectivity losses). The second case takes only into account that the union of the communication graphs over uniform intervals of time remains connected (uniformly connected in mean). This framework if further extended and applied for the case of multiple UAVs in [37] where the communications graph is disconnected during some interval of time or even fails to be connected for the entire duration of the mission. Moreover, flight tests of a coordinated road-search mission scenario are shown. A framework that takes into account the topology of the communication links, the communication channel model, and the cost of exchanging information is proposed in [2]. Following the cooperative control architecture presented in [2] and exploiting the virtual target approach for trajectory optimization of UAVs proposed in [33], this chapter addresses a decentralized multi-vehicle control structure for a set of UAVs, where the vehicles dynamics and communication topology constraints are taken into account.

The control approach used here is a sampled-data model predictive control (MPC) architecture. Sampled-data MPC schemes and its stabilizing properties have been studied in [9, 16, 18, 26]. Regarding path-following via MPC, there are several recent works on the subject [4, 8, 11, 12, 30, 38]. The control of multiple vehicles by MPC schemes has been reported in [10, 19, 29]. The MPC technique, in addition to its well-known capacity of handling constraints and optimization of performance characteristics, is particularly suited to the multi-vehicle control since, even when a communication link becomes temporarily unavailable, breaking the feedback loop, an already computed open-loop plan can be implemented.

In this chapter, we propose a sampled-data MPC architecture to solve the decentralized cooperative path-following (CPF) problem of multiple UAVs. We refer the reader to [20] for a discussion on the centralized and decentralized approaches. In decentralized CPF, multiple vehicles are required to follow pre-specified paths at nominal speed profiles (that may be path dependent) while keeping a desired possibly time-varying geometric formation pattern and considering that each vehicle receives information about its neighbors. In the proposed framework, we exploit the potential of optimization-based control strategies with significant advantages on explicitly addressing input and state constraints and on the ability to allow the minimization of meaningful cost functions. We show how stability of the overall strategy can be guaranteed and, moreover, we are able to guarantee a prescribed rate of exponential convergence to the desired solution. We provide numerical computations to show the effectiveness of the proposed MPC cooperative path-following system.

The rest of the chapter is organized as follows. In Sect. 7.2, we introduce the UAV model and formulate the path-following problem using the VTV approach. Then, a formal definition of the cooperative path-following control problem is given. In Sect. 7.3, we first propose the MPC law for a single UAV and then we extend it for a set of UAVs. In Sect. 7.4, we provide numerical computations that illustrate the effectiveness of the proposed MPC-CPF architecture.

7.2 Problem Formulation

This section formulates the cooperative path-following (CPF) control problem for multiple UAVs, in which a fleet of n_v UAVs is tasked to converge to and to follow a set of desired geometric paths under a specified formation and velocity profile. In this chapter, the problem of cooperative trajectory generation is not addressed. It is assumed that the desired geometric paths for each vehicle are given such that each path is conveniently parameterized by a single variable $\gamma^{[i]}$, $i = 1, \ldots, n_v$, and with the restriction that when the fleet is in the desired formation, the parameterization variables satisfy the equality constraint $\gamma^{[1]} = \gamma^{[2]} = \cdots = \gamma^{[n_v]}$. The next subsections describe the adopted UAV model, the path-following, coordination, and CPF problem statements.

7.2.1 Constrained UAV Model

This section presents a planar aerial vehicle model in coordinated flight, based on a simplified model widely used in the literature, see, e.g., [6, 22], but extended as in [15] to accommodate the effects of the roll dynamics and include the roll rate as a control input. In this case, the planar UAV motion, [33], can be described by

$$\begin{aligned} \dot{x} &= v \cos \psi, \\ \dot{y} &= v \sin \psi, \\ \dot{\psi} &= \frac{g \tan \phi}{v}, \\ \dot{v} &= u_1, \\ \dot{\phi} &= u_2, \end{aligned} \tag{7.1}$$

where (x, y) is the longitudinal and lateral position with respect to an inertial global frame, ψ is the heading angle, ϕ is the roll angle, v is the airspeed, and g denotes the gravity acceleration. The longitudinal acceleration $\dot{v}$ and the roll rate $\dot{\phi}$ are the control inputs u_1 and u_2, respectively.

Due to physical limitations or security specifications, state and input constraints are imposed on the model as follows:

$$\begin{aligned} v_{\min} \leq v &\leq v_{\max}, \\ |\phi| &\leq \phi_{\max}, \\ |u_1| &\leq u_{1\max}, \\ |u_2| &\leq u_{2\max}, \end{aligned} \tag{7.2}$$

where the constraint parameters used in the simulation section are based on the ones given in [29], i.e., $v_{\min} = 18\,\text{m/s}$, $v_{\max} = 25\,\text{m/s}$, $\phi_{\max} = 24°$, $u_{1\max} = 0.2\,\text{m/s}^2$, $u_{2\max} = 28\,\text{deg/s}$.

7.2.2 Error Dynamics and Path-Following Formulation

Following the approach developed in [33], this section introduces a virtual target vehicle (VTV) that is constrained to move along a given desired geometric path $(x_d(\gamma), y_d(\gamma))$ parameterized by the variable $\gamma \in \mathbb{R}$, and defines the error vector between the position of the VTV and the actual UAV through the use of a Serret–

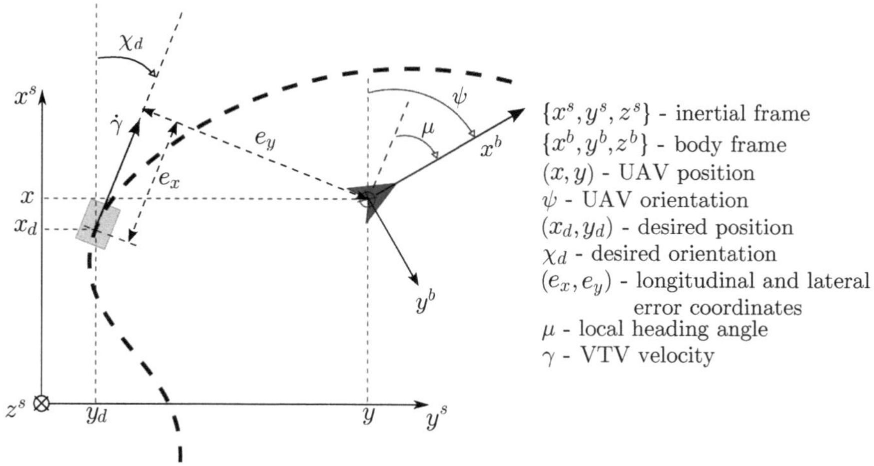

Fig. 7.1 Tracking error coordinates. The *bold triangle* and the *light rectangle* indicate the real UAV and the VTV, respectively. The *bold dash line* indicates the desired path

Frenet frame, which can be viewed as the body frame of the VTV that moves according to a "convenient velocity", effectively yielding an extra control input.

Figure 7.1 captures the setup, where the dashed line denotes the desired path, and the light rectangle represents the VTV. The desired course heading χ_d and the curvature σ_d are related by

$$\begin{aligned} x_d' &= \cos \chi_d \\ y_d' &= \sin \chi_d \\ \chi_d' &= \sigma_d \end{aligned} \tag{7.3}$$

where the prime symbol denotes the first derivative of a variable with respect to the parameterization variable γ. Note that one way to obtain the desired geometric path $(x_d(\gamma), y_d(\gamma))$ is to solve (7.3) with a given initial condition $(x_d(0), y_d(0), \chi_d(0))$ and specified input signal $\sigma_d(\gamma)$.

The coordinates of the UAV can be defined with respect to the position of the VTV, [33]. In particular, defining the longitudinal and lateral error coordinates e_x and e_y, the local heading angle $\mu = \psi - \chi_d$ and the VTV's velocity as an additional control input $u_3 = \dot{\gamma}$, the nonlinear system (7.1) can be written with respect to the new set of coordinates $(\mathbf{x}, \mathbf{u}) = \left(e_x, e_y, \mu, v, \phi, \gamma, u_1, u_2, u_3\right)$:

$$\begin{aligned}
\dot{e}_x &= v\cos\mu - (1 - e_y\sigma_d)u_3,\\
\dot{e}_y &= v\sin\mu - e_x\sigma_d u_3,\\
\dot{\mu} &= \frac{g\tan\phi}{v} - \sigma_d u_3,\\
\dot{v} &= u_1,\\
\dot{\phi} &= u_2,\\
\dot{\gamma} &= u_3.
\end{aligned} \tag{7.4}$$

A detailed description of this derivation, including references to some of the related literature, is given in [33].

With this framework, the path-following problem for a single UAV can be formulated as follows.

Problem 7.1 (*Path-following*) Consider a given UAV described by (7.4) subject to the constraints in (7.2), and let $(x_d(\gamma), y_d(\gamma))$ be a given desired spatial path, and $v_d(\gamma)$ a given desired speed assignment, both parameterized by $\gamma \in \mathbb{R}$. Under feasibility assumption, design feedback control laws for the longitudinal acceleration u_1, the roll rate u_2, and the rate of progression u_3 of the VTV along the desired path such that the errors e_x, e_y, and μ converge to zero as $t \to \infty$ and the parameter γ satisfies the speed assignment $\dot{\gamma} \to v_d(\gamma)$ as $t \to \infty$.

7.2.3 Coordination and Cooperative Path-Following Formulation

Consider now a fleet of n_v vehicles and for each vehicle $i \in \mathcal{N} = \{1, \ldots, n_v\}$, it is given a desired geometric path $(x_d^{[i]}, y_d^{[i]})$. Suppose that each vehicle is endowed with a path-following controller, which means that eventually they will converge to their respective VTVs. Then, one can conclude that if the n_v virtual target vehicles asymptotically synchronize, the UAVs asymptotically reach a desired formation because the desired paths were conveniently parameterized to satisfy the formation constraints. In this context, the parametric variable $\gamma^{[i]}$ represents a measure of the position of the ith VTV and is said to be the coordination state. To enforce the formation, we formulate a consensus problem by extending the definition of the desired speed profile to

$$v_d^{[i]} = v_L(\gamma^{[i]}) + u_c^{[i]}$$

where v_L is the (common) formation speed assignment, and the correction speed $u_c^{[i]}$ is viewed as an input control signal that should vanish when the vehicles are in formation. To reach consensus the UAVs will have to exchange the coordination states. Further, we consider explicitly the fact that communications do not occur in a continuous manner. To this end, let $t_k^{[i]}$, $k \geq 0$, be the instants of time that data is transmitted to the ith UAV by a set of neighbor vehicles denoted by $\mathcal{N}_{t_k}^{[i]}$. The coordination problem can be formulated as follows.

Problem 7.2 (*Coordination*) Assume that for each vehicle $i \in \mathcal{N}$ at $t = t_k^{[i]}$, the variables $\gamma^{[i]}$ and $\gamma^{[j]}, j \in \mathcal{N}_{t_k}^{[i]}$, are available. Derive a control law for $u_c^{[i]}$ such that, for all $i, j \in \mathcal{N}$, $(\gamma^{[i]} - \gamma^{[j]})$ and $u_c^{[i]}$ converge to zero as $t \to \infty$.

Note that $u_c^{[i]}$ is a piecewise continuous signal and is only updated (with new information) at discrete instants of time $t = t_k^{[i]}$.

Equipped with the above formulations we can now state the problem of cooperative path-following, which is a combination of the two previously stated problems.

Problem 7.3 (*Cooperative Path-following*) Consider a fleet of n_v UAVs each one with dynamics described by (7.4) and subject to the constraints in (7.2), a set of desired geometric paths $(x_d^{[i]}, y_d^{[i]})$ parameterized by $\gamma^{[i]}$, $i \in \mathcal{N}$, and a formation speed assignment v_L. Suppose that the UAVs are supported by an inter-vehicle communication network, and, therefore, for each vehicle $i \in \mathcal{N}$ at $t = t_k^{[i]}$, the variables $\gamma^{[i]}$ and $\gamma^{[j]}, j \in \mathcal{N}_{t_k}^{[i]}$, are available. Design feedback control laws for the longitudinal acceleration and roll rate of each vehicle such that

1. the corresponding path-following errors e_x, e_y, and μ of each UAV converge to zero as $t \to \infty$, and
2. for each pair of vehicles $i, j \in \mathcal{N}$, the coordination errors $(\gamma^{[i]} - \gamma^{[j]})$ and the speed errors $(v_d^{[i]} - v_L)$ converge to zero as $t \to \infty$.

7.3 MPC-CPF Control System

In this section, we propose a MPC cooperative path-following controller for a set of UAVs. We start with the design of a MPC path-following controller for a single UAV using the virtual target approach introduced in the previous section. Then, we propose a coordination controller that adjusts the speed of the virtual target at each sampling time, thus exploiting the MPC scheme. Finally, the overall MPC cooperative path-following system is proposed and discussed.

7.3.1 MPC Path-Following for a Single UAV

We now address the design of the MPC law to solve the path-following problem for a single UAV. To this end, we first consider the following open-loop optimal control problem:

Given a pair $(t_k, z) \in \mathbb{R} \times \mathbb{R}^n$, a desired curve $(\mathbf{x}_d(\cdot), \mathbf{u}_d(\cdot))$, a horizon length $T > 0$, and positive definite weighting matrices Q, R, and P_1 find the optimal pair $\big(x^*([t_k, t_k + T]), u^*([t_k, t_k + T])\big)$ that solves

$$\begin{aligned}
\min_{\mathbf{x}(\cdot),\mathbf{u}(\cdot)} \quad & \frac{1}{2}\int_{t_k}^{t_k+T} \left(\|\mathbf{x}(t)-\mathbf{x}_d(t)\|_Q^2+\|\mathbf{u}(t)-\mathbf{u}_d(t)\|_R^2\right) dt + \frac{1}{2}\|\mathbf{x}(t_k+T)-\mathbf{x}_d(t_k+T)\|_{P_1}^2 \\
\text{subject to} \quad & \text{dynamic constraints (7.4)}, \ \forall t \in [t_k, t_k+T], \\
& \text{state and input constraints (7.2)}, \ \forall t \in [t_k, t_k+T], \\
& x(t_k) = z.
\end{aligned} \tag{7.5}$$

The desired curve $(\mathbf{x}_d(\cdot), \mathbf{u}_d(\cdot))$ is chosen as follows. The desired kinematic coordinates, e_{xd}, e_{yd} and μ_d, are set to zero. The desired speed, v_d, and the desired geometric path, parametrized by γ_d, are assigned, see Problem 7.1. The desired roll angle and control inputs, ϕ_d, u_{1d} and u_{2d}, are set to zero (in order to minimize the control effort). The desired control input u_{3d} is set to be equal to the desired speed. The optimal control trajectory $u^*(\cdot)$ minimizes the deviations of the actual trajectory from the desired curve according to the quadratic cost function used in (7.5).

It is worth noting that we formulate Problem 7.3 using a sampled-data MPC approach in which the vehicle model, the state and control constraints, and the cost function are described in continuous time. Next, we discuss stability and convergence requirements, and, then, we address the optimal control problem (7.5) using an iterative algorithm for solving continuous-time optimal control problem (where the optimization problem is not transcribed into a discrete optimization problem). Exploiting a projection operator approach, [21], we are able to compute approximate solutions of the Problem 7.3 in an efficient way (e.g., avoiding numerical instabilities) using continuous-time second-order approximations of the cost function.

7.3.1.1 Stability and Exponential Rate of Decay

We now show that, for a conveniently selected set of design parameters—the matrices Q, R, P_1 of the objective function and the horizon T-we can guarantee not only stability, but also a prescribed rate of exponential decay. The rate of exponential decay is related to the design parameters and can, therefore, under some limitations, be chosen.

We start by defining the new states and control, which are deviations from the desired behavior, to be $\bar{x}(t) = \mathbf{x}(t) - \mathbf{x}_d(t)$ and $\bar{u}(t) = \mathbf{u}(t) - \mathbf{u}_d(t)$. We rewrite the dynamics as

$$\dot{\bar{x}}(t) = \bar{f}(\bar{x}(t), \bar{u}(t)) \tag{7.6}$$

$$\bar{u}(t) \in \bar{U} \tag{7.7}$$

$$\bar{x}(t) \in \bar{X} \tag{7.8}$$

$$\bar{x}(0) = x_0 \tag{7.9}$$

and its linearization

$$\dot{\bar{x}}(t) = A\bar{x}(t) + B\bar{u}(t)$$

where $A = \bar{f}_x(0, 0)$ and $B = \bar{f}_u(0, 0)$. We define $M\mathbb{B}$ as the ball around the origin containing all possible initial states, i.e., $x_0 \in M\mathbb{B}$.

We assume the following hypotheses.

H1 The linearization of the system around the origin is stabilizable, i.e., the unstable modes of the pair (A, B) are controllable.

H2 There exists a trajectory solving (7.6)–(7.9) that can be bounded by a trajectory of some stable linear system. More precisely, there exist positive scalars M, M_1 and α such that for any initial state $x_0 \in M\mathbb{B}$ we can find a control $\tilde{u} : [t_0, \infty] \to \mathbb{R}^m$ satisfying (7.6)–(7.9) and

$$\|\bar{x}(t; x_0, \tilde{u})\|^2 \leq M_1 \|x_0\|^2 \mathrm{e}^{-\alpha t}$$

Given the nature of the system we have in hands, we expect small deviations from the nominal trajectory, and therefore the previous assumptions for this case hold. We note that nonholonomic vehicles are generally not adequately modeled by a linearized system (the linearization of the system around the origin is in general not stabilizable). On the other hand, the relative distances and relative velocities between vehicles following a similar trajectory already have a stabilizable linearization, see [19] for a discussion on this issue.

Define the "maximal vector field speed" $\bar{S} := \max\{ \|\bar{f}(\bar{x}, \bar{u})\| : \bar{x} \in M\mathbb{B},\ \bar{u} \in \bar{U} \}$. Such a maximum exists since the sets involved are compact and $\bar{f}$ is continuous.

Consider the following result from [17] which establishes stability and a given minimum exponential rate of decay.

Theorem 7.1 *Select a scalar $\bar{\mu} > 0$ such that the pair $(A + \bar{\mu}I, B)$ is stabilizable. There exist a set of design parameters (T, Q, R, P_1) and a scalar $\bar{k} > 1$ such that the trajectory resulting from applying the MPC strategy starting at t_0 satisfies the exponential convergence rate*

$$\|\bar{x}(t + t_0)\| \leq \sqrt{\frac{4\lambda_{\max}(P_1)}{\rho\lambda_{\min}(Q)}} \|\bar{x}(t_0)\| \mathrm{e}^{-\frac{\bar{\mu}}{\bar{k}}t}, \quad \forall t \geq T,$$

where $\rho = \min\{T, \frac{\|\bar{x}(t_0)\|}{2\bar{S}}\}$.

Again following [17], to choose design parameters (the horizon T, the matrices Q, R, P_1, and the scalar $\bar{k}$) guaranteeing a prescribed rate of exponential stability we can follow the following steps.

1. Choose the desired rate $\bar{\mu}$ conditioned to $(A + \bar{\mu}I, B)$ being stabilizable.
2. Choose the scalar $\bar{k} > 1$.
3. Choose some positive definite matrices $\hat{R}$ and $\hat{Q}$.
4. Find P to solve the algebraic Riccati equation

$$(A + \bar{\mu}I)^T P + P(A + \bar{\mu}I) - PB\hat{R}^{-1}B^T P + \hat{Q} = 0.$$

5. Let $Q = \hat{Q} + 2\bar{\mu}P$.
6. Let $P_1 = \bar{k}P$.
7. Choose $R \geq 0$ such that:

   a. $\hat{R} - R \geq 0$,
   b. there exists a scalar $M_2 > 0$ such that $\alpha > 0$ and $\tilde{u}$ of H2 satisfy

$$\|\tilde{u}(t)\|_R^2 \leq M_2 \|x_0\|^2 e^{-\alpha t}.$$

   (Always true if $R = 0$.)

It remains to choose the horizon T. In [17] a way to compute a conservative bound is given. We highlight that the larger we choose $\bar{\mu}$ the faster the rate of convergence. Moreover, the selection of $\bar{k}$ implies the following compromise: (i) the nearer it is of 1, the faster will be the convergence rate $\bar{\mu}/\bar{k}$, (ii) the smaller $\bar{k}$, the longer will be the horizon T. We refer the reader to [17] for more details.

7.3.1.2 Numerical Solution to the Optimal Control Problem

Here, we propose to solve the optimal control problem (7.5) using the projection operator-based Newton method for trajectory optimization (PRONTO), [21]. This method is a direct method based for solving continuous-time optimal control problems. Using a projection operator that maps a state-control curve (e.g., a desired curve) onto the trajectory manifold, the constrained optimization problem (7.5) can be converted into an unconstrained one, so that a Newton-based descent method can be used.

We recall that a trajectory is a (state-input) curve $\xi = (\mathbf{x}(\cdot), \mathbf{u}(\cdot))$ defined on $L_\infty[0, T]$ such that

$$\dot{\mathbf{x}}(t) = f(\mathbf{x}(t), \mathbf{u}(t)),$$

for all $t \in [t_k, t_k + T]$, where f is a (sufficiently smooth) map, [23]. Consider now the cost functional

$$h(\xi) = \frac{1}{2}\int_{t_k}^{t_k+T} \left(\|\mathbf{x}(t) - \mathbf{x}_d(t)\|_Q^2 + \|\mathbf{u}(t) - \mathbf{u}_d(t)\|_R^2\right) dt + \frac{1}{2}\|\mathbf{x}(t_k + T) - \mathbf{x}_d(t_k + T)\|_{P_1}^2$$

and denote by $\mathscr{T}$ the manifold of bounded trajectories $\xi = (\mathbf{x}(\cdot), \mathbf{u}(\cdot))$ on $[t_k, t_k + T]$.

Based on the idea developed in [24], we use a barrier function relaxation to handle the constraints. We rewrite the constraints (7.2) in the equivalent form

$$c_j = \left(\frac{2\tilde{x} - (\tilde{x}_{\max} + \tilde{x}_{\min})}{\tilde{x}_{\max} - \tilde{x}_{\min}}\right)^2 - 1 \leq 0 \quad j = 1, \ldots, n_c,$$

where $\tilde{x}$ can be a state-input variable, c_j is a smooth function that enforces the constraint $\tilde{x}_{\min} \leq \tilde{x}(t) \leq \tilde{x}_{\max}$, and n_c is the number of constraints (i.e., $n_c = 4$). For instance, for $\tilde{x} = u$, we have the constraint $u_{\min} \leq u(t) \leq u_{\max}$. For a given (state-input) trajectory $\xi = (\mathbf{x}(\cdot), \mathbf{u}(\cdot))$, a barrier functional can be defined as

$$b_\delta(\xi) = \int_{t_k}^{t_k+T} \left(\sum_{j=1}^{n_c} \beta_\delta(-c_j(\mathbf{x}(t), \mathbf{u}(t))) \right) dt,$$

where

$$\beta_\delta(\tilde{x}) = \begin{cases} -\log \tilde{x}, & \tilde{x} > \delta \\ -\log \delta + \frac{1}{2}\left[\left(\frac{\tilde{x}-2\delta}{\delta}\right)^2 - 1 \right], & \tilde{x} \leq \delta \end{cases}.$$

Using the barrier functional defined above, the relaxed version of problem (7.5) is given by

$$\min_{\xi \in \mathscr{T}} \left(h(\xi) + \varepsilon b_\delta(\xi) \right), \tag{7.10}$$

for some $\varepsilon > 0$. Using the projection operator defined in [21] to locally parameterize the trajectory manifold, we convert the constrained optimization problem (7.10) into one of minimizing the unconstrained functional

$$g_{\varepsilon,\delta}(\xi) = h(\mathscr{P}(\xi)) + \varepsilon b_\delta(\mathscr{P}(\xi)). \tag{7.11}$$

The `PRONTO`, [21], is used to optimize the functional (7.11), as part of a continuation method to seek an approximate solution to (7.5). The strategy is to start with a reasonably large ε and δ. Then, for the current ε and δ, the problem

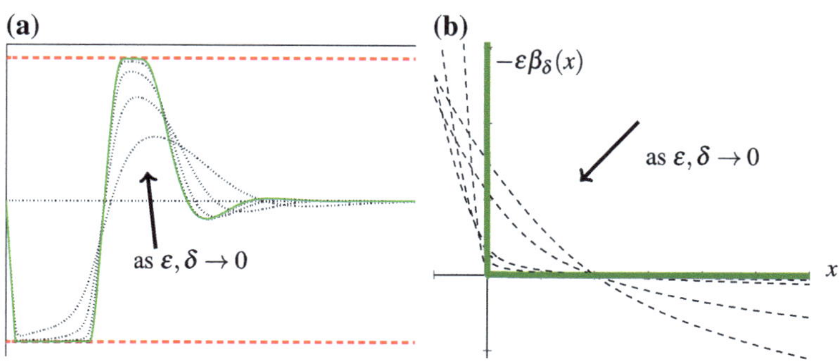

Fig. 7.2 Trajectory exploration strategy for a fixed prediction horizon. Start with a feasible trajectory and solve the relaxed optimal control problem. At the ith step, update the initial trajectory with the previous optimal trajectory and decrease the barrier functional parameters, see the approximate barrier function in (**b**), to compute feasible intermediate optimal trajectories, see the *light dot black lines* in (**a**). The intermediate optimal trajectories approach the optimal trajectory (*solid green line* in (**a**)) for the constrained problem

$$\min g_{\varepsilon,\delta}(\xi)$$

is solved using the PRONTO method starting from the current trajectory. An illustration of the strategy is shown in Fig. 7.2. An important feature of the optimization process is its ability to handle state and input constraints. Thus the (temporary) optimal trajectories, light dot lines in Fig. 7.2a, are strictly feasible as the constraints parameters are reduced (the optimization process works in an interior point fashion).

Following a sampled-data MPC approach, the optimal control input $u^*(\cdot)$ is repeatedly computed at the discrete-time samples $\mathbb{T} := \{t_0, t_1, \dots\}$ with $z = x(t_k)$, and where $t_{k+1} > t_k$ and $t_{k+1} - t_k < T$ for $k = 0, 1, \dots$ The sampled-data MPC control law only uses $u^*(\cdot)$ in the interval $t \in [t_k, t_{k+1})$ being the remaining interval $t \in [t_{k+1}, T]$ discarded, that is,

$$u(t) = u^*(t), \quad t \in [t_k, t_{k+1}), \ \forall t_k \in \mathbb{T}.$$

7.3.2 Coordination

From the adopted CPF setup, to achieve the specified formation, the UAVs only have to exchange the coordination states (which are the path-parameterizations) over the supported communication network to reach agreement on these states. From the theory of consensus of distributed systems and assuming that each vehicle receives information from its neighbors continuously, a natural choice for the coordination control law would be (the so-called neighboring rule, see, e.g., [2])

$$u_c^{[i]} = -\kappa \sum_{j \in \mathcal{N}^{[i]}} \left(\gamma^{[i]} - \gamma^{[j]}\right), \tag{7.12}$$

where κ is a positive scalar, $\gamma^{[i]}$ is related to the position and orientation of the ith VTV, see (7.3), and $(\gamma^{[i]} - \gamma^{[j]})$ is the coordination error. To reduce the communication rate and lift the continuously transmissions, we consider instead that each vehicle relies on their estimates that are updated at the transmission times $t = t_k$. In this case, we propose the following modification:

$$u_c^{[i]} = -\kappa \sum_{j \in \mathcal{N}^{[i]}} \left(\hat{\gamma}_i^{[i]} - \hat{\gamma}_j^{[i]}\right), \tag{7.13}$$

where $\hat{\gamma}_\ell^{[i]}$ is an estimate of $\gamma^{[\ell]}$ running on agent i. The estimators of the coordination states in each UAV run open-loop most of the time but are reset once new data is received from the supported network to correct the state estimates. More precisely, each estimator at the ith UAV is described by

- For $t_k^{[i]} \leq t < t_{k+1}^{[i]}$

$$\dot{\hat{\gamma}}_\ell^{[i]} = f_\ell^{[i]}\left(\hat{\gamma}_j^{[i]}, \cdot\right) \quad \forall \ell \in \{i\} \cup \mathcal{N}_{t_k}^{[i]}$$

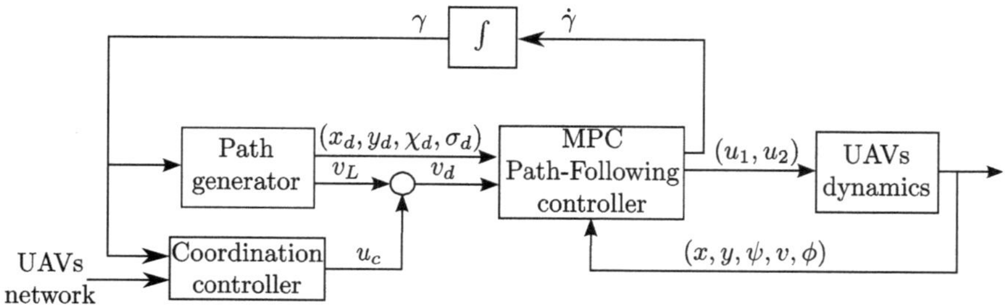

Fig. 7.3 Overall control system architecture

- For $t = t_{k+1}^{[i]}$

$$\hat{\gamma}_\ell^{[i]} = \gamma^{[\ell]}, \quad \forall \ell \in \{i\} \cup \mathcal{N}_{t_{k+1}}^{[i]}$$

where the vector fields $f_\ell^{[i]}(\hat{\gamma}_j^{[i]}, \cdot)$ could be in general functions of the corresponding coordinated state estimate and other states. A simple choice is to make $f_\ell^{[i]}(\hat{\gamma}_j^{[i]}, \cdot) = 0$, which means that (7.13) is just a piecewise continuous sample and hold discretization of (7.12).

7.3.3 Overall MPC Cooperative Path-Following System

Following closely the approach in [2, 20], Fig. 7.3 shows the overall MPC cooperative path-following control architecture of each vehicle, which consists of three interconnected subsystems: the path-generator, the MPC path-following controller, and the coordination controller, all supported by the communication system. Next, we briefly describe the control architecture.

The path-generator system is not addressed in this chapter, but its role is to provide to the MPC path-following controller the desired path. The coordination controller has the responsibility to output a correction speed signal u_c to adjust the speed of the virtual target about their nominal value so as to synchronize its position with the other virtual vehicles and achieve, indirectly, vehicle coordination. Note that the coordination is achieved by resorting to a decentralized control law whereby the exchange of data among the vehicles is kept at a minimum. In the consensus literature (see, e.g., [13, 20, 27]), there exist several algorithms and conditions under which the coordination is achieve, even when the topology of communication links among the vehicles is time-varying. For the simple neighboring rule proposed, it is shown in [20] that consensus is achieved in the case where the communication graph is intermittently connected with brief connectivity losses, and where the union of the communication graphs over uniform intervals of time remains connected (uniformly connected in mean). Regarding the stability of the overall system, it is shown in [2, 20] that the system obtained by putting together the path-following and the vehicle coordination can be either a feedback interconnection or a cascade of two input-to-state stable (ISS) systems by assuming that individually each block is ISS. Stability and convergence properties of the resulting interconnected system can be then formally concluded through a small gain theorem-based argument.

7.4 Simulation Results

In this section we provide numerical computations showing the effectiveness of the MPC cooperative path-following introduced in the previous section. First, we apply the MPC path-following controller to one vehicle ($n_v = 1$) based on the straight line maneuver. We highlight some features of the computed trajectory. Second, we illustrate the performance of the MPC cooperative path-following, when applied to a group of three UAVs ($n_v = 3$).

7.4.1 MPC Path-Following

Similar to the computations shown in [33], we set as desired curve the straight line shown in Fig. 7.4a, see dash-dot line. Given the initial position $(x, y) = (-10, 20)$ and orientation $\psi = 45°$ of the actual vehicle, the goal is to approach the desired straight line path with a desired velocity, $v_d = 19$ m/s, along it. We run the MPC strategy by setting the following weighting matrices $Q = diag([0.01,\ 0.01,\ 0.1,\ 1.0,\ 0.1,\ 1e-09])$, $R = diag([1.0,\ 1.0,\ 0.0001])$, and $P_1 = diag([0.003,\ 0.003,\ 3.8,\ 1.0,\ 1.4,\ 1e-05])$. Note that the VTV velocity is weighted lightly thus giving the optimization the necessary freedom to track the states. We make use of a relatively short update time $(t_{k+1} - t_k) = 1$ s and a fixed horizon time $T = 15$ s.

Figure 7.4 shows the trajectory of the UAV. It is worth noting that such a maneuver allows us to highlight an important aspect of the MPC path-following: the VTV's velocity allows us to *improve the convergence of the actual path to the desired one (that is main feature of the VTV approach)*. In fact, due to the initial conditions of the actual vehicle, see Fig. 7.4a, the VTV assumes a negative velocity value, see Fig. 7.4f. The VTV goes backward along the straight line thus approaching (almost instantaneously) the point on the desired path at minimum distance from the actual vehicle. It is interesting to note in Fig. 7.4e that the actual vehicle starts the maneuver by applying the maximum roll angle. At the same time, the vehicle decelerates to increase the yaw rate, see Fig. 7.4d. Thus, decreasing the velocity allows the vehicle to increase the yaw rate, see (7.1), thus improving the convergence of the actual path to the desired straight line path. At about $t = 4$ s, the actual vehicle reaches the point with zero longitudinal tracking error e_x, see Fig. 7.4b, and maximum lateral tracking error e_y, see Fig. 7.4c. Immediately after, the actual vehicle turns right by applying positive roll angle, see Fig. 7.4e. The actual vehicle crosses the desired straight line at about $t = 11$ s and, in order to regain the desired position, it turns to the left by

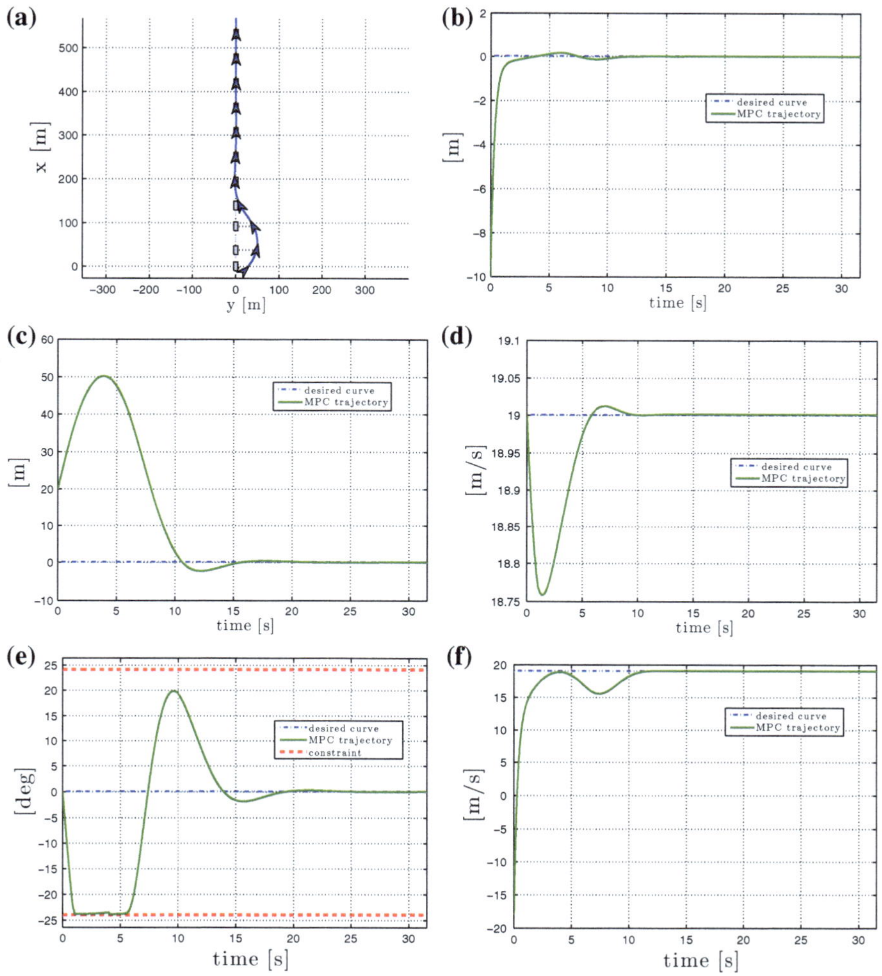

Fig. 7.4 MPC path-following: *straight line* maneuver. The desired (*dash-dot line*) and the UAV (*solid line*) trajectories are shown. The path, the longitudinal, and lateral tracking error profiles are shown in (**a**), (**b**), and (**c**), respectively. The velocity, roll angle, and virtual target velocity profiles are shown in (**d**), (**e**), and (**f**), respectively (constraints in *dash line*). **a** x-y. **b** e_x. **c** e_y. **d** v. **e** ϕ. **f** u_3

applying a negative roll angle. Now the VTV (slightly) decelerates and the actual vehicle (slightly) accelerates thus approaching the desired straight line path.

7.4.2 MPC Cooperative Path-Following

Based on the straight line maneuver of the previous numerical computation, in Figs. 7.5 and 7.6 we illustrate the performance of the MPC cooperative path-following con-

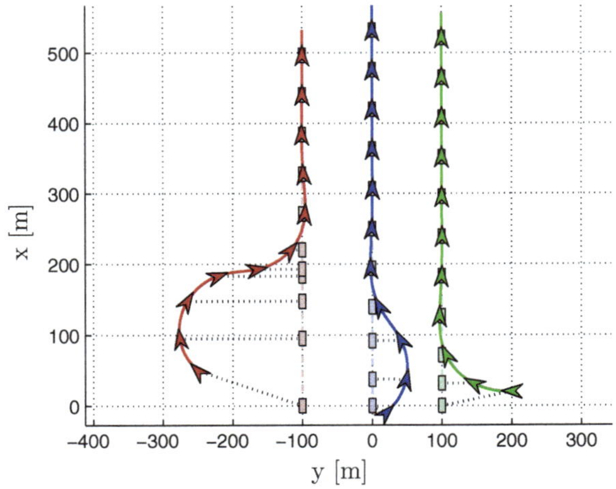

Fig. 7.5 MPC path-following of 3 UAVs (no coordination). The *bold triangle* and the *light rectangle* indicate the real UAV and the VTV, respectively. UAV 1 in *blue*, UAV 2 in *green*, and UAV 3 in *red*

troller proposed, when applied to a group of three homogenous UAVs ($n_v = 3$). The vehicles are required to (i) follow their own straight line and (ii) keep a formation pattern that consists of having them aligned along a common horizontal line (i.e., same inertial longitudinal coordinate). We consider the following scenario: UAV 1 is allowed to communicate with UAVs 2 and 3, but the latter two do not communicate between themselves directly. The initial positions of the three vehicles are UAV 1, $(x^{[1]}, y^{[1]}) = (-10, 20)$ and orientation $\psi^{[1]} = 45°$, UAV 2, $(x^{[2]}, y^{[2]}) = (-10, 20)$ and orientation $\psi^{[2]} = -90°$, and UAV 3, $(x^{[3]}, y^{[3]}) = (50, -250)$ and orientation $\psi^{[3]} = -45°$. Similar to the previous computation, the desired velocity is 19 m/s, $(t_{k+1} - t_k) = 1$ s, and $T = 15$ s. A controller gain, see (7.13), of $\kappa = 0.2$ was found to work quite well (i.e., it provides a fast convergence of the coordination error to zero).

Figure 7.5 shows the trajectories of the UAVs when the coordination is disabled. Similar to the previous computation, the MPC path-following allows the three vehicles to approach their own desired straight line path. Once again, the VTV's velocity allows us to improve the convergence of the actual path to the desired one. As instance, this is evident from the trajectory performed by the UAV 3, see Fig. 7.5: at about $x = 190$ m, the VTV is "waiting" the actual vehicle while it is performing a maneuver at the maximum of its capabilities.

Figure 7.6 shows the trajectories of the UAVs when the coordination is taken into account. Figure 7.6b–d show the coordination error. Figure 7.6e–g show the VTV's velocities. More precisely, from Fig. 7.6a it can be seen that synchronization takes places after a transient behavior of the controlled system. Thus, the coordination error $(\gamma^{[i]} - \gamma^{[j]})$ approaches zero, the input control of the coordination controller (see Fig. 7.3) vanishes, and the VTV's velocities as well as the actual vehicles' speed stabilize (after a transient time due to the different initial conditions of the three vehicles) around the desired velocity. It is worth noting that the trajectories performed by the UAVs are slightly different from the ones performed without coordination. This is evident, as instance, from the trajectory of the UAV 3 (red vehicle in Figs.7.5 and 7.6a).

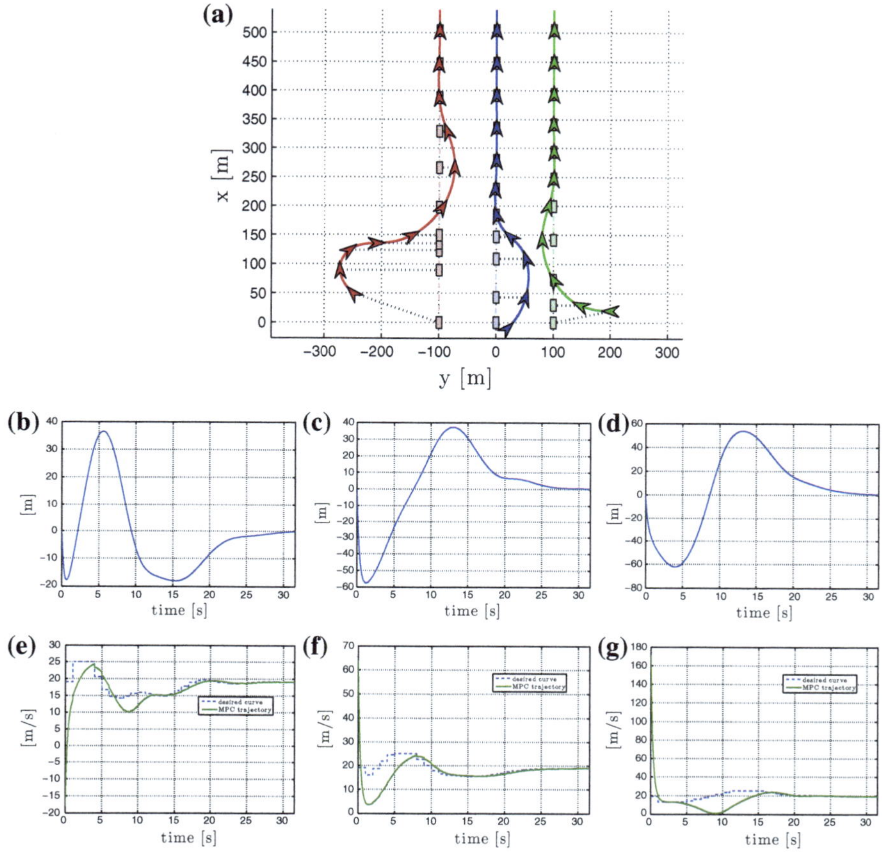

Fig. 7.6 MPC cooperative path-following of 3 UAVs. **a** $x^{[i]}$-$y^{[i]}$. UAV 1 in blue, UAV 2 in green and UAV 3 in red. **b** $\gamma^{[1]} - \gamma^{[2]}$. **c** $\gamma^{[1]} - \gamma^{[3]}$. **d** $\gamma^{[2]} - \gamma^{[3]}$. **e** $u_3^{[1]}$. **f** $u_3^{[2]}$. **g** $u_3^{[3]}$

7.5 Conclusion

In this chapter we proposed a sampled-data MPC architecture to solve the decentralized cooperative path-following problem of multiple UAVs. The MPC controller takes into account the UAVs dynamics and the communication topology constraints. Based on the virtual target approach and nonlinear optimal control techniques, we provided a MPC controller to solve the cooperative path-following problem of multiple UAVs. The MPC framework used guarantees exponential stability with a rate of convergence that can be prescribed. The coordination is obtained by exploiting the speed of the virtual targets and the MPC scheme (i.e., the desired speed is updated at each sampling time). A key property of the proposed architecture is that it explicitly handles state-input constraints and communication topology constraints in a decentralized fashion. We provided numerical computations showing the effectiveness of the proposed architecture.

Future directions of research include the (i) implementation of a vehicle/obstacle collision avoidance technique, as the one proposed in [25], (ii) development of the path-generator system for a general class of desired paths, and (iii) study and analysis of stringent communication constraints and non-constant update time. Finally, the efficiency of the proposed cooperative path-following controller should be demonstrated throughout experimental tests on a real UAV platform.

Acknowledgments All the authors gratefully acknowledge the support of FCT under the projects “Advanced control of a fleet of heterogeneous autonomous vehicles” (PESSOA program). The first author acknowledges the FCT-investigator grant to support his contract with FEUP. The last three authors acknowledge the support of the FCT grant “Incentivo/EEI/UI0147/2014.” The first four authors also acknowledge the support of FCT under the projects PEst-OE/EEI/UI0147/2014, and the R&D unit SYSTEC - UID/EEA/00147/2013.

References

1. A.P. Aguiar, J.P. Hespanha, Trajectory-tracking and path-following of underactuated autonomous vehicles with parametric modeling uncertainty. IEEE Trans. Autom. Control **52**(8), 1362–1379 (2007)
2. A.P. Aguiar, A.M. Pascoal, Coordinated path-following control for nonlinear systems with logic-based communication, in *IEEE Conference on Decision and Control* (2007), pp. 1473–1479
3. A.P. Aguiar, J.P. Hespanha, P.V. Kokotović, Performance limitations in reference tracking and path following for nonlinear systems. Automatica **44**(3), 598–610 (2008)
4. A. Alessandretti, A.P. Aguiar, C.N. Jones, Trajectory-tracking and path-following controllers for constrained underactuated vehicles using model predictive control, in *IEEE European Control Conference* (2013), pp. 1371–1376
5. E.P. Anderson, R.W. Beard, T.W. McLain, Real-time dynamic trajectory smoothing for unmanned air vehicles. IEEE Trans. Control Syst. Technol. **13**(3), 471–477 (2005)
6. F. Bayer, J. Hauser, Trajectory optimization for vehicles in a constrained environment, in *IEEE Conference on Decision and Control* (2012), pp. 5625–5630
7. R. Bencatel, P. Kabamba, A. Girard, Perpetual dynamic soaring in linear wind shear. AIAA J. Guid. Control Dyn. **37**(5), 1712–1716 (2014)
8. A.C.D. Caldeira, F.A.C.C. Fontes, Model predictive control for path-following of nonholonomic systems, in *Proceedings of the 10th Portuguese Conference in Automatic Control* (2010), pp. 374–379
9. H. Chen, F. Allgöwer, A quasi-infinite horizon nonlinear model predictive control scheme with guaranteed stability. Automatica **34**(10), 1205–1217 (1998)
10. W.B. Dunbar, R.M. Murray, Distributed receding horizon control for multi-vehicle formation stabilization. Automatica **42**(4), 549–558 (2006)
11. T. Faulwasser, R. Findeisen, Nonlinear model predictive path-following control, in *Nonlinear Model Predictive Control*. Lecture Notes in Control and Information Sciences, vol. 384 (Springer, Berlin, 2009), pp. 335–343
12. T. Faulwasser, R. Findeisen, Nonlinear model predictive control for constrained output path following. Technical report, EPFL-ARTICLE-200852, (submitted to IEEE) (2015), http://arxiv.org/abs/1502.02468
13. J.A. Fax, R.M. Murray, Graph laplacians and stabilization of vehicle formations, in *IFAC World Congress* (Barcelona, Spain, 2002)

14. J.A. Fax, R.M. Murray, Information flow and cooperative control of vehicle formations. IEEE Trans. Autom. Control **49**(9), 1465–1476 (2004)
15. G. Flores, I. Lugo-Cárdenas, R. Lozano, A nonlinear path-following strategy for a fixed-wing MAV, in *IEEE International Conference on Unmanned Aircraft Systems* (2013), pp. 1014–1021
16. F.A.C.C. Fontes, A general framework to design stabilizing nonlinear model predictive controllers. Syst. Control Lett. **42**(2), 127–143 (2001)
17. F.A.C.C. Fontes, R.B. Vinter, Nonlinear model predictive control: specifying rates of exponential stability without terminal state constraints, in *IFAC Meeting on Advanced Control of Chemical Processes* (2000), pp. 821–826
18. F.A.C.C. Fontes, L. Magni, E. Gyurkovics, Sampled-data model predictive control for nonlinear time-varying systems: stability and robustness, in *Assessment and Future Directions of Nonlinear Model Predictive Control*. Lecture Notes in Control and Information Systems, vol. 358 (Springer, Berlin, 2007), pp. 115–129
19. F.A.C.C. Fontes, D.B.M.M. Fontes, A.C.D. Caldeira, Model predictive control of vehicle formations, in *Optimization and Cooperative Control Strategies*. Lecture Notes in Control and Information Sciences, vol. 381 (Springer, Berlin, 2009), pp. 371–384
20. R. Ghabcheloo, A.P. Aguiar, A.M. Pascoal, C. Silvestre, I. Kaminer, J.P. Hespanha, Coordinated path-following in the presence of communication losses and time delays. SIAM J. Control Optim. **48**(1), 234–265 (2009)
21. J. Hauser, A projection operator approach to the optimization of trajectory functionals, in *IFAC World Congress* (Barcelona, Spain, 2002)
22. J. Hauser, R. Hindman, Aggressive flight maneuvers, in *IEEE Conference on Decision and Control* (1997), pp. 4186–4191
23. J. Hauser, D.G. Meyer, The trajectory manifold of a nonlinear control system, in *IEEE Conference on Decision and Control* (1998), pp. 1034–1039
24. J. Hauser, A. Saccon, A barrier function method for the optimization of trajectory, in *IEEE Conference on Decision and Control* (2006), pp. 864–869
25. A.J. Hausler, A. Saccon, A.M. Pascoal, J. Hauser et al., Cooperative AUV motion planning using terrain information, in *OCEANS-Bergen, 2013 MTS/IEEE* (2013), pp. 1–10
26. A. Jadbabaie, J. Yu, J. Hauser, Unconstrained receding horizon control of nonlinear systems. IEEE Trans. Autom. Control **46**(5), 776–783 (2001)
27. I. Kaminer, E. Xargay, V. Cichella, N. Hovakimyan, A.M. Pascoal, A.P. Aguiar,V. Dobrokhodov, R. Ghabcheloo, Time-critical cooperative path following of multiple UAVs: casestudies, in *Advances in Estimation, Navigation, and Spacecraft Control* (Springer, 2015), pp. 209–233
28. D.R. Nelson, D.B. Barber, T.W. McLain, R.W. Beard, Vector field path following for miniature air vehicles. IEEE Trans. Robot. **23**(3), 519–529 (2007)
29. I. Prodan, S. Olaru, R. Bencatel, J.B. De Sousa, C. Stoica, S.-I. Niculescu, Receding horizon flight control for trajectory tracking of autonomous aerial vehicles. Control Eng. Pract. **21**(10), 1334–1349 (2013)
30. I. Prodan, S. Olaru, F.A.C.C. Fontes, C. Stoica, S-I Niculescu, A predictive control-based algorithm for path following of autonomous aerial vehicles, in *IEEE Conference on Control Applications* (2013), pp. 1042–1047
31. A. Ratnoo, P.B. Sujit, M. Kothari, Adaptive optimal path following for high wind flights, in *IFAC World Congress* (Milano, Italy, 2011)
32. W. Ren, R.W. Beard, Trajectory tracking for unmanned air vehicles with velocity and heading rate constraints. IEEE Trans. Control Syst. Technol. **12**(5), 706–716 (2004)
33. A. Rucco, A.P. Aguiar, J. Hauser, Trajectory optimization for constrained UAVs: a virtual target vehicle approach, in *IEEE International Conference on Unmanned Aircraft Systems* (Denver, CO, USA, 2015)
34. A. Ryan, J.K. Hedrick, A mode-switching path planner for UAV-assisted search and rescue, in *IEEE Conference on Decision and Control* (2005), pp. 1471–1476
35. S. Spedicato, G. Notarstefano, H.H. Bülthoff, A. Franchi, Aggressive maneuver regulation of a quadrotor UAV, in *The 16th International Symposium on Robotics Research* (Singapore, 2013)

36. P.B. Sujit, S. Saripalli, J. Sousa, Unmanned aerial vehicle path following: a survey and analysis of algorithms for fixed-wing unmanned aerial vehicles. IEEE Control Syst. Mag. **34**(1), 42–59 (2014)
37. E. Xargay, I. Kaminer, A. Pascoal, N. Hovakimyan, V. Dobrokhodov, V. Cichella, A.P. Aguiar, R. Ghabcheloo, Time-critical cooperative path following of multiple UAVs over time-varying networks. AIAA J. Guid. Control Dyn. **36**(2), 499–516 (2013)
38. S. Yu, X. Li, H. Chen, F. Allgöwer, Nonlinear model predictive control for path following problems. Int. J. Robust Nonlinear Control **25**(8), 1168–1182 (2015)

Chapter 8
Predictive Control for Path-Following. From Trajectory Generation to the Parametrization of the Discrete Tracking Sequences

Ionela Prodan, Sorin Olaru, Fernando A.C.C. Fontes, Fernando Lobo Pereira, João Borges de Sousa, Cristina Stoica Maniu and Silviu-Iulian Niculescu

Abstract This chapter discusses a series of developments on predictive control for path following via a priori generated trajectory for autonomous aerial vehicles. The strategy partitions itself into offline and runtime procedures with the assumed goal of moving the computationally expensive part into the offline phase and of leaving only tracking decisions to the runtime. First, it will be recalled that differential flatness represents a well-suited tool for generating feasible reference trajectory.

I. Prodan (✉)
Laboratory of Conception and Integration of Systems (LCIS EA 3747),
Université Grenoble Alpes, 26902 Valence, France
e-mail: ionela.prodan@lcis.grenoble-inp.fr

S. Olaru · C. Stoica Maniu
Laboratory of Signals and Systems, CentraleSupélec-CNRS-Université Paris-Sud, Université Paris-Saclay, 3 Rue Joliot Curie, 91192 Gif-sur-Yvette, France
e-mail: sorin.olaru@centralesupelec.fr

C. Stoica Maniu
e-mail: cristina.maniu@centralesupelec.fr

S.-I. Niculescu
Laboratory of Signals and Systems (L2S, UMR CNRS 8506),
CentraleSupélec-CNRS-Université Paris-Sud, Université Paris-Saclay, 3 Rue Joliot Curie,
91192 Gif-sur-Yvette, France
e-mail: silviu.niculescu@lss.supelec.fr

F.A.C.C. Fontes · F. Lobo Pereira
SYSTEC—Faculty of Engineering, Porto University and ISR-Porto,
4200-465 Porto, Portugal
e-mail: faf@fe.up.pt

F. Lobo Pereira
e-mail: flp@fe.up.pt

J. Borges de Sousa
LSTS—Faculty of Engineering, Porto University, 4200-465 Porto, Portugal
e-mail: jtasso@fe.up.pt

S. Olaru et al. (eds.), *Developments in Model-Based Optimization and Control*,
Lecture Notes in Control and Information Sciences 464,
DOI 10.1007/978-3-319-26687-9_8

Next, an optimization-based control problem which minimizes the tracking error for the nonholonomic system is formulated and further enhanced via path following mechanisms. Finally, possible changes of the selection of sampling times along the path and their impact on the predictive control formulation will be discussed in detail.

Keywords Model predictive control (MPC) · Differential flatness · Trajectory tracking · Path following · Autonomous aerial vehicles

8.1 Introduction

There are many situations in control and coordination of dynamical systems for which a trajectory has to be generated a priori in order to provide a reference for a tracking control problem, [1, 3, 30]. These control applications are often difficult to handle in embedded solutions (complex dynamics, difficult real-time constraints, short sampling times, limited computational resources, and the like). For all these reasons, it is essential to push as many of the reference management and control tasks offline. Therefore, enforcing the computationally demanding effort of trajectory generation for an offline stage leaves for the runtime only the relatively straightforward trajectory tracking, [14, 20, 23]. Moreover, having a priori feasible reference trajectory implies that we may offer guarantees of performance for the overall system, [22, 26, 30].

Another challenging problem frequently used in control is path following which allows dynamical (nonlinear) systems to follow a predefined path specified by points, lines, and the like. The clear specification of the difference between the following two close notions is of most importance: trajectory tracking and path following. The latter provides a desired time-independent route for the vehicle, while for the trajectory tracking the reference is represented by a function of time. Both problems have their strengths and weaknesses depending on the general control objectives. For example, we may consider that the former has the advantage of providing simultaneously both feasible input and state variables for the corresponding system. However, a disadvantage would be its time dependence, which often imposes an additional constraint on the real-time functioning. This is to be compared with the reference path which remains time-independent and, as such, provides a certain flexibility for tracking.

There is a wealth of work in the literature on trajectory tracking and path following algorithms. From an optimization-based control viewpoint, a widely used technique for solving tracking problems is model predictive control (MPC) (see, for instance, [15, 25] for an overview of MPC, and [12] for MPC of nonholonomic systems) due to its ability to handle control and state constraints, while offering good performance specifications. For example, [17] uses a predictive guidance controller for an autonomous UAV and a fault detection filter for taking into account the disturbances. Mixed-integer programming (MIP) techniques combined with receding horizon strategy were useful for coordinating the efficient interaction of multiple UAVs in scenarios with many sequential tasks and tight timing constraints (see, [16, 27]). Furthermore, some works investigate the capability of nonlinear MPC for tracking control. Among these contributions, [18] formulates a nonlinear MPC algorithm combined with the gradient descent method for trajectory tracking, and [13] proposes a two-layer control scheme composed by a nonlinear and a linear predictive controller for a group of nonholonomic vehicles moving in formation. Reference [21] proposes as well a gradient-based optimization algorithm for trajectory generation for aircraft avoidance maneuvers where concepts like polar sets and gauge function are used to partition the feasible region in convex partitions. The combination of MPC and path following has been previously addressed in [4, 11], and [35].

The combination of MPC with flatness represents a challenging combination in the current state of the art by allowing real-time control, trajectory generation, and robustness by using set-theoretic methods, [22]. In the present work, we choose to use one of the few generic tools, those based on differential flatness for constructing a reference trajectory. Then, we propose a trade-off between trajectory and path tracking. We pre-compute a feasible trajectory but we use it as a path by considering the velocity along it as the solution of an optimization problem. By allowing this degree of freedom on how fast we move along the path, we actually increase the flexibility and robustness of the problem, while at the same time guaranteeing a feasible path.

The present chapter is motivated mainly by our previous work [22–24], where a flat trajectory was generated and further used as a reference in an output tracking MPC problem. The results were also implemented on real UAVs. This work extends the optimization-based control approach and the path following versus trajectory tracking discussions previously presented. More specifically, the original contributions are the following:

- greater flexibility of trajectory tracking by allowing a variable speed along the path in order to decrease the sensitivity to disturbances and perturbations;
- discretization and the linearization along the reference trajectory are adapted to the variable speed profile by using variable sampling intervals; and
- simulations results over a high-order unmanned aerial vehicle (UAV) model are provided.

The chapter is organized as follows. Section 8.2 introduces the prerequisites needed for trajectory generation: flat trajectory and its parametrization together with proof of concepts examples. Also, the principles of the key background underlying optimization-based control are briefly introduced. While Sect. 8.3 details the control part of the trajectory tracking problem, Sect. 8.4 presents its reconfiguration as a path. Section 8.5 presents illustrative simulation results for an UAV system, and Sect. 8.6 concludes the paper.

Notation

Let $x(k+1)$ denote the value of x at time instant $k+1$, predicted upon the information available at time $k \in \mathbb{N}$. The length of the prediction horizon is denoted by N_p, and the time step within prediction horizon is denoted by s. We write $R \succ 0$ and $R \succeq 0$ to denote that R is a positive definite and semidefinite matrix, respectively. The discretization step is denoted by T_e.

8.2 Prerequisites

This section presents some general details on flat trajectories and the optimization-based control principles.

8.2.1 Flat Trajectory

Consider the nonlinear continuous time-invariant system:

$$\dot{\mathbf{x}}(t) = f(\mathbf{x}(t), \mathbf{u}(t)), \tag{8.1}$$

where $\mathbf{x}(t) \in \mathbb{R}^{n_x}$ is the state vector and $\mathbf{u}(t) \in \mathbb{R}^{n_u}$ is the input vector.

Definition 8.1 The system (8.1) is called differentially flat if there exists a flat output $\mathbf{z}(\tau) \in \mathbb{R}^{n_u}$ such that the states and inputs can be algebraically expressed in terms of $\mathbf{z}(\tau)$ and a finite number of its higher order derivatives:

$$\mathbf{x}(\tau) = \Phi_0(\mathbf{z}(\tau), \dot{\mathbf{z}}(\tau), \ldots, \mathbf{z}^{(q)}(\tau)), \tag{8.2}$$

$$\mathbf{u}(\tau) = \Phi_1(\mathbf{z}(\tau), \dot{\mathbf{z}}(\tau), \ldots, \mathbf{z}^{(q)}(\tau)), \tag{8.3}$$

where $\mathbf{z}(\tau) = \gamma(\mathbf{x}(\tau), \mathbf{u}(\tau), \dot{\mathbf{u}}(\tau), \cdots, \mathbf{u}^{(q)}(\tau))$ and $q \in \mathbb{N}$ represents the maximum order of $\mathbf{z}(\tau)$ arising in the problem (see also Fig. 8.1 for a general view on differential flatness concept). □

Remark 8.1 Note that, in (8.2)–(8.3), $\tau \in \mathbb{R}$ is a scalar parameter which can be assimilated to $t \in \mathbb{R}$ in (8.1), but will be used as a decision variable interpreted as a virtual time when the reference tracking problem is casted into path following problem. □

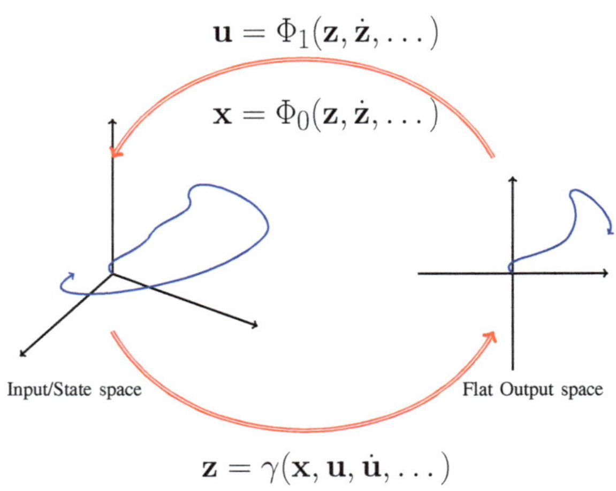

Fig. 8.1 Differentially flat systems

Remark 8.2 For any system admitting a flat description, the number of flat outputs equals the number of inputs [19]. In the case of linear systems [29], the flat differentiability (existence and constructive forms) is implied by the controllability property. □

The most important aspect of flatness is that it reduces the problem of trajectory generation to finding an adequate flat output solving an algebraic system of equalities and inequalities. This means choosing $\mathbf{z}(t)$ such that, via mappings $\Phi_0(\cdot)$, $\Phi_1(\cdot)$, various constraints on state and inputs are verified. The flat output may be itself difficult to compute. The usual solution (also followed here) is to parameterize $\mathbf{z}(t)$ by using a set of smooth basis functions $\Lambda^i(t)$:

$$\mathbf{z}(t) = \sum_{i=1}^{N} c_i \Lambda^i(t), \quad c_i \in \mathbb{R}. \tag{8.4}$$

The number of basis functions directly depends on the number of constraints imposed onto the dynamics [33].

There are multiple choices for the basis functions $\Lambda^i(t)$ in (8.4). The polynomial basis $\lambda^i(t) = t^i$ is a well-known choice but suffers from numerical deficiencies: the number of functions depends on the trajectory constraints and on the degree of the derivatives appearing in the state and input parametrizations, [6, 10, 31]. Another choice is the *Bésier basis functions* [28], they mitigate the numerical difficulties but their degree still depends on the order of derivatives that appear, [9, 34]. *B-spline* basis functions represent an alternative well suited to flatness parametrization due to their ease of enforcing continuity. Moreover, their degree depends only up to which derivative is needed to ensure continuity. This basis will be used in the simulation results presented in this chapter. For details on B-spline functions and their applications, the interested reader is referred to recent research works, [7, 31, 32].

8.2.2 Principles of Optimization-Based Control

The optimization-based control refers to the control design that optimizes a given criterion by using methods that generate optimal control laws whose parameters are such that a certain desired property, such as stability or robustness, is fulfilled. This is a broad definition which actually can cover the classical optimal control, the LMI-based techniques, MPC, or interpolation-based techniques. We provide in this chapter, a trajectory tracking MPC problem for which a control action $\mathbf{u}(k)$ for a given state $\mathbf{x}(k)$ is obtained from the control sequence $\mathrm{u} \triangleq \{\mathbf{u}(k), \mathbf{u}(k+1), \ldots, \mathbf{u}(k+N_p-1)\}$ as the result of the optimization problem:

$$u^* = \arg\min_{\mathrm{u}} \left\{ V_f(\mathbf{x}(k+N_p), \mathrm{r}(k+N_p)) + \sum_{s=1}^{N_p-1} V(\mathbf{x}(k+s), \mathbf{u}(k+s), \mathrm{r}(k+s)) \right\}, \tag{8.5}$$

subject to:

$$\begin{cases} \mathbf{x}(k+s+1) = f(\mathbf{x}(k+s), \mathbf{u}(k+s)), \quad s = 0 : N_p - 1, \\ h(\mathbf{x}(k+s), \mathbf{u}(k+s), \mathrm{r}(k+s)) \leq 0, \quad s = 1 : N_p - 1, \end{cases} \tag{8.6}$$

over a finite horizon N_p. Here, $V_f(\cdot, \cdot) : \mathbb{R}^{n_x} \times \mathbb{R}^{n_x} \rightarrow \mathbb{R}_+$ represents the terminal cost function, $V(\cdot, \cdot, \cdot) : \mathbb{R}^{n_x} \times \mathbb{R}^{n_u} \times \mathbb{R}^{n_x} \rightarrow \mathbb{R}$, the cost at stage s function, $f(\cdot, \cdot)$ describes the evolution of the systems' trajectory, $h(\cdot, \cdot, \cdot)$ the constraints in a general (input state parameters) form, being $\mathrm{r}(\cdot) \in \mathbb{R}^{n_x}$ a vector of time-varying parameters which includes the reference trajectory along the prediction horizon.

Therefore, our main objective being the design of a predictive control strategy, a reference trajectory needs to be available beforehand at least for a finite prediction window. Hereinafter, we use flatness concepts previously presented in order to provide flat states and inputs of the nonlinear finite-time optimization problem (8.5) at the pre-design stage (trajectory generation), which can be updated in real-time in order to allow rescheduling and target moves. Finally, the MPC optimization problem and the a priori generated reference are linked through an optimization block which adapts the speed of tracking, being this actually the main contribution of the paper and thus it will be our main focus in the forthcoming section.

Remark 8.3 Note that, the nonlinear systems used in a wide class of practical applications are differentially flat. For the cases where the specifications of flat outputs are not possible, other strategies for trajectory generation can be employed (see, for example, [1, 5]). □

8.3 Optimization-Based Trajectory Tracking

This section starts by presenting the linearization procedure of the system model which will be further used in the optimization-based control design for trajectory tracking.

8.3.1 Linearization

Hereinafter, we choose to discretize and linearize the dynamics (8.1) along the flat trajectory (for the construction details regarding the discretization method, linearization points along the flat trajectory and the like, the reader can consult our previous work, [22]). For the time discretization via Euler explicit method, we compute a first-order approximation of the state of the system at a time later. The one at the current time is as follows:

$$\mathbf{x}(k+1) = \mathbf{x}(k) + T_e \cdot f(\mathbf{x}(t), \mathbf{u}(t))|_{t=k \cdot T_e}, \tag{8.7}$$

where T_e is the discretization step. Furthermore, for the discretized model of the nonlinear system (8.1) defined as:

$$\mathbf{x}(k+1) = f^d(\mathbf{x}(k), \mathbf{u}(k)), \tag{8.8}$$

we consider a collection of points along the reference trajectory in which we pre-compute linear approximations of (8.8):

$$\mathbb{L} \triangleq \{l^j = (\mathbf{x}^j, \mathbf{u}^j), \quad j = 0 : N_l\}, \tag{8.9}$$

with N_l the number of chosen linearization points. For a given point $l^j \in \mathbb{L}$ we consider the following Taylor decomposition:

$$f^d(\mathbf{x}(k), \mathbf{u}(k)) = f^d(\mathbf{x}^j, \mathbf{u}^j) + \mathbf{A}_j(\mathbf{x}(k) - \mathbf{x}^j) + \mathbf{B}_j(\mathbf{u}(k) - \mathbf{u}^j) + \beta_j(\mathbf{x}(k), \mathbf{u}(k)), \tag{8.10}$$

where the matrices $\mathbf{A}_j \in \mathbb{R}^{n_x \times n_x}$ and $\mathbf{B}_j \in \mathbb{R}^{n_x \times n_u}$ are defined by

$$\mathbf{A}_j = \frac{\partial f^d}{\partial \mathbf{x}}|_{(\mathbf{x}^j, \mathbf{u}^j)}, \qquad \mathbf{B}_j = \frac{\partial f^d}{\partial \mathbf{u}}|_{(\mathbf{x}^j, \mathbf{u}^j)} \tag{8.11}$$

and $\beta_j(\mathbf{x}(k), \mathbf{u}(k)) \in \mathbb{R}^{n_x}$ represents the terms of the Taylor decomposition of rank greater than 1 (i.e., the nonlinear residue of the linearization):

$$\beta_j(\mathbf{x}(k), \mathbf{u}(k)) = f^d(\mathbf{x}(k), \mathbf{u}(k)) - f^d(\mathbf{x}^j, \mathbf{u}^j) - \mathbf{A}_j(\mathbf{x}(k) - \mathbf{x}^j) - \mathbf{B}_j(\mathbf{u}(k) - \mathbf{u}^j), \tag{8.12}$$

for all $j = 0, \ldots, N_l$. Therefore, the system (8.8) is linearized in $l^j \in \mathbb{L}$ as follows:

$$\mathbf{x}(k+1) = f_j^d(\mathbf{x}(k), \mathbf{u}(k)) \triangleq \mathbf{A}_j\mathbf{x}(k) + \mathbf{B}_j\mathbf{u}(k) + d_j, \tag{8.13}$$

where $f_j^d(\mathbf{x}(k), \mathbf{u}(k)) = f^d(\mathbf{x}(k), \mathbf{u}(k)) - \beta_j(\mathbf{x}(k), \mathbf{u}(k))$, and the affine constant terms $d_j \in \mathbb{R}^{n_x}$ are given by

$$d_j = f^d(\mathbf{x}^j, \mathbf{u}^j) - \mathbf{A}_j\mathbf{x}^j - \mathbf{B}_j\mathbf{u}^j, \tag{8.14}$$

for all $j = 0, \ldots, N_l$.

Remark 8.4 Note that, in general, the linearization and the discretization operations can be interchanged and even mixed in order to obtain a discrete-time linear system which best approximates the sampling of the continuum time dynamics. □

Furthermore, for selecting between the predefined linearization points (8.9) for the current input/state values, we partition the state space into a collection of *Voronoi cells*:

$$\mathcal{V}_j = \left\{(\mathbf{x}, \mathbf{u}) : ||(\mathbf{x}, \mathbf{u}) - (\mathbf{x}^j, \mathbf{u}^j)|| \leq ||(\mathbf{x}, \mathbf{u}) - (\mathbf{x}^i, \mathbf{u}^i)||, \ \forall i \neq j\right\}, \tag{8.15}$$

where each cell consists of all points whose linearization error around point $(\mathbf{x}^j, \mathbf{u}^j)$ is lower than the one with respect to any other point $(\mathbf{x}^i, \mathbf{u}^i)$ from $\mathbb{L}$, with $i, j = 0, \ldots, N_l$. This allows a practical criterion for the selection of the linearization point during runtime:

$$\text{if} \quad (\mathbf{x}(k), \mathbf{u}(k)) \in \mathcal{V}_j \quad \text{then} \quad \mathbf{x}(k+1) = f_j^d(\mathbf{x}(k), \mathbf{u}(k)), \quad \forall\, j = 0, \ldots, N_l. \tag{8.16}$$

It is worth mentioning that, for a given error norm, the Voronoi decomposition is unique (by its geometrical properties) and, as such, it offers a generic design tool for any localization of the linearization points. The drawback is that this criterion is purely geometric and does not take into account the dynamical properties of the model. This disadvantage can be mitigated by two practical procedures: increasing the number of linearization points, and computing the maximal linearization error (see [8] for a discussion on the accuracy of the linearization and its impact in the design of stabilizing control laws). Since $\beta_j(\mathbf{x}, \mathbf{u}) = f^d(\mathbf{x}, \mathbf{u}) - f_j^d(\mathbf{x}, \mathbf{u})$, it follows that the linearization error is related to the topology of its corresponding cell, $\mathcal{V}_j$: For all $k = 0, 1, \ldots$, we have

$$||\beta_j(\mathbf{x}(k), \mathbf{u}(k))|| \leq \max_{(\mathbf{x},\mathbf{u}) \in \mathcal{V}_j} ||f^d(\mathbf{x}, \mathbf{u}) - f_j^d(\mathbf{x}, \mathbf{u})||. \tag{8.17}$$

Basically, a Voronoi decomposition with decreasing volume of the cells leads to an increasing quality of the PWA approximation for the function (8.8).

The following remarks are in order.

Remark 8.5 An a priori computation of the linearization (8.11), (8.12), and (8.14) in all feasible combinations of inputs and states is difficult to handle. As such, we prefer to select the linearization points (8.9) along the flat trajectory under the assumption (to be verified along the system evolution) that the real trajectory will stay in the corresponding validity domain (Voronoi cell), and thus the chosen linearization points will remain relevant to the problem at hand. □

Remark 8.6 Here, we have chosen the linearization points equidistantly along the reference trajectory. This choice is acceptable as long as the trajectory tracking error is contained by similar uncertainty bounds over the associated Voronoi cells. Adaptive curve sampling can be employed via a different parametrization scheme in order to select these points. For example, the selection of linearization points can be seen as an optimization problem where the goal is to position the points in such a way as to minimize the linearization errors $\beta_j(\mathbf{x}(k), \mathbf{u}(k))$ in (8.10). Such an approach becomes relevant when the control problem is specified in a high-dimensional space and an automatic implementation of the scheme is required. □

Remark 8.7 Note that, a detailed exposition on the procedure for selecting between the linearization points can be found in [22]. □

Next, an optimization problem is formulated in a predictive control framework. It includes the minimization of the system tracking error since the nominal trajectory is generated by taking into account the state and input constraints, but the real vehicle state may not follow exactly the reference flat trajectory. Hence, the system will be controlled in real time to remain adequately close to the reference trajectory over a finite-time horizon in the presence of constraints by using the available information.

8.3.2 *Real-Time Control*

We consider the recursive construction of an optimal open-loop control sequence $\mathrm{u} = \{\mathbf{u}(k), \mathbf{u}(k+1), \ldots, \mathbf{u}(k+N_p-1)\}$ over a *finite* constrained receding horizon, which leads to a feedback control policy by the effective application of the first control action as system input (see also Fig. 8.2 which illustrates the general trajectory tracking mechanism):

$$u^* = \arg\min_{\mathrm{u}} \sum_{s=0}^{N_p-1} \left(||\mathbf{x}(k+s) - \mathbf{x}^{ref}(k+s)||_{\mathbf{Q}} + ||\mathbf{u}(k+s) - \mathbf{u}^{ref}(k+s)||_{\mathbf{R}} \right), \tag{8.18}$$

subject to the set of constraints:

$$\begin{cases} \mathbf{x}(k+s+1) = \mathbf{A}_j\mathbf{x}(k+s) + \mathbf{B}_j\mathbf{u}(k+s) + r_j, \; s = 0 : N_p - 1, \\ \quad \mathbf{x}(k+s) \in \mathcal{X}, \; s = 1 : N_p - 1, \\ \quad \mathbf{u}(k+s) \in \mathcal{U}, \; s = 1 : N_p - 1, \\ \qquad \mathbf{x}(k) = \mathbf{x}_p(k), \end{cases} \tag{8.19}$$

for some $j \in \{1 : N_l\}$. Here, $\mathbf{x}_p(k)$ denotes the state of the plant measured at instant k, the matrices $\mathbf{Q} = \mathbf{Q}^T \succeq 0$, $\mathbf{R} \succ 0$ are weighting matrices and N_p denotes the length of the prediction horizon.

The solution of the optimization problem (8.18)–(8.19) needs to satisfy the dynamical constraints, expressed by the equality constraints in (8.19). At the same time, other security or performance specifications can be added to the system trajectory. These physical limitations (velocity and bank control inputs) are stated in terms of pointwise hard constraints on both the state variables, and input control action as

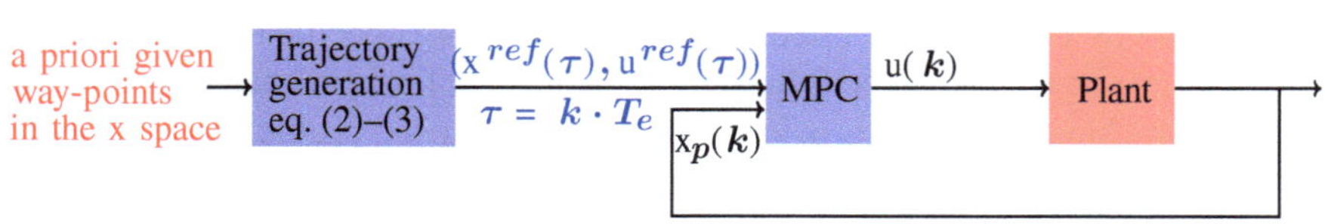

Fig. 8.2 Trajectory tracking mechanism using MPC

detailed by the set inclusion constraints in (8.19). Practically, the closed and compact sets $\mathcal{X}, \mathcal{U}$ denote in a compact formulation the magnitude constraints on states and inputs, respectively. In the following, all these sets are supposed to be polytopic (and, thus, bounded) and to contain the reference control process. This means that $\mathbf{x}^{ref}(k) \in \mathcal{X}$ and $\mathbf{u}^{ref}(k) \in \mathcal{U}$.

Stability constraints can be considered by adding to problem (8.18)–(8.19) a terminal set-terminal cost arguments, the particularity being the time-varying nature of the prediction model.

8.4 Optimization-Based Path Following

This section recalls first the difference between path and trajectory tracking. The former only provides a desired route which may or may not be feasible when taking into account the dynamics of the vehicle or the input and state constraints. At each point in time, the latter provides a pair of reference state and input and thus guarantees the feasibility of the problem. While superior to deal with real-time requirements, the trajectory tracking problem is more challenging. Besides the increased difficulty in generating a trajectory rather than a path there is also the issue of time dependence, which may imply the infeasibility of the real-time optimization problem (8.18)–(8.19).

The synchronization of the absolute time t in (8.1) with the virtual time τ in (8.2)–(8.3), which parametrizes the flat trajectory, forces a constant "velocity" in the sense that the state has to track the current values of the reference. If for some reason (disturbances, unforeseen obstacles), the vehicle "lags", the tracking controller has forced the vehicle to remain in a reachable domain around the current values of the reference trajectory, otherwise the closed-loop control design is compromised. Conversely, if the vehicle is slightly ahead the reference trajectory, it may perform "complex" maneuvers (with an inefficient energy use) when trying to be "in sync" with the reference. This highlights the need for a mechanism which the vehicle can "decelerate" or "accelerate" as desired along the reference trajectory, by adjusting the virtual time flow with respect to the real controller time. Specifically, if for some reason the vehicle remains constantly behind it is not reasonable to follow the trajectory with a constant or an increasing tracking error which can lead to infeasibility in predictive control terms. The alternative will be to reconfigure the trajectory in terms of a path to be followed.

In this section, we propose a formal relaxation of the trajectory tracking scheme. This is done by choosing the optimal point (in the sense specified above) at each sampling time on the path to be followed. Subsequently, further enhancement of the trajectory tracking scheme is provided by varying the speed profile along the path, thus increasing the flexibility and robustness of the problem while guaranteeing the path feasibility.

8.4.1 Selection of the Initial Point of the Reference Trajectory

The basic idea is to find the point and the associated time value τ_c on the predefined trajectory, which is the closest one to the current position of the vehicle $\mathbf{x}(k)$:

$$\tau_c = \arg \min_{\tau \in [\tau_{k-1}, t_f]} ||\mathbf{x}^{ref}(\tau) - \mathbf{x}(k)||_2^2. \tag{8.20}$$

Note that it is assumed implicitly in (8.20) that $\tau_k \geq \tau_{k-1}$ which means that the only possibility for the vehicle is to advance forward along the path (rather than go backward).

In order to better explain the proposed strategy, Fig. 8.3 illustrates a reference trajectory (depicted in blue), which passes through some a priori given waypoints (denoted by the green dots), and the optimal time obtained by solving (8.20), for which the current position is the closest one from the reference trajectory. If the real-time position and the one in virtual time are synchronized, then the vehicle is forced to track a point which is behind, whereas the optimal choice is to reorient the vehicle to track the reference using the optimal time τ_c obtained as in (8.20). Note that the same figure illustrates where to look for the optimal time by using solid blue and dashed blue to denote the past and the future, respectively, along the trajectory.

Once the optimal time along the reference is found, an optimization-based control problem similar to (8.18) is implemented (see also a similar mechanism illustrated in Fig. 8.2):

$$u^* = \arg \min_{\mathrm{u}} \sum_{s=0}^{N_p-1} \big(||\mathbf{x}(k+s) - \mathbf{x}^{ref}(\tau_c(k) + s \cdot T_e)||_{\mathbf{Q}} + ||\mathbf{u}(k+s) - \mathbf{u}^{ref}(\tau_c(k) + s \cdot T_e)||_{\mathbf{R}}\big), \tag{8.21}$$

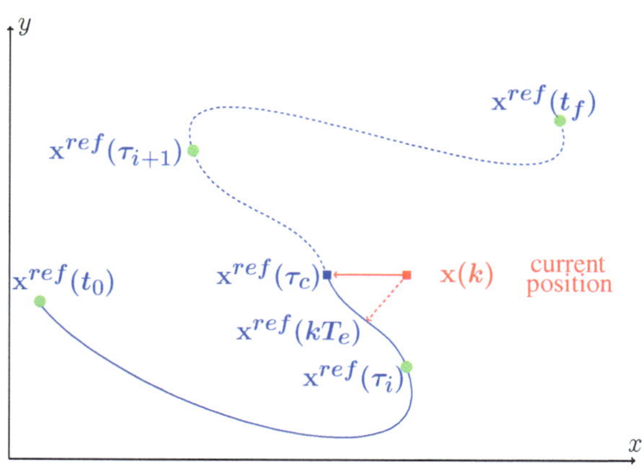

Fig. 8.3 Optimal time for which the current position is the closest from the reference trajectory

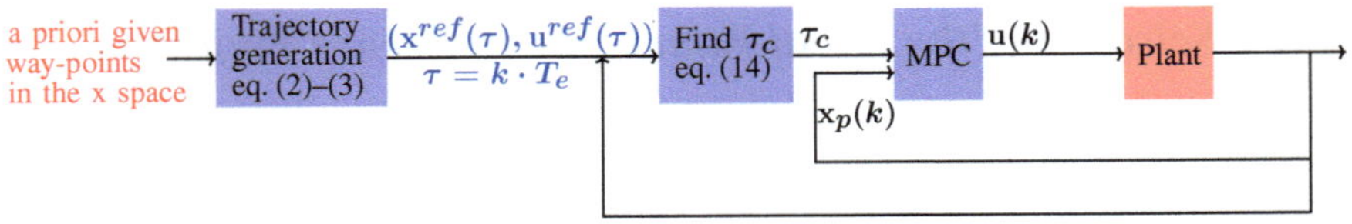

Fig. 8.4 Conversion to path following mechanism using MPC

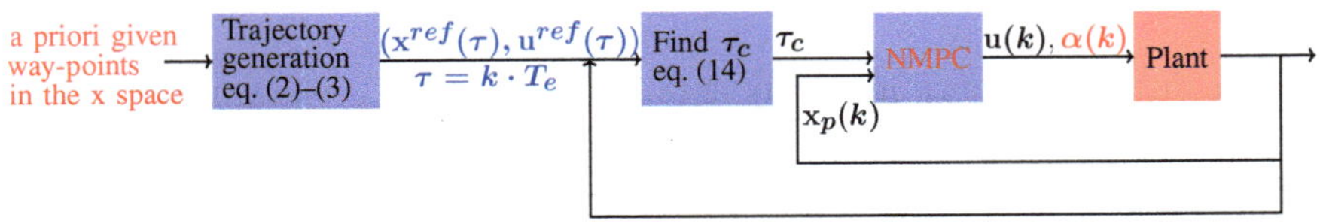

Fig. 8.5 Conversion to path following mechanism using NMPC

subject to the set of constraints (8.19), with T_e the sampling time as in (8.7) and τ_c the optimal time obtained as the result of the optimization (8.20) (Fig. 8.4).

Note that the optimization problem (8.21) retains its linear structure, while (8.20) is the simplest nonlinear optimization problem that can be formulated since it involves only one decision variable τ subject to the constraints in (8.2)–(8.3).

8.4.2 Selection of the Speed Profile Along the Path

In the following, we go further in the design of the trajectory tracking scheme by introducing a scalar term α to adjust the speed along the trajectory. This is equivalent to a decorrelation of the flow along the virtual time by linear acceleration ($\alpha > 1$) or deceleration ($\alpha < 1$). Therefore, the optimization problem (8.21) is reformulated as follows

$$(u^*, \alpha^*) = \arg\min_{\mathrm{u},\alpha} \sum_{s=0}^{N_p-1} \left(||\mathbf{x}(k+s) - \mathbf{x}^{ref}(\tau_{T_e}(s;k,\alpha(k)))||_{\mathbf{Q}} + ||\mathbf{u}(k+s) \right.$$
$$\left. - \mathbf{u}^{ref}(\tau_{T_e}(s;k,\alpha(k)))||_{\mathbf{R}} \right), \tag{8.22}$$

subject to the set of constraints (8.19), where $\tau_{T_e}(s;k,\alpha(k)) = \tau_c(k) + s\alpha(k)T_e$, with T_e being the sampling time as in (8.7), τ_c the optimal time obtained as in (8.20), and α the time-varying speed profiling factor. Now, the structure of the optimization problem (8.21) becomes nonlinear, and, thus, more complex, but still simple in the sense that, it involves only one additional dimension, the decision variable α, which permits to adjust the way the path is followed. Figure 8.5 illustrates the path following mechanism using a nonlinear model predictive control (NMPC) strategy.

8.4.3 Linearization with Varying Speed Profile

In Sect. 8.3.1, we introduced a general linearization procedure for the system. Following the previous sections, the goal is to compare the system trajectory with the reference trajectory in a discrete set of time instances when the reference trajectory is followed at varying speed. To do that, we consider here non-equidistant consecutive time instants $\{t_k\}_{k\geq 0}$ such that:

$$t_{k+1} = t_k + \alpha(k)T_e, \quad k \in \mathbb{N}, \quad \alpha(k) \in (0, 1], \tag{8.23}$$

where T_e is the nominal sampling period.

As with (8.7), we consider Euler explicit method for the discretization:

$$\mathbf{x}(t_{k+1}) = \mathbf{x}(t_k) + \alpha(k)T_e f(\mathbf{x}(t_k), \mathbf{u}(t_k)), \quad k \in \mathbb{N}. \tag{8.24}$$

Next, for the linearization procedure, the collection of linearization points along the reference trajectory is redefined by:

$$\mathbb{L} \triangleq \{l^j = (\mathbf{x}^j, \mathbf{u}^j, \alpha^j), \quad j = 1, 2, \ldots, N_l\}, \tag{8.25}$$

with the additional component α^j, whose nominal value is chosen to be $\alpha^j = 1$, corresponding to the reference trajectory at the nominal speed. Furthermore, the linearization around a point $l^j \in \mathbb{L}$ yields the following dynamics:

$$f(\mathbf{x}(t_k), \mathbf{u}(t_k)) \simeq f(\mathbf{x}^j, \mathbf{u}^j) + f_x(\mathbf{x}^j, \mathbf{u}^j)(\mathbf{x}(t_k) - \mathbf{x}^j) + f_u(\mathbf{x}^j, \mathbf{u}^j)(\mathbf{u}(t_k) - \mathbf{u}^j) \tag{8.26}$$

and

$$\begin{aligned} \mathbf{x}(t_{k+1}) \simeq\ & \mathbf{x}(t_k) + \alpha^j T_e f_x(\mathbf{x}^j, \mathbf{u}^j)\mathbf{x}(t_k) + \alpha^j T_e f_u(\mathbf{x}^j, \mathbf{u}^j)\mathbf{u}(t_k) \\ & + \alpha(k)T_e(f(\mathbf{x}^j, \mathbf{u}^j) - f_x(\mathbf{x}^j, \mathbf{u}^j)\mathbf{x}^j - f_u(\mathbf{x}^j, \mathbf{u}^j)\mathbf{u}^j), \end{aligned} \tag{8.27}$$

which can be rewritten as:

$$\begin{aligned} \mathbf{x}(t_{k+1}) \simeq\ & (I + \alpha^j T_e f_x(\mathbf{x}^j, \mathbf{u}^j))\mathbf{x}(t_k) + \alpha^j T_e f_u(\mathbf{x}^j, \mathbf{u}^j)\mathbf{u}(t_k) \\ & + \alpha(k)T_e(f(\mathbf{x}^j, \mathbf{u}^j) - f_x(\mathbf{x}^j, \mathbf{u}^j)\mathbf{x}^j - f_u(\mathbf{x}^j, \mathbf{u}^j)\mathbf{u}^j). \end{aligned} \tag{8.28}$$

Note that (8.27) is a bilinear model and a further approximation ($\alpha(k) = \alpha^j$ in the second and third terms) leads to the linear model (8.28).

As with Sect. 8.3.1, we denote $\mathbf{x}(k+1) = f^d(\mathbf{x}(k), \mathbf{u}(k), \alpha(k))$, thus, $\mathbf{x}(t_k)$ and $\mathbf{u}(t_k)$ become $\mathbf{x}(k)$ and $\mathbf{u}(k)$, respectively. The discrete-time linearized model can be written as:

$$\mathbf{x}(k+1) = \mathbf{A}_j\mathbf{x}(k) + \mathbf{B}_j^1\mathbf{u}(k) + \mathbf{B}_j^2\alpha(k), \tag{8.29}$$

where

$$\mathbf{A}_j = I + \alpha^j T_e f_x(\mathbf{x}^j, \mathbf{u}^j), \tag{8.30}$$

$$\mathbf{B}_j^1 = \alpha^j T_e f_u(\mathbf{x}^j, \mathbf{u}^j), \tag{8.31}$$

$$\mathbf{B}_j^2 = T_e(f(\mathbf{x}^j, \mathbf{u}^j) - f_x(\mathbf{x}^j, \mathbf{u}^j)\mathbf{x}^j - f_u(\mathbf{x}^j, \mathbf{u}^j)\mathbf{u}^j). \tag{8.32}$$

8.4.4 Real-Time Control

The optimization problem (8.22) is now reformulated as follows:

$$(u^*, \alpha^*) = \arg\min_{\mathbf{u},\alpha} \sum_{s=0}^{N_p-1} \Big(||\mathbf{x}(k+s) - \mathbf{x}^{ref}(\tau(k+s))||_{\mathbf{Q}} + ||\mathbf{u}(k+s) - \mathbf{u}^{ref}(\tau(k+s))||_{\mathbf{R}} + ||\alpha(k+s) - 1|| \Big), \tag{8.33}$$

subject to the set of constraints:

$$\begin{cases} \mathbf{x}(k+s+1) = \mathbf{A}_j \mathbf{x}(k+s) + \mathbf{B}_j^1 \mathbf{u}(k+s) + \mathbf{B}_j^2 \alpha(k+s), \; s = 0 : N_p - 1, \\ \tau(k+s+1) = \tau(k+s) + \alpha(k) T_e, \; s = 0 : N_p - 1, \\ \mathbf{x}(k+s) \in \mathcal{X}, \; s = 1 : N_p - 1, \\ \mathbf{u}(k+s) \in \mathcal{U}, \; s = 1 : N_p - 1, \\ \mathbf{x}(k) = \mathbf{x}_p(k), \\ \tau(k) = \tau_c(k), \end{cases} \tag{8.34}$$

for the appropriate index $j \in \{1, \dots, N_l\}$ corresponding to the active Voronoi cell and τ_c the optimal time obtained as in (8.20) and α the speed profile factor.

Remark 8.8 Note that adapting the discretization step via a time-varying coefficient α and a constant prediction horizon N_p leads to a time-varying prediction horizon on the absolute continuous timescale. Indeed, in continuous time the prediction time will be $N_p T_e \sum_{s=0}^{N_p-1} \alpha(s)$. In order to mitigate this phenomenon, the MPC prediction horizon, N_p can be adapted, by choosing

$$N_p(k) = \min\{n \in \mathbb{N} : nT_e \sum_{s=0}^{N_p-1} \alpha(s) \geq N_p\}. \qquad \square$$

Finally, let us wrap-up in Algorithm 1 the mechanism implemented based on the theoretical elements previously presented.

Algorithm 8.1 Path following optimization-based control problem

Input: Specify a collection of waypoints
-construct the flat trajectory as in (8.2)–(8.3), passing through the waypoints;
-choose a collection of linearization points $\mathbb{L}$ (8.19);
-construct the PWA function as in (8.23);
for $k = 1 : k_{max}$ **do**
  -measure the current state of the plant $\mathbf{x}_p(k)$;
  -find the optimal time τ_c by solving (8.20);
  -select the linearization point $l^j \in \mathbb{L}$ and consequently the pair $(\mathbf{A}_j, \mathbf{B}_j^1, \mathbf{B}_j^2)$ as in Remark 8.7;
  -find the optimal control action u^* and the speed profile factor α by solving (8.27);
  -apply to the plant the first value of the control sequence $u^*(k)$ during the time $\alpha^*(k)T_e$;
end

8.5 Simulation Example for an UAV System

In this section, we start with the case of a 2D 3-DOF model (8.35) of an Unmanned Aerial Vehicle (UAV) in which the autopilot forces coordinated turns (zero side-slip) at a fixed altitude:

$$\begin{aligned} \dot{x}(t) &= v_a(t)\cos\Psi(t) + d_x, \\ \dot{y}(t) &= v_a(t)\sin\Psi(t) + d_y, \\ \dot{\Psi}(t) &= \tfrac{g\ \tan\Phi(t)}{v_a(t)}. \end{aligned} \tag{8.35}$$

The state variables are represented by the position $(x(t), y(t))$ and the heading (yaw) angle $\Psi(t) \in [0, 2\pi]$ rad, which we denote as $\mathbf{x}(t) = [x^T(t)y^T(t)\Psi^T(t)]^T$. The input signals are the airspeed $v_a(t)$ and the bank (roll) angle $\Phi(t)$, respectively, denoted as $\mathbf{u}(t) = [v_a^T(t)\Phi^T(t)]^T$. Also, the airspeed and the bank angle are regarded as the autopilot pseudocontrols. Furthermore, we assume a small angle of attack and that the autopilot provides a higher bandwidth regulator for the bank angle, making its dynamics negligible when compared to the heading dynamics. Also, in (8.35), d_x and d_y represent the wind velocity components on the x and y axes. The dynamical model of the vehicle corresponds to a nonholonomic system, which is completely controllable (under the natural assumption that the velocity is different from zero), but it cannot make instantaneous turns in certain directions.

We take as flat output the position components of the state, $\mathbf{z}(\tau) = [z_1(\tau)\, z_2(\tau)]^T = [x(t)y(t)]^T$, which permits to compute the remaining variables:

$$\mathbf{x}^{ref}(\tau) = \left[z_1(\tau)\, z_2(\tau) \arctan\left(\frac{\dot{z}_2(\tau)}{\dot{z}_1(\tau)}\right)\right]^T, \tag{8.36}$$

$$\mathbf{u}^{ref}(\tau) = \left[\sqrt{\dot{z}_1^2(\tau) + \dot{z}_2^2(\tau)}\ \arctan\left(\frac{1}{g}\frac{\ddot{z}_2(\tau)\dot{z}_1(\tau) - \dot{z}_2(\tau)\ddot{z}_1(\tau)}{\sqrt{\dot{z}_1^2(\tau) + \dot{z}_2^2(\tau)}}\right)\right]^T, \tag{8.37}$$

where $\tau \in [t_0, t_f]$ is a scalar parameter which can be assimilated to t in (8.35), but will be used as a decision variable interpreted as a virtual time when the reference tracking problem is recast in a path following problem.

Note that, in the heading component of the state variable appears first-order derivatives of the flat outputs, while in the roll angle input appears the second-order derivatives of the flat outputs. Hence, for obtaining a smooth state and input (i.e., their derivatives are continuous) it follows that the B-spline parametrization has to have at least degree 4. Further, the linearized model was used for the control part of the trajectory tracking problem.

Next, for testing the proposed trajectory tracking method and its reformulation into a path following problem we use an extended model of (8.35) (for low-level control of an UAV [2]) with 12 states in a 6-DOF simulation. More precisely, the 12-state model includes the positions (x [m], y [m], z [m]), the velocities (v_x [m/s], v_y [m/s], v_z [m/s]), the roll, pitch and yaw angles (ϕ [rad], θ [rad], ψ [rad]), and the angular rates (p [rad/s], q [rad/s], and r [rad/s]), all measured along body frame coordinates, X, Y, and Z. The simulations also incorporate perturbations like the wind with an intensity bounded by some reasonable values (e.g., a maximum speed of 8 m/s).

The objective here is to force the UAV to track 6 given waypoints (denoted as red dots in the forthcoming simulations). The following data and tuning parameters were used for the simulations:

- the list of waypoints: $\{(500, 200, 150), (450, -250, 150), (0, -350, 150), (-350, 0, 150), (-350, 300, 150), (-200, 500, 150), (50, 450, 150), (400, 0, 150)\}$.
- the sampling time is $T_e = 0.01$ s;
- constraints on the input: the velocity $v_a \in [18,\ 25]$ m/s, the bank angle $\Phi \in [-0.43, 0.43]$ rad and the wind components d_x, d_y with $\|\left[d_x\ d_y\right]\|_2 \leq 8$ m/s;
- small variations on the velocity and bank command are admitted: the rate of change of v_a is limited to the maximum acceleration the aircraft can produce, i.e., $0.1 \sim 0.2\,\text{m/s}^2$; the variation of Φ is limited to 0.04 rad/s;
- the weights matrices are: $\mathbf{Q} = [10e1\,0\,0;\,0\,10e1\,0;\,0\,0\,0.1]$, $\mathbf{R} = 10e4 \cdot [1\,0\,0;\,0\,1]$;
- the prediction horizon is $N_p = 7$.

First, we add the current position of the UAV to the list of waypoints. Next, by using the theoretical results presented in Sect. 8.2.1 and (8.36)–(8.37), we generate a flat trajectory starting from the current position of the UAV and passing through the given waypoints. Figure 8.6 illustrates the a priori generated flat trajectory and the heading angle, whereas Fig. 8.7 depicts the control input signals and their derivatives. The linearized model is used for the control part of the trajectory tracking problem with the above-mentioned tuning parameters.

Second, by considering the trajectory tracking mechanism detailed in Fig. 8.2 and solving (8.18)–(8.19) with no wind conditions we obtain good tracking performance as illustrated in 3D in Fig. 8.8a. In green dashed line, we represent the UAV actual motion, and, in magenta dashed line, the path projection on the ground.

However, sometimes it may be the case that, under different wind conditions, the UAV may track the trajectory with an increasing tracking error as proved by the

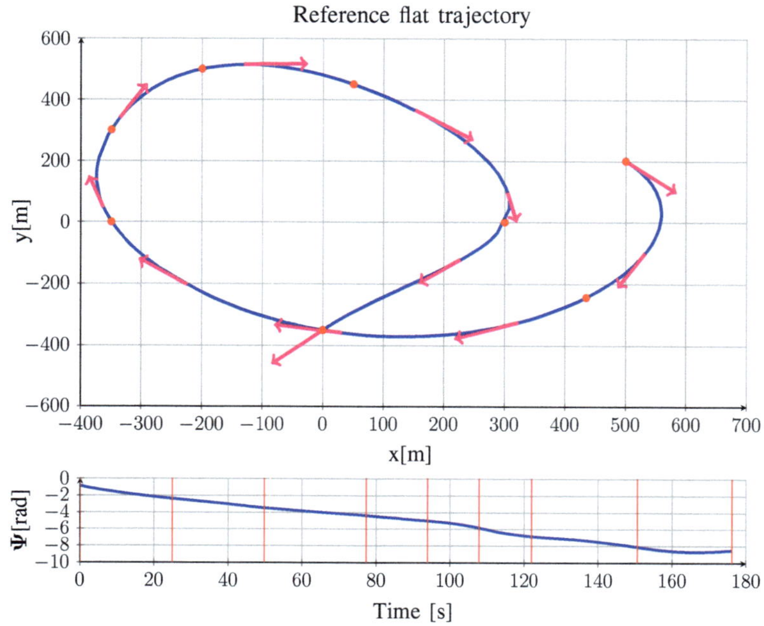

Fig. 8.6 Reference trajectory passing through the waypoints and the corresponding heading angle

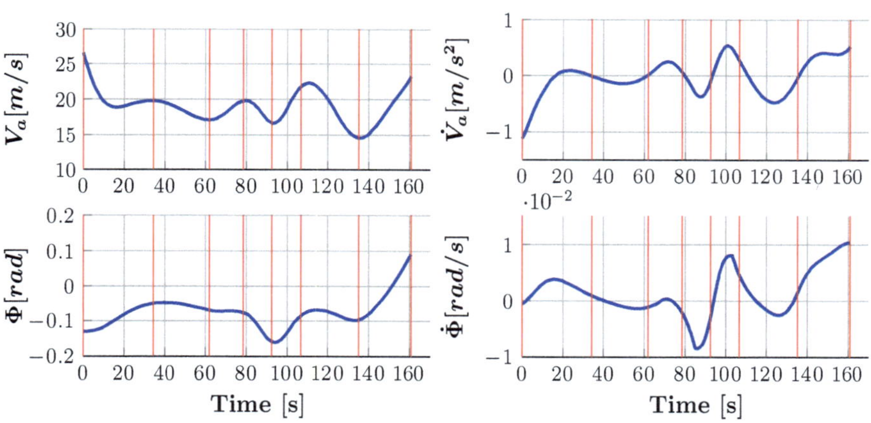

Fig. 8.7 Velocity and roll angle control signals and their derivatives

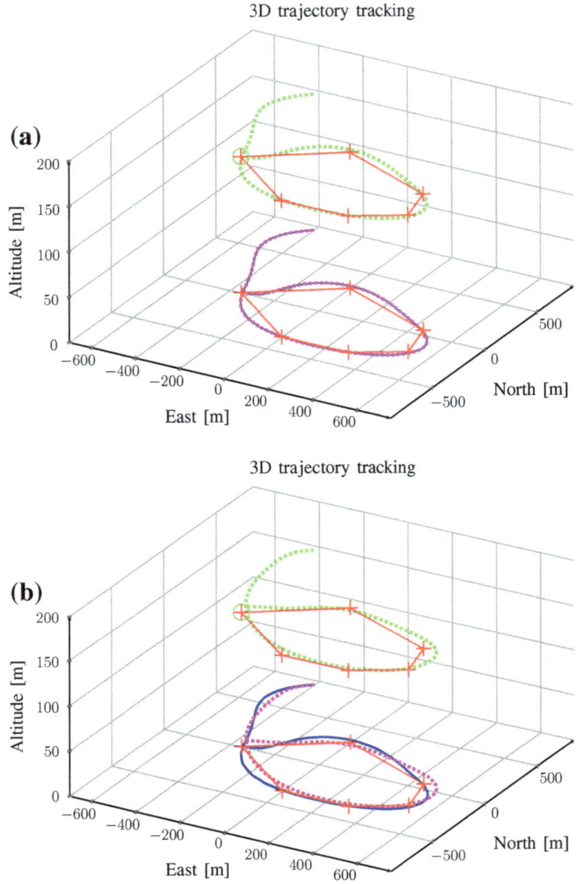

Fig. 8.8 Trajectory tracking: **a** UAV actual motion with no wind conditions. **b** UAV actual motion with wind conditions

simulation scenario with a wind velocity of 5[m/s] from East depicted in Fig. 8.8b in blue continuous line.

To deal with these type of situations, the trajectory tracking problem is reformulated via a path following problem. Equation (8.20) is useful in various scenarios. For instance, we can consider the initial time instant when the real trajectory is far away from the reference. Also, we can consider it during the runtime whenever we need to reinitialize the reference time τ: whenever the real trajectory steers too far away from the reference, we have to recalculate the best time τ_c as addressed in (8.20). The scalar α permits to shrink or expand the sampling period and, in this way, to adjust the speed of the virtual reference vehicle. Note that the use of a speed profile as in (8.22) shows no discernible difference with respect to the simpler method (8.21). Starting from a different initial position $(-30, -800, 150)$, Fig. 8.9 illustrates the performance of the proposed path following mechanism.

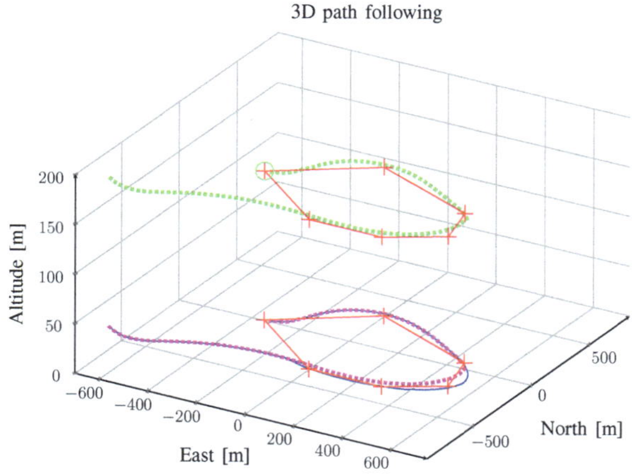

Fig. 8.9 Path following: UAV actual motion with wind conditions

8.6 Conclusions

This chapter presents a predictive control strategy for path following of autonomous aerial vehicles. The strategy is decomposed in a two-stage procedure, where a reference trajectory was pre-specified using differential flatness formalism and than an optimization-based control problem is formulated for minimizing the tracking error for the vehicle. The discussions provided highlight the advantages of reconfiguring the generated feasible trajectory in terms of a path along with the structural properties of the resulting optimization-based control problem. By allowing a variable speed along the path the control problem sensitivity to disturbances and perturbations is decreased. Moreover, the discretization and the linearization along the reference trajectory are adapted to the variable speed profile by using time-varying sampling intervals. Some simulation examples for the control of autonomous aerial vehicles are presented in order to illustrate the proposed approach.

Acknowledgments All the authors acknowledge the support of FCT/ANR PESSOA Project "Advanced control of a fleet of heterogeneous autonomous vehicles." F.A.C.C. Fontes, F. Lobo Pereira, and J. Borges de Sousa acknowledge the support of LSTS - Laboratory of Underwater Systems and Technologies of FEUP. F.A.C.C. Fontes and F. Lobo Pereira acknowledge the support of FCT underlying the funding of the R& D Unit SYSTEC-Research Center for Systems and Technologies.

References

1. A.P. Aguiar, J.P. Hespanha, Trajectory-tracking and path-following of underactuated autonomous vehicles with parametric modeling uncertainty. IEEE Trans. Autom. Control **52**(8), 1362–1379 (2007)
2. R. Bencatel, M. Faied, J. Sousa, A.R. Girard, Formation Control with Collision Avoidance (2011), pp. 591–596

3. M. Burger, K.Y. Pettersen, Smooth transitions between trajectory tracking and path following for single vehicles and formations, in *Estimation and Control of Networked Systems*, (2010), pp. 115–120
4. A.C.D. Caldeira, F.A.C.C. Fontes. Model predictive control for path-following of nonholonomic systems in *Proceedings of the 10th Portuguese Conference in Automatic Control* (Coimbra, Portugal, 2010), pp. 374–379
5. C.L. Castillo, W. Moreno, K.P. Valavanis, Unmanned helicopter waypoint trajectory tracking using model predictive control, in *Proceedings of the IEEE Mediterranean Conference on Control and Automation* (2007), pp. 1–8
6. M. Daniel, J.C. Daubisse, The numerical problem of using bé zier curves and surfaces in the power basis. Comput. Aided Geom. Des. **6**(2), 121–128 (1989)
7. J.A. De Doná, F. Suryawan, M.M. Seron, J. Lévine, A flatness-based iterative method for reference trajectory generation in constrained NMPC, in *Nonlinear Model Predictive Control*, vol. 384, LNCIS, ed. by L. Magni, D.M. Raimondo, F. Allgöwer (Springer, Heidelberg, 2009), pp. 325–333
8. L. Fagiano, M. Canale, M. Milanese, Set membership approximation of discontinuous NMPC laws, pp. 8636–8641
9. G.E. Farin, Curves and Surfaces for Computer-Aided Geometric Design: A Practical Guide. Morgan Kaufmann, 5th edn (2001)
10. R.T. Farouki, V.T. Rajan, On the numerical condition of polynomials in bernstein form. Comput. Aided Geomet. Des. **4**(3), 191–216 (1987)
11. T. Faulwasser, B. Kern, R. Findeisen, Model Predictive Path-following for Constrained Nonlinear Systems, pp. 8642–8647
12. F.A.C.C. Fontes, A general framework to design stabilizing nonlinear model predictive controllers. Syst. Control Lett. **42**(2), 127–143 (2001)
13. F.A.C.C. Fontes, D. Fontes, A. Caldeira, *Model predictive control of vehicle formations* (Optimization and Cooperative Control, Strategies, 2009), pp. 371–384
14. A. Grancharova, E.I. Grøtli, D. Ho, T.A. Johansen, Uavs trajectory planning by distributed MPC under radio communication path loss constraints. J. Intell. Robotic Syst. 1–20 (2014)
15. L. Grüne, J. Pannek, Nonlinear Model Predictive Control (Springer, 2011)
16. J. How, E. King, Y. Kuwata, Flight demonstrations of cooperative control for UAV teams, in *AIAA 3rd Unmanned Unlimited Technical Conference, Workshop and Exhibit* (2004), pp. 20–23
17. T. Keviczky, G.J. Balas, Software-enabled receding horizon control for autonomous unmanned aerial vehicle guidance. J. Guid. Control Dyn. **29**(3), 680–694 (2006)
18. H.J. Kim, D.H. Shim, S. Sastry, Nonlinear model predictive tracking control for rotorcraft-based unmanned aerial vehicles, in *Proceedings of the 2002 American Control Conference*, vol. 5 (IEEE, 2002) pp. 3576–3581
19. J. Lévine, Analysis and control of nonlinear systems: A flatness-based approach (Springer, Hidelberg, 2009)
20. D. Mellinger, N. Michael, V. Kumar, Trajectory generation and control for precise aggressive maneuvers with quadrotors. Int. J. Robotics Res. 027–36 (2012)
21. R.B. Patel, P.J. Goulart, Trajectory generation for aircraft avoidance maneuvers using online optimization. J. Guid. Control Dyn. **34**(1), 218–230 (2011)
22. I. Prodan, R. Bencatel, S. Olaru, J. Sousa, C. Stoica, S.I. Niculescu, Predictive control for autonomous aerial vehicles trajectory tracking, in *Proceedings of the IFAC Nonlinear Model Predictive Control Conference* (Noordwijkerhout, The Netherlands, 23–27 August 2012), pp. 508–513
23. I. Prodan, S. Olaru, R. Bencatel, J. Sousa, C. Stoica, S.I. Niculescu, Receding horizon flight control for trajectory tracking of autonomous aerial vehicles. Control Eng. Pract. **21**(10), 1334–1349 (2013)
24. I. Prodan, S. Olaru, F.A.C.C. Fontes, C. Stoica, S.-I. Niculescu, A predictive control-based algorithm for path following of autonomous aerial vehicles, in *2013 IEEE International Conference on Control Applications (CCA)* (IEEE, 2013), pp. 1042–1047

25. J.B. Rawlings, D.Q. Mayne, Model Predictive Control: Theory and Design (2009)
26. A.P. Schoellig, F.L. Mueller, R. D'Andrea, Optimization-based iterative learning for precise quadrocopter trajectory tracking. Auton. Robots **33**(1–2), 103–127 (2012)
27. T. Schouwenaars, M. Valenti, E. Feron, J. How, Implementation and flight test results of milp-based uav guidance, in *Aerospace Conference* (IEEE, 2005), pp. 1–13
28. L.L. Schumaker, Spline Functions: Basic Theory (Cambridge University Press, 2007)
29. H. Sira-Ramírez, S. Agrawal, *Differential Flatness* (Marcel Dekker, New York, 2004)
30. F. Stoican, I. Prodan, D. Popescu, Flat trajectory generation for way-points relaxations and obstacle avoidance, in *The 23st IEEE Mediterranean Conference on Control and Automation* (2015), pp. 729–734
31. F. Suryawan, Constrained trajectory generation and fault tolerant control based on differential flatness and B-splines. Ph.D thesis, School of Electrical Engineering and Computer Science (The University of Newcastle, Australia, 2012)
32. F. Suryawan, J. De Dona, M. Seron, Methods for trajectory generation in a magnetic-levitation system under constraints, in *18th Mediterranean Conference on Control and Automation* (2010), pp. 945–950
33. J.H. Wilkinson, *The Algebraic Eigenvalue Problem*, vol. 155 (Oxford University Press, 1965)
34. F. Yamaguchi, F. Yamaguchi, *Curves and Surfaces in Computer Aided Geometric Design* (Springer, Berlin, 1988)
35. S. Yu, X. Li, H. Chen, F. Allgower, Nonlinear model predictive control for path following problems. In In Proceedings of the IFAC Nonlinear Model Predictive Control Conference (Noordwijkerhout, The Netherlands, 23–27 August 2012), pp. 508–513

Chapter 9
Formation Reconfiguration Using Model Predictive Control Techniques for Multi-agent Dynamical Systems

Minh Tri Nguyen, Cristina Stoica Maniu, Sorin Olaru and Alexandra Grancharova

Abstract The classical objective for multiple agents evolving in the same environment is the preservation of a predefined formation because it reinforces the safety of the global system and further lightens the supervision task. One of the major issues for this objective is the *task assignment problem*, which can be formulated in terms of an optimization problem by employing set-theoretic methods. In real time the agents will be steered into the defined formation via task (re)allocation and classical feedback mechanisms. The task assignment calculation is often performed in an offline design stage, without considering the possible variation of the number of agents in the global system. These changes (i.e., including/excluding an agent from a formation) can be regarded as a typical fault, due to some serious damages on the components or due to the operator decision. In this context, the present chapter proposes a new algorithm for the dynamical task assignment formulation of multi-agent systems in view of real-time optimization by including *fault detection and isolation* capabilities. This algorithm allows to detect whether there is a fault in the global multi-agent system, to isolate the faulty agent and to integrate a recovered/healthy agent. The proposed methods will be illustrated by means of a numerical example with connections to multi-vehicle systems.

M.T. Nguyen (✉) · C. Stoica Maniu · S. Olaru
Laboratory of Signals and Systems (L2S, UMR CNRS 8506), CentraleSupélec-CNRS-Université Paris-Sud, Université Paris-Saclay, 3 rue Joliot Curie, 91192 Paris, Gif-sur-Yvette cedex, France
e-mail: minhtri.nguyen@supelec.fr

C. Stoica Maniu
e-mail: cristina.stoica@supelec.fr

S. Olaru
e-mail: sorin.olaru@supelec.fr

A. Grancharova
Department of Industrial Automation, University of Chemical Technology and Metallurgy, Bul. Kl. Ohridksi 8, 1756 Sofia, Bulgaria
e-mail: alexandra.grancharova@abv.bg

S. Olaru et al. (eds.), *Developments in Model-Based Optimization and Control*, Lecture Notes in Control and Information Sciences 464,
DOI 10.1007/978-3-319-26687-9_9

Keywords Multi-agent dynamical systems · Fault detection and isolation · Set-membership theory · Tight formation control

9.1 Introduction

A multi-agent system (MAS) is a group of multiple intelligent (decision-making) agents interacting within an environment subject to constraints. A dynamical multi-agent system is first a MAS but each agent in this group is a relatively independent subsystem characterized by a *dynamical equation*. MAS actually receives considerable attention due to flourishing of various applications which require studying the best policy to control the entire MAS interpreted as a formation. In fact, the control of a dynamical MAS is translated in terms of supervising the interaction between the agents and further defining the best strategy of control suitable to a common objective.

Besides the mission planning performance, the mission safety becomes an intensive research field for MAS applications. In this context, a supplementary fault diagnosis layer is required to detect and isolate the possible faults appearing during the mission. Thus, recently designing a fault detection and isolation (FDI) strategy becomes a highly required priority for MAS. Many studies in the literature have been conducted on this topic and various results were obtained. Precisely in [1–3] the authors have developed a set of FDI filters to detect the actuator faults in the presence of large environmental disturbance and then the faulty functioning of MAS is recovered by applying the Markov chain theory. Other works, [4–6], have used model-based fault detection to generate residual signals for MAS.

Recently, results have been reported on the application of set-theoretic and optimization tools for MAS control, notably [7]. Furthermore, these tools were also used to design FDI schemes based on the separation between different functioning modes [8]. The faults treated in this work are principally sensor faults ([9, 10]) and actuator faults ([11, 12]) for mono-agent dynamical systems. Other recent results concerning the application of set-theoretic tools to design a fault tolerant control framework on MAS are presented in [13–15].

The main idea of this chapter is to build on the resent results in [13, 14] and present a unified approach for the use of set-theoretic tools to design a centralized fault tolerant control (FTC) for an homogeneous MAS. The aim is on the one hand, to supervise the interaction of the agents in the global MAS and on the other hand, to detect if some agents from exterior try to integrate the current global system. For all of these cases, the priority is to preserve the formation and thus the design can be considered to be placed at a supervisory level. The main contribution is twofold:

- First, the proposed centralized FDI scheme based on set-theoretic methods for MAS is able to detect if an agent is faulty[1] and if this fault falls in a serious

[1] Actuator faults are considered further.

category, to eliminate the faulty agent from the team (and automatically from the formation).

- Second, it can be used as a threshold to detect intruders. The FDI step is subsequently completed by a reconfiguration step to calculate a new optimal configuration for the global system. After finding the optimal formation, a classical control action is designed to steer and keep the MAS into this new formation, with respect to the collision avoidance constraints between the agents.

The design of such centralized FDI scheme is based mainly on the quality of the communication tasks between the agents. This requires that the communication graph of the global MAS is *fully connected*, i.e., any agent can send its information (e.g., position, speed...) to all the agents in the global system and also receive the information from all these agents. Moreover, we assume that there is no degradation in the information exchanged between the agents due to the large disturbance of the environment or due to the delay of communication.

The structure of this chapter is organized as follows: Section 9.2 presents a brief resume of the necessary elements from previous work. Section 9.3 formulates the tracking problem for a MAS completed by few challenges which inspire the current work. At the end of this section, we describe two scenarios of functioning in the presence of faults and present a corresponding FDI framework with the associated reconfiguration step. The treatment of these two cases will be detailed, respectively, in Sects. 9.4 and 9.5. Section 9.6 proposes an example to illustrate the performance of the new FDI algorithm for a MAS composed of three agents. This MAS is subject to two faulty scenarios above. Finally, some concluding remarks and perspectives are mentioned in Sect. 9.7.

General definitions and notation: In order to use set-theoretic concepts, we introduce next a series of basic notions allowing us to link the dynamical systems to static geometrical sets in the state space. In order to formalize the present results and describe the appropriate framework, it should be mentioned that the considered dynamical systems are linear, time invariant and described by discrete-time models. The uncertainties are introduced in the model via bounded (additive) disturbances.

Given two sets $\mathcal{A}$ and $\mathcal{B}$, the operator $\mathcal{A} \oplus \mathcal{B} = \{a + b | a \in \mathcal{A}, b \in \mathcal{B}\}$ denotes their *Minkovski sum set* and $cv\,(\mathcal{A}, \mathcal{B})$ denotes their convex hull set.

Consider a bounded polyhedral set $\mathcal{S} \subset \mathbb{R}^n$, its closure is denoted by $cl(\mathcal{S})$ and $C(\mathcal{S}) = cl(\mathcal{S}) \backslash \mathcal{S}$ is the complement of $\mathcal{S}$.

With respect to the vector operator, $|\cdot|$ denotes the *element-wise absolute value.* $\|\mathbf{x}\|_{\mathbf{Q}} = \mathbf{x}^T \mathbf{Q} \mathbf{x}$ denotes the value of the *quadratic norm* of vector $\mathbf{x} \in \mathbb{R}^n$, with $\mathbf{Q} = \mathbf{Q}^\top \succ 0$ a *weighting symmetric positive definite matrix.* $\mathbf{I}_n \in \mathbb{R}^{n \times n}$ denotes the unitary matrix of dimension n.

$\mathbb{N} = \{0, 1, 2 \ldots\}$ is the set of all natural numbers. In the present study, $\mathbb{N}_{[1,N]} \triangleq \{1, 2, \ldots, N\}$, with $N \in \mathbb{N}$, will denote the index of each agent in the MAS. We use $\mathcal{N}_E \in \mathbb{N}_{[1,N]}$ to denote the subset containing the indices of the eliminated agents due to the fault occurrence. Hence the complement set of the faulty agents' indices is $\mathcal{N}_R = \mathbb{N}_{[1,N]} \backslash \mathcal{N}_E$.

In this paper two notions of receding horizons will be used:

- *Prediction horizon* denoted by N_p—will be employed as a finite-time window in the future;
- *Fault monitoring horizon* denoted by N_m—a finite-time window in the recent past, used to characterize the time to detect a fault.

$\check{\mathbf{x}}_i(k)$ denotes the *one-step predicted state* of the ith agent.

$\bar{\mathbf{x}}_i$ denotes the *target position* of the ith agent in the case that the common reference of the entire MAS reduces to the origin. Its role will be detailed in the next subsection.

$\check{\mathbf{x}}_i(k)$ denotes the *trajectory reference* of the ith agent, once one configuration for the entire system is determined.

Definition 9.1 ([16]) Consider an autonomous linear discrete time-invariant system $\mathbf{x}(k+1) = \mathbf{A}\mathbf{x}(k) + \mathbf{w}(k)$, with the matrix $\mathbf{A}$ assumed to be a Schur matrix. A set $\mathcal{S}$ is called *robustly positive invariant* (RPI) for this system, if $\mathbf{A}\mathbf{x}(k) + \mathbf{w}(k) \in \mathcal{S}$ for all $\mathbf{x}(k) \in \mathcal{S}$, $\mathbf{w}(k) \in \mathcal{W}$, which is equivalent to:

$$\mathbf{A}\mathcal{S} \oplus \mathcal{W} \subseteq \mathcal{S} \tag{9.1}$$

Lemma 9.1 *([17]) Consider the system $\mathbf{x}(k+1) = \mathbf{A}\mathbf{x}(k) + \mathbf{w}(k)$, with the matrix $\mathbf{A}$ assumed to be a Schur matrix and a nonnegative vector $\mathbf{w}(k)$ such that $|\mathbf{w}(k)| \leq \bar{\mathbf{w}}$, $\forall \mathbf{w}(k) \in \mathcal{W} \subset \mathbb{R}^n$. Let $\mathbf{A} = \mathbf{V}\mathbf{J}\mathbf{V}^{-1}$ be the Jordan decomposition of $\mathbf{A}$. Then the set*

$$\mathcal{S} = \{\mathbf{x} \in \mathbb{R}^n : |\mathbf{V}^{-1}\mathbf{x}| \leq (\mathbf{I} - |\mathbf{J}|)^{-1}|\mathbf{V}^{-1}\bar{\mathbf{w}}|\} \tag{9.2}$$

is robustly invariant with respect to the system dynamics.

9.2 Background in Multi-agent Formation Control

The notions presented in this section are based on the results in [18]. In order to elaborate a contribution in the next sections, we need to recall three main elements introduced and discussed in [18] respectively:

- Safety region construction which is the basic characterization for one agent beside its dynamical equation.
- The optimization-based framework to obtain an optimal formation.
- The control action for the tracking mission for a formation in the fault-free functioning.

9.2.1 Robust Tube-Based Safety Region of an Agent

Consider the global formulation of a MAS Σ composed of N heterogeneous agents:

$$\Sigma : \mathbf{x}(k+1) = \mathbf{A}\mathbf{x}(k) + \mathbf{B}\mathbf{u}(k) \tag{9.3}$$

where $\mathbf{x} = [\mathbf{x}_1^T \ \mathbf{x}_2^T \dots \mathbf{x}_N^T]^T \in \mathbb{R}^{Nn}$ and $\mathbf{u} = [\mathbf{u}_1^T \ \mathbf{u}_2^T \dots \mathbf{u}_N^T]^T \in \mathbb{R}^{Nm}$ denote respectively the collective state and input vector of the global system Σ. The dynamics are fully decoupled and the matrices $\mathbf{A} = diag(\mathbf{A}_1, \mathbf{A}_2, \dots, \mathbf{A}_N) \in \mathbb{R}^{Nn \times Nn}$ and $\mathbf{B} = diag(\mathbf{B}_1, \mathbf{B}_2, \dots, \mathbf{B}_N) \in \mathbb{R}^{Nn \times Nm}$ collect all the matrices corresponding to each agent by juxtaposition.

The nominal dynamics of each agent i is described by a LTI discrete-time model:

$$\Sigma_i : \mathbf{x}_i(k+1) = \mathbf{A}_i\mathbf{x}_i(k) + \mathbf{B}_i\mathbf{u}_i(k), \quad \forall i \in \mathbb{N}_{[1,N]} \tag{9.4}$$

where $\mathbf{x}_i \in \mathbb{R}^n$ and $\mathbf{u}_i \in \mathbb{R}^m$ are the ith agent's nominal state vector and input vector, respectively. In the presence of disturbances, the model of the ith agent is considered to account for it additively:

$$\tilde{\mathbf{x}}_i(k+1) = \mathbf{A}_i\tilde{\mathbf{x}}_i(k) + \mathbf{B}_i\tilde{\mathbf{u}}_i(k) + \mathbf{w}_i(k), \quad \forall i \in \mathbb{N}_{[1,N]} \tag{9.5}$$

where $\mathbf{w}_i \in \mathcal{W}$ is the disturbance vector, $\tilde{\mathbf{x}}_i \in \mathbb{R}^n$ and $\tilde{\mathbf{u}}_i \in \mathbb{R}^m$ are the ith agent's state and input vector, respectively. In the present work it is assumed that the set $\mathcal{W} \subset \mathbb{R}^n$ is bounded and contains the origin in its interior. If the robust control input vector in (9.5) accounts for a nominal control action and a linear disturbance rejection term:

$$\tilde{\mathbf{u}}_i(k) = \mathbf{u}_i(k) + \mathbf{K}_i \left(\tilde{\mathbf{x}}_i(k) - \mathbf{x}_i(k)\right) \tag{9.6}$$

then, by denoting $\mathbf{e}_i = \tilde{\mathbf{x}}_i - \mathbf{x}_i$ as the tracking error of the ith agent, the following expression is obtained:

$$\mathbf{e}_i(k+1) = (\mathbf{A}_i + \mathbf{B}_i\mathbf{K}_i)\mathbf{e}_i(k) + \mathbf{w}_i(k) \tag{9.7}$$

The pairs $(\mathbf{A}_i, \mathbf{B}_i)$ are assumed to be stabilizable. The stabilizability assumption of the pairs $(\mathbf{A}_i, \mathbf{B}_i)$ guarantees the existence of the feedback gain $\mathbf{K}_i \in \mathbb{R}^{m \times n}$ which stabilizes (9.7). Applying Lemma 9.1 for the tracking error equation, a RPI set $\mathcal{S}_i$ can be constructed (see [7]), ensuring that the tracking error $\mathbf{e}_i(k) \in \mathcal{S}_i$ at each time instant if $\mathbf{e}_i(0) \in \mathcal{S}_i$. The set $\mathcal{S}_i$ is considered as a basic geometrical structure for the safety region around the ith agent. Furthermore, although the real state $\tilde{\mathbf{x}}_i$ is unknown due to disturbances $\mathbf{w}_i$, its trajectory is always bounded by the parameterized tube:

$$\mathcal{S}_i(\mathbf{x}_i(k)) = \{\mathbf{x}_i(k)\} \oplus \mathcal{S}_i \tag{9.8}$$

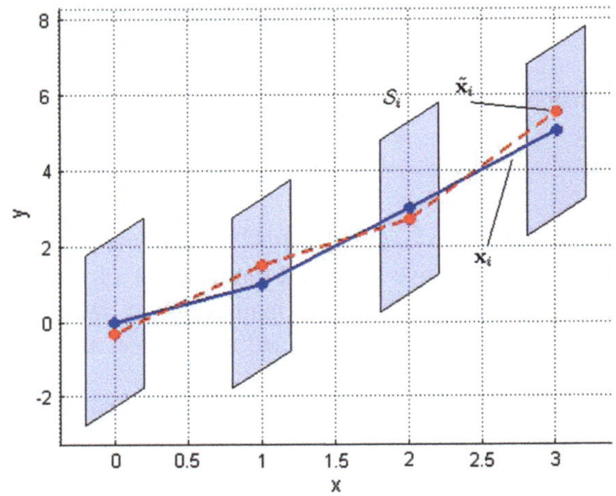

Fig. 9.1 Safety region of an agent

Therefore, the nominal dynamics (9.4) together with its robust tube-based safety region are used to characterize the behavior of an agent. As illustrated in Fig. 9.1, for the ith agent, its real state $\tilde{\mathbf{x}}_i$ (red line) is always bounded in a tube $\mathcal{S}(\mathbf{x}_i)$ composed of its nominal state $\mathbf{x}_i$ (blue line) and its safety region $\mathcal{S}_i$.

9.2.2 *Minimal Formation*

A minimal formation of the MAS system Σ as presented in [7] is defined as an ideal configuration where all the considered agents are as close as possible to a common reference, which is the reference of the formation center. This formation is defined as the optimal solution $\bar{\mathbf{x}}^*$ of the following problem:

$$\bar{\mathbf{x}}^* = \underset{\bar{\mathbf{u}}^*}{\text{argmin}} \sum_{i=1}^{N} \|\bar{\mathbf{x}}_i\| \tag{9.9a}$$

subject to:

$$\bar{\mathbf{x}}_i - \bar{\mathbf{x}}_j \notin (-\mathcal{S}_i) \oplus \mathcal{S}_j \tag{9.9b}$$

$$\bar{\mathbf{x}}_i = \mathbf{A}_i\bar{\mathbf{x}}_i + \mathbf{B}_i\bar{\mathbf{u}}_i \tag{9.9c}$$

with $\forall i, j \in \mathbb{N}_{[1,N]}, i \neq j$. Here, $\bar{\mathbf{x}}_i$ indicates the error between the state of the ith agent and the origin (which represents the common reference). The expression (9.9b) denotes the collision avoidance constraints, while (9.9c) emphasizes that $\bar{\mathbf{x}}_i$ is determined as a static equilibrium point according to the dynamical constraints. The problem (9.9) is a nonconvex problem and due to (9.9b) can be casted in the

Mixed Integer Programming (MIP) class with the associated solvers and resolution [19]. The solution given by MIP is the set of target positions for a tight equilibrium formation Σ. Moreover, solving (9.9) requires the full information of all agents in the global system, hence its calculation is centralized and thus it is computed in a centralized way.

9.2.3 Centralized Tracking Reference

Once the optimal formation is determined, it will be preserved along the *common reference*[2] $\mathbf{x}_{ref}$ (depicted by a red line in Fig. 9.2 for the example of a MAS composed of three agents). Hence, the target trajectory of the ith agent is denoted by:

$$\begin{aligned} \breve{\mathbf{x}}_i(k) &= \mathbf{x}_{ref}(k) + \bar{\mathbf{x}}_i^* \\ \breve{\mathbf{u}}_i(k) &= \mathbf{u}_{ref}(k) + \bar{\mathbf{u}}_i^* \end{aligned} \tag{9.10}$$

This trajectory is associated to the dynamical equation:

$$\breve{\mathbf{x}}_i(k+1) = \mathbf{A}_i\breve{\mathbf{x}}_i(k) + \mathbf{B_i}\breve{\mathbf{u}}_i(k) \tag{9.11}$$

This can be verified by observing that the pairs $[\bar{\mathbf{x}}_i^*, \bar{\mathbf{u}}_i^*]$ obtained by solving (9.9) are always static. In others words, they are used to represent the offset between the time-varying common reference $[\mathbf{x}_{ref}(k), \mathbf{u}_{ref}(k)]$ and the reference of each agent $[\breve{\mathbf{x}}_i(k), \breve{\mathbf{u}}_i(k)]$. Hence we obtain

$$\begin{aligned} \breve{\mathbf{x}}_i(k+1) &= \mathbf{x}_{ref}(k+1) + \bar{\mathbf{x}}_i^* \\ &= \mathbf{A}_i\mathbf{x}_{ref}(k) + \mathbf{B}_i\mathbf{u}_{ref}(k) + \bar{\mathbf{x}}_i^* \\ &= \mathbf{A}_i\breve{\mathbf{x}}_i(k) + \mathbf{B}_i\breve{\mathbf{u}}_i(k) - \mathbf{A}\bar{\mathbf{x}}_i^* - \mathbf{B}_i\bar{\mathbf{u}}_i^* + \bar{\mathbf{x}}_i^* \\ &= \mathbf{A}_i\breve{\mathbf{x}}_i(k) + \mathbf{B}_i\breve{\mathbf{u}}_i(k) \end{aligned}$$

where the last expression exploits the static equality $\bar{\mathbf{x}}_i^* = \mathbf{A}_i\bar{\mathbf{x}}_i^* + \mathbf{B}_i\bar{\mathbf{u}}_i^*$.

The main purpose of the formation control remains the design of a closed-loop control scheme so that the MAS's states track the common reference which can be interpreted as a feedforward signal:

$$\mathbf{x}_{ref}(k+1) = \mathbf{A}_{ref}\mathbf{x}_{ref}(k) + \mathbf{B}_{ref}\mathbf{u}_{ref}(k) \tag{9.12}$$

[2]The common reference is the reference of the formation center.

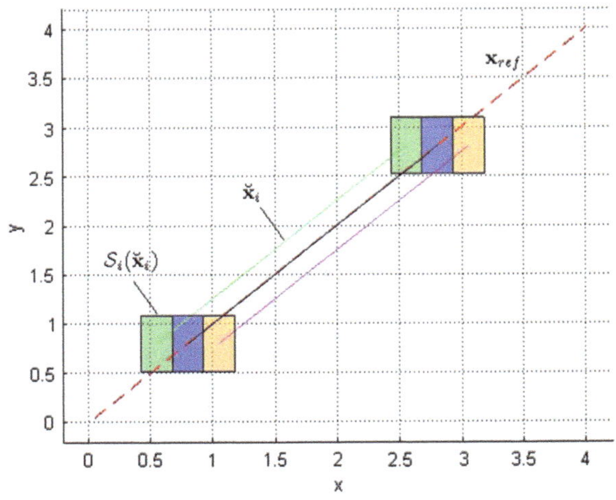

Fig. 9.2 Formation of three agents

with $\mathbf{A}_{ref} = \mathbf{A}_i$ for an homogeneous MAS. With this purpose, the following *centralized model predictive control* (MPC) action is designed:

$$\mathbf{u}^*(k) = \underset{\mathbf{u}^*}{\operatorname{argmin}} \sum_{t=k}^{k+N_p-1} \left(\|\mathbf{z}(t)\|_{\mathbf{Q}} + \|\mathbf{v}(t)\|_{\mathbf{R}} \right) + \left\|\mathbf{z}(k+N_p)\right\|_{\mathbf{P}} \tag{9.13a}$$

subject to:

$$\mathbf{z}(t) = \mathbf{x}(t) - \check{\mathbf{x}}(t) \tag{9.13b}$$
$$\mathbf{v}(t) = \mathbf{u}(t) - \check{\mathbf{u}}(t) \tag{9.13c}$$
$$\mathbf{x}(t+1) = \mathbf{A}\mathbf{x}(t) + \mathbf{B}\mathbf{u}(t) \tag{9.13d}$$
$$\check{\mathbf{x}}(t+1) = \mathbf{A}\check{\mathbf{x}}(t) + \mathbf{B}\check{\mathbf{u}}(t) \tag{9.13e}$$
$$\mathbf{x}_i(t+1) - \mathbf{x}_j(t+1) \notin -\mathcal{S}_i \oplus \mathcal{S}_j \tag{9.13f}$$

with $\mathbf{u}^* = \left[\mathbf{u}^{*\top}(k), \ldots, \mathbf{u}^{*\top}(k+N_p-1)\right]^\top$ and $t \in \mathbb{N}_{[k,k+N_p-1]}$, $\forall i,j \in \mathbb{N}_{[1,N]}$, $i \neq j$.

Alternative formulations enforcing the stability can be employed by adding terminal constraints to (9.13f) but this deserves additional considerations with respect to the length of the prediction horizon and feasible domain. Their presentation is beyond the scope of the present work which focuses on the monitoring and fault detection of the MAS formation (supposed to run on a properly designed tracking control mechanism). We note that the prediction horizon has to be long enough in order to allow convergence. Note also that (9.13) does not include static input-state limitations and the feasibility of the anticollision constraint (9.13f) can be handled via reachability analysis or viability theory whenever these constraints are considered as "hard", [20]. Such centralized control structure requires that any agent has to send its information (e.g., position, speed...) to all agents in the global system and also receive the information from all these agents. The computational time for each prediction depends on the dimension of the MAS.

9.3 Problem Statement

9.3.1 *Different Results in Reference Tracking of MAS*

The control task of a MAS is a challenging one, especially when addressing the mission safety. Given a dynamical MAS, the mission safety is defined as the achievement of a common tracking goal, while ensuring that the interaction between the agents does not damage the structural organization of the MAS. More precisely, all agents have to track a given reference within a predefined formation and integrate the collision avoidance constraints while fulfilling the mission objective. The framework (9.9) can give a minimal configuration for the system but this predefined configuration does not adapt to the real-time evolution of MAS. In case of changing the number of agents in MAS, typically when an agent leaves definitely its team,[3] due to a serious fault or due to the operator decision, or when some agents from exterior try to join the current MAS, clearly a fixed configuration is not suitable. Moreover, having agents that leave the formation may have a disastrous impact if a subgroup of agent follows such a faulty element.

9.3.2 *Interest and Challenge*

Our interest is to reinforce the safety of a MAS during a tracking mission. This chapter proposes a centralized framework for fault monitoring and the functioning of the global system in the presence of damages that are acceptable. This requirement becomes challenging whenever the supervision level has to integrate the environmental disturbance and cumbersome computation load due to the dimension of Σ. The most important feature is that this supervision/monitoring framework has to be suitable to the established control objectives. More precisely, the implementation of a fault tolerant supplementary layer can introduce some loss of performance during the operation of the global system.

The theoretic base is offered by the use of set-theoretic tools. Based on the knowledge of the agents dynamics and their safety region construction, we will present in the next sections a novel *Fault Detection and Isolation* (FDI) layer for the global system. This framework is considered as an adaptive threshold with respect to the

[3]In practice, it may even become even adversary with respect to the team but such behavior is not considered here. In the following, all the intruders are considered as cooperative and their inclusion is automatically granted to the formation subject to reconfiguration.

disturbance and furthermore based on the predicted state of Σ. We consider here only additive disturbances to simplify the presentation. The FDI layer for the fault monitoring is followed by a reconfiguration step which aims to recover the control structure with respect to the remaining healthy agents in the system. This FDI layer is installed on each agent, based on the communication between this agent and all the remaining agents in the global system. Concerning the changing of the number of agents in the MAS, we consider in the following two notable cases.

First, an agent belonging to the Σ can suffer some damages on its components which decrease the performance of functioning. This is often translated to a faulty behavior relative to the common healthy behavior of the other agents in Σ (due to the impact in the common prediction model). The anomalies can also come from the decision of the central operator in order to isolate this agent out of the current formation. For this case, an FDI layer has to be introduced to characterize the behavior of each agent allowing to detect if an agent has some anomalies in its behavior and if the anomalies are characterized as serious enough to subsequently isolate this agent.

Second, during the mission, some agents positioned outside of the formation can make maneuvers in order to integrate to the current formation. This scenario is critical because it can lead to safety loss. A FDI layer which is based on set-theoretic methods is further used as a threshold to detect these *intruders*. In order to guarantee the safety of the formation, this faulty case can be decomposed into two phases. First, once detected, the status *healthy* of the agent has to be validated. This validation step needs a validation time to ensure the integration objective of the intruders. Second, after being validated as healthy, a suitable reconfiguration step will be effectuated. It authorizes the inclusion of a novel index (of the intruder) to the global system Σ and then reconfigure accordingly the formation at the next iteration. The new formation has to be a set of optimal positions to which the current agents and the new ones may converge while fulfilling the anticollision constraints. These two scenarios will be described in the next two sections.

9.4 Outgoing Agent Case Study

9.4.1 Fault Detection and Isolation for Outgoing Agents

In this section, the FDI layer is designed in order to detect and eliminate the faulty agents from the current formation. After the elimination step, the formation will be reconfigured based on the healthy (remaining) subset of agents. A certain time window has to be predefined to validate the faulty status of an outgoing agent. Two levels of faults are considered:

- *Quarantined Faulty Agent*: The faulty agent suffers anomalies in its dynamic behavior relative to the healthy behavior of its neighbors. These anomalies make the agent's behavior different from its nominal (predicted) dynamics, but this still cannot prove that this agent is faulty, because the anomalies may be issued from the environmental disturbance or local decisions made at the control level, see the problem (9.13). For this reason, we need to set a certain time window of length N_m to validate the fault. If it is maintained more than N_m time steps, this agent will be certified as faulty.

- *Certified Faulty Agent*: This case is similar to the previous case but the impact of the decision is important in the global formation. The certification is done whenever the system in quarantine presents a state which is largely different from the remaining agents, practically outside of the current formation envelope. In this case, the faulty status has to be reported as soon as possible to eliminate the faulty agent and reconfigure the formation.

In the sequel, for brevity, $\tilde{\mathbf{x}}_i(k)$ denotes the real state of the ith agent, as specified in Sect. 9.2.

9.4.1.1 Quarantined Faulty Agent Detection

In order to determine the functioning mode (Healthy or Faulty) of an agent, a set of N residuals will be used, one for each agent. Each residual is defined as:

$$\mathbf{r}_i(k) = \tilde{\mathbf{x}}_i(k) - \check{\mathbf{x}}_i(k), \quad \text{with } i \in \mathbb{N}_{[1,N]} \tag{9.14}$$

with $\check{\mathbf{x}}_i(k)$ denoting the one-step predictable state of the ith agent. The value of $\check{\mathbf{x}}_i(k)$ is obtained by using the nominal dynamics (9.4) and the last available state $\mathbf{x}_i(k-1)$, i.e.,

$$\check{\mathbf{x}}_i(k) = \mathbf{A}_i\check{\mathbf{x}}_i(k-1) + \mathbf{B}_i\mathbf{u}_i^*(k-1), \quad \text{with } i \in \mathbb{N}_{[1,N]} \tag{9.15}$$

where $\mathbf{u}_i^*(k-1)$ is the control action of the ith element of the optimal solution of (9.13) at time instant $k-1$.

If there is no fault, then $\mathbf{r}_i(k) \in \mathcal{S}_i$. Hence, the safety region $\mathcal{S}_i$ is also the set $\mathcal{R}_i^H$ which characterizes the Healthy functioning of the ith agent:

$$\mathcal{R}_i^H = \mathcal{S}_i, \quad \text{with } i \in \mathbb{N}_{[1,N]} \tag{9.16}$$

We consider a set denoted $\mathcal{R}_i^F$ to characterize the Faulty functioning and $\mathcal{R}_i^{H\to F}$ to characterize the Healthy-to-Faulty transition functioning. These sets must respect the following detectability set-separation condition:

$$\begin{cases} \mathcal{R}_i^H \cap \mathcal{R}_i^{H\to F} = \emptyset \\ \mathcal{R}_i^H \cap \mathcal{R}_i^F = \emptyset \end{cases}, \quad \text{with } i \in \mathbb{N}_{[1,N]} \tag{9.17}$$

If one of these separations does not hold, then the FDI mechanism will not be able to ensure one-step detection but can engage in a monitoring procedure [8].

In the present framework, we consider only the critical faults, like leaving the formation due to serious faults. The fault identification is not considered in this paper. This means that once the fault occurs, the residual $\mathbf{r}_i(k)$ jumps out of $\mathcal{R}_i^H$ and transits to $\mathcal{R}_i^F$. The condition of FDI (9.17) is thus simplified to:

$$\mathcal{R}_i^H \cap \mathcal{R}_i^F = \emptyset, \quad \text{with } i \in \mathbb{N}_{[1,N]} \tag{9.18}$$

A candidate set which satisfies the condition (9.18) is the complement set of $\mathcal{R}_i^H$, i.e., $\mathcal{R}_i^F = C(\mathcal{R}_i^H)$.

9.4.1.2 Faulty Agent Certification

The previous FDI scheme is used to detect and quarantine an agent which exhibits a fault. We propose here another set construction to detect whenever the fault becomes critical. The threshold set for this case is parameterized by a set

$$\check{\mathcal{S}}(\mathbf{x}_{ref}(k)) = cv\left(\bigcup \mathcal{S}(\check{\mathbf{x}}_i(k))\right), \quad \forall i \in \mathcal{N}_R \tag{9.19}$$

Here, $\mathbf{x}_{ref}$ is the common reference of the global system Σ (see Sect. 9.2.3) and $\check{\mathbf{x}}_i(k)$ is the one-step predicted state of the ith agent. We recall that $\mathcal{N}_R$ denotes the indices set of the remaining agents in Σ. In words, the set (9.19) describes an envelope of formation evolving under healthy behavior.

This threshold set $\check{\mathcal{S}}(\mathbf{x}_{ref}(k))$ is defined as the convex hull of the one-step predicted position of all agents around the reference. This description is similar to obtaining a tube-based construction centered by the common reference $\mathbf{x}_{ref}(k)$, i.e.,

$$\check{\mathcal{S}}(\mathbf{x}_{ref}(k)) = \left\{\mathbf{x}_{ref}(k)\right\} \oplus \check{\mathcal{S}} \tag{9.20}$$

This set can be called the *monitoring tube* for the formation Σ.

9.4.2 Reconfiguration—Outgoing Agents Case Study

This section proposes a reconfiguration mechanism which is activated when an agent is certified faulty. It uses the definition of the two faulty cases from the Sect. 9.4.1. This reconfiguration step is performed (and is enabled) after the FDI step.

First, the nature of the fault has to be determined. At each iteration k, we check if the residual signal $\mathbf{r}_i(k)$ belongs to $\mathcal{R}_i^H$ or not. If $\mathbf{r}_i(k) \notin \mathcal{R}_i^H$, the respective agent will be labeled as quarantined faulty agent. After N_m iterations corresponding to the fault monitoring horizon (N_m is used to characterize the time to detect a fault), if $\mathbf{r}_i(k+N_m) \notin \mathcal{R}_i^H$, the ith agent is certified faulty and subsequently will be eliminated from the team.

Otherwise, $\tilde{\mathbf{x}}_i(k) \notin \check{\mathcal{S}}(\mathbf{x}_{ref}(k))$ and the ith agent is immediately certified faulty and it will be eliminated without passing by the quarantine stage.

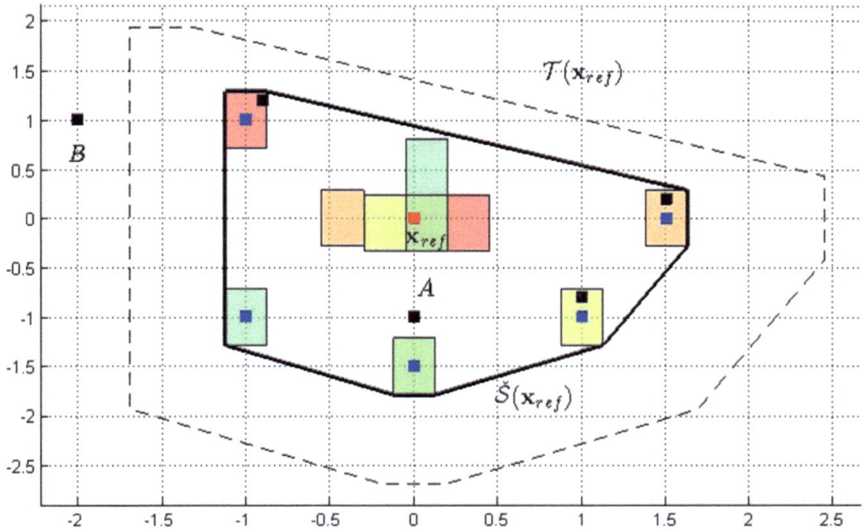

Fig. 9.3 Faulty cases in the interior of the formation

After characterizing the fault nature, the reconfiguration layer will be activated. Thus the formation is further reconfigured at the next iteration for the remaining healthy agents.

We illustrate in Fig. 9.3 two faulty cases presented above for a group of five agents. These five agents are at the initial stage in a minimal formation centered by

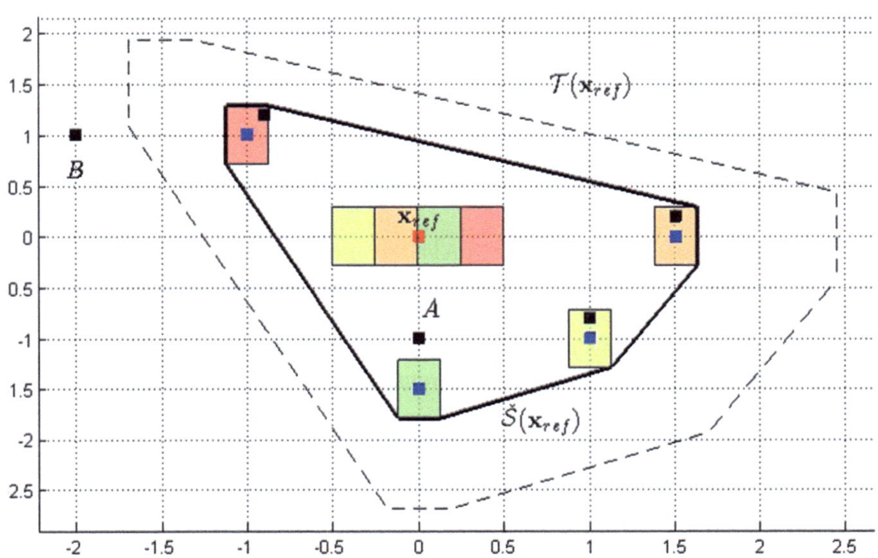

Fig. 9.4 Reconfigured formation after elimination of B

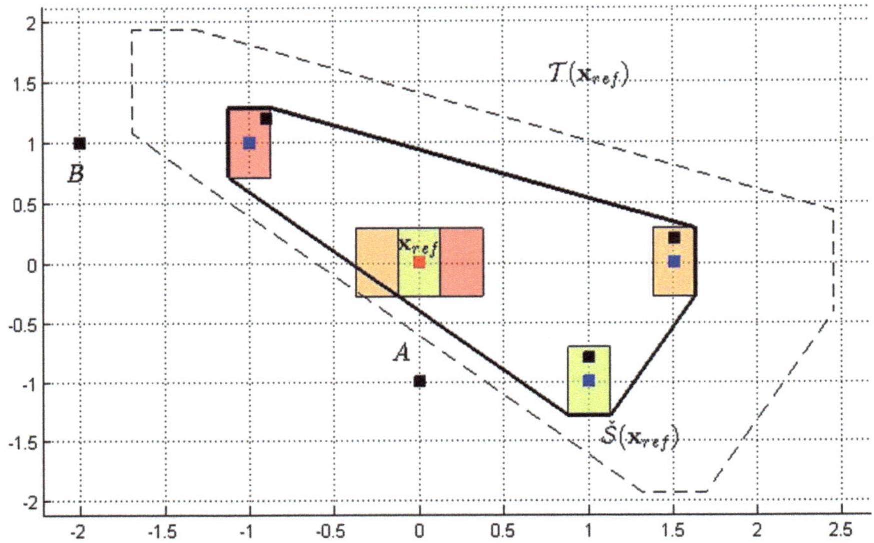

Fig. 9.5 Reconfigured formation after elimination of A

the common reference $\mathbf{x}_{ref}$ (red square). Their nominal states $\mathbf{x}_i$ are presented by the blue squares and their real states $\tilde{\mathbf{x}}_i$ are denoted by the black squares. We describe the elimination tube set $\check{\mathcal{S}}(\mathbf{x}_{ref}(k))$ by the black lined set. One agent is quarantined (the A point). The other agent is certified faulty (the B point) and eliminated and a reconfiguration takes place for a MAS of four agents (see Fig. 9.4). If the quarantined status is maintained for N_m steps, then the agent A will be also eliminated and the formation is reconfigured for the three remaining agents, thus a new formation (of three agents) is depicted in Fig. 9.5.

9.4.3 Algorithm for the Outgoing Agents Scenario

All the above ideas are incorporated in Algorithm 1 for the task assignment of the $\mathcal{N}_R$ healthy subset of agents. This algorithm is executed at each sampling time. A set of timers will be activated in order to count the time steps when $\mathbf{r}_i(k) \notin \mathcal{R}_i^H$. Each timer is associated with one agent in Σ.

The functioning mode of each agent is represented by:

$$status_i = \begin{cases} 1 & \text{if Healthy} \\ 0 & \text{if Faulty} \end{cases}$$

Algorithm 9.1 Task assignment reconfiguration in case of elimination of faulty agent

```
   Data: current state x(k), residual set R_i^H
   Result: Minimal reconfigured formation at sample time k
1  - construct Š(x_ref) for N_R;
2  for i ∈ N_R do
3      if x̃_i(k) ∉ Š(x_ref(k)) then
4          status_i := 0;
5          N_R := N_R \ {i};
6      else
7          - calculate residual r_i(k) = x̃_i(k) − x̌_i(k);
8          if timer_i = 1 then
9              if r_i(k) ∉ R_i^H then
10                 if t = N_m then
11                     status_i := 0;
12                     N_E := N_E ∪ {i};
13                 else
14                     t := t + 1;
15                 end
16             else
17                 timer_i := 0;
18                 status_i := 1;
19                 t := 0;
20             end
21         else
22             if r_i(k) ∉ R_i^H then
23                 timer_i := 1;
24                 status_i := 1;
25                 t := 1;
26             else
27                 timer_i := 0;
28                 status_i := 1;
29                 t := 0;
30             end
31         end
32     end
33 end
34 - solve (9.9) for N_R;
```

The status (activated/deactivated) of each timer is described by the corresponding element in the vector $timer = [timer_1 \; timer_2 \ldots timer_N]^\top$, with

$$timer_i = \begin{cases} 1 & \text{if activated} \\ 0 & \text{if deactivated} \end{cases}$$

At each sampling time, the current state of all agents in Σ will be collected and also their residual sets $\mathcal{R}_i^H$. A loop of calculation is activated for all agents (line 2–33). For each agent, if $\tilde{\mathbf{x}}_i(k) \notin \check{\mathcal{S}}(\mathbf{x}_{ref}(k))$ then it will be eliminated immediately, else the verification necessary for the quarantine status is activated (line 3–6). In this case, its residual signal $\mathbf{r}_i(k) = \tilde{\mathbf{x}}_i(k) - \check{\mathbf{x}}_i(k)$ is calculated (line 7) and the corresponding timer will be activated (line 8–31). During the activation of its timer, if $\mathbf{r}_i(k) \in \mathcal{R}_i^H$ then the agent is healthy and its timer is deactivated (line 26–30). Else after N_m sampling time, if $\mathbf{r}_i(k) \notin \mathcal{R}_i^H$ this agent will be certified faulty and it will be eliminated (line 10–12). The reconfiguration step is activated and the end of the loop and it considers just the remaining healthy agents (line 34).

9.5 Incoming-Agent Case Study

Section 9.4 presented our FDI framework to detect and isolate the faulty agent in the interior of the formation. Here we study a different situation when the agents from exterior integrate the formation.

9.5.1 Fault Detection and Isolation—Incoming Agent Case Study

Apart from the faulty agent detection and isolation, we propose here a FDI scheme to preserve safety when some new agents coming from the exterior of the current formation join the system. The main purpose is to detect whether/when an agent tries to join the formation, while guaranteeing the safety of the global system during the integration process.

For this purpose of detection of intrusion, based on the elimination tube set $\check{\mathcal{S}}$ as defined in (9.19), we introduce a new set $\mathcal{T}$, called *intruder detection tube* which encircles the one-step forward predicted formation, e.g.,

$$\mathcal{T}(\mathbf{x}_{ref}(k)) = \{\mathbf{x}_{ref}(k)\} \oplus \alpha\check{\mathcal{S}} \tag{9.21}$$

This set $\mathcal{T}$ is constructed to bound the monitoring tube $\check{\mathcal{S}}(\mathbf{x}_{ref}(k))$. In (9.21) $\alpha > 1$ is a scaling factor which accounts for the *visibility region* in the neighborhood of the formation. Its value can be adjusted in order to enlarge or reduce the scope of detection. Similar to the description of $\check{\mathcal{S}}(\mathbf{x}_{ref}(k))$, $\mathcal{T}(\mathbf{x}_{ref}(k))$ represents a tube set centered in the common reference $\mathbf{x}_{ref}(k)$.

This set is used to detect if some agents from exterior try to join the current formation of Σ. If this integration effort is accepted by the supervision decision level, the following reconfiguration step has to be activated.

9.5.2 Reconfiguration—Incoming Agent Case Study

Let consider an ith agent as not taken into account by Σ, i.e., $i \notin \mathcal{N}_R$. When this agent approaches the formation of Σ, if $\tilde{\mathbf{x}}_i(k) \in \mathcal{T}(\mathbf{x}_{ref}(k))$, a timer will be activated. After N_m iterations (for brevity we take the same monitoring horizon in the case of incoming or outgoing agents, although a different length can be adopted if necessary), if $\tilde{\mathbf{x}}_i(k+N_m) \in \mathcal{T}(\mathbf{x}_{ref}(k+N_m))$, then the integration is accepted. The index of this agent will be taken into account in the global task allocation and minimal formation, i.e., $\mathcal{N}_R = \{i\} \cup \mathcal{N}_R$ and a new configuration will be generated for the new subset $\mathcal{N}_R$. A natural question can rise about the need of a monitoring window for the incoming agents. The reason for its inclusion resides in the undesired effects of coupling the outgoing and incoming agents monitoring. Indeed, in the case of an outgoing agent, its fault can be considered concomitantly as a potential incoming behavior and the status of the respective agent will be indiscernible.

Here, in order to improve the safety of the integration process, the new formation obtained from solving (9.9) has to be an optimal formation but admissible from the current position of all agents in $\mathcal{N}_R$.

We illustrate in Fig. 9.6 the faulty case presented above for a group of three agents. These three agents have to be in a minimal formation centered by the common reference $\mathbf{x}_{ref}$ (red square). Their nominal states $\mathbf{x}_i$ are presented by the blue squares and their real states $\tilde{\mathbf{x}}_i$ are denoted by the black squares. We describe the monitoring

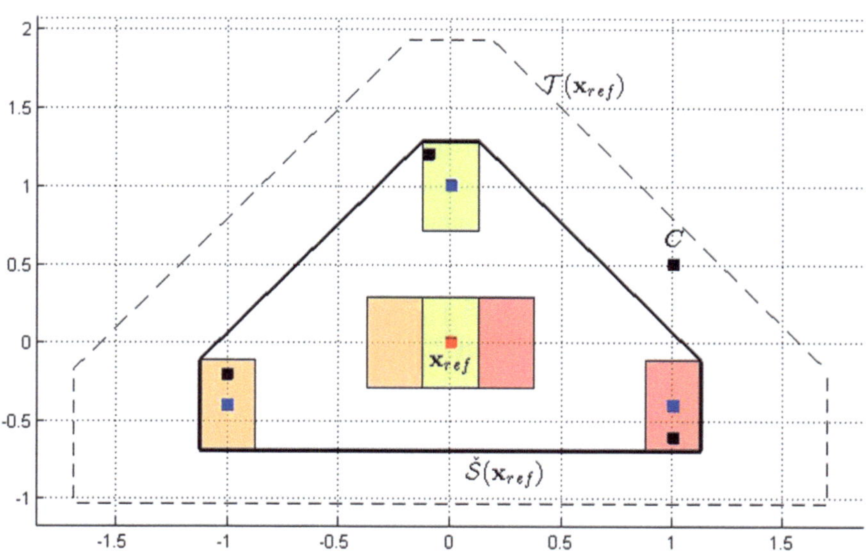

Fig. 9.6 Detection of an agent outside of the current formation

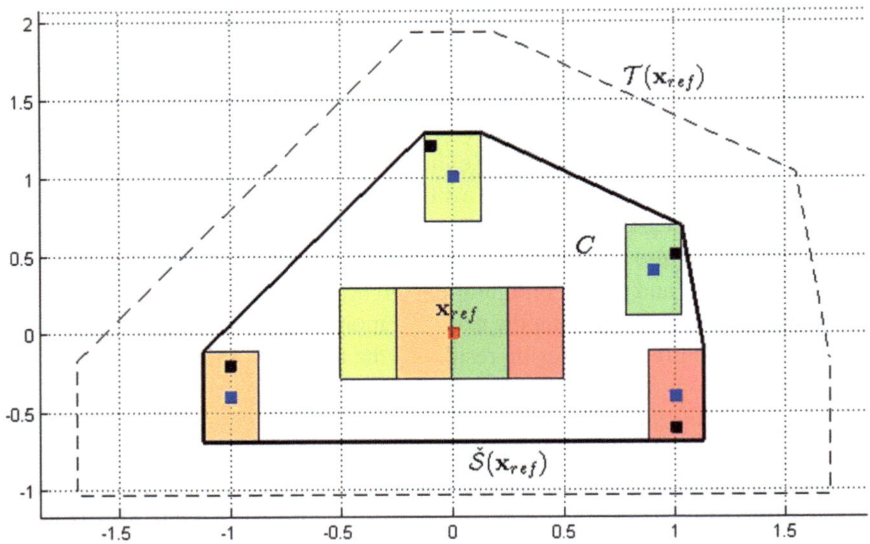

Fig. 9.7 Integration accepted

tube $\check{\mathcal{S}}(\mathbf{x}_{ref}(k))$ by a black line set and the intruder detection tube $\mathcal{T}(\mathbf{x}_{ref}(k))$ by a dash line set. One agent is trying to join the current group (the C point). When its integration is accepted, the formation is reconfigured for the four actual agents, thus a new formation is shown in Fig. 9.7.

9.5.3 Algorithm for the Incoming Agent Scenario

The main ideas in this scenario are resumed in Algorithm 2. We reuse the notation concerning the timer in Sect. 9.4.3. It is important to note that such a timer is activated for an unique agent from outside and there is no confusion between the incoming and outgoing agents. Moreover, we need to consider the following valuation of the status of the agent which approaches the formation:

$$status_i = \begin{cases} 1 & \text{if accepted} \\ 0 & \text{if denied} \end{cases}$$

This means that the integration of an exterior agent is accepted if and only if its status binary variable is 1, else it is rejected.

At each sampling time, the current real state of the intruder agent $\tilde{\mathbf{x}}_i(k)$ will be collected and also the sets $\check{\mathcal{S}}(\mathbf{x}_{ref})$, $\mathcal{T}(\mathbf{x}_{ref})$, $\mathcal{N}_R$ of the current formation. If $\tilde{\mathbf{x}}_i(k) \in \mathcal{T}(\mathbf{x}_{ref}(k))$ and there is no timer activated for its integration, then the activation will take place (line 15–18). During the activation of its timer, if $\tilde{\mathbf{x}}_i(k) \notin \mathcal{T}(\mathbf{x}_{ref}(k))$ then the timer is deactivated and the integration request is denied (line 19–23). Else after N_m sampling time, if $\tilde{\mathbf{x}}_i(k) \in \mathcal{T}(\mathbf{x}_{ref}(k))$ the integration request is accepted (line 2–5). The reconfiguration step is activated at the end of the Algorithm for the healthy agents in the new subset $\mathcal{N}_R$ (line 25).

Algorithm 9.2 Task assignment reconfiguration in case of integration of agent from exterior

Data: $\tilde{\mathbf{x}}_i(k)$,$\check{\mathcal{S}}(\mathbf{x}_{ref})$, $\mathcal{T}(\mathbf{x}_{ref})$, $\mathcal{N}_R$
Result: Minimal reconfigured formation at sample time k

1 **if** $timer_i = 1$ **then**
2   **if** $\tilde{\mathbf{x}}_i(k) \in \mathcal{T}(\mathbf{x}_{ref}(k))$ **then**
3   **if** $t = N_m$ **then**
4             $status_i := 1$;
5             $\mathcal{N}_R := \mathcal{N}_R \cup \{i\}$;
6   **else**
7             $t := t + 1$;
8   **end**
9   **else**
10         $timer_i := 0$;
11         $status_i := 0$;
12         $t := 0$;
13   **end**
14 **else**
15   **if** $\tilde{\mathbf{x}}_i(k) \in \mathcal{T}(\mathbf{x}_{ref}(k))$ **then**
16         $timer_i := 1$;
17         $status_i := 0$;
18         $t := 1$;
19   **else**
20         $timer_i := 0$;
21         $status_i := 0$;
22         $t := 0$;
23   **end**
24 **end**
25 - solve (9.9) for $\mathcal{N}_R$;

9.6 Illustrative Example

In this section, a numerical example is presented in order to illustrate the results obtained by applying Algorithms 1 and 2 on a MAS Σ composed of $N = 3$ homogeneous agents. The common dynamics is $\mathbf{x}_i(k+1) = \begin{bmatrix} -0.2 & 0.5 \\ 0.2 & 0.71 \end{bmatrix} \mathbf{x}_i(k) + \begin{bmatrix} 0.71 & 0 \\ 0 & 0.22 \end{bmatrix} \mathbf{u}_i(k) + \mathbf{w}_i,\ i \in \{1, 2, 3\}$. The safety region is constructed by using the pole placement technique. The chosen poles to construct their safety region are [0.2; 0.5]. The disturbance is bounded, i.e., $|w_i| \leq [0.2\,0.2]^\top$. For the MPC controller used in (9.13), the weighting matrices are $\mathbf{Q} = 100\mathbf{I}_{Nn}$, $\mathbf{R} = \mathbf{I}_{Nm}$. The prediction horizon is $N_p = 3$ and the monitoring horizon chosen to validate the elimination of a faulty agent from the formation and also the integration of an exterior agent in the current formation is $N_m = 3$.

The feedforward common reference is generated by a MPC reference controller, included an integral tracking error as cost function and the weighting matrices chosen are $\mathbf{Q}_r = 10\mathbf{I}_n$, $\mathbf{R}_r = 0.01\mathbf{I}_m$.

$$\mathbf{u}^*_{ref}(k) = \underset{\mathbf{u}^*_{ref}(k),\ldots,\mathbf{u}^*_{ref}(k+N_p-1)}{\operatorname{argmin}} \sum_{t=k}^{k+N_p} \|\mathbf{z}(t)\|_{\mathbf{Q}_r} + \sum_{t=k+1}^{k+N_p-1} \|\mathbf{v}(t)\|_{\mathbf{R}_r} \tag{9.22a}$$

subject to:

$$\mathbf{z}(t) = \mathbf{x}_{ref}(t) - \mathbf{r}(k),\ t \in \mathbb{N}_{[k,k+N_p]} \tag{9.22b}$$

$$\mathbf{v}(t) = \mathbf{u}_{ref}(t) - \mathbf{u}_{ref}(t-1),\ t \in \mathbb{N}_{[k+1,k+N_p-1]} \tag{9.22c}$$

$$\mathbf{x}_{ref}(t+1) = \mathbf{A}_{ref}\mathbf{x}_{ref}(t) + \mathbf{B}_{ref}\mathbf{u}_{ref}(t),\ t \in \mathbb{N}_{[k,k+N_p]} \tag{9.22d}$$

We recall here that Σ is homogeneous. Thus the model dynamics implemented in this MPC layer is the common dynamics of all agents, hence $\mathbf{A}_{ref} = \mathbf{A}_i$ with $i \in \{1, 2, 3\}$.

As illustrated in Fig. 9.8, the global MAS Σ pursuits a periodic trajectory illustrated by the black line. At the beginning (the A point), Σ is composed of three agents, with $\mathcal{N}_R = \{1, 2, 3\}$. The task assignment gives the initial admissible optimal formation for Σ. This formation is preserved with the common reference $\mathbf{x}_{ref}$, and the trajectories evolve in a tube centered at the $\mathbf{x}_{ref}(k)$ with its associated monitoring tube $\breve{\mathcal{S}}(\mathbf{x}_{ref}(k))$ and its intruder detection tube $\mathcal{T}(\mathbf{x}_{ref}(k))$. The red lines present the set of reference trajectory $\breve{\mathbf{x}}_i$ of the agents.

At $k = 13$ (the B point), the 3rd agent is subject to an actuator fault, then it stops. The remaining agents continue to track their reference trajectory with respect to the last configuration. When $\mathbf{x}_3 \notin \breve{\mathcal{S}}(\mathbf{x}_{ref})$ (the C point), the 3rd agent is certified faulty, and then the formation is reconfigured for the two remaining agents. The new MAS is denoted by Σ^*, with $\mathcal{N}_R = \{1, 2\}$. Two new sets $\breve{\mathcal{S}}^*(\mathbf{x}_{ref}(k))$ and $\mathcal{T}^*(\mathbf{x}_{ref}(k))$ are calculated for Σ^*.

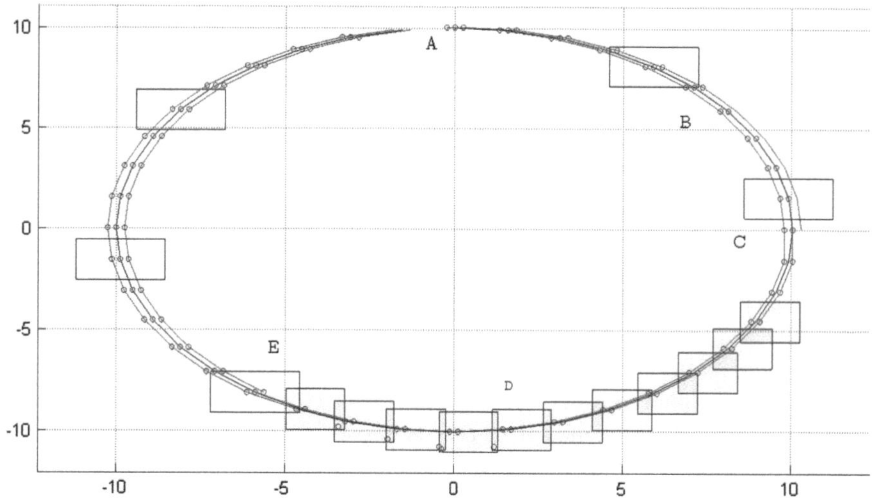

Fig. 9.8 Illustrative example

At $k = 25$ (the D point), the 4th agent from outside exhibits an incoming trajectory with respect to Σ^*. When $\mathbf{x}_4(k) \in \mathcal{T}^*(\mathbf{x}_{ref}(k))$ for more than N_m time steps (the E point), the integration of the 4th agent is accepted. Thus $\mathcal{N}_R = \{1, 2, 4\}$ and the formation will be reconfigured for three healthy agents.

9.7 Conclusion

This chapter uses set-theoretic methods as a basic tool to design a supervision layer in order to maintain the functioning of a global MAS under two particular faulty situations: elimination of a faulty agent from the formation and addition of an external agent to the current formation. In order to serve this fault detection purpose, set-theoretic tools are employed to construct offline the safety region of the agents and further to combine them online in order to build a threshold set for fault detection and isolation.

In order to deal with computational constraints, a decentralized approach, where the optimization problem is solved at the agent's level will be considered in the future work. In this approach, both fault detection and isolation techniques, also reconfiguration techniques have to be constructed in a decentralized manner. Furthermore, this have to be suitable to the multi-agent typical problem, for example, the tracking mission, collision avoidance, etc.

We emphasize that the problem mentioned in the main contribution was already simplified by imposing a series of assumptions. In particular, all of the frameworks above are presented based on the nominal dynamics of the global MAS. It means that the impact of multiplicative uncertainties is not considered. These perturbations in the context of MAS are various.

There can be the disturbances issued from the working environment which cause the delay in the communication channel or more seriously degradation of the information exchanged between the agents. These can lead to the degradation in the communication graph of the MAS. In this context, the robustness of the proposed framework seems interesting to study and specifically the relationship with the fault detection decision-making. The main purpose is to robustify the functioning of the global system by compensating the impact of these disturbances. Moreover, the link between the FDI layer and the control action applied to the global system needs to be further clarified and thus the theoretic aspects have to be developed in order to offer a clear picture on the severity of collision constraints after reconfiguration. Another research avenue which considers fault tolerant control for a heterogeneous system in a wider sense will be welcomed. This case is interesting and challenging because it is relative to the constraints on the dynamics of each agent. This can affect strongly the tracking objective and, more importantly, the generation of a consensus of the global system to an optimal configuration.

Acknowledgments The authors acknowledge the support of PHC RILA Project "Robust Distributed Model Predictive Control of Medium- and Large-Scale Systems" and PHC PESSOA "Advanced control of a fleet of heterogeneous autonomous vehicles".

References

1. N. Meskin, K. Khorasani, Actuator fault detection and isolation for a network of unmanned vehicles. IEEE Trans. Autom. Control **54**(4), 835–840 (2009)
2. N. Meskin, K. Khorasani, A geometric approach to fault detection and isolation of continuous-time markovian jump linear systems. IEEE Trans. Autom. Control **55**(6), 1343–1357 (2010)
3. N. Meskin, K. Khorasani, A hybrid fault detection and isolation strategy for a network of unmanned vehicles in presence of large environmental disturbances. IEEE Trans. Control Syst. Technol. **18**(6), 1422–1429 (2010)
4. S. Stankovic, N. Illic, Z. Djurovic, M. Stankovic, K.H. Johansson, Consensus based overlapping decentralized fault detection and isolation, in *Proceedings of the Conference on Control and Fault Tolerant Systems*, Nice (2010)
5. G. Antonelli, F. Arrichiello, F. Caccavale, A. Marino, A decentralized controller-observer scheme for multi-agent weighted centroid tracking. IEEE Trans. Autom. Control **58**(5), 1310–1316 (2013)
6. P. Kempker, A. Ran, J. van Schuppen, A formation flying algorithm for autonomous underwater vehicles, in *Proceedings of the 50th IEEE CDC*, Orlando, Florida (2011), pp. 1293–1298
7. I. Prodan, S. Olaru, C. Stoica, S. Niculescu, Predictive control for tight group formation of multi-agent systems, in *Proceedings of the IFAC World Congress*, Milan, Italy (2011)
8. F. Stoican, S. Olaru, *Set-Theoretic Fault-Tolerant Control in Multisensor Systems* (Wiley, New York, 2013)
9. F. Stoican, S. Olaru, G. Bitsoris, Controlled invariance-based fault detection for multisensory control systems. Control Theory Appl., IET **7**(4), 606–611 (2013)
10. S. Olaru, J.A. De Dona, M. Seron, F. Stoican, Positive invariant sets for fault tolerant multisensor control schemes. Int. J. Control **83**(12), 2622–2640 (2010)

11. M.M. Seron, J.A. De Dona, Fault tolerant control using virtual actuators and invariant-set based fault detection and identification, in *Proceedings of the 48th IEEE CDC and 28th CCC*, Shanghai (2009)
12. G. Franze, F. Tedesco, D. Famularo, Actuator fault tolerant control: a set-theoretic approach, in *Proceedings of the 51st IEEE CDC*, Maui, Hawaii (2012)
13. M. Nguyen, C. Stoica Maniu, S. Olaru, A. Grancharova, About formation reconfiguration for multi-agent dynamical systems, in *Proceedings of the Automatics and Informatics*, Sofia, Bulgaria (2014)
14. M. Nguyen, C. Stoica Maniu, S. Olaru, A. Grancharova, Fault tolerant predictive control for multi-agent dynamical systems: formation reconfiguration using set-theoretic approach, in *Proceedings of the Control, Decision and Information Technologies*, Metz, France (2014), pp. 417–422
15. A. Rosich, H. Voos, M. Darouach, Cyber-attack detection based on controlled invariant sets, in *Proceedings of the ECC*, Strasbourg, France (2014), pp. 2176–2181
16. F. Blanchini, S. Miani, *Set-Theoretic Methods in Control* (Birkhauser, Boston, 2007)
17. E. Kofman, H. Haimovich, M.M. Seron, A systematic method to obtain ultimate bounds for perturbed systems. Int. J. Control **80**(2), 167–178 (2007)
18. I. Prodan, Commande sous contraintes de systemes dynamiques multi-agents, Ph.D. thesis, Supelec, 2012
19. I. Prodan, F. Stoican, S. Olaru, S.I. Niculescu, Mixed-integer representations in control design: Mathematical foundations and applications, Springer (2015)
20. J.-P. Aubin, *Viability Theory* (Springer Science & Business Media, Berlin, 2009)

Part IV
Applications of Optimization-Based Control and Identification

Chapter 10
Optimal Operation of a Lumostatic Microalgae Cultivation Process

Sihem Tebbani, Mariana Titica, George Ifrim, Marian Barbu and Sergiu Caraman

Abstract This chapter proposes the optimization of batch microalgae cultures in artificially lighted photobioreactors. The strategy consists in controlling the incident light intensity so that the microalgae growth rate is maximized. Two approaches were developed and compared. In the first one, the ratio between the incident light intensity and the cell concentration (light-to-microalgae ratio) is optimized, either offline or online, and then maintained at its optimal value. In the second approach, the cells growth rate is maintained at its optimal value by means of nonlinear model predictive controller (NMPC). The proposed control strategies are illustrated and their efficiency is assessed, in simulation, for *Chlamydomonas reinhardtii* batch cultures. The proposed lumostatic operation strategies are shown to lead to a higher cell productivity and to a more efficient light utilization in comparison to conventional constant light operation approach.

Keywords Optimization · Nonlinear model predictive control · Bioprocess · Microalgae · Photobioreactor · Lumostatic operation · Light-to-microalgae ratio · Photoautotrophy · Photosynthetic microorganism

---

S. Tebbani (✉)
Laboratory of Signals and Systems, CentraleSupélec-CNRS Univ Paris Sud, Université Paris-Saclay, 3 Rue Joliot-Curie, F91192 Gif Sur Yvette, France
e-mail: sihem.tebbani@centralesupelec.fr

M. Titica
GEPEA, CNRS, Université de Nantes, UMR-CNRS 6144, Bd. de L'Université, CRTT-BP 406, 44600 Saint-nazaire Cedex, France
e-mail: mariana.titica@univ-nantes.fr

G. Ifrim · M. Barbu · S. Caraman
"Dunărea de Jos" University of Galati,
Domnească Street, no. 47, 800008 Galaţi, Romania
e-mail: george.ifrim@ugal.ro

M. Barbu
e-mail: marian.barbu@ugal.ro

S. Caraman
e-mail: sergiu.caraman@ugal.ro

S. Olaru et al. (eds.), *Developments in Model-Based Optimization and Control*,
Lecture Notes in Control and Information Sciences 464,
DOI 10.1007/978-3-319-26687-9_10

10.1 Introduction

The microalgae use photosynthesis in order to convert the sunlight energy into chemical energy required for the conversion of CO_2 into organic compounds. As a result of their ability to biosynthesize various components with high nutritional values, genera such as *Arthrospira* (*Spirulina*) are used for human nutrition, while certain microalgae are used in aquaculture and early juvenile stages of crustaceans and fish. The microalgae are also attractive for environmental applications such as greenhouse gas bio-mitigation, [18, 24], and wastewater treatment processes, [4, 16]. The production of biofuels from microalgae (third generation biofuels) is a very sound idea leading to numerous fundamental and applied researches, [5, 21]. Lately, increased attention was conferred to photoautotrophic[1] H_2 production with microalgae, seen as a novel source of sustainable and renewable energy, without greenhouse gas emissions or environmental pollution, [12].

Despite their widely recognized potential, industrial applications are still marginal, mainly because of the difficulty to propose adapted cultivation systems, called photobioreactors (PBR), that permits to reach sufficient productivity (with increased photosynthetic efficiency). Indeed, the specific biological needs of photosynthetic microorganisms consist in light transparent systems with adapted geometries, maximizing light energy collection and its supply to the cells. To fulfill these requirements PBRs have increased surface-to-volume ratio, resulting in specific designs (tubular, planar (flat panel), and column (vertical and inclined column)). They could be naturally or artificially lighted, depending on the application. These reactors are operated in continuous[2] or discontinuous mode.[3] Efficient operation of PBRs poses specific challenges in terms of process control. In all cases, considering light as a "substrate" for growth, its distribution inside the bulk governs the performances of the production system. Biomass itself has an "inhibitor" effect on its growth because an increase in biomass concentration diminishes the light availability; in contrast, when biomass concentration is too low, an excess of light energy will inhibit growth. The process could be optimized by imposing light gradient into the bulk so that the light energy absorption to be maximized. This type of operated reactor is called lumostat. While in artificially lighted reactors, the incident light intensity could be controlled, in outdoor conditions the light intensity varies all along the day; specific control strategies have to be implemented.

This chapter concerns the optimal control of a batch cultivation of microalgae in a photobioreactor, illuminated with artificial light. More specifically, the objective is to optimize the supplied light energy so that the bioprocess efficiency is maximized by maximizing light use. When light energy is in excess, the microalgae growth declines, especially in the early stage of the culture when cells concentration is low. On the other hand, when light intensity is too low inside the photobioreactor, the

---

[1] Uses light as source of energy and the only carbon source is CO_2.

[2] Continuous addition and withdrawal of medium during the culture.

[3] No addition or removal of medium during the culture. It is also called batch mode.

productivity also declines. In addition, to ensure that the production of microalgae is economically competitive, the light energy should be reduced as much as possible, while ensuring high efficiency.

In this chapter, the problem of the optimization of the batch culture by controlling the light supply is addressed. The objective is to maximize the efficiency of the light use while maximizing the quantity of the produced biomass in a given duration. The control of the light energy is a recent concern in the scientific community, [14, 17]. The objective is to improve further the bioprocess efficiency by means of advanced control strategies. Two strategies are proposed and their performances are compared.

The first strategy consists in considering the light-to-microalgae ratio. This quantity was used in the literature and was optimized empirically (considering a constant ratio). In this chapter, the ratio will be optimized by deriving an optimization problem, that will be solved either offline (constant ratio) or online (leading to a time-varying ratio).

In the second approach, the specific growth rate will be controlled so that it is maintained at its optimal value. Since the bioprocess model is highly nonlinear, a model predictive controller will be developed. These two strategies will be compared in simulation, highlighting the advantages and drawbacks of each method.

In order to illustrate the efficiency of the proposed strategies, the microalga *Chlamydomonas reinhardtii* is considered. This unicellular green microalga can produce hydrogen and thus is studied as a candidate for biofuel production.

The chapter is organized as follows. In Sect. 10.2, the studied bioprocess and its modeling are presented. Then, the lumostatic culture mode is detailed in Sect. 10.3 and compared to a conventional operation at constant incident light intensity. Two approaches were investigated and developed: the control of the light-to-microalgae ratio (Sect. 10.4), and the control of the growth rate (Sect. 10.5). For each strategy, simulation results are presented and discussed. Finally, conclusions are stated in Sect. 10.6.

10.2 Process Description and Modeling

10.2.1 Process Description

In many applications, a two-stage process is implemented consisting first in producing high biomass concentrations, before inducing the production of the metabolite of interest. It is the case with *Chlamydomonas reinhardtii*, which is able to produce molecular hydrogen in a clean way, under anoxic conditions, obtained and maintained by combining high density cultivation conditions coupled with the modulation of the incident light, [9]. This study focuses on the first biomass production phase, with the aim of proposing optimized production protocols increasing the efficiency of light use. It considers the batch cultivation of photosynthetic microorgansims in photobioreactor, under artificial light and in standard autotrophic conditions. In these

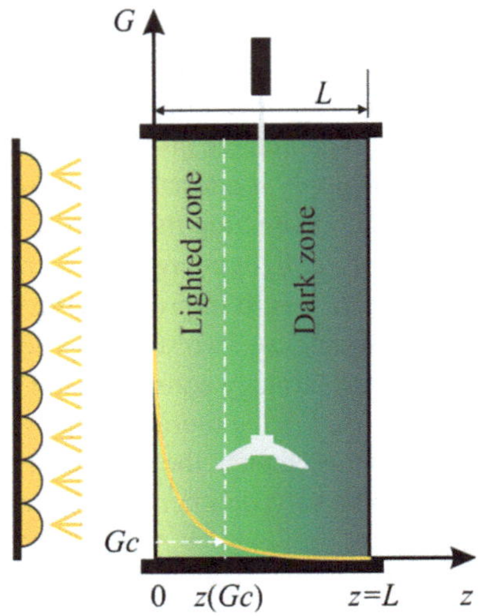

Fig. 10.1 Schematic representation of the studied photobioreactor

conditions, all mineral nutrients (nitrogen, sulfate, phosphate, and micronutrients) required for growth are introduced at the beginning in excess, while the inorganic carbon source (CO_2) is supplied continuously in order to maintain optimum pH during cultivation. In this way, light becomes the sole factor controlling growth. Under artificial light conditions, the photobioreactors are generally exposed to constant incident light intensities, provided by a LED panel or other light sources, applied on their surface. Owing to absorption and scattering by cells, the light effectively received by cells is attenuated and thus heterogeneously distributed into the culture (the light intensity decreases along the depth). The attenuation of light is induced by pigments of photosynthetic cells which absorb the photonic energy and the self-shading phenomenon, and depends on various factors such as photobioreactor geometry, lighting source position, hemispherical incident flux density, emission spectrum, optical properties of culture medium, biomass concentration, etc., [22]. Since the photosynthetic growth is governed by the received light, a local photosynthetic growth response can be defined at each location in the bulk. It results in using a particular class of models which are able to return local photosynthetic responses, as described below. The considered photobioreactor in this study is rectangular, flat panel, lighted from one side with a LED panel (Fig. 10.1).

10.2.2 Process Modeling

10.2.2.1 Mass Balance Model

Dynamic models of microalgal growth are mainly represented by a set of ordinary differential equations issued from mass (and possibly energy) balance of major components of the process. Assuming that the culture broth is completely homogeneous, i.e., the concentration of microalgae is the same in all the points of the reactor, the biomass concentration dynamics in batch culture will be represented by:

$$\frac{dC_X(t)}{dt} = r_X \tag{10.1}$$

where C_X is the biomass concentration, and r_X represents the mean biomass volumetric growth rate in the photobioreactor. The growth rate is mainly function of cell and nutrients concentrations, pH, temperature, and of the incident light flux. Its modeling is detailed hereafter.

10.2.2.2 Kinetic Modeling of Photosynthetic Growth

The mean biomass volumetric growth rate r_X in (10.1) is the result of the biomass increase by photosynthesis in chloroplast (anabolism process) and its partial degradation by respiration in mitochondria (catabolism process):

$$r_X = \mu_p C_X(t) - \mu_s C_X(t) \tag{10.2}$$

where μ_p and μ_s describe the photosynthetic and respiration rates, respectively. In this study, μ_s has been assumed to be constant and not affected by the evolution of the culture medium, [12]. Under the assumption that all nutrients are introduced in excess, the photosynthetic rate, μ_p, is a function of the light received by the cell, characterized by the available light inside the culture. In a more general formulation, other nutrient limitations and/or inhibitions can be considered, using appropriate kinetic relations, [2, 12].

A Haldane type model, [10], has been proposed here to represent the local growth rate dependency on light, describing growth inhibition at high irradiance, and limitation at low irradiance (see Fig. 10.2). The specific growth rate is in this case expressed as:

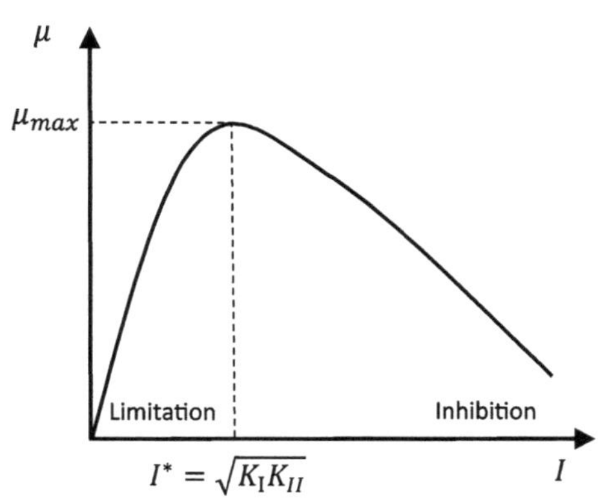

Fig. 10.2 Growth rate evolution w.r.t. light intensity (Haldane model)

$$\mu = \mu_0 \frac{I}{K_I + I + \frac{I^2}{K_{II}}} \quad (10.3)$$

where I is the spectral irradiance, K_I the irradiance half-saturation constant, and K_{II} the irradiance inhibition constant. μ_0 is a kinetic parameter representing the ideal maximum growth rate. It is related to the real maximum specific growth rate by the relation:

$$\mu_{\max} = \frac{\mu_0}{1 + 2\sqrt{K_I/K_{II}}}. \quad (10.4)$$

Indeed, the optimal growth rate in a Haldane model is obtained for $I^* = \sqrt{K_I K_{II}}$, [10]. The Haldane model (10.3) is thus applied at each location in the reactor. The mean volumetric rate is then obtained by integration of these local responses along reactor volume. For a cultivation system with one-dimensional light attenuation (flat panel, see Fig. 10.1), this consists in a simple integration along the depth of culture (z, between 0 and L):

$$\mu_p = \mu_0 \frac{1}{L} \int_0^L \frac{G(z)}{K_I + G(z) + \frac{G^2(z)}{K_{II}}} dz \quad (10.5)$$

where G is the spectral irradiance.

10.2.2.3 Radiative Model

As expressed before, the computation of the volumetric rate of growth μ_p is based on the knowledge of the available light inside the culture broth. Aside optical properties and concentration of cells, the light gradient strongly depends on the reactor geometry. In the case of flat panel photobioreactors lighted on one side, as considered here, an analytical solution of irradiance distribution is obtained [22]:

$$G(z) = 2q_0 \frac{(1+\alpha)\, e^{\delta(L-z)} - (1-\alpha)\, e^{-\delta(L-z)}}{(1+\alpha)^2 e^{\delta L} - (1-\alpha)^2 e^{-\delta L}} \quad (10.6)$$

with $\delta = C_X\sqrt{E_a(E_a + 2bE_s)}$, the two-flux extinction coefficient, and $\alpha = \sqrt{E_a/(E_a + 2bE_s)}$, the linear scattering modulus. E_a and E_s are optical parameters and represent the mass absorption and the mass scattering coefficients, and b is the backward scattering fraction. q_0 represents the hemispherical incident light flux (or incident light intensity, or photons flux density, PFD, as commonly named in photobioreactors studies, [22]). C_X represents the biomass concentration inside the photobioreactor, z is the depth of the culture, and L is the depth of the photobioreactor (Fig. 10.1). The optical parameters can be determined spectrophotometrically for any given photosynthetic organism, and depends on microorganism shape and size, as well as pigment content. Their values depend also on the incident light flux, [23]. In a first approach, their variation has been neglected.

10.2.3 Sensors and Measurements

A certain number of reliable sensors for online measurements exist, most of them being common in any biotechnological process, [13]. Aside usual sensors measuring incident radiance, temperature, pH, measurements related to the O_2 production rate and carbon dioxide uptake rate provide reliable online estimation of photosynthetic growth, especially in mineral nonlimiting conditions.

In particular, the oxygen is produced through water photolysis and is partially consumed through respiration; its kinetic rate r_{O_2} is, therefore, proportional to the photosynthetic growth rate, based on overall stoichiometry of the microorganism under study (which characterizes the yields of conversion of substrates into biomass and related products), as follows:

$$r_{O_2} = Y_{O_2/X}\, r_X. \tag{10.7}$$

In other words, the oxygen production rate of the cell is proportional to its growth rate. The conversion yield $Y_{O_2/X}$ is constant in standard nonlimiting autotrophic conditions. For *Chlamydomonas reinhardtii*, its value has been determined in batch cultivations in, [12].

The oxygen production rate (r_{O_2}) could be online estimated from dissolved oxygen data, knowing the oxygen transfer rate from the liquid to the gaseous phase. The mass balance for the dissolved oxygen assuming well-mixed liquid phase can be expressed as:

$$\frac{dC_{O_2}}{dt} = r_{O_2} + K_La(C^*_{O_2} - C_{O_2}) \tag{10.8}$$

where $C^*_{O_2}$ and C_{O_2} are the saturation and the actual concentrations of dissolved oxygen in the liquid phase, K_La is the overall volumetric mass transfer coefficient for oxygen. K_La strongly depends on photobioreactor geometry and aeration conditions, [15]. The term $K_La(C^*_{O_2} - C_{O_2})$ is known as oxygen transfer rate (OTR).

Other method uses gaseous oxygen analyzer to measure the oxygen concentration of the gas leaving the reactor. OTR can be determined from gas oxygen mass balance on the reactor as the difference between oxygen flow rate leaving the reactor and the oxygen flow rate entering in the system (if enriched air is bubbled into the culture broth).

Whatever the method for measuring the oxygen production rate, Eq. (10.7) allows to estimate online the volumetric photosynthetic growth rate. The estimated r_X will be denoted $\hat{r}_X$.

The measurement of the oxygen production rate is more informative, than a biomass measurement (via an optical sensor, for example). It allows to estimate with better accuracy the specific growth. Thus, from the estimated global growth rate $\hat{r}_X$, the estimated biomass concentration $\hat{C}_X$ can be also determined using the relation:

$$\frac{\mathrm{d}\hat{C}_X}{\mathrm{d}t} = \hat{r}_X. \tag{10.9}$$

In this study, Euler integration scheme is used to solve (10.9), since the sampling time is small in comparison to the system dynamics. Indeed, the sampling time is chosen about 10 min since the bioprocess dynamics is very slow (characteristic time about hours) and the response time of oxygen sensor is less than 1 min.

10.2.4 Overall Photosynthetic Growth Model

The model of the considered system is made of an ordinary differential equation:

$$\frac{\mathrm{d}C_X}{\mathrm{d}t} = r_X(C_X, q_0) \tag{10.10}$$

where C_X is the state variable and r_X, the model dynamics, is given by (10.2), (10.5), and (10.6). The control input is the incident light flux $u = q_0$, and the output is the estimated volumetric growth rate $y = \hat{r}_X$ (from r_{O_2}, the oxygen production rate measurements, using (10.7)). The sampled output is available every 10 min (the chosen sampling time is $T_s = 10$ min).

The considered model parameters were identified on experimental data in [12] and [15]. They are reported in Table 10.1.

Table 10.1 Model parameters

	Parameter	Value	Unit
Kinetic model	E_a	172	$m^2\,kg^{-1}$
	E_b	870	$m^2\,kg^{-1}$
	b	$8 \cdot 10^{-4}$	–
Radiative model	μ_0	0.14	h^{-1}
	K_I	120	$\mu mol\,m^{-2}\,s^{-1}$
	K_{II}	500	$\mu mol\,m^{-2}\,s^{-1}$
	μ_s	0.013	h^{-1}
Oxygen model	$Y_{O_2/X}$	1.42	g O_2/g biomass
	K_La	0.9	h^{-1}
	$C^*_{O_2}$	$4.85 \cdot 10^{-3}$	$g\,L^{-1}$
PBR[a] geometry	V	1.47	L
	L	0.04	m

[a]Photobioreactor

The process of photoautotrophic growth of microalgae in photobioreactor is thus very complex. The models used to represent the system evolution are strongly non-linear, characterized by parametric and model uncertainties and usually time varying.

10.3 Lumostatic Operating Mode

Microalgae culture in artificially lighted photobioreactors are usually supplied by a constant light flux. In this case, the incident light intensity is set at a high level, so that the growth rate is maximized (by ensuring a high average light flux inside the photobioreactor). However, at the early stage of the culture, when the cell concentration is low, there is a risk of photoinhibition (because of light excess). Consequently, the bioprocess productivity decreases, and possibly, the lag phase is longer because of cells' stress. If a low level of light flux is applied during all the culture to prevent photoinhibition phenomenon, the adverse effect occurs: with the growth of the cells, light limitation increases, leading also to a decrease of the bioprocess productivity.

Figure 10.3 illustrates the evolution of spectral irradiance G inside the photobioreactor, for different cell concentrations. This reflects the evolution of light distribution all along a batch (where the biomass concentration progressively increase). Notice that based on (10.2), it exists a radiant light energy (called compensation point, G_c) for which photosynthesis compensates the respiration ($r_X = 0$). Based on this value, the culture broth can be partitioned in two zones: one with positive growth (light zone) and one with negative growth (dark zone).

It was shown in [23] that maximal productivity will be obtained when the light intensity at the backside of the PBR equals G_c (i.e., no dark zone). However, this condition could be obtained only at low biomass concentrations. At high biomass concentrations, when the PBR is operated in batch mode, very high incident light intensity is needed to achieve this goal.

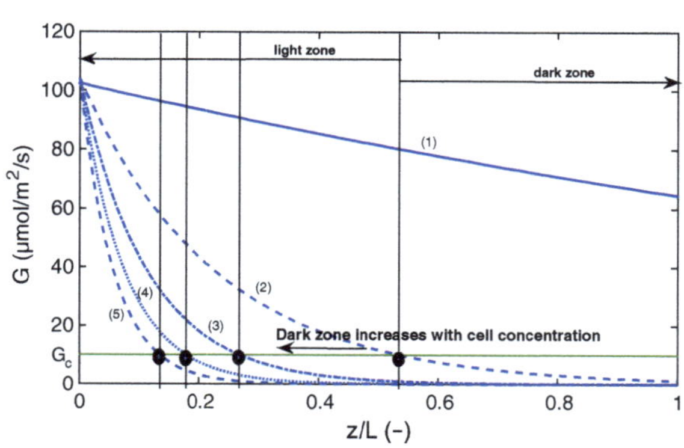

Fig. 10.3 Light distribution inside the culture for different cell concentrations: *(1)* $C_X = 0.05$ g/L, *(2)* $C_X = 0.5$ g/L, *(3)* $C_X = 1$ g/L, *(4)* $C_X = 1.5$ g/L, *(5)* $C_X = 2$ g/L

Recent studies aim at improving the batch culture by optimizing the light supply so that the phenomena of photolimitation and photoinhibition are reduced, [3, 8, 14, 17]. It was concluded that an incremental light supply reduces energy consumption compared to a conventionally operated PBR, for the same cell production. The light intensity supply is adjusted continuously at an optimal level that depends on the cell concentration, so that the cell growth is maximized. This culture strategy is called the *lumostatic operation mode* or *lumostat*.

In the literature, the incident light intensity is usually chosen as the control input, but several possibilities for the controlled outputs are proposed. The main approaches involve the specific light uptake rate, [6, 19], (and its particular formulation named light-to-microalgae ratio, [14]); and the cell growth rate, [3, 17]. In this study, these two approaches will be investigated and compared (in Sects. 10.4 and 10.5).

10.3.1 Control of the Light-to-Microalgae Ratio

The specific light uptake rate, q_e, represents the total amount of energy absorbed by the culture divided by the total microalgae present in the culture. It can be defined as:

$$q_e = \frac{(q_0 - q_L)}{C_X} \frac{A}{V} \tag{10.11}$$

where q_L is the outgoing light energy.

In batch mode, the ratio A/V is constant. Its value is optimized by an appropriate design of the PBR. Thus, the specific light uptake rate can be simplified to the light-to-microalgae ratio (i.e., q_0/C_X, the only varying term in (10.11)). q_L value is neglected; indeed it becomes zero even at low biomass concentration (see Fig. 10.3).

In this study, we will assume that the optimal lumostatic operation of the culture corresponds to maintaining the amount of absorbed light energy proportional to the cell concentration. In other words, the light intensity is increased proportionally to the cell concentration. This corresponds to maintaining the light-to-microalgae ratio constant during the entire culture. The value of this variable should be optimized to ensure high performance of the bioprocess. Indeed, if this ratio is too high, the light intensity can reach quickly its upper bounds when cell concentration increases. On the contrary, if its value is too low, lower cell concentration will be obtained, leading to longer culture duration.

10.3.2 Lumostatic Versus Constant Light Operations

In this section, cultures operated in batch for constant light and lumostatic operation modes (with constant light-to-microalgae ratio) are compared. The efficiency of light

Table 10.2 Simulation conditions

Parameter	Value	Unit
$C_X(t_0)$	0.02	$\mathrm{g\,L^{-1}}$
$q_{0\min}$	100	$\mathrm{\mu mol\,m^{-2}\,s^{-1}}$
$q_{0\max}$	800	$\mathrm{\mu mol\,m^{-2}\,s^{-1}}$
t_f	150	h

utilization has been evaluated as the biomass yield on light energy, expressed as dry weight of produced biomass per amount of photons absorbed:

$$Y_{X/q_0}(t) = \frac{\left(C_X - C_{X_0}\right) V}{A \int_0^t q_0 \mathrm{d}t} \tag{10.12}$$

where C_{X0} is the initial biomass concentration (concentration at initial time t_0, after inoculation).

A culture with conditions reported in Table 10.2 is studied. The photobioreactor is assumed to be illuminated with a constant light flux of $q_0 = 800\,\mathrm{\mu mol\,m^{-2}\,s^{-1}}$ during 150 h. The LED panel can deliver a varying light flux between $q_{0\min}$ and $q_{0\max}$. The lower bound, $q_{0\min}$, ensures to have a minimum level of light regardless the operating conditions of the culture, to prevent photolimitation. The upper bound, $q_{0\max}$, depends on the characteristics of the light source used in the experimental setup. The model parameters used for the simulation are reported in Table 10.1.

For this culture, an incremental light intensity is applied to reach $q_0 = 800$ $\mathrm{\mu mol\,m^{-2}\,s^{-1}}$ (here equal to the upper bound). To achieve this, the light flux is increased by maintaining the light-to-microalgae ratio constant and equal to $R = 5845\,\mathrm{\mu mol\,m\,s^{-1}\,kg^{-1}}$. This value is the solution of an optimization problem where the criterion to be maximized is the cell concentration at the final time. This optimization problem and its solution determination will be detailed in Sect. 10.4.1, but it is used in this section to highlight the benefits of the lumostatic strategy. Hereafter, the ratio unit will be omitted to simplify notation.

Figure 10.4 illustrates the temporal evolution of the cell concentration, the biomass yield on light energy, the specific growth rate ($\mu = r_X/C_X$), and the incident light intensity for the two batch cultures. In the case of lumostatic operation mode, the light flux is increased gradually from its minimum value until it reaches its upper bound (at time $t_{\max} = 35.6$ h). Then, the light intensity is maintained at its upper bound. This strategy yields to a higher final cell concentration than the conventional constant light culture (an increase of about 4.8 % at the final time), and higher biomass per mole of supplied photon energy (up to 25 %). These results are summarized in Table 10.3. In addition, the specific growth rate μ is higher at the early stage of the culture for the lumostatic culture than the constant light one (see Fig. 10.4). However, once the light flux reaches its upper bound at time $t_{\max}$, the growth rate becomes lower than the conventional strategy. Indeed, from time $t_{\max}$ onwards, even if the two cultures are similar (constant light cultures with the same incident light flux), the

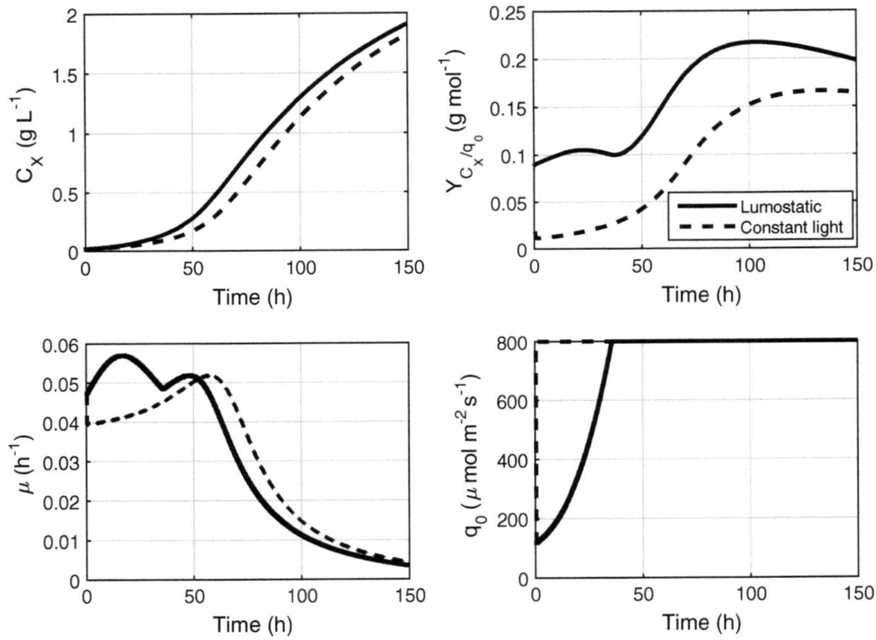

Fig. 10.4 Comparison of lumostatic and constant light batch cultures

Table 10.3 Results with a constant light and a constant light-to-microalgae ratio strategies, t_{max} is the time when the control input reaches its upper bound for the lumostatic culture

Strategy	t_{max} (h)	$C_X(t_{max})$ (g.L^{-1})	$Y_{C_X\|q_0}(t_{max})$ (g.mol^{-1})	$C_X(t_f)$ (g.L^{-1})	$Y_{C_X\|q_0}(t_f)$ (g.mol^{-1})
Constant light	0	0.08	0.03	1.81	0.16
Lumostatic	35.6	0.13 (62 %)[a]	0.10 (**233 %**)[a]	1.90 (4.9 %)[a]	0.20 (**25 %**)[a]

[a] Variation from constant light strategy to lumostatic strategy

cell concentrations are different (higher for the lumostatic culture). Consequently, the growth rate of the lumostatic culture becomes lower than the one of the conventional culture. Nevertheless, significant improvements are obtained with lumostatic strategy, mainly in terms of yield use efficiency over the entire culture duration.

Lumostatic operation leads to better performances than conventional constant light cultures. It is obvious that the superiority of the lumostatic mode is amplified in the case of higher upper bound of q_0.

As mentioned previously, the value of the light-to-microalgae ratio is a key parameter that conditions the bioprocess efficiency. In the following section, it will be optimized so that the cell production is maximized and its efficiency is evaluated in the case of model parameters' uncertainties.

10.4 Optimization of the Light-to-Microalgae Ratio

10.4.1 *Offline Optimization of the Ratio*

As highlighted in Sect. 10.3, the lumostatic cultivation mode leads to higher performances in comparison to a constant incident light one. More specifically, the light-to-microalgae ratio is maintained constant in order to increase the light availability in the bulk. This section will focus on the optimization of the lumostatic mode by controlling the light-to-microalgae ratio. In order to improve further the bioprocess efficiency, the ratio is optimized so that the biomass concentration at the final time is maximized. This light-to-microalgae ratio optimization is solved offline, based on a model of the bioprocess. The derived optimization problem is a constrained unidimensional one and can be expressed as:

$$\max_{R \in \mathbb{R}^+} C_X(t_f) \tag{10.13}$$

$$\text{s. t.} \begin{cases} q_0(t) = R C_X(t) \\ q_{0\min} \leq q_0(t) \leq q_{0\max} \\ \dfrac{\mathrm{d}C_X(t)}{\mathrm{d}t} = r_X(C_X(t), q_0(t)),\ C_X(t_0) = C_{X0} \end{cases}$$

with $t \in [t_0, t_f]$, t_0 and t_f are the initial and final times, respectively. C_X and C_{X0} are the biomass concentration and its initial value, respectively. The incident light intensity, the control input of the bioprocess, is assumed to be bounded.

The formulation of the problem given in (10.13) helps calculating the optimal ratio that maximizes the biomass growth, while ensuring that the light intensity respects the bounds constraints. The optimization problem (10.13) could be further transformed into an unconstrained one, by removing constraints on the light intensity and to simply bound the control input during the model dynamics integration. Since it is a unidimensional unconstrained maximization problem, it can be solved by a Golden Section Search algorithm, [11]. The transformation of problem (10.13) into an unconstrained optimization problem is possible because of the optimality properties of the optimal solution of (10.13). Indeed, since the ratio is constant, the light intensity temporal evolution will be proportional to the cell concentration one. Consequently, the higher the cell concentration is, the higher the corresponding light intensity will be, reminding that the optimal cell concentration trajectory has an increasing behavior. The limitation on the light intensity (linked to the light panel maximal capacity) limits the possible applied light intensity, so it can be reached at an instant either previous to or after the final time, depending on the photoinhibition phenomenon occurrence.

Once the solution of problem (10.13), denoted as R_{opt}, is determined, it is applied online to the real system as depicted by Fig. 10.5. From the measurement of dissolved oxygen concentration, the cell concentration is calculated as detailed in Sect. 10.2.3.

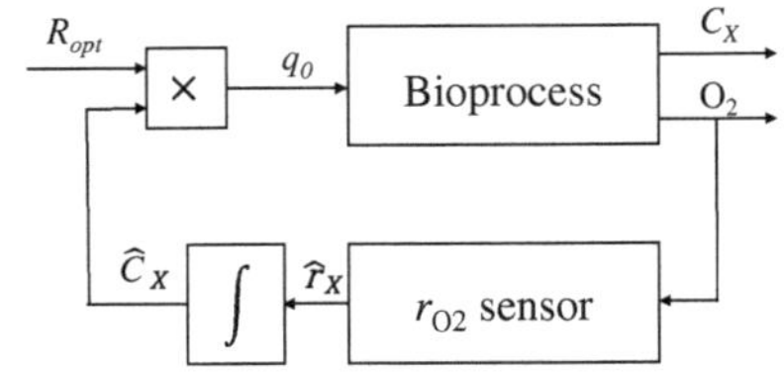

Fig. 10.5 Bioprocess control with a fixed constant ratio

Then, at each instant time t_k, the incident light intensity to be applied to the photobioreactor is calculated online as follows:

$$q_0(t_k) = \begin{cases} q_{0\min}, \text{ if } R_{opt}\hat{C}_X(t_k) < q_{0\min} \\ q_{0\max}, \text{ if } R_{opt}\hat{C}_X(t_k) > q_{0\max} \\ R_{opt}\hat{C}_X(t_k), \text{ otherwise} \end{cases} \tag{10.14}$$

The light intensity $q_0(t_k)$ is set piecewise constant over the time interval $[t_k, t_{k+1}[$:

$$q_0(t) = q_0(t_k)\,\forall t \in [t_k, t_{k+1}[. \tag{10.15}$$

The control law is detailed by Algorithm 10.1.

Algorithm 10.1 Constant optimal ratio

1. *Determine R_{opt} by solving Problem (10.13)*
2. *For each time index k:*
   a. *Measure r_{O_2} and calculate $\hat{C}_X$ from (10.7) and (10.9)*
   b. *Calculate q_0 with (10.14) and apply it to the bioprocess with a zero-order hold*
   c. *Go to step 2.*

10.4.2 Online Optimization of the Ratio

In the previous section, the light-to-microalgae ratio is calculated offline, using a model of the microalgae growth. However, when applied to the real photobioreactor, the determined solution can be suboptimal because of model uncertainties. In order to improve the lumostatic cultivation efficiency, an alternative approach consists in optimizing online the ratio as depicted by Fig. 10.6.

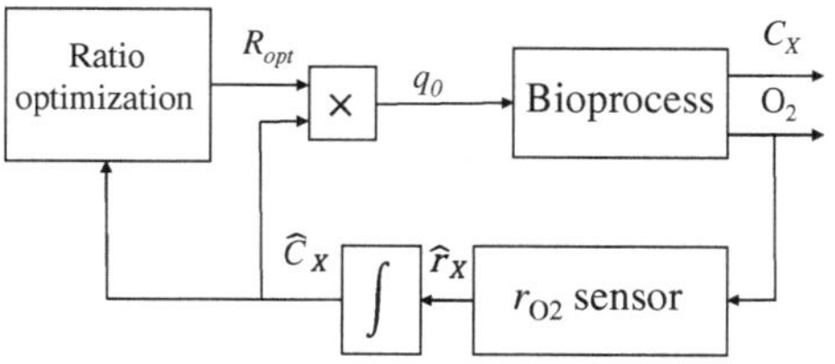

Fig. 10.6 Bioprocess control with an online optimized ratio

Thus, at each time index k, the ratio is determined by solving a problem similar to (10.13), over the time interval $[t_k, t_f]$, and with the current state value as initialization:

$$\max_{R \in \mathbb{R}^+} C_X(t_f) \qquad (10.16)$$

$$\text{s. t. for } t \in [t_k, t_f] \begin{cases} q_O(t) = R C_X(t) \\ q_{O\min} \le q_O(t) \le q_{O\max} \\ \dfrac{\mathrm{d}C_X(t)}{\mathrm{d}t} = r_X(C_X(t), q_O(t)),\ C_X(t_k) = \hat{C}_X(t_k). \end{cases}$$

Algorithm 10.2 Time-varying optimal ratio

1. *For* $k = 0$*, apply* q_O *given by the offline optimization procedure (Sect. 10.4.1)*
2. *For each time index k:*

   a. *Measure* r_{O_2} *and calculate* $\hat{C}_X$ *from (10.7) and (10.9)*
   b. *Update* R_{opt} *by solving Problem (10.16) over time interval* $[t_k, t_f]$
   c. *Calculate* q_O *with (10.14) and apply* q_O *to the bioprocess with a zero-order hold*
   d. *Go to step 2.*

As detailed in the above algorithm, this new approach induces solving an optimization problem at each time step. Nevertheless, the computation burden is still limited since this problem can be formulated as a unidimensional one (that could be solved by a Golden Section Search algorithm, [11], as previously). In addition, in the case of small model uncertainties, the optimal ratio is close to the value determined offline (Sect. 10.4.1). The search interval can be then chosen according to the ratio optimal value determined by the offline optimization procedure. Last, it should be reminded that the sampling time for this application is quite large (about 10 min), allowing solving, online, complex optimization problems.

The update of the optimal ratio value is performed at each time index. If the computation burden becomes critical, the simplest way to carry out this control strategy is to consider two sampling times: the ratio is updated with a large sampling time (typically 1–2 h), whereas the light intensity is calculated with a small sampling time (e.g., 10 min), so that the growth of the cells is taken into consideration appropriately. Indeed, to keep the ratio between the light intensity and the biomass concentration approximatively constant, the light intensity should be increased frequently so that it follows the exponential growth of microalgae. The use of a first-order hold for the control input may improve further the results, but this solution is not adequate for experimental implementation for this process.

Finally, it should be mentioned that the online optimization-based approach leads to a time-varying light-to-microalgae ratio since its value is updated at each time step. The ratio at each time index indirectly takes into account model mismatch (from the actual biomass concentration). An improvement of the efficiency of this control law could be to include an adaptive identification of the model, so that the online optimization procedure uses an accurate up-to-date model of the bioprocess.

10.4.3 Simulation Results

In this section, the performances obtained by the two previously presented control strategies are tested and compared in simulation. The simulation conditions are those given by Table 10.2. The values of model parameters considered in this simulation study are given by Table 10.1. The sampling time is chosen constant and equals 10 min in accordance with the probes constant time (Sect. 10.2.3). Measurements are corrupted by a centered Gaussian white noise with standard deviation of 10 %. The real system is assumed to have the same mathematical structure as the model, but with different parameter values. In order to simplify the study, we will consider a mismatch on the value of the specific growth rate μ_0 only. Three cases are studied hereafter (where μ_0^r denotes the real maximal specific growth rate):

- **Case a**: a lower growth of microalgae than expected ($\mu_0^r = 0.09\,\mathrm{h}^{-1}$)
- **Case b**: a higher growth than expected ($\mu_0^r = 0.16\,\mathrm{h}^{-1}$)
- **Case c**: the growth is low at the early stage of the culture (corresponding to a lag phase of the microalgae growth) and then becomes higher than expected at the end of the culture (corresponding to an exponential phase of the microalgae growth). This model of growth rate translates the adaptation of microorganisms at the beginning of the culture (up to 2 days) to the cultivation conditions [14]. In our case, the following specific growth rate was considered:

$$\begin{cases} \mu_0^r = 0.09\,\mathrm{h}^{-1} \text{ if } t_0 \le t < 24\,\mathrm{h} \\ \mu_0^r = 0.16\,\mathrm{h}^{-1} \text{ if } 24\,\mathrm{h} \le t \le t_f. \end{cases} \tag{10.17}$$

The results obtained for these cases and for the two control strategies are given in Figs. 10.7, 10.8, and 10.9. The control law that maintains a constant ratio optimized offline (Sect. 10.4.1) is referred to as *constant ratio* (in dashed line), whereas the controller where the ratio is updated at each time step (Sect. 10.4.2) is referred to

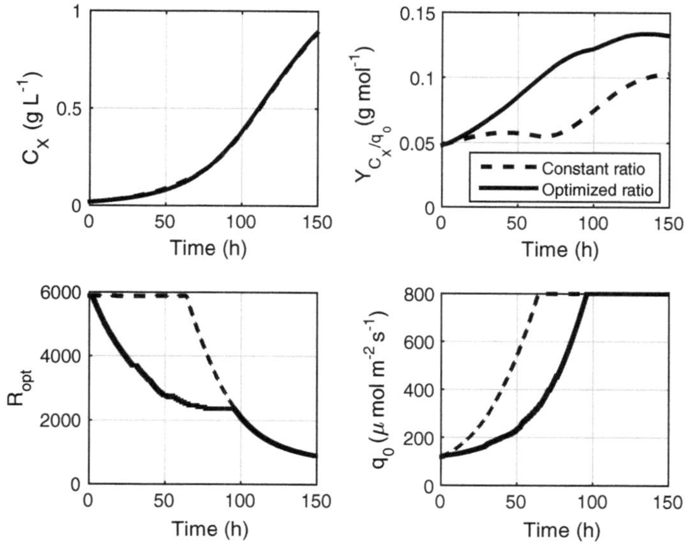

Fig. 10.7 Simulation results of the control of the light-to-microalgae ratio (**Case a**)

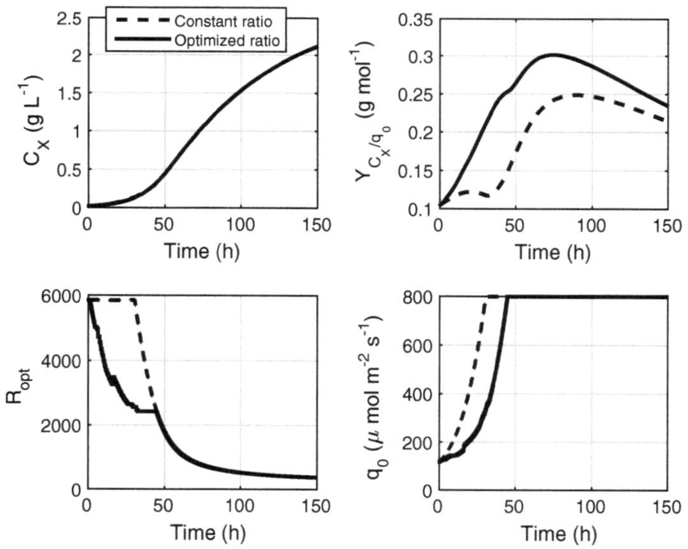

Fig. 10.8 Simulation results of the control of the light-to-microalgae ratio (**Case b**)

as *optimized ratio* (in full line). These figures compare the temporal evolution of biomass concentration, yield, ratio, and light intensity obtained under controlled light intensity with both control laws. These results are also summarized in Table 10.4.

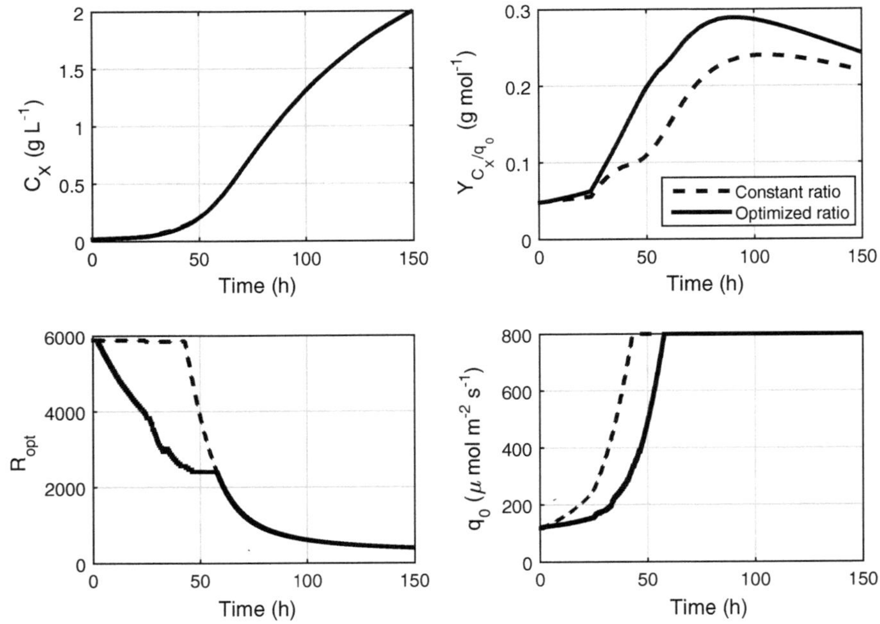

Fig. 10.9 Simulation results of the control of the light-to-microalgae ratio (**Case c**)

Table 10.4 Results with the control of the light-to-microalgae ratio ($t_{\max}$ is the time when the control input reaches its upper bound) and variation of the results of online optimization compared with the offline approach

	Optimization approach	$t_{\max}$ (h)	$C_X(t_{\max})$ (g.L^{-1})	$Y_{C_X\vert q_0}(t_{\max})$ (g.mol^{-1})	$C_X(t_f)$ (g.L^{-1})	$Y_{C_X\vert q_0}(t_f)$ (g.mol^{-1})
Case a	Offline	63.6	0.13	0.05	0.88	0.10
	Online	95.5	0.33	0.12	0.89	0.13
	Variation	+50 %	+153 %	**+140 %**	1 %	**+30 %**
Case b	Offline	30.1	0.13	0.21	2.11	0.21
	Online	44.5	0.32	0.25	2.11	0.23
	Variation	+47 %	+153 %	**+19 %**	<0.01 %	**+9.5 %**
Case c	Offline	42.8	0.13	0.09	1.99	0.22
	Online	57.6	0.33	0.22	1.99	0.24
	Variation	+34 %	+153 %	**+144 %**	<0.01 %	**+9 %**

From Figs. 10.7, 10.8, and 10.9, it can be observed that the optimal ratio variation can be divided into two phases:

- Phase 1: from the start of the cultivation until the time $t_{\max}$ when the control input q_0 reaches its upper bound $q_{0\max}$.
- Phase 2: from time $t = t_{\max}$ onwards. The control input is constant and equals its upper bound ($q_0(t) = q_{0\max}$).

In the first phase, the ratio is either constant (for the first control strategy), or time-decreasing (when it is updated online), whereas its evolution is given by $q_{0_{\max}}/\hat{C}_X(t)$ in the second phase.

The benefit of the control of the ratio is mainly present in the first phase since in the second one, the ratio value is dictated by the cell growth. An improvement could be to shorten the batch culture and to end it at time $t_{\max}$ instead of time t_f. At the end of the batch culture, either the microalgae can be harvested, or the bioprocess operation switches to a continuous mode by applying a flow rate through the reactor (fresh medium is continuously added and reactor fluid is continuously removed) with fresh medium. In the latter case, microalgae or high-added value products can be produced over long time periods.

From Table 10.4, it can be noticed that the online optimization of the ratio leads to a higher biomass yield on light energy in all cases (up to 30 % over the entire culture duration). The biomass production is slightly the same with both control strategies.

For the online optimization approach, the incident light irradiance reaches its upper bound later in comparison to the constant ratio strategy (between 34 and 50 % of increase of this duration). The ratio is indeed adjusted online by optimizing the light absorption by microalgae. It can be observed that the best performances are obtained in the case of overestimated growth rate (**case a** and early stage of **case c**). The control strategy adapts the ratio to the real characteristics of microalgae growth, avoiding photoinhibition by limiting light excess. On the contrary, when the growth rate is underestimated (**case b** and last stage of **case c**), the control law takes advantage of the microalgae higher growth rate to increase further the bioprocess performance.

Remark: the lumostatic strategy (with a constant ratio) was compared to the constant light strategy for these three cases of model parameter uncertainty, assessing the superiority of the lumostatic approach, similarly to the results in the nominal case in Sect. 10.3.2.

Controlling the light-to-microalgae ratio by updating its reference value online leads to high performance of the cultivation process. It helps reducing the light utilization by adjusting the light intensity depending on the real cell growth. The implementation of this control strategy is quite simple, with a reduced computation load even if it involves the solution of an optimization problem at each time index.

10.5 Predictive Control of Microalgae Growth

10.5.1 Control of the Growth Rate

Previously, the control strategy consists in optimizing the light-to-microalgae ratio either offline or online. Another strategy consists in maintaining the growth rate at an optimal value, using the incident light intensity as the control input.

In [17], this control problem was addressed by means of a feedforward controller. The control strategy consists of two steps. First, the optimal average light intensity in

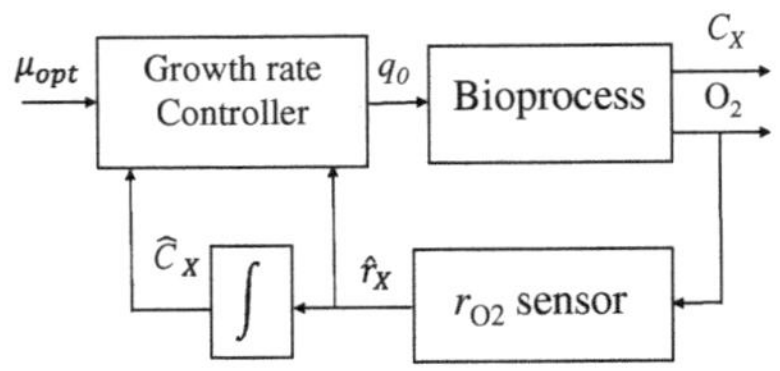

Fig. 10.10 Scheme of the control of microalgae growth rate

the photobioreactor that maximizes the growth rate is empirically determined via an optimal search procedure using Brent's method. Then, the incident light intensity is generated from this optimal average light irradiance and from the measured biomass concentration.

In this study, the knowledge of the growth model structure will be used to determine the optimal growth rate. Then, a controller will be designed to track this optimal value.

As detailed in Sect. 10.2.2.2, the growth model considered in this study follows a Haldane kinetics that takes into consideration the light limitation and inhibition phenomena on the cell growth. In the case of this growth kinetics, the maximum specific growth rate, $\mu_{\max}$, is related to μ_0 by relation (10.4).

The optimal growth rate, μ_{opt}, that can be achieved in the photobioreactor is then obtained from this maximal value and by taking into account respiration phenomenon (Sect. 10.2.2.2):

$$\mu_{opt} = \mu_{\max} - \mu_s. \tag{10.18}$$

The control strategy presented hereafter consists in maintaining the growth rate at this optimal value, by appropriately adjusting the incident light intensity. The controller structure is depicted by Fig. 10.10.

10.5.2 Predictive Control Strategy

Since the bioprocess model is nonlinear with respect to control and states variables, a nonlinear model predictive controller (NMPC) is selected. The advantage of this kind of control law is that it can be easily used for nonlinear systems, especially for the tracking of a reference trajectory for system outputs. In addition, it can take into account constraints on the control input and/or state trajectory. The principle of the NMPC law is to create an anticipated effect, based on the prediction of the future behavior of the system. The control inputs are derived from the minimization, online, using this prediction, of an appropriate cost function under operating constraints, [20]. The minimization of the tracking error over a moving horizon determines the optimal sequence of the control inputs. Since microalgae growth dynamics is highly nonlinear, the NMPC control strategy allows achieving high performances for the bioprocess operation, [1, 24].

This strategy is applied to the studied system where the criterion to be minimized is the tracking error of growth rate μ, over the prediction interval of length N; and the control inputs are the incident light intensities over this interval. The optimization problem to be solved at each time index k to determine the sequence of optimal control inputs is expressed as:

$$\begin{aligned}\min_{\mathbf{Q}_k} \sum_{j=0}^{N-1} & \left(\mu_{ref}(t_{k+j}) - \mu(t_{k+j})\right)^2 + \lambda\left(q_0(t_{k+j}) - q_0(t_{k+j-1})\right)^2 \\ & + \left(\mu_{ref}(t_N) - \mu(t_N)\right)^2 \\ s.t.\ & q_{0\min} \leq q_0(t_{k+j-1}) \leq q_{0\max} \text{ for } j = 1 \ldots N\end{aligned} \tag{10.19}$$

with k the time index, N the prediction horizon length, μ_{ref} and μ are the reference and real ratio, respectively, and λ is a weighting factor on the control variation. The second term in the criterion (10.19) allows smoothing the control evolution. The last term in the criterion, the so-called terminal cost, is added to guarantee stability of the algorithm. Nevertheless, it should be pointed out that the stability is not a critical issue in the case of this application. Indeed, the culture duration is fixed and the evolution of the state variables is limited (it depends on the maximal growth capacity of the cells). The prediction horizon length N and the weighting factor λ are tuning parameters of the controller. Their choice is dictated by a trade-off between computation load and the accuracy of the prediction of the future behavior of the system (and consequently of the controller performance). In (10.19), $\mathbf{Q}_k$, the optimization variable, is the vector of control inputs over the prediction horizon, defined as:

$$\mathbf{Q}_k = (q_0(t_k), \ldots, q_0(t_{k+N-1})). \tag{10.20}$$

The control inputs are set piecewise constant over each interval $[t_{k+j}, t_{k+j-1}[$, $j = 0 \ldots N$, similarly to (10.15). The criterion (10.19) involves the prediction of the growth rate $\mu(t_{k+j})$, derived from the growth model (10.2):

$$\mu(t_{k+j}) = \frac{r_X(C_X(t_{k+j}), q_0(t_{k+j}))}{C_X(t_{k+j})}. \tag{10.21}$$

The prediction of biomass concentrations $C_X(t_{k+j})$ is obtained by the integration of dynamics (10.1), starting from $\hat{C}_X(t_k)$ and applying the control sequence $\mathbf{Q}_k$. The global volumetric growth rate r_X in (10.21) is given by model (10.2).

The reference value of the growth rate, μ_{ref}, is set equal to μ_{opt} (defined by (10.18)). However, because of model mismatch between the real system and the prediction model, a tracking error will be induced. A simple way to reduce this error consists in correcting the reference value by considering the error between predicted and real value at the current time index k:

$$\mu_{ref}(t_{k+j}) = \mu_{opt} + \left(\frac{r_X(\hat{C}_X(t_k), q_0(t_k))}{\hat{C}_X(t_k)} - \frac{\hat{r}_X(t_k)}{\hat{C}_X(t_k)}\right), \quad j = 1 \ldots N \tag{10.22}$$

where $\hat{r}_X(t_k)$ and $\hat{C}_X(t_k)$ are measured global volumetric growth rate and measured cell concentration, respectively (from measurement of r_{O_2} as detailed in Sect. 10.2.3), r_X is given by (10.2).

The optimization problem (10.19) is solved at each time index using the determined control inputs sequence at the previous time step as initial guess. The first control input of the optimal sequence is applied until the next time index, as per the moving horizon principle. This optimization problem could be solved either by a Sequential Quadratic Programming technique [11] or by an interior-reflective Newton method [7]. The latter approach is the most suitable since the problem (10.19) is a nonlinear least-squares problem with only bounds constraints on the optimization variables (no nonlinear constraints on the optimization variables). The controller algorithm is summarized by Algorithm 10.3.

Algorithm 10.3 Predictive control of the specific growth rate

1. *At initial time* t_0, *apply* $q_0(t_0) = q_{0\min}$ *with a zero-order hold*
2. *For each time index k:*

    a. *Measure* r_{O_2} *and determine* $\hat{r}_X$ *and* $\hat{C}_X$ *from (10.7) and (10.9)*
    b. *Update* μ_{ref} *with (10.22)*
    c. *Solve Problem (10.19) and determine the optimal sequence* $\boldsymbol{Q}_k^*$
    d. *Apply the first value of the optimal sequence* $\boldsymbol{Q}_k^*$ *to the bioprocess and to the model, with a zero-order hold*
    e. *Go to step 2.*

10.5.3 Simulation Results

The performance of the predictive controller to track the optimal value of the specific growth rate is tested in simulation, for the same case study as in Sect. 10.4.3 (**case a** for an underestimated growth, **case b** for overestimated growth, and **case c** for varying growth rate). The predictive controller performance is compared to the one obtained with application of an online optimized ratio (presented in Sect. 10.4.2). The same simulation conditions as in Sect. 10.4.3 are considered. The prediction horizon was chosen equal to 1 h ($N = 6$) and the weighting factor λ was fixed to normalize the two terms in the criterion (10.19) and to give the tracking error more importance than the smoothing of the control evolution. The results presented hereafter are obtained with $\lambda = 0.1\mu_{ref}$. For the prediction model, the optimal specific growth rate given by (10.18) equals $0.057\,\mathrm{h}^{-1}$. The predictive controller will regulate the specific growth rate of the microalgae to this value. It should be mentioned that the real optimal specific growth rates for the mismatched models are 0.032 and $0.067\,\mathrm{h}^{-1}$ for **case a** and **case b**, respectively. Figures 10.11, 10.12, and 10.13 illustrate the temporal evolution

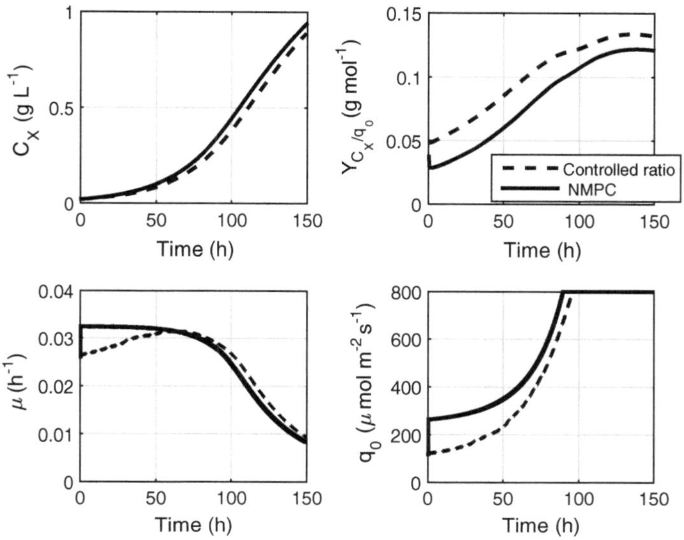

Fig. 10.11 Simulation results of the control of the growth rate (**Case a**)

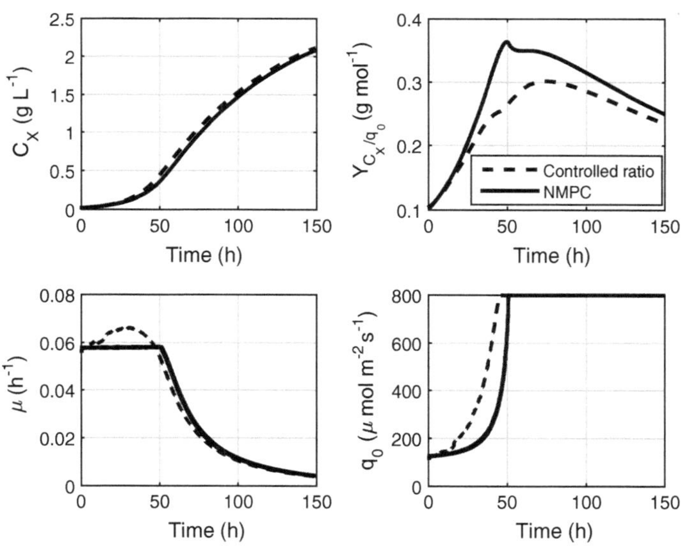

Fig. 10.12 Simulation results of the control of the growth rate (**Case b**)

of the biomass concentration, the biomass yield on light energy, the specific growth rate, and the incident light energy, obtained with the application of the predictive controller and of the online optimized ratio strategy. These results are summarized in Table 10.5.

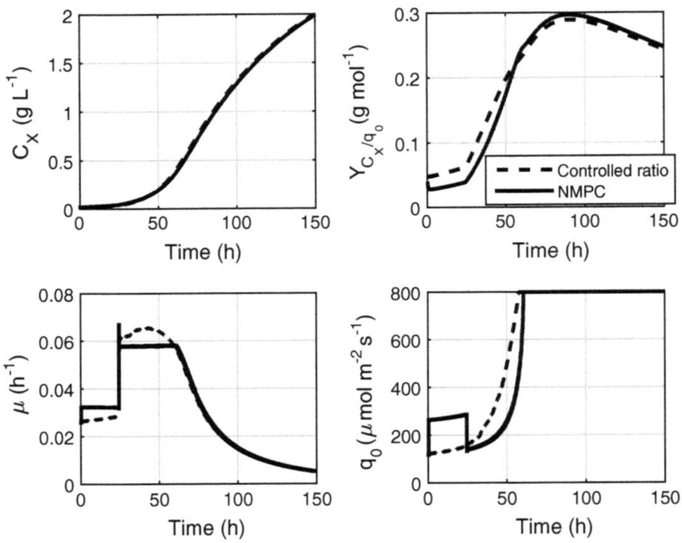

Fig. 10.13 Simulation results of the control of the growth rate (**Case c**)

Table 10.5 Results with the control of the growth rate (t_{max} is the time when the control input reaches its upper bound) and comparison to the results with the control of the ratio

	Controlled output	t_{max} (h)	$C_X(t_{max})$ (g.L^{-1})	$Y_{C_X\|q_0}(t_{max})$ (g.mol^{-1})	$C_X(t_f)$ (g.L^{-1})	$Y_{C_X\|q_0}(t_f)$ (g.mol^{-1})
Case a	Ratio	95.5	0.33	0.12	0.89	0.13
	μ	89.5	0.33	0.1	0.93	0.12
	variation	−6.2 %	−0.9 %	**−16 %**	+4.5 %	**−7.6 %**
Case b	Ratio	44.5	0.32	0.25	2.11	0.23
	μ	50	0.37	0.36	2.08	0.25
	variation	+12 %	+15 %	**+44 %**	−1.5 %	**+8.7 %**
Case c	Ratio	57.6	0.33	0.22	1.99	0.24
	μ	60.5	0.36	0.24	1.98	0.25
	variation	+5 %	+9 %	**+9 %**	−0.5 %	**4 %**

From the above-mentioned figures, it can be observed that the growth rate evolution can be divided into two phases:

- Phase 1: from the start of the cultivation until the time t_{max} when the control input q_0 reaches its upper bound q_{0max}. In this phase the growth rate is quasi-constant.
- Phase 2: from time $t = t_{max}$ onwards. The control input equals its upper bound ($q_0(t) = q_{0max}$) and consequently the growth rate decreases. Indeed, since the cell concentration increases with time, the light availability inside the reactor decreases, leading to a decrease in the growth rate.

In the case of an underestimated growth (**case a**, Fig. 10.11), the predictive controller regulates μ at a lower value than expected. Indeed, in this case, the maximal specific growth rate that can be reached by the microalgae equals $0.032\,h^{-1}$. Thus, the controller maintains the system at its optimal growth capacity, leading to a higher biomass growth than the one obtained by an online optimized ratio. Nevertheless, this increase of the biomass growth induces an increase of the applied light intensity, and thus to a decrease of the yield (of $-7.6\,\%$ at the final time), in comparison to the results obtained with the control of the ratio.

On the contrary, when the optimal growth rate is attainable, the predictive controller succeed to regulate the biomass growth rate to the desired value (**case b**, Fig. 10.12). In this case, the yield is higher than the one obtained with the control of the ratio, with a slight decrease of the biomass growth. The bioprocess is not operated at its optimal capacity (in this case, the real maximal specific growth rate equals $0.067\,h^{-1}$), but the global performance is good (an increase of about 44 % of the yield at t_{max} in comparison to the ratio control approach).

In the last case (**case c**, Fig. 10.13), the predictive controller regulates the growth rate either at a lower level than the reference when the latter is not attainable (early stage of the culture), or to the desired level when the real optimal value is higher than expected. The yield increases up to 9 % in comparison to the control of the ratio approach at time t_{max} (Table 10.5).

Globally, the predictive strategy improves the performance of the microalgae cultivation in comparison to the control of the ratio. Nevertheless, it suffers from a higher sensitivity to the model accuracy than the previous method, more particularly to the value of the maximal growth rate. An effort for the determination of the real maximal growth rate of the microalgae is required, either prior to the batch culture [17], or by an online identification of the model parameters.

10.6 Conclusion

The optimization of the batch cultivation of microalgae was considered, by focusing the study on the lumostatic operation mode. In this mode, the light intensity is controlled so that the cell growth and the light energy absorption are maximized. It yields to higher performances than the classical operation at constant incident light intensity (up to 25 % of increase of the yield and about 5 % of cell concentration ate the final time in simulation for a *Chlamydomonas reinhardtii* culture). In order to improve further the lumostatic mode performance, two optimization-based approaches were proposed. First, the control of the light-to-microalgae ratio was developed. In this case, the incident light intensity is increased following the cell growth. The online optimization of the ratio leads to at least 9 % of increase of the biomass production. In the second approach, a predictive controller is proposed to regulate the specific growth rate to its optimal value. The performance of the lumostatic mode is improved in comparison to the online optimized ratio strategy. Nevertheless, simulations highlighted the sensitivity of this approach to the model accuracy. This strategy should be further investigated to improve its performances. From the simulation study, the control of the ratio presents the best compromise between performance, computation load, and robusteness with respect to model mismatch. These strategies' efficiency should be demonstrated through experimental implementation so that their performances are assessed when applied to a real bioprocess.

Acknowledgments Authors wish to acknowledge support from the Romanian French research cooperation project PACBIO and from the French ANR Program "Algo-H2" ANR-10-BIOE-0004.

References

1. J. Abdollahi, S. Dubljevic, Lipid production optimization and optimal control of heterotrophic microalgae fed-batch bioreactor. Chem. Eng. Sci. **84**, 619–627 (2012)
2. O. Bernard, Hurdles and challenges for modelling and control of microalgae for CO_2 mitigation and biofuel production. J. Process Control **21**, 1378–1389 (2011)
3. X. Chen, Q.Y. Goh, W. Tan, I. Hossain, W.N. Chen, R. Lau, Lumostatic strategy for microalgae cultivation utilizing image analysis and chlorophyll a content as design parameters. Bioresour. Technol. **102**, 6005–6012 (2011)
4. P. Chevalier, D. Proulx, P. Lessard, W.F. Vincent, J. de la Noue, Nitrogen and phosphorus removal by high latitude mat-forming cyanobacteria for potential use in tertiary wastewater treatment. J. Appl. Phycol. **12**, 105–112 (2000)
5. Y. Chisti, Biodiesel from microalgae. Biotechnol. Adv. **25**, 294–306 (2007)
6. S.L. Choi, I.S. Suh, C.-G. Lee, Lumostatic operation of bubble column photobioreactors for Haematococcus pluvialis cultures using a specific light uptake rate as a control parameter. Enzyme Microb. Technol. **33**, 403–409 (2003)
7. T.F. Coleman, Y. Li, An interior, trust region approach for nonlinear minimization subject to bounds. SIAM J. Optim. **6**, 418–445 (1996)
8. P. Das, J.P. Obbard, Incremental energy supply for microalgae culture in a photobioreactor. Bioresour. Technol. **102**, 2973–2978 (2011)
9. B. Degrenne, J. Pruvost, M. Titica, H. Takache, J. Legrand, Kinetic modeling of light limitation and sulphur deprivation effects in the induction of hydrogen production with Chlamydomonas reinhardtii Part II: Definition of Model-Based Protocols and Experimental Validation. Biotechnol. Bioeng. **108**, 2288–2299 (2011)
10. D. Dochain (ed.), *Automatic Control of Bioprocesses* (ISTE and Wiley, London, 2008)
11. R. Fletcher, *Practical Methods of Optimization*, 2nd edn. (Wiley-Interscience, New York, 1987)
12. S. Fouchard, J. Pruvost, B. Degrenne, M. Titica, J. Legrand, Kinetic modeling of light limitation and sulfur deprivation effects in the induction of hydrogen production with Chlamydomonas reinhardtii: Part I. Model development and parameter identification. Biotechnol. Bioeng. **102**, 232–245 (2009)
13. I. Havlik, P. Lindner, T. Scheper, K.F. Reardon, On-line monitoring of large cultivations of microalgae and cyanobacteria. Trends Biotechnol. **31**, 406–414 (2013)
14. G. Ifrim, M. Titica, L. Boillereaux, S. Caraman, Feedback linearizing control of light-to-microalgae ratio in artificially lighted photobioreactors, in *Computer Applications in Biotechnology Conference* (Mumbai, India, 2013)
15. G.A. Ifrim, M. Titica, G. Cogne, L. Boillereaux, J. Legrand, S. Caraman, Dynamic pH model for autotrophic growth of microalgae in photobioreactor: a tool for monitoring and control purposes. AIChE J. **60**, 585–599 (2014)
16. L. Jiang, S. Luo, X. Fan, Z. Yang, R. Guo, Biomass and lipid production of marine microalgae using municipal wastewater and high concentration of CO_2. Appl. Energy **88**, 3336–3341 (2011)

17. R. Kandilian, T.-C. Tsao, L. Pilon, Control of incident irradiance on a batch operated flat-plate photobioreactor. Chem. Eng. Sci. **119**, 99–108 (2014)
18. M.K. Lam, K.T. Lee, A.R. Mohamed, Current status and challenges on microalgae-based carbon capture. Int. J. Greenh. Gas Control **10**, 456–469 (2012)
19. H.-S. Lee, Z.-H. Kim, S.-E. Jung, J.-D. Kim, C.-G. Lee, Specific light uptake rate can be served as a scale-up parameter in photobioreactor operations. J. Microbiol. Biotechnol. **16**, 1890–1896 (2006)
20. D.Q. Mayne, J.B. Rawlings, C.V. Rao, P.O.M. Scokaert, Constrained model predictive control: stability and optimality. Automatica **36**, 789–814 (2000)
21. A. Pandey (ed.), *Biofuels: Alternative Feedstocks and Conversion Processes* (Academic Press, San Diego, 2011)
22. L. Pottier, J. Pruvost, J. Deremetz, J.F. Cornet, J. Legrand, C.G. Dussap, A fully predictive model for one-dimensional light attenuation by Chlamydomonas reinhardtii in a torus reactor. Biotechnol. Bioeng. **91**, 569–582 (2005)
23. H. Takache, G. Christophe, J.F. Cornet, J. Pruvost, Experimental and theoretical assessment of maximum productivities for the microalgae Chlamydomonas reinhardtii in two different geometries of photobioreactors. Biotechnol. Prog. **26**, 431–440 (2010)
24. S. Tebbani, F. Lopes, R. Filali, D. Dumur, D. Pareau, Nonlinear predictive control for maximization of CO_2 bio-fixation by microalgae in a photobioreactor. Bioprocess Biosyst. Eng. **37**, 83–97 (2014)

Chapter 11
Bioprocesses Parameter Estimation by Heuristic Optimization Techniques

Dorin Şendrescu, Sihem Tebbani and Dan Selişteanu

Abstract This work presents a bioprocesses parameter estimation method based on heuristic optimization approaches. The identification problem is formulated as a multimodal numerical optimization problem in a high-dimensional space. Then, the optimization problem is split in simpler sub-problems that require fewer computational resources. The main results are obtained using genetic algorithms (GA) and particle swarm optimization (PSO) methods. One applies three global-search metaheuristic algorithms for numerical optimization: two variants of PSO and one type of genetic algorithm. The estimation procedures are applied for identification of a bacterial growth model associated with the enzymatic catalysis where reaction kinetics is described by Monod and Haldane models. The performances of the proposed methods are analysed by numerical simulations. The simulation results indicate that the proposed metaheuristic algorithms are effective and efficient, and demonstrate that the applied techniques exhibit a significant performance improvement over classical optimization methods.

Keywords Parameter estimation · Biotechnology · Particle swarm optimization · Genetic algorithms

D. Şendrescu (✉) · D. Selişteanu
Department of Automation and Electronics, University of Craiova, Craiova, Romania
e-mail: dorins@automation.ucv.ro

D. Selişteanu
e-mail: dansel@automation.ucv.ro

S. Tebbani
Laboratory of Signals and Systems, CentraleSupélec-CNRS-Université Paris-Sud, Université Paris-Saclay, Control Department, 3 rue Joliot-Curie, F91192 Gif-sur-Yvette, France
e-mail: sihem.tebbani@centralesupelec.fr

S. Olaru et al. (eds.), *Developments in Model-Based Optimization and Control*, Lecture Notes in Control and Information Sciences 464,
DOI 10.1007/978-3-319-26687-9_11

11.1 Introduction

The biotechnology is one of the fields that over the last decades has a high development. Biotechnology applications can be found especially in agriculture, in food industry, in medicine and pharmaceutical processes, in waste treatment processes, etc. Due to its advantages, the control of industrial bioprocesses is an important practical problem attracting wide attention. A frequent and important challenge in control of such living processes is finding an accurate model of the system, [2]. The bioprocesses are highly nonlinear and their kinetic parameters are usually badly or inadequately known. This problem becomes of great importance in complex systems where critical oscillations (functioning) and instability of the process must be avoided. Parameters characterizing the internal behaviour of biotechnological systems are usually not directly accessible to measurement and their attainment is usually approached indirectly as a parameter estimation problem, [3]. In this work a dynamic model describing the internal structure of a biotechnological system is formulated and a parameter estimation method based on heuristic optimization algorithms is designed.

In recent years, a progress has been made in the area of continuous-time system identification, [14, 22]. Even if most physical systems are naturally continuous, a much more attention has been paid to parameter estimation of discrete-time systems, mainly because they are better suited for numerical implementations. Continuous-time identification makes possible a more direct link to the physical properties and operation of the underlying systems and the direct estimation of physical parameters which have a clear significance. The most common approach for parameter estimation of linear or nonlinear systems is the use of prediction error identification methods (PEM), [15]. In this category falls the well-known least squares methods or the maximum likelihood methods. In this approach, identification consists of minimization of a scalar-valued function of the model parameters. In general, this function cannot be minimized by analytical methods so, the solution has to be found by iterative, numerical techniques. In classical approach the most used are the quasi-Newton methods and interior point algorithms. The main drawback of these nonlinear parameter optimization techniques is that they are often unreliable, e.g., they give no guarantee of converging to a global minimum. Indeed, they converge to a local minimum that depends on the algorithm initialization. The increasing computational power of personal computers and microcontrollers allowed the implementation of several optimization algorithms inspired from natural phenomena. Examples of these algorithms include the simulated annealing, [13], genetics algorithms (GA), [10], or ant colony optimization, [5], algorithms. Particle swarm optimization (PSO), [12], is among these nature inspired algorithms and it is inspired by the ability of birds flocking to find food that they have no previous knowledge of its location. Every member of the swarm is affected by its own experience and its neighbours' experiences. Although the idea behind PSO is simple and can be implemented by two lines of programming code, the emergent behaviour is complex and hard to completely characterize, [11]. GAs are inspired on principles of natural selection and genetics. GAs encode the decision variables in the sets of solutions called chromosomes, formed by different parts (genes) with some values (alleles). Once a problem is encoded in chromosomes and a fitness measure for selecting good solutions (usually the objective function value) has been chosen, the population can start to evolve.

The most important approaches (based on heuristic optimization) for the yield and kinetic coefficients estimation of biotechnological systems make use of the state transformations based on the general structure, [3]. The main drawback of these methods is the computation time which increase exponentially with the population size, [20, 23]. In this paper we propose a multi-step identification method based on particle swarm optimization and genetic algorithms techniques for these classes of biotechnological systems considering that the unknown parameters can appear in rational relations with measured variables. The principle of multi-step method is to split general estimation problem in several simpler sub-problems that require a reduced population size for optimization.

This chapter is an extended work of the research achieved in [19]. The chapter is organized in the following way. The nonlinear dynamical model of a bacterial growth process associated with the enzymatic catalysis is given in Sect. 11.2. Section 11.3 presents the identification method using the particle swarm optimization and genetic algorithms techniques. Some numerical simulations are presented in Sect. 11.4, and conclusions in Sect. 11.5.

11.2 Nonlinear Dynamical Model of Bacterial Growth Bioprocesses

A process that takes place in a bioreactor can be described as a set of m biochemical reactions involving n components (with $n \geq m$). The global dynamics can be represented by the following dynamical state-space model [2], assuming a well-stirred bioreactor:

$$\frac{\mathrm{d}\xi}{\mathrm{d}t} = K \cdot \Phi(\xi, t) - D\xi + F - Q \tag{11.1}$$

where $\xi \in \mathbb{R}^{n \times 1}$ is the components concentrations, K is a matrix containing the yield coefficients, Φ is the reaction rate vector, D is the dilution rate (medium feed flow rate over effective volume), F the vector of feeding rates and Q the vector of rates of removal of the components.

This model describes the behaviour of an entire class of biotechnological processes and is referred to as the general dynamical state-space model of this class of bioprocesses, [11]. In (11.1), the term $K \cdot \Phi(\xi, t)$ is the rate of consumption and/or production of the components in the reactor, i.e., the reaction kinetics and the term $-D\xi + F - Q$ represents the exchange with the environment. The strongly nonlinear character of the model (11.1) is given by the reaction kinetics. In many situations, the yield coefficients, the structure and the parameters of the reaction rates are partially known or unknown. Many of the evolved control methods for these kind of systems—like sliding mode control, [18], robust or adaptive control, [1]—are based on good initial estimates of the yield and kinetic parameters.

The identification scheme proposed in this chapter is illustrated using bacterial growth model associated with the enzymatic catalysis. The simplified reaction scheme is given by two reactions, [18, 19]:

$$S + O \xrightarrow{\mu_1} X + P_1, \tag{11.2}$$

$$S + X \xrightarrow{\mu_2} X + P_2. \tag{11.3}$$

In the reactions (11.2)–(11.3), S is the substrate, X the biomass, O is the dissolved oxygen, P_1 and P_2 are the biosynthesis products and μ_1 and μ_2 are the specific growth rates and are function of the concentrations of the components. The simplified dynamical model of the bioprocess with the reaction schemes (11.2)–(11.3) can be obtained by using the mass balance of the components, considering that the reaction takes place in a fed-batch bioreactor:

$$\frac{\mathrm{d}S}{\mathrm{d}t} = (-k_1\mu_1(S,O,X,P_1,P_2) - k_2\mu_2(S,O,X,P_1,P_2))X - DS + F_1 \tag{11.4}$$

$$\frac{\mathrm{d}O}{\mathrm{d}t} = -k_3\mu_1(S,O,X,P_1,P_2)X - DO + F_2 \tag{11.5}$$

$$\frac{\mathrm{d}X}{\mathrm{d}t} = \mu_1(S,O,X,P_1,P_2)X - DX \tag{11.6}$$

$$\frac{\mathrm{d}P_1}{\mathrm{d}t} = k_4\mu_1(S,O,X,P_1,P_2)X - DP_1 \tag{11.7}$$

$$\frac{\mathrm{d}P_2}{\mathrm{d}t} = \mu_2(S,O,X,P_1,P_2)X - DP_2. \tag{11.8}$$

The dynamical model is expressed as a set of differential equations, in terms of the concentrations of components. In (11.4)–(11.8), all concentrations are in g/L, $k_i, i = 1\ldots4$ are the positive yield coefficients (in g/g), F_1 is the input feed rate (in g/L/h) and $F_1 = D * S_{in}$ with D the so-called dilution rate (in h^{-1}) and S_{in} the concentration of influent substrate (in g/L) and F_2 the oxygen supply flow (in g/L/h). The model (11.4)–(11.8) can be expressed in the matrix form:

$$\begin{bmatrix} \dot{S} \\ \dot{O} \\ \dot{X} \\ \dot{P}_1 \\ \dot{P}_2 \end{bmatrix} = \begin{bmatrix} -k_1 & -k_2 \\ -k_3 & 0 \\ 1 & 0 \\ k_4 & 0 \\ 0 & 1 \end{bmatrix} \begin{bmatrix} \mu_1(S,O,X,P_1,P_2)X \\ \mu_2(S,O,X,P_1,P_2)X \end{bmatrix} - D \begin{bmatrix} S \\ O \\ X \\ P_1 \\ P_2 \end{bmatrix} + \begin{bmatrix} F_1 \\ F_2 \\ 0 \\ 0 \\ 0 \end{bmatrix}. \tag{11.9}$$

The next notations will be used: the state vector is $\xi = [S\,O\,X\,P_1\,P_2]^T = [\xi_1\,\xi_2\,\xi_3\,\xi_4\,\xi_5]^T$, the vector of reaction rates (the reaction kinetics) is $\Phi(\xi) =$

$[\,\mu_1(\xi)X\ \mu_2(\xi)X\,]^T$, K is the matrix of yield coefficients, $F = [\,F_1\ F_2\ 0\ 0\ 0\,]^T$ is the vector of feeding rates. With these notations, the system can be compactly written in the form (11.1).

The most difficult task for the construction of the dynamical model is the modelling of the reaction kinetics. The form of kinetics is complex, nonlinear and in many cases unknown. In our study one considers that reaction rates are given by the Monod law

$$\mu_1(\xi) = \mu_1^* \frac{S}{K_{M_1} + S}, \tag{11.10}$$

and the Haldane kinetic model

$$\mu_2(\xi) = \mu_2^* \frac{S}{K_{M_2} + S + S^2/K_i} \tag{11.11}$$

where K_{M_1}, K_{M_2} are Michaelis–Menten constants (in g/L), μ_1^*, μ_2^* represent maximum specific growth rates (in h^{-1}) and K_i is the inhibition constant (in g/L) .

For simplicity, we shall denote the plant parameters by the vector:

$$\theta = [\,\theta_1\ \theta_2\ \theta_3\ \theta_4\ \theta_5\ \theta_6\ \theta_7\ \theta_8\ \theta_9\,]^T \tag{11.12}$$

where:

$$\begin{aligned} &\theta_1 = k_1;\ \theta_2 = k_2;\ \theta_3 = k_3;\ \theta_4 = k_4; \\ &\theta_5 = \mu_1^*;\ \theta_6 = \mu_2^*;\ \theta_7 = K_{M_1};\ \theta_8 = K_{M_2};\ \theta_9 = K_i. \end{aligned} \tag{11.13}$$

Because the dilution rate D can be externally modified, it will be considered as the first component of the input vector

$$u = \begin{bmatrix} u_1\ u_2\ u_3 \end{bmatrix}^T. \tag{11.14}$$

The other two components of u are S_{in} and F_2 so,

$$u_1 = D;\ u_2 = S_{in};\ u_3 = F_2. \tag{11.15}$$

Remark 11.1 In the following, one considers that the full system state is measurable.

11.3 Parameter Estimation using Heuristic Optimization

At the beginning of parameter estimation, the input and output data are known and the real system parameters are assumed as unknown. The full state vector is considered to be measured and the objective is to determine the values of model parameters from these data. The identification problem is formulated in terms of an optimization problem in which the error between an actual physical measured response of the system and the simulated response of a parameterized model is minimized. The estimation of the system parameters is achieved as a result of minimizing the error function by the heuristic algorithm.

11.3.1 Problem Statement

Consider the following n-dimensional nonlinear system:

$$\frac{d\xi(t)}{dt} = f(\xi,\ t,\ \theta) \tag{11.16}$$

where

$\xi \in \mathbb{R}^n$ is the state vector,
$\theta \in \mathbb{R}^m$ is the unknown parameters vector,
f is a given nonlinear vector function.

To estimate the unknown parameters in (11.16), a parameter identification system is defined as follows:

$$\frac{d\hat{\xi}(t)}{dt} = f(\hat{\xi},\ t,\ \hat{\theta}) \tag{11.17}$$

where

$\hat{\xi} \in \mathbb{R}^n$ is the estimated state vector,
$\hat{\theta} \in \mathbb{R}^m$ is the estimated parameters vector.

The objective function defined as the mean squared errors between real and estimated responses for a set of samples is considered as fitness of estimated model parameters:

$$W = \frac{1}{N+M} \sum_{j=1}^{M} \sum_{k=1}^{N} \left(\xi_j^k - \hat{\xi}_j^k\right)^2 \tag{11.18}$$

s.t. relation (17) is fulfilled.

where M is the number of measurable states, N is the data length used for parameter identification, whereas ξ_j^k and $\hat{\xi}_j^k$ are the real and estimated values of state j at time k, respectively.

This objective function is difficult to minimize because there are many local minima, due for example to measurement noises, and the global minimum has a very narrow domain of attraction. Our goal is to determine the system parameters, using heuristic optimization algorithms in such a way that the value of W is minimized, approaching zero as much as possible. Because this optimization problem requires great computational resources, a multi-step approach was used (that will be presented in Sect. 11.3.4).

Remark 11.2 In the following, one analyses the noise-free case and the case when an additive measurement noise is included in model (11.16).

11.3.2 Overview of Basic PSO Algorithms

PSO algorithms have gained much attention and wide applications in different fields due to their effectiveness in addressing difficult optimization issues, as well as simplicity of implementation and ability to fast converge to a reasonably good solution. PSO is a population-based heuristic global optimization technique, first introduced by Kennedy and Eberhart, [12], and referred to as a swarm intelligence technique. It is motivated from the simulation of social behaviour of animals such as bird flocking, fish schooling and swarm. In this algorithm, the population is called a swarm, and the trajectory of each particle in the search space is controlled through the medium of a term called "velocity", according to its own flying experience and swarm experience in the search space. Mathematical description of basic PSO and some important variants is presented in the following.

Candidate solutions of a population called particles coexist and evolve simultaneously based on knowledge sharing with neighbouring particles. Each particle represents a potential solution to the optimization problem and it has a fitness value decided by optimal function. Supposing search space is D-dimensional, each individual is treated as a particle in the D-dimensional search space. The position and rate of position change for ith particle can be represented by D-dimensional vector, $x_i = (x_{i1}, x_{i2}, \ldots, x_{iD})$ and $v_i = (v_{i1}, v_{i2}, \ldots, v_{iD})$, respectively. The best position previously visited by the ith particle is recorded and represented as $p_i = (p_{i1}, p_{i2}, \ldots, p_{iD})$, called *pbest*. The swarm best position previously visited by all the particles in the population is represented as $p_g = (p_{g1}, p_{g2}, \ldots, p_{gD})$, called *gbest*. Then particles search their best position, which are guided by swarm information p_g and their own information p_i. Each particle modifies its velocity to find a better solution (position) by applying its own flying experience (i.e., memory of the best position found in earlier flights) and the experience of neighbouring particles (i.e., the best solution found by the population). Each particle position is evaluated by using fitness function and updates its position and velocity according to the following equations:

$$
\begin{aligned}
v_i^{k+1} &= \omega \cdot v_i^k + c_1 r_1 (pbest_i^k - x_i^k) + c_2 r_2 (gbest_i^k - x_i^k) \\
x_i^{k+1} &= x_i^k + v_i^{k+1}
\end{aligned}
\tag{11.19}
$$

where k is iteration number, ω is inertia weight, c_1 and c_2 are two acceleration coefficients regulating the relative velocity towards local and global best position, r_1 and r_2 are two random numbers from interval [0, 1]. Many effects have been made over the last decade to determinate the inertia weight. Various studies have shown that under certain conditions convergence is guaranteed to a stable equilibrium point. These conditions include:

$$\omega > \frac{c_1 + c_2}{2} - 1 \text{ and } 0 < \omega < 1. \tag{11.20}$$

The technique originally proposed was to bound velocities so that each component of v_i is kept within the range $[-V_{\max}, +V_{\max}]$.

Unfortunately, this simple form of PSO suffers from the premature convergence problem, which is particularly true in complex problems since the interacted information among particles in PSO is too simple to encourage a global search. Many efforts have been made to avoid the premature convergence. One solution is the use of a constriction factor to ensure convergence of the PSO, introduced in [4]. Thus, the expression for velocity has been modified as:

$$\begin{aligned} v_i^{k+1} &= h \cdot [v_i^k + c_1 r_1 (pbest_i^k - x_i^k) + c_2 r_2 (gbest_i^k - x_i^k)] \\ x_i^{k+1} &= x_i^k + v_i^{k+1} \end{aligned} \tag{11.21}$$

where h represents the constriction factor and is defined as

$$h = \frac{2}{\left|2 - \alpha - \sqrt{\alpha^2 - 4\alpha}\right|} \tag{11.22}$$

$$\alpha = c_1 + c_2 > 4. \tag{11.23}$$

In this variant of the PSO algorithm, h controls the magnitude of the particle velocity and can be seen as a dampening factor. It provides the algorithm with two important features, [6]. First, it usually leads to faster convergence than standard PSO. Second, the swarm maintains the ability to perform wide movements in the search space. The constriction PSO has the potential to avoid being trapped in local optima while possessing a fast convergence.

It is shown that a larger inertia weight tends to facilitate the global exploration and a smaller inertia weight achieves the local exploration to fine-tune the current search area. The best performance could be obtained by initially setting ω to some relatively high value (e.g., 0.9), which corresponds to a system where particles perform extensive exploration, and gradually reducing ω to a much lower value (e.g., 0.4), where the system would be more dissipative and exploitative and would be better at homing into local optima. In [21], a linearly decreased inertia weight ω over time is proposed, where ω is given by the following equation:

$$\omega = (\omega_i - \omega_f) \cdot \frac{k_{\max} - k}{k_{\max}} + \omega_f \tag{11.24}$$

where ω_i, ω_f are starting and final values of inertia weight, respectively; $k_{\max}$ is the maximum number of the iteration and k is the current iteration number. It is generally taken that starting value $\omega_i = 0.9$ and final value $\omega_f = 0.4$.

On the other hand, in [17] was introduced a PSO with time-varying acceleration coefficients. The improvement has the same motivation and the similar techniques as the adaptation of inertia weight. In this case, the cognitive coefficient c_1 is decreased linearly and the social coefficient c_2 is increased linearly over time as follows:

$$c_1 = (c_{1f} - c_{1i}) \cdot \frac{k_{\max} - k}{k_{\max}} + c_{1i} \tag{11.25}$$

$$c_2 = (c_{2f} - c_{2i}) \cdot \frac{k_{\max} - k}{k_{\max}} + c_{2i} \tag{11.26}$$

where c_{1i} and c_{2i} are the initial values of the acceleration coefficients c_1 and c_2; c_{1f} and c_{2f} are the final values of the acceleration coefficients c_1 and c_2, respectively. Usually, $c_{1i} = 2.5$; $c_{2i} = 0.5$; $c_{1f} = 0.5$ and $c_{2f} = 2.5$. Considering known all the states of the nonlinear system (11.16) at the sampling moments $k * T_S$ (T_S = sampling period), the PSO algorithm has the following steps:

Algorithm 11.1 PSO algorithm

1. *Initialize a population of particles with random positions and velocities on D dimensions in search space.*
2. *For each particle, evaluate the desired optimization fitness function* (11.18) *in D variables.*
3. *Compare particle's fitness evaluation with its pbest. If the current value is better than pbest, then set pbest equal to the current value x_i in D-dimensional space.*
4. *Identify the particle in swarm with the best success so far, and assign its index to the variable gbest.*
5. *Change the velocity and position of the particle according to the equations* (11.21).
6. **if** *a criterion is met (usually a sufficiently good fitness or a maximum number of iterations)*

   then *Stop;*
   else
   go to Step 2.

11.3.3 Overview of Genetic Algorithms

Implementation of a genetic algorithm (GA) begins with a population of random chromosomes. The algorithm then evaluates these structures and allocates reproductive opportunities such that chromosomes which represent a better solution to the problem are given more chance to "reproduce". In selecting the best candidates, new fitter offspring are produced and reinserted, and the less fit removed. In using operators such as crossover and mutation the chromosomes exchange their characteristics. The suitability of a solution is typically defined with respect to the current population. GA techniques have a solid theoretical foundation, based on the Schema Theorem. GAs are often viewed as function optimizers, although the range of problems to which they have been applied is broad, including: pattern discovery, signal processing and training neural networks, [9, 16]. To illustrate the working process of genetic algorithm, the steps to realize a basic GA are listed:

Algorithm 11.2 GA algorithm

1. *Represent the problem variable domain as a chromosome of fixed length; choose the size of the chromosome population N, the crossover probability* P_c *and the mutation probability* P_m.
2. *Define a fitness function to measure the performance of an individual chromosome in the problem domain. The fitness function establishes the basis for selecting chromosomes that will be mated during reproduction.*
3. *Randomly generate an initial population of size N.*
4. *Calculate the fitness of each individual chromosome.*
5. *Select a pair of chromosomes for mating from the current population. Parent chromosomes are selected with a probability related to their fitness. High fit chromosomes have a higher probability of being selected for mating than less fit chromosomes.*
6. *Create a pair of offspring chromosomes by applying the genetic operators.*
7. *Place the created offspring chromosomes in the new population.*
8. *Repeat Step 5 until the size of the new population equals that of initial population, N.*
9. *Replace the initial (parent) chromosome population with the new (offspring) population.*
10. *Go to Step 4, and repeat the process until the termination criterion is satisfied.*

A GA is an iterative process. Each iteration is called a generation. A typical number of generations for a simple GA can range from 50 to over 500. A common practice is to terminate a GA after a specified number of generations and then examine the best chromosomes in the population. If no satisfactory solution is found, then the GA is restarted, [7, 8]. Indeed, multiple runs of the optimization algorithm are performed to get a good estimation of its performance. Depending on this performance, tuning methods can be applied to obtain a good behaviour of the algorithm, [8].

11.3.4 Implementation of Multi-step Parameter Estimation

In the following, a multi-step parameter estimation version is proposed and an optimal set of yield and kinetic parameter values of the bacterial growth process is obtained. In order to implement a heuristic optimization-based technique, the model of bacterial growth process obtained from the macroreactions schemes is used, translated into the parameter identification system framework represented in (11.9).

To facilitate the application of the proposed parameter estimation strategy, the time derivatives of the states from model (11.9) must be reconstructed. Because the measured data are usually very few, an interpolation method is necessary to find intermediate values of the states, which are actually the biological parameters of the process, i.e., the concentrations of reactants. Such situations with a small number of experimental measurements are typical for many bioprocesses. Indeed, there is a lack for reliable and non-expensive sensors for this kind of systems. Measurements are usually performed offline, with a large sampling time (for example, the determination of the concentration of biomass is usually performed by cell counting with a microscope). Ideally speaking, the online measurements (in each sampling moment) for each concentration are necessary. However, these online measurements are achieved with expensive instrumentation, or there are no such fast sensors for some concentrations. Thus, the infrequent offline measurements are preferred. To facilitate the achievement of an accurate estimation of model parameters, we need the interpolation of these measured data, which allows us to obtain the unavailable data between adjacent measurement points (i.e., to estimate the unavailable data needed to calculate model predictions between these measurement points).

Remark 11.3 From mathematical point of view, a discussion about the interpolation technique can be done. Many authors use the linear interpolation, with advantages related to rapidity and simple implementation. However, the linear interpolation is not very precise. Another disadvantage is that the interpolant is not differentiable at the points where the value of the function is known. Therefore, we propose a cubic interpolation method that is the simplest method that offers true continuity between the measured data. A cubic Hermite spline or cubic Hermite interpolator is a spline where each piece is a third-degree polynomial specified in Hermite form: that is, by its values and first derivatives at the end points of the corresponding domain interval. Cubic Hermite splines are typically used for interpolation of numeric data specified at given argument values $t_1, t_2, \ldots, t_n$ to obtain a smooth continuous function. The Hermite formula is applied to each interval (t_k, t_{k+1}) separately. The resulting spline will be continuous and will have continuous first derivative.

The time derivatives of the states are approximated using central differences:

$$\frac{d\xi(t_k)}{dt} \approx \frac{\xi(t_k + T_S) - \xi(t_k - T_S)}{2T_S} \tag{11.27}$$

Table 11.1 The sub-problems solved by using the multi-step approach

Sub-problem	Estimated parameters	Number of state equation
P1	μ_1^*, K_{M1}	3
P2	μ_2^*, K_{M2}, K_i	5
P3	k_4	4
P4	k_3	2
P5	k_1, k_2	1

where T_S represents the sampling period. In this approximation, the error is proportional with the sampling interval (a smaller sampling period will give a smaller approximation error).

Because a 9-dimensional optimization problem (for parameters defined by (11.12) and (11.13)) that must be solved for simultaneous estimation of all unknown parameters requires great computational resources, a multi-step approach was used. So, the problem was split in five simpler problems that are solved sequentially until all 9 parameters are found. These problems are denoted by P1, P2,…, P5 and the corresponding resulted parameters are presented in Table 11.1. There are also presented the state equations involved in the corresponding sub-problem. For example, the problem P1 corresponds to the third equation from system (11.4)–(11.8) (that represents time evolution of the biomass) and only two parameters must be estimated in this case: μ_1^* and K_{M1}. The heuristic optimization algorithm is used to minimize the sum of the square errors between measured and estimated data:

$$W_{P1} = \sum_{k=1}^{N} \left(\xi_3(k \cdot T_S) - \hat{\xi}_3(k \cdot T_S) \right)^2 \qquad (11.28)$$

$$\text{s.t. } \hat{\xi}_3((k+1) \cdot T_S) = \hat{\xi}_3(k \cdot T_S) + T_S \cdot \frac{\mu_1^* \cdot \xi_1(k \cdot T_S) \cdot \xi_3(k \cdot T_S)}{K_{M1} + \xi_1(k \cdot T_S)}. \qquad (11.29)$$

The main purpose of this decomposition in simpler problems is to speed up the estimation procedure and a good performance is obtained if the number of parameters that must be estimated in one sub-problem is as small as possible (in this case, a small number of particles is necessary). The structure presented in Table 11.1 is the simplest and is based on the assumption that full system state is measurable but other combinations of sub-problems can be developed. For example, parameters estimated in sub-problem P1 (that characterize the specific growth rate μ_1) can be obtained from any of the first four state equations, and the parameters estimated in sub-problem P2 (that characterize the specific growth rate μ_2) can be obtained from state equations (11.1) or (11.5).

Remark 11.4 Parameters estimated in sub-problems P1, P2 and P5, i.e., (μ_1^*, K_{M_1}, μ_2^*, K_{M_2}, K_i, k_1, k_2), can be obtained simultaneously from the first state equation using a single sub-problem. In this case, it is not necessary that states 3 and 5 to be measurable.

11.4 Simulation Results

The efficacy of our approach is shown by numerical simulations on an interval of 30 h. The model given by relation (11.9) was integrated using a fourth-order Runge–Kutta routine with a sampling period of 1 min (T_S =1 min) and with initial conditions: IC = [6 2 1 0.06 0.008] g/L. The optimization algorithms were implemented on a computer with Intel Core i5, 64-bit, 3.3 GHz processor.

The influence of sampling period, type of the optimization algorithm and of noisy measurements is analysed. To compare statistical performances of the different approaches the empirical normalized mean square error (NMSE) was used, that is defined as:

$$NMSE = \frac{1}{m}\sum_{j=1}^{m} NMSE(\hat{\theta}_j) \tag{11.30}$$

with $NMSE(\hat{\theta}_j) = \left(\frac{\hat{\theta}_j - \theta_j^*}{\theta_j^*}\right)^2$, where m is the number of estimated parameters, $\hat{\theta}_j$ is the jth element of the estimated parameter vector, while the '$*$' superscript denotes the true value of the parameter. The estimation method was tested using three optimization algorithms (for a sampling period $T_S = 1$ min and noise free measurements):

- **Type 1**: PSO algorithm based on relation (11.21) with h defined by relation (11.22) $c_1 = c_2 = 2.1$;
- **Type 2**: PSO algorithm based on relation (11.19) with ω, c_1, c_2 defined by relation (11.24) and (11.25) and $\omega_i = 0.9, \omega_f = 0.4, c_{1i} = 2.5, c_{2i} = 0.5, c_{1f} = 0.5, c_{2f} = 2.5$;
- **Type 3**: genetic algorithm.

The results are presented in Table 11.2.

In order to study the sensitivity of the estimation method to the sampling period of PSO algorithm (**Type 1**) the following sampling periods were used:

$$T_S \in \{1\,\text{min},\ 10\,\text{min},\ 30\,\text{min}\}.$$

The results are presented in Table 11.3.

The influence of noisy measurements (using also PSO algorithm—Type 1) was analysed using a zero mean gaussian white noise with following signal-to-noise ratio (SNR):

$$SNR = \{50dB, 40dB, 30dB\}.$$

The results of these simulations are presented in Table 11.4. The simulations for optimization algorithm were performed with a number of particles between 15 and

Table 11.2 Influence of optimization algorithm type

	Estimated coefficients			
	"Real value"	Type of optimization algorithm		
		Type 1	Type 2	Type 3
θ_1	1	1.0166	0.9288	1.0542
θ_2	2	1.9486	2.2733	1.6852
θ_3	1	0.9999	0.9949	1.0006
θ_4	1.5	1.5000	1.5000	1.5027
θ_5	1	1.0020	1.0555	0.9881
θ_6	2	1.9370	1.9419	2.1539
θ_7	1	1.0028	1.1559	0.9563
θ_8	20	18.9822	19.5968	20.5016
θ_9	10	9.3692	10.0865	9.5672
NMSE	0	9.45e-04	0.0012	0.0320

Table 11.3 Influence of the sampling period (Optimization algorithm—Type 1)

	Estimated coefficients			
	"Real value"	Sampling period		
		$T_S = 1\,\text{min}$	$T_S = 10\,\text{min}$	$T_S = 30\,\text{min}$
θ_1	1	1.0011	0.9846	1.0831
θ_2	2	1.9980	2.1402	1.6398
θ_3	1	1.0000	0.9993	1.0046
θ_4	1.5	1.5000	1.4946	1.5045
θ_5	1	0.9999	0.9961	0.9673
θ_6	2	2.0237	1.9874	2.0709
θ_7	1	0.9999	0.9959	0.8817
θ_8	20	20.2339	20.7856	19.7722
θ_9	10	9.9841	9.8054	11.0265
NMSE	0	3.13e-05	2.21e-04	0.003

100. All the presented results are obtained for a number of 30 particles. For a greater number of particles the accuracy of the estimates was not better.

The parameter estimates are very good, results given in Tables 11.2–11.4 showing that the performance of the algorithm deteriorates in the following cases:

- at the increasing of sampling period;
- when ω, c_1, c_2 are kept constant;
- at high level of noise.

Also, the PSO algorithms lead to better results than GA.

Table 11.4 Noise influence (Optimization algorithm—Type 1)

	Estimated coefficients			
	"Real value"	Noise level		
		SNR = 50 dB	SNR = 40 dB	SNR = 30 dB
θ_1	1	1.0821	0.9253	0.9666
θ_2	2	1.5984	2.1694	2.0836
θ_3	1	1.0041	0.9959	0.9923
θ_4	1.5	1.5010	1.5010	1.4771
θ_5	1	0.9866	1.0763	0.9854
θ_6	2	2.0342	1.9723	1.7536
θ_7	1	0.9564	1.2388	1.0069
θ_8	20	20.1248	19.7956	17.6142
θ_9	10	10.1156	11.0008	7.0000
NMSE	0	2.83e-04	0.0081	0.1364

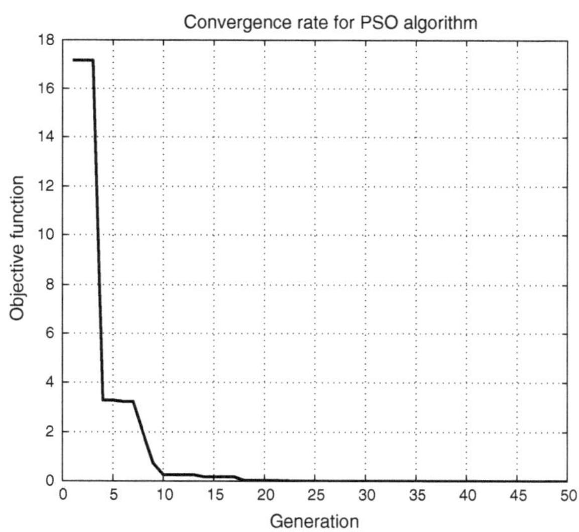

Fig. 11.1 Convergence rate for PSO algorithm in sub-problem P1 case

In Figs. 11.1 and 11.2 are presented the convergence rates for Type 1 (PSO) and Type 3 (GA) algorithms for sub-problem P1 using a sampling period of 1 min. The algorithms have a good convergence rate, the optimization tolerance being achieved after 20 iterations in PSO case and 40 iterations in GA case.

As was specified in Sect. 11.3.4, the main advantage of the multi-step approach is the superior speed to estimate the parameters using the same optimization algorithm. In Table 11.5, a comparison between multi-step approach (called in the following **Method 1**) and simultaneously estimation of all parameters (**Method 2**) is given. The simulations were performed using a **Type 1** algorithm (PSO with constriction

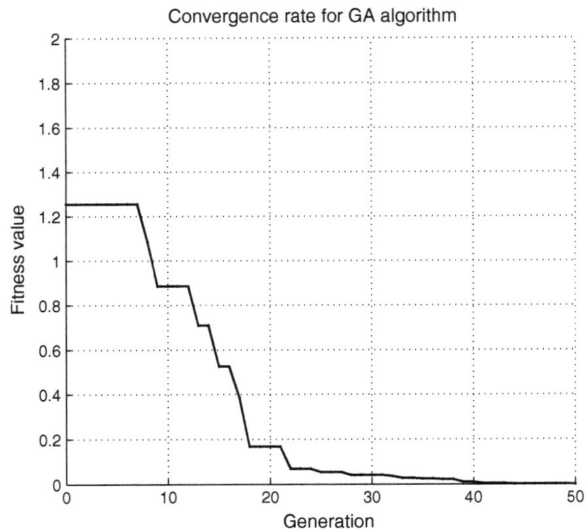

Fig. 11.2 Convergence rate for GA algorithm in sub-problem P1 case

factor), noise-free case with sampling period $T_S = 1\,\mathrm{min}$. For solving optimization sub-problems in **Method 1** a number of 20 particles were used, and for **Method 2**, 80 particles were used.

Remark 11.5 As it can be seen in Table 11.5, **Method 1** is five times faster than **Method 2**. Also, the estimation results are slightly better in **Method 1** case. These can be explained by the fact that optimization problem in **Method 2** is more complex with many local minima.

Table 11.5 Influence of optimization algorithm type

	Estimated coefficients		
	"Real value"	Type of optimization algorithm	
		Method 1	Method 2
θ_1	1	1.0011	1.0476
θ_2	2	1.9979	1.6885
θ_3	1	0.9999	0.9590
θ_4	1.5	1.5000	1.4582
θ_5	1	0.9999	0.9523
θ_6	2	2.1323	2.0724
θ_7	1	1.0000	1.0282
θ_8	20	21.3394	20.2341
θ_9	10	8.3762	11.1848
NMSE	0	0.0039	0.0053
Simulation time		89 s	461 s

11.5 Conclusions

This chapter presents an offline estimation procedure based on heuristics for global optimization for identification of yield and kinetics coefficients in a microbial growth model associated with the enzymatic catalysis bioprocess. The identification problem is formulated in terms of an optimization problem in which the error between an actual physical measured response of the system and the simulated response of a parameterized model is minimized. This function is multimodal and classical iterative methods fail to find the global optimum. The estimation of the system parameters is achieved as a result of minimizing the error function by the PSO and genetic algorithms. The optimization problem is split in simpler sub-problems that require fewer computational resources. The simulations evaluated the effects of sampling period, some basic variants of PSO and GA algorithms and of the noisy measurements. The proposed strategy can still converge to accurate results even in the presence of measurement noise, as illustrated by the numerical study. The obtained dynamical model of the bacterial growth process is accurate and can contribute to the development of model-based applications, which lead to high productivity and better quality products. The proposed estimation approach can also be applied to other bioprocesses belonging to the nonlinear class considered in the present study.

Acknowledgments This work was supported by UEFISCDI, project ADCOSBIO no. 211/2014, PN-II-PT-PCCA-2013-4-0544.

References

1. S.F. Azevedo, P. Ascencio, D. Sbarbaro, An adaptive fuzzy hybrid state observer for bioprocesses. IEEE Trans. Fuzzy Syst. **12**, 641–651 (2004)
2. G. Bastin, D. Dochain, *On-line Estimation and Adaptive Control of Bioreactors* (Elsevier, New York, 1990)
3. L. Chen, Modelling, Identifiability and Control of Complex Biotechnological Systems, Ph.D. thesis, Universit Catholique de Louvain, Belgium (1992)
4. M. Clerc, J. Kennedy, The particle swarm: explosion, stability, and convergence in a multidimensional complex space. IEEE Trans. Evol. Comput. **6**, 58–73 (2002)
5. A. Colorni, M. Dorigo, V. Maniezzo, Distributed optimization by ant colonies. in Proceedings of ECAL91—European Conference on Artificial Life, Elsevier Publishing (1991)
6. R.C. Eberhart, Y. Shi, Comparing inertia weights and constriction factors in Particle Swarm Optimization. Proceedings of the Congress on Evolutionary Computating, 84–88 (2000)
7. A.E. Eiben, J. Smith, *Introduction to Evolutionary Computing* (Springer, London, 2003)
8. A.E. Eiben, S.K. Smit, Parameter tuning for configuring and analyzing evolutionary algorithms. Swarm Evol. Comput. **1**, 19–31 (2011)
9. R.L. Haupt, S. EHaupt, *Practical Genetic Algorithms* (Wiley, Hoboken, 2004)
10. J.M. Holland, *Adaptation in Natural and Artificial Systems* (University of Michigan Press, Ann Arbor, 1975)
11. J. Kennedy, Particle swarms: optimization based on sociocognition, in *Recent Developments in Biologically Inspired Computing* (The Idea Group Inc, 2004)
12. J. Kennedy, R.C. Eberhart, Particle swarm optimization, in *Proceedings of International Conference on Neural Networks* (1995), pp. 1942–1948

13. S. Kirkpatrick, C.D. Gelatt, M.P. Vecchi, Optimization by simulated annealing. Science **220**, 671–680 (1983)
14. L.M. Li, S.A. Billings, Continuous-time non-linear system identification in the frequency domain. Intern. J. Control **74**, 1052–1061 (2001)
15. L. Ljung, *System Identification—Theory for the User*, 2nd edn. (Prentice-Hall, Upper Saddle River, 1999)
16. M. Mitchell, *An Introduction to Genetic Algorithms* (MIT press, Cambridge, 1998)
17. A. Ratnaweera, S.K. Halgamure, H.C. Watson, Self-organizing hierarchical particle swarm optimizer with time-varying acceleration coefficients. IEEE Trans Evol. Comput. **8**, 240–255 (2004)
18. D. Selisteanu, E. Petre, V. Răsvan, Sliding mode and adaptive sliding-mode control of a class of nonlinear bioprocesses. Intern. J. Adapt. Control Signal Process. **21**, 795–822 (2007)
19. D. Sendrescu, M. Roman, Parameter Identification of Bacterial Growth Bioprocesses using Particle Swarm Optimization. 9th Asian Control Conference (ASCC), Istanbul, Turkey, June 23–26 (2013)
20. B. Shen, C. Liu, J. Ye, E. Feng, Z. Xiu, Parameter identification and optimization algorithm in microbial continuous culture. Appl. Math. Model. **36**, 585–595 (2012)
21. Y. Shi, R.C. Eberhart, Empirical study of particle swarm optimization, in *Proceedings of the 1999 IEEE congress on Evolutionary Computation* (IEEE Press, 1999), pp. 1945–1950
22. E. Walter, L. Pronzato, *Identification of Parametric Models from Experimental Data* (Springer, Berlin, 1997)
23. J. Zhai, J. Ye, L. Wang, E. Feng, H. Yin, Z. Xiu, Pathway identification using parallel optimization for a complex metabolic system in microbial continuous culture. Nonlinear Anal.: Real World Appl. **12**, 2730–2741 (2011)

Chapter 12
Real-Time Experimental Implementation of Predictive Control Schemes in a Small-Scale Pasteurization Plant

Albert Rosich and Carlos Ocampo-Martinez

Abstract This chapter proposes three closed-loop control topologies based on model predictive control (MPC) for a small-scale pasteurization plant. The topologies are designed taking into account the role of the predictive controller within the loop: (i) as supervisor control for the computation of the references for regulatory controllers, (ii) as unique controller within the closed loop and (iii) acting simultaneously as supervisor and regulatory controllers together with other regulatory controllers. All control designs have been applied in real time to a test bench station, then experimental results are both presented and discussed. The main advantages and drawbacks for each topology are presented for the regulation of the temperature of the output product, while the energy consumption of the overall system is minimized.

Keywords Model predictive control · Pasteurization plant · Industrial processes constrained control modelling and identification · Real application · PID controllers · Control topologies · Optimization-based control

12.1 Introduction

Along the last decades, model predictive control (MPC) has had a significant impact on industrial control engineering. Its implementation in process industry is justified by its capabilities of handling multi-variable control problems in a natural form, while taking into account actuator limitations and other physical and operational constraints, [8, 20]. Given the computational burden associated to the optimization problem solved online when an MPC controller is implemented, the use of this control

A. Rosich
Flanders Mechatronics Technology Centre, Celestijnenlaan 300,
3001 Leuven, Belgium
e-mail: arosich@fmtc.be

C. Ocampo-Martinez (✉)
Automatic Control Department, Universitat Politècnica de Catalunya,
Institut de Robòtica i Informàtica Industrial (CSIC-UPC),
Llorens i Artigas, 4-6, 08028 Barcelona, Spain
e-mail: cocampo@iri.upc.edu

S. Olaru et al. (eds.), *Developments in Model-Based Optimization and Control*,
Lecture Notes in Control and Information Sciences 464,
DOI 10.1007/978-3-319-26687-9_12

technique was preferred for control architectures formed by two levels: the former, as a supervisory/management level, where set-points for regulatory controllers at the lower level are computed, and the latter, as the proper regulatory level, where the control actions are applied to the dynamical system. Notice that the regulatory controllers attempt to hide the non-linear behaviours of the systems, allowing the supervisory controller to use a simpler control-oriented model, [21]. Thus, MPC controllers, acting as supervisory ones, are based then on those simpler models and hence their associated optimization problems get less computational burden (see, e.g., [6, 17] and references therein).

Nevertheless, relevant technological advances during the last years make possible the real-time implementation of MPC controllers based on more complex and large dynamical models. Therefore, several possibilities arise when implementing real-time predictive controllers for industrial processes. The possible topologies can be such as the aforementioned two-level schemes, where the MPC acts as a supervisory controller, topologies where the MPC is the unique controller within the closed loop, and topologies where there is a convenient combination of the MPC controller with a twofold function and *classical* regulatory controllers (such as PIDs), interacting altogether.

On the other hand, a pasteurization system involves typical behaviours of industrial processes, where considering complex dynamical models with nonlinearities imply important challenges when a suitable controller should be designed. In that sense, some previous modelling approaches and control schemes have been already proposed. The so-called *divide-and-conquer* technique for modelling the system is applied in [9], where the input–output mathematical model of the system is obtained from the decomposition of the plant in functional subsystems. Other non-linear models from the whole system and/or some subsystems are obtained in [1, 12]. Regarding its control, in [5] it is proposed a scheme based on PID with Smith predictor in order to compensate delays when some temperatures are regulated. In [10], a dynamic matrix control (DMC) is designed and implemented using the system models proposed in [9], where the delay and the energy reduction of the system were not improved. The regulation of both water and milk temperatures by using MPC is recently reported in [16], where transient behaviours have been suitably handled with respect to other control techniques such as cascade generic model control.

According to the previous discussions, this chapter performs the design and implementation of three control topologies based on MPC, where the temperature of the output product in a small-scale pasteurization plant is regulated, while the energy consumption required to this end is simultaneously reduced. Therefore, the main contribution of this chapter is not only the suitable design of controllers and topologies in order to satisfy the control objectives fulfilling system constraints, but also the real-time experimental implementation of those MPC-based topologies in the real system and the analysis of the performance results in order to highlight the advantages and disadvantages for each topology. This analysis aims at motivating the use of MPC controllers interacting within the existing industrial topologies, where it is well known the hegemony of the PID controllers.

The mathematical models of the pasteurization plant are properly obtained from the experimental data, [13], but the deep description of the system identification procedures for the corresponding subsystems is out of the scope of this chapter. None of the models previously reported in the literature were considered for the study performed in this chapter since the real pilot plant considered here is quite different with respect to those used in [1, 9, 16], among others. Final results where the energy consumption comparison is performed are taken into account with respect to a control scheme based only on PID controllers properly tuned by using well-known existing tools, [2, 4, 19]. In any case, these PID controllers have been accurately tuned by using the available tools in MATLAB©, achieving a proper response and avoiding the unfair comparison between topologies based on the possibly wrong PID tunings.

The remainder of the chapter is structured as follows. In Sect. 12.2, the pasteurization plant test bench is described. Section 12.3 presents the general statement of the MPC problems considered in this chapter. Besides the mathematical models, control problem formulations and main experimental results for each one of the three proposed topologies are presented and discussed. Section 12.4 performs the discussion of the results presented throughout the chapter. Finally, in Sect. 12.5 the main conclusions are drawn.

12.2 System Description

A process plant trainer for control purposes is used in this chapter as a real benchmark to test the different proposed control topologies. Specifically, the small-scale pasteurization plant PCT23 MKII from Armfield is used, [3]. Only those parts of the system that are relevant to the chapter are described in this section (see Fig. 12.1).

The system emulates an industrial high-temperature short-time (HTST) pasteurization process. In this process, the goal is to heat and keep the product, which is usually a liquid, at a predetermined temperature for a minimum time, typically for bacteriological purposes. This is achieved by circulating the heated liquid through a holding tube that delays the product stream.

A water heating unit is available in order to provide the necessary heat to the product. This unit consists of a water pump that circulates the hot water through a heat exchanger, and a hot water tank equipped with a temperature sensor (T_2) and an electrical resistor. The water heat is transferred to the product inside the first phase of the heat exchanger. A temperature sensor (T_3) at the heat exchanger outlet is used to ensure that the product has gained the desired temperature. The product is always pumped at a constant flow velocity in order to guarantee that it remains inside the holding tube at a constant pasteurization temperature (adiabatic phase of the process) for the minimum required time. With this end in view, a PID controller is implemented to regulate the product flow by means of a pump (feeding the product into the plant) and a flow metre located before the holding tube. This control loop will be considered fixed and out of the study carried out in this chapter. A temperature sensor (T_1) at the output of the holding tube is used to monitor the product temperature after the

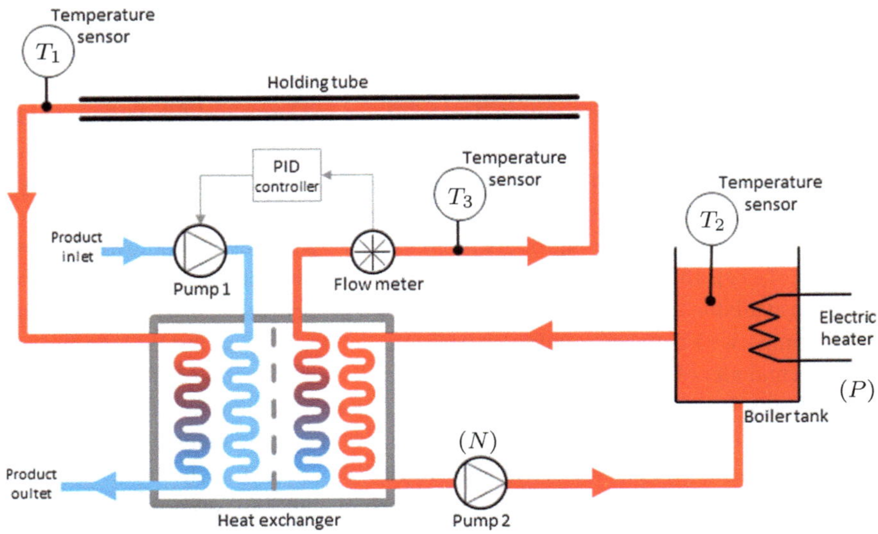

Fig. 12.1 Pasteurization plant diagram

pasteurization process. Finally, the product is cooled in the second phase of the heat exchanger, where residual heat is transferred to the inlet product.

In summary, from a control point of view, the pasteurization plant can be seen as a multiple-input and multiple-output (MIMO) system with electric heater power, P, and water pump speed, N, as inputs, and temperatures, T_1, T_2, and T_3 as outputs. The control objectives are twofold: *i)* the temperature T_1 must follow a predefined profile (tracking control problem), and *ii)* the energy consumption should be minimized.

It is considered that the pasteurization plant will be operated around the working point defined by

$$\begin{aligned} P^o &= 290\,\text{W}, \quad N^o = 65\,\%, \quad T_1^o = 55.5\,^\circ\text{C}, \\ T_2^o &= 66\,^\circ\text{C}, \quad T_3^o = 56\,^\circ\text{C}, \end{aligned} \tag{12.1}$$

which will be used to obtain the linear models required throughout the chapter. The design and simulation of the controllers have been performed in MATLAB© R2012b by using Tomlab Optimization Software [7] in an Intel Xeon CPU E31225 - 4 cores, 3.10 GHz and 4 GB RAM. The discretization of all dynamics as well as the implementation of the MPC controllers for experimentation were performed using a sampling time of 1 s. In particular, the dynamical model of the pump N was obtained by selecting a lower sampling time (0.5 s) given the fast dynamics shown by this element compared with the remainder processes.

12.3 Predictive Control Schemes

12.3.1 MPC Problem Statement

Given that the proposed controllers consider dynamical models around the working point (12.1), they are expressed in the discrete-time state-space linear form

$$\mathbf{x}(k+1) = \mathbf{A}\,\mathbf{x}(k) + \mathbf{B}\,\mathbf{u}(k), \tag{12.2a}$$

$$\mathbf{y}(k) = \mathbf{C}\,\mathbf{x}(k), \tag{12.2b}$$

where $\mathbf{x} \in \mathcal{X} \subseteq \mathbb{R}^{n_x}$, $\mathbf{u} \in \mathcal{U} \subseteq \mathbb{R}^{n_u}$ and $\mathbf{y} \in \mathcal{Y} \subseteq \mathbb{R}^{n_y}$ correspond to the vector of system states, the vector of input signals and the vector of measured outputs, respectively, and $k \in \mathbb{Z}_+$ denotes the discrete time. **A**, **B** and **C** are the system matrices of suitable dimensions. In the sequel, identified models obtained as transfer functions are conveniently expressed by their equivalent controllable realizations in state space as in (12.2). Moreover, let[1]

$$\hat{\mathbf{u}}(k) \triangleq \big(\mathbf{u}(0|k), \ldots, \mathbf{u}(H_p - 1|k)\big) \tag{12.3}$$

be the sequence[2] of input signals over a fixed-time prediction horizon H_p. Notice that (12.3) depends on the initial condition $\mathbf{x}(0|k) \triangleq \mathbf{x}(k)$. Therefore, the design of the different MPC controllers for the proposed control topologies/schemes in this chapter is based on Problem 12.1.

Problem 12.1 [*MPC Design*] The MPC design is based on the solution of the open-loop optimization problem (OOP)

$$\min_{\{\hat{\mathbf{u}}(k)\in\mathcal{U}^{H_p},\, \hat{\boldsymbol{\xi}}(k)\in\mathbb{R}^{H_p}\}} J(\mathbf{x}(k), \hat{\mathbf{u}}(k), \hat{\boldsymbol{\xi}}(k)), \tag{12.4a}$$

subject to

$$\mathbf{x}(i+1|k) = \mathbf{A}\mathbf{x}(i|k) + \mathbf{B}\mathbf{u}(i|k), \quad \forall\, i \in [0, H_p - 1], \tag{12.4b}$$

$$\mathbf{y}(i|k) = \mathbf{C}\mathbf{x}(i|k), \quad \forall\, i \in [0, H_p - 1], \tag{12.4c}$$

$$\mathbf{x}(k) \in \mathbb{R}^{n_x}, \quad \forall\, k, \tag{12.4d}$$

$$\mathbf{u}(i|k) \in \mathcal{U}, \quad \forall\, i \in [0, H_p - 1], \tag{12.4e}$$

$$\mathbf{G_1}\mathbf{y}(i|k) + \mathbf{G_2}\,\xi(i|k) \leq \mathbf{g}, \quad \forall\, i \in [0, H_p - 1], \tag{12.4f}$$

[1] Here, $m(k+i|k)$ denotes the prediction of the variable m at time $k+i$ performed at k. For instance, $x(k+i|k)$ denotes the prediction of the system state, starting from its initial condition $\mathbf{x}(0|k) = \mathbf{x}(k)$.

[2] In the sequel, the notation $\hat{\mathbf{z}}$ means a sequence of vectorial elements of suitable dimensions.

where $J(\cdot) : \mathcal{U}^{n_u H_p} \times \mathbb{R}^{H_p} \mapsto \mathbb{R}$ in (12.4a) is the cost function, H_p denotes the *prediction horizon* and $\mathbf{G_1}$, $\mathbf{G_2}$ and $\mathbf{g}$ are matrices of suitable dimensions. Moreover, $\xi \in \mathbb{R}$ is the *slack* variable for softening the output constraints (12.4f), and $\hat{\boldsymbol{\xi}}(k) \triangleq \left(\xi(0|k), \ldots, \xi(H_p - 1|k)\right) \in \mathbb{R}^{H_p}$. Notice that, in this chapter, expression in (12.4d) means the unconstrained nature of the system states. Assuming that the OOP (12.4) is feasible, i.e., $\mathbf{u}(k) \neq \emptyset$, there will be an optimal solution for the sequence of control inputs

$$\mathbf{u}(k)^* \triangleq \left(\mathbf{u}(0|k)^*, \mathbf{u}(1|k)^*, \ldots, \mathbf{u}(H_p - 1|k)^*\right), \tag{12.5}$$

and then, according to the receding horizon philosophy, $\mathbf{u}^*(0|k)$ is applied to the system, while the whole process is repeated for the next time instant $k \in \mathbb{Z}_+$. □

The following subsections present and discuss three closed-loop control schemes based on different roles of the MPC controllers wherein the main control objectives need to be accomplished. Aspects such as the model used, the controller design and the corresponding experimental results are explained for each considered topology.

12.3.2 Topology 1: MPC as a Supervisory Controller

The pasteurization plant described in Sect. 12.2 can be *simply* controlled by using a PID-based control scheme, [15, 18, 22]. A typical approach to track the temperature T_1 is to implement two PID controllers in cascade, [3]. The inner control loop, denoted here as PID_{T_3}, regulates T_3 by manipulating the pump velocity, N, whereas the outer loop, denoted here as PID_{T_1} provides the reference for the inner-loop controller. On the other side, another control loop, denoted here as PID_{T_2} is typically implemented to regulate the temperature of the water heating unit T_2. The PID_{T_2} controller manipulates the power of the electrical heater, P, in order to maintain the temperature T_2 at a certain constant and high value in order to guarantee that enough energy can be transferred to the product. Hence, no energy saving is considered.

In this section, this standard PID-based configuration is used. Nevertheless, the reference T_2^r, provided to regulate the temperature T_2, is supervised by an MPC (see Fig. 12.2) with the objective of saving energy. For the sake of space and because it does not involve any additional difficulty, the implementation of the PID controllers is not here presented. Instead, the chapter is focused on the supervisory MPC design.

12.3.2.1 System Identification and Control-Oriented Model

To accomplish the proposed control configuration, suitable models of the pasteurization plant are needed. In particular, two transfer functions are identified. These transfer functions relate the measured outputs T_1 and T_2 to their corresponding input

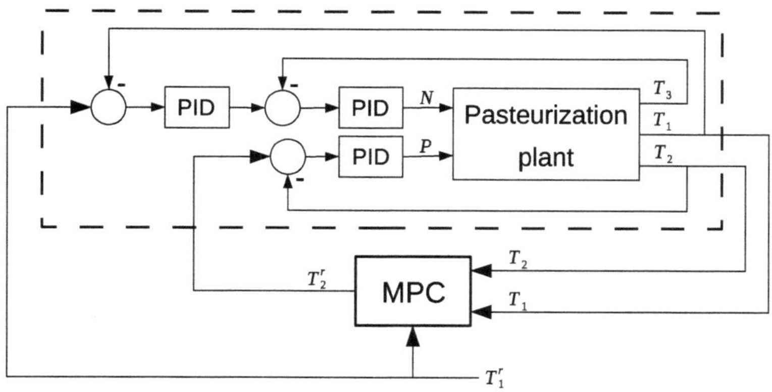

Fig. 12.2 Topology 1: MPC as a supervisory controller

references, T_1^r and T_2^r. By applying parametric identification methods (mainly based on least squares) to different model structures, [14], the following transfer functions are obtained:

$$\frac{T_1(z)}{T_1^r(z)} = \frac{0.0008008}{z^{30} - 1.963z^{29} + 0.9639z^{28}} \tag{12.6}$$

and

$$\frac{T_2(z)}{T_2^r(z)} = \frac{0.00358z}{z^2 - 1.937z + 0.9406}, \tag{12.7}$$

where z is the z-transform variable. Observe that the transfer function (12.6) has order 30 because the transportation delay induced by the holding tube. In addition, it is worth noting that, since the PID control loops are considered in the plant modelling, the two input/output pairs can be decoupled.

Models (12.6) and (12.7) are validated with real data from the pasteurization plant obtaining satisfactory results. In Fig. 12.3 a comparison between the real system behaviour and model-obtained temperatures is depicted.

12.3.2.2 Control Problem Setup

In order to design the corresponding MPC, the constraints and the cost function in Problem 12.1 should be defined. For the proper system operation, T_2 must always be greater than T_1. According to the performed experiments, this temperature difference between T_2 and T_1 should be greater than $D = 11.8\,°\text{C}$. The value of D is considered as a design parameter, which was determined by iterative simulations in order to have a safety temperatures difference taking into account the variations given by the devices dynamics and signal noises. Given that the system is operated around the working point (12.1), this difference can be conveniently adapted through a soft constraint of the form

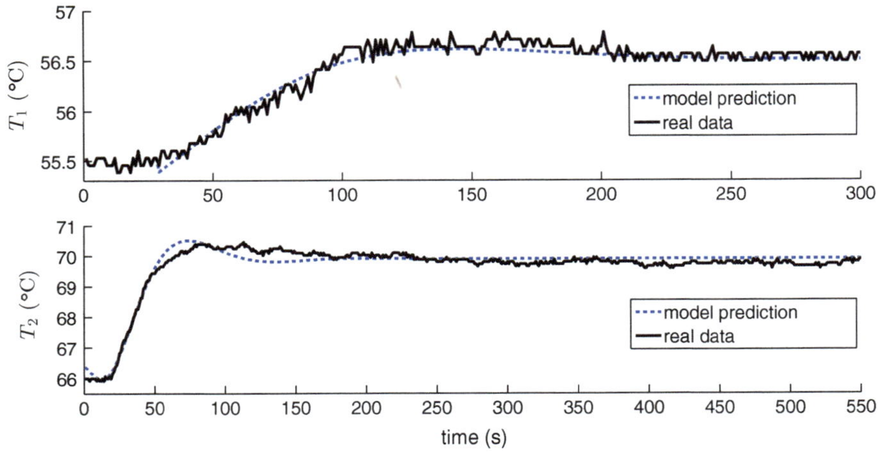

Fig. 12.3 Responses from the models (12.6)–(12.7) and comparison with real data

$$T_2(k) - T_1(k) \geq D + \xi(k), \tag{12.8}$$

where $\xi \in \mathbb{R}$ allows to avoid infeasibility of the OOP in (12.4). On the other hand, looking at the energy consumption of the actuators, the water pump consumes 35 VA for flows between 100–300 ml/min, [23], while the heater resistor consumes about 300 W. Therefore, minimization of the energy consumption is done through the heater resistor since its consumption is significantly greater than the energy consumption of the pump. Hence, T_2^r should be minimized. Finally, the minimization of the slew rate, $\Delta T_2^r(k) \triangleq T_2^r(k) - T_2^r(k-1)$, is also considered in order to reduce oscillations in T_2. Thus, the OOP (12.4) for this topology is defined as

$$\min_{\{\hat{\mathbf{T}}_2^{\mathbf{r}}(k),\, \hat{\boldsymbol{\xi}}(k)\}} \left\| \hat{\mathbf{T}}_2^{\mathbf{r}}(k) \right\|_{\mathbf{W}_1}^2 + \left\| \hat{\mathbf{\Delta T}}_2^{\mathbf{r}}(k) \right\|_{\mathbf{W}_2}^2 + \left\| \hat{\boldsymbol{\xi}}(k) \right\|_{\mathbf{W}_3}^2, \tag{12.9a}$$

subject to

$$\mathbf{x}(i+1|k) = \mathbf{A}_1 \mathbf{x}(i|k) + \mathbf{B}_1 \begin{bmatrix} T_1^r(i|k) \\ T_2^r(i|k) \end{bmatrix}, \tag{12.9b}$$

$$\begin{bmatrix} T_1(i|k) \\ T_2(i|k) \end{bmatrix} = \mathbf{C}_1 \mathbf{x}(i|k), \tag{12.9c}$$

$$T_2(i|k) - T_1(i|k) \geq D + \xi(i|k), \tag{12.9d}$$

for all $i \in [0, H_p - 1]$, where

$$\hat{\mathbf{T}}_2^{\mathbf{r}}(k) = (\mathbf{T}_2^r(0|k), \ldots, \mathbf{T}_2^r(H_p - 1|k)),$$
$$\hat{\mathbf{\Delta T}}_2^{\mathbf{r}}(k) = (\mathbf{\Delta T}_2^{\mathbf{r}}(0|k), \ldots, \mathbf{\Delta T}_2^{\mathbf{r}}(H_p - 1|k)),$$

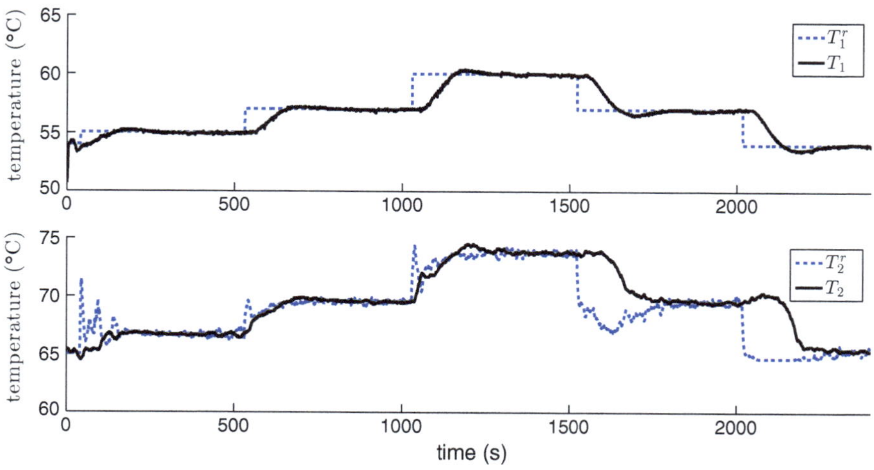

Fig. 12.4 Controlled temperatures, T_1 and T_2, with Topology 1

and W_i, $i = 1, 2, 3$, are the weighting matrices needed to prioritize the different control goals within the multi-objective cost function. Notice that, in general, $\mathbf{W}_i = \omega_i \mathbf{I}$, where $\mathbf{I}$ is the identity matrix of suitable dimensions and $\omega_i \in \mathbb{R}$. Here, the prediction model in (12.9b)–(12.9c) is derived from (12.6) and (12.7).

The MPC controller has been implemented in MATLAB© and tested on the pasteurization plant. The controller has been experimentally tuned for the weight values $\omega_1 = \omega_2 = 0.1$, $\omega_3 = 10$, whereas a prediction horizon $H_p = 35$ has been set.[3] It should be note that, in this case, shorter horizons degrade the closed-loop performance, while a larger horizon does not improve the results and moreover takes longer in solving the optimization problem (12.9). The controlled temperatures from the real plant (i.e., T_1 and T_2) are shown in Fig. 12.4 together with their corresponding references, where T_2^r is in this case the control variable. Observe that the MPC tries to keep T_2 as lower as possible and fulfil constraint (12.8) at the same time. This is particularity difficult when T_1 decreases (e.g., time $k = 1500$ s and $k = 2000$ s) since T_2 should be reduced by the MPC, accordingly. However, there is no actuator to reduce T_2 and the water tank is really cooled by dissipating the heat to the atmosphere.

12.3.3 Topology 2: MPC as Unique Controller

Another possibility to control the pasteurization plant is a holistic approach by means of one single MPC controller that directly operates the actuators from the system measurements (see Fig. 12.5).

---

[3]In this chapter, it is supposed that the prediction horizon H_p and the control horizon H_u have the same length in order to have more degrees of freedom when computing the optimal control action at each time instant $k \in \mathbb{Z}_+$.

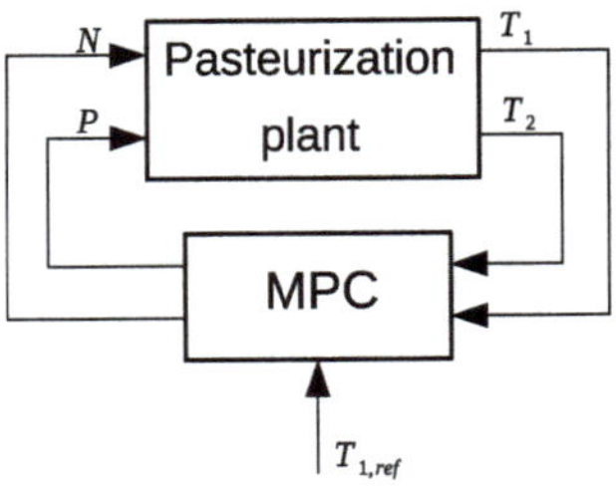

Fig. 12.5 Topology 2: MPC as unique controller

12.3.3.1 System Identification and Control-Oriented Model

A complete model of the pasteurization plant is needed for this topology. However, due to the complexity of the system, it is not feasible to identify the whole plant at once. Instead, the different subsystems are identified and modelled separately.

Holding tube subsystem: From a thermodynamic point of view, the holding tube can be modelled as a single-input and single-output system, where temperature T_3 is the input and temperature T_1 is the output. By experimentation, the following discrete transfer function is obtained:

$$\frac{T_1(z)}{T_3(z)} = \frac{0.2231}{z^{30} - 0.7649z^{29}}. \tag{12.10}$$

The measurement of T_3 has been used to validate the model. In Fig. 12.6, the responses of T_1 obtained by applying the same measurement of T_3 in both, the real plant and the model (12.10) are compared.

Hot water tank subsystem: The dynamics of the temperature T_2 in the water tank of the heating unit system are derived from first principles of thermodynamics [11]. Specifically, the following differential equations is used:

$$c_w \frac{dT_2(t)}{dt} = P(t) - c_s F(k)(T_{out}(t) - T_{in}(t)) - k_l(T_2(t) - T_a(t)), \tag{12.11}$$

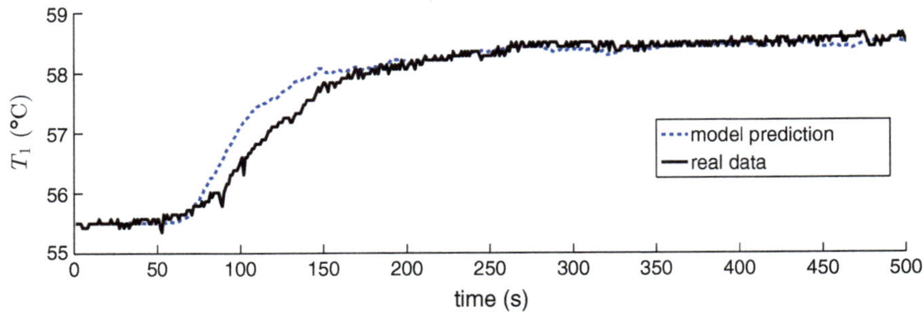

Fig. 12.6 Response from the model of the holding tube

where c_w is the heat capacity coefficient of the water mass in the tank, P is the power provided by the electrical resistor, c_s the specific heat capacity of the water, F is the water flow given by the pump, T_{out} and T_{in} are the output and input water temperatures, k_l is the heat loss coefficient and T_a is the atmospheric temperature. Note that the water flow is proportional to the pump speed, therefore $F = \alpha_1 N$. For the sake of simplicity, it is assumed that the output water temperature is the same as the measured water tank temperature, i.e., $T_{out} = T_2$, and also that the T_{in}, which cannot be measured, is proportional to T_3, i.e., $T_{in} = \alpha_2 T_3$. In addition, T_a is assumed to be known and constant.

The unknown parameters of (12.11) are properly identified (by using parametric identification based on least squares, [13, 14]) and then the model is discretized and linearized around the working point in (12.1). The resulting discrete linear model for the hot water tank system is

$$\begin{aligned} T_2(k+1) =& 9.98 \times 10^{-1} T_2(k) + 1.2 \times 10^{-4} P(k) \\ &- 4.26 \times 10^{-4} N(k) - 1.64 \times 10^{-4} T_3(k). \end{aligned} \tag{12.12}$$

Temperature T_2 from both, real plant and model (12.12) are shown in Fig. 12.7. In this case, the resistor power P and pump speed N are slightly modified from the working point in (12.1), while input temperature T_3 is taken from the real plant.

Heat exchanger subsystem: The first phase of the heat exchanger is here modelled. For simplicity, the input product temperature is assumed to be known and constant. Therefore, the dynamics of the output product temperature, T_3, are only directly affected by the pump speed, N. It should be noted that the temperature T_3 is indirectly affected by both T_2 and the pasteurized product temperature T_1 since the latter circulates through the second phase of the heat exchanger, which is in contact with the first phase. However, in order to keep the simplicity of the model, these side effects on T_3 are not taken into account. They can be seen as unknown disturbances.

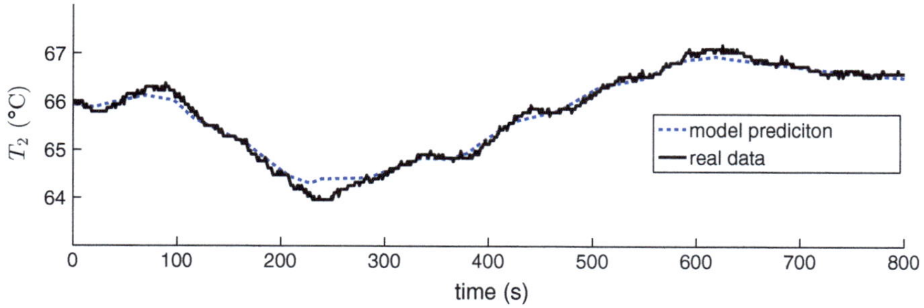

Fig. 12.7 Response from the model of the hot water tank

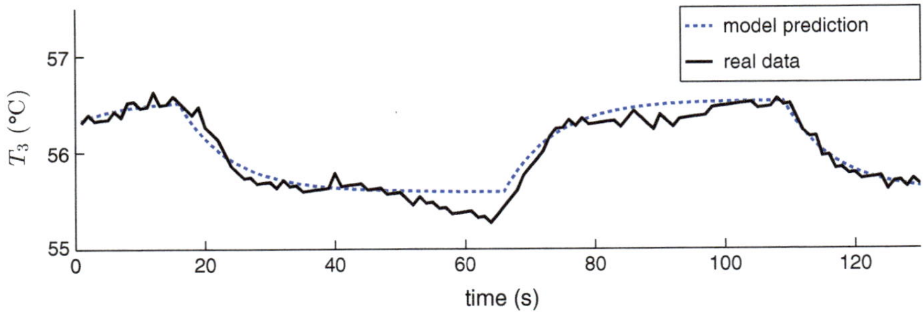

Fig. 12.8 Response from the model of the heat exchanger

The obtained model is represented in discrete transfer function as

$$\frac{T_3(z)}{N(z)} = \frac{0.01141}{z - 0.8796}. \tag{12.13}$$

The real and the model-predicted temperature T_3 are compared in Fig. 12.8.

12.3.3.2 Control Problem Setup

As in the Topology 1, the control objectives remain the same, remarking the priority on the reduction of the energy consumption. In this case, the management of the power dissipated by the heater resistor determines the performance of the closed-loop control. Now, the MPC controller must compute the direct control actions to the actuators, while performing the tracking task with T_1. This task implies the minimization of the tracking error $e_{T_1}(k) \triangleq T_1(k) - T_1^r(k)$, while the control actions are minimized and smoothed, and the softening of the operational constraint (12.8) is penalized. Thus, the OOP (12.4) for this topology is defined as

$$\min_{\{[\hat{\mathbf{N}}(k)\ \hat{\mathbf{P}}(k)]^T,\, \hat{\xi}(k)\}} \left\|\hat{\mathbf{P}}(k)\right\|_{\mathbf{W}_1}^2 + \left\|\begin{bmatrix}\Delta\hat{\mathbf{N}}(k)\\ \Delta\hat{\mathbf{P}}(k)\end{bmatrix}\right\|_{\mathbf{W}_2}^2 + \left\|\hat{\xi}(k)\right\|_{\mathbf{W}_3}^2 + \left\|\hat{\mathbf{e}}_{\mathbf{T}_1}(k)\right\|_{\mathbf{W}_4}^2, \tag{12.14a}$$

subject to

$$\mathbf{x}(i+1|k) = \mathbf{A}_2\mathbf{x}(i|k) + \mathbf{B}_2\begin{bmatrix}N(i|k)\\ P(i|k)\end{bmatrix}, \tag{12.14b}$$

$$\begin{bmatrix}T_1(i|k)\\ T_2(i|k)\end{bmatrix} = \mathbf{C}_2\mathbf{x}(i|k), \tag{12.14c}$$

$$N(i|k) \in [-40, 15], \quad P(i|k) \in [-0.3, 1.3], \tag{12.14d}$$

$$T_2(i|k) - T_1(i|k) \geq D + \xi(i|k), \tag{12.14e}$$

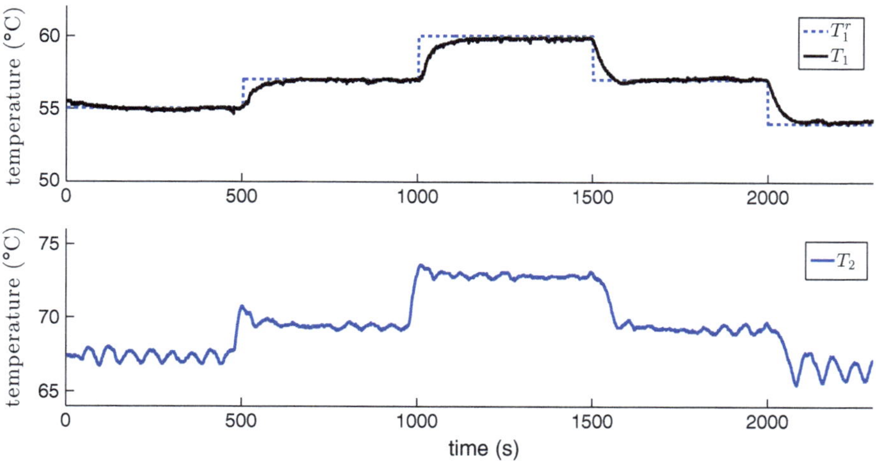

Fig. 12.9 Controlled temperatures, T_1 and T_2, with Topology 2

for all $i \in [0, H_p - 1]$, where

$$\hat{\mathbf{P}}(k) = (\mathbf{P}(0|k), \ldots, \mathbf{P}(H_p - 1|k)),$$
$$\hat{\mathbf{\Delta N}}(k) = (\Delta N(0|k), \ldots, \Delta N(H_p - 1|k)),$$
$$\hat{\mathbf{\Delta P}}(k) = (\Delta P(0|k), \ldots, \Delta P(H_p - 1|k)),$$
$$\hat{\mathbf{e}}_{\mathbf{T_1}}(k) = (\mathbf{e}_{\mathbf{T_1}}(0|k), \ldots, \mathbf{e}_{\mathbf{T_1}}(H_p - 1|k)),$$

and W_4 is the weighting matrix related to the tracking error of T_1. Notice here that W_2 is a block diagonal matrix, whose elements are matrices $\Omega_2 = \begin{bmatrix} \Omega_{21} & 0 \\ 0 & \Omega_{22} \end{bmatrix}$. In this case, the prediction model (12.14b)–(12.14c) comes from merging of the equivalent controllable realizations of (12.10), (12.12) and (12.13). Moreover, input constraints (12.14d) should be stated since

- both the dead zone of the pump (which is up to 25 % of its operating range) and the definition of a safety range below 80 % also of its operating range (over this value some elements can deteriorate rapidly) should be taken into account[4]; and
- the heating power of the resistor, P, can take values in the range [0, 1.6] kW. Notice that values in (12.14d) consider the working point (12.1).

The MPC controller was implemented with the trial-and-error weight values $\omega_1 = 2$, $\omega_{21} = 10^{-3}$, $\omega_{22} = 10^{-2}$, $\omega_3 = 10^2$ and $\omega_4 = 10^3$. Moreover, $H_p = 35$.

Results obtained by applying this topology are depicted in Fig. 12.9. As before, the temperature T_2 ensures enough transfer heating to the product (i.e., constraint (12.8) is satisfied). In addition, now the delay caused by the holding tube is compensated.

[4]Notice that the pump speed N is given in percentage with respect to the maximal speed of the corresponding pump according to the device specifications.

On the other hand, although it is not evident in the figure, the steady-state error in T_1 is not null. This is mainly due to two reasons: *i)* the models, especially the heat exchanger model in (12.13), are not accurate and *ii)* the pump actuator range is limited to 80 % which degrades the controller performance. It is well known that it can be corrected by using an integrator-in-series configuration but it was not implemented here due to the small amount of such error.

12.3.4 Topology 3: MPC and PID

The last control topology presented in this chapter is a combination of the previous ones. A PID controller, denoted as PID_{T_3}, is used to control T_3 by means of the pump speed N, whereas the MPC provides the set-point for PID_{T_3} and drives directly the electric resistor at the same time (see, Fig. 12.10). In this topology, it is intended that a simple PID controller is more suitable for controlling T_3 because fast dynamics are present in both the temperature T_3 and the pump speed N. Furthermore, by using PID_{T_3} controller, the effect of the unknown disturbances over T_3 are mitigated which means that the model used in the MPC becomes more accurate.

12.3.4.1 System Identification and Control-Oriented Model

The models used in this topology can easily be derived by combining the previous models. For instance, the dynamics of T_3 driven by T_3^r can straightforwardly be obtained by computing the corresponding PID closed-loop transfer function with (12.13). Therefore, no new models are needed for this topology.

12.3.4.2 Control Problem Setup

The control design for this topology combines an MPC coping with the regulation of P, together with PID_{T_3} that manages N but which receives its reference from

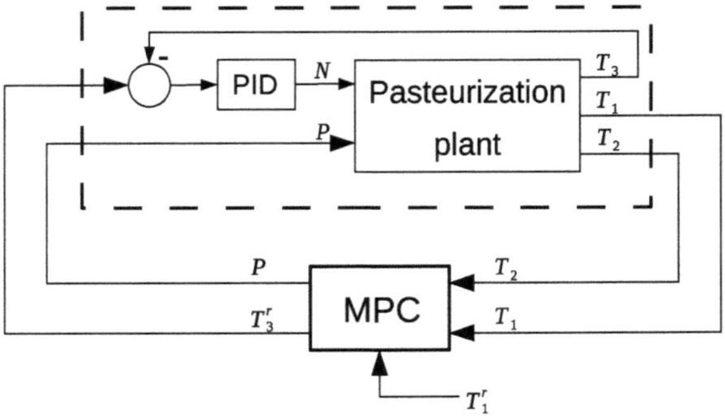

Fig. 12.10 Topology 3: MPC and PID

the MPC controller. Given the control objectives and aforementioned operational constraints, the OOP (12.4) for this topology is defined as

$$\min_{\{[\mathbf{P}(k)\ \mathbf{T_3^r}(k)]^T\ \xi(k)\}} \|\mathbf{P}(k)\|^2_{\mathbf{W_1}} + \left\|\hat{\Delta \mathbf{T_3^r}}(k)\right\|^2_{\mathbf{W_2}} + \left\|\hat{\xi}(k)\right\|^2_{\mathbf{W_3}} + \left\|\mathbf{e_{T_1}}(k)\right\|^2_{\mathbf{W_4}}, \quad (12.15a)$$

subject to

$$\mathbf{x}(i+1|k) = \mathbf{A_3}\mathbf{x}(i|k) + \mathbf{B_3}\begin{bmatrix} P_{i|k} \\ T_3^r(i|k) \end{bmatrix}, \quad (12.15b)$$

$$\begin{bmatrix} T_1(i|k) \\ T_2(i|k) \end{bmatrix} = \mathbf{C_2}\mathbf{x}(i|k), \quad (12.15c)$$

$$P(i|k) \in [-0.3, 1.3], \quad T_3^r(i|k) \in [-2, 7], \quad (12.15d)$$

$$T_2(i|k) - T_1(i|k) \geq D + \xi(i|k), \quad (12.15e)$$

for all $i \in [0, H_p - 1]$, where

$$\hat{\Delta \mathbf{T_3^r}}(k) = (\Delta \mathbf{T_3^r}(0|k), \ldots, \Delta \mathbf{T_3^r}(H_p - 1|k)),$$

and the constraints for input T_3^r are given considering a small range of variation around the system working point (12.1). The prediction model (12.15b)–(12.15c) is obtained for this topology by merging the controllable realizations of (12.10) and (12.12). Notice here that the induction to a smooth behaviour of T_3^r implies less oscillations of T_2. For the implementation of this control topology with the real system, the tuning parameters are set to $\omega_1 = 10^{-5}$, $\omega_2 = 300$, $\omega_3 = 1$, $\omega_4 = 10$, and $H_p = 35$.

In Fig. 12.11, the temperature responses from the pasteurization plant under the control Topology 3 are shown. In this case, the delay is compensated and the null steady-state is achieved. This is because the MPC is now used to compensate the delay by providing a suitable reference to PID_{T_3}, and this in turn provides the perfect tracking in steady-state despite the model inaccuracies. Moreover, the model inaccuracies alleviation also provides a better performance in terms of T_2. Observe that now the fluctuations in the temperature T_2 have been significantly reduced. However, as far as T_1 goes away from its working point, an undesired overshoot appears in its transient dynamics. It is also produced by the new resultant behaviour of T_2, which also experiments bigger overshoots. This phenomena might be avoided by an accurate tuning of the cost function in (12.15a) (conveniently increasing the prioritization of $\hat{\Delta \mathbf{T_3^r}}$).

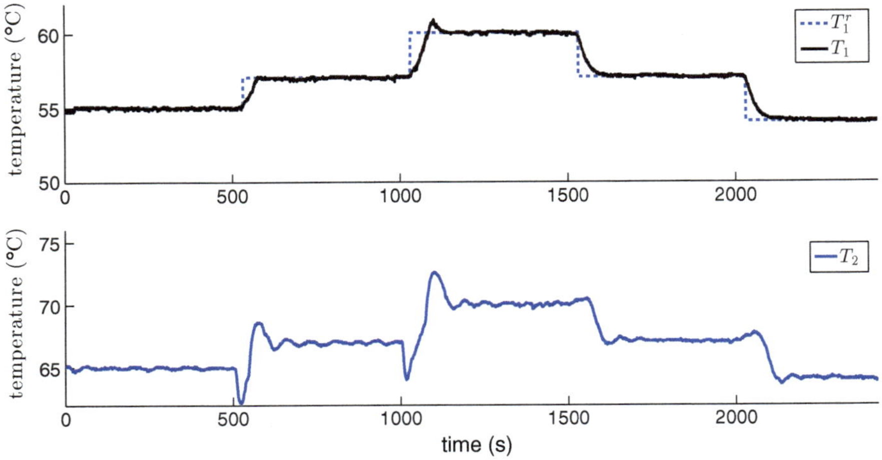

Fig. 12.11 Controlled temperatures, T_1 and T_2, with Topology 3

12.4 Results Discussion

The different proposed control topologies are here compared by means of three performance indices. The values of these indices for each topology are displayed in Table 12.1.

The first index is the MSE (Mean Square Error) between temperature T_1 and its reference. This index indicates how accurate the tracking is performed. In this case, Topology 1 presents the worst value. This is mainly due to the fact that the holding tube delay is not compensated. On the other hand, the disturbance effect mitigation accomplished by the PID-based control loop in Topology 3 improves the MSE in front of Topology 2.

Another performance index is the settling time for temperature T_1. Short settling times are crucial in pasteurization processes, since all the product obtained during the settling time is rejected because it does not reach the proper temperature. Topology 1 presents again the worst result while Topology 3 is significantly better that the others. This is because Topology 3 takes advantage of the fast PID-based control loop.

Third index takes into account the energy consumption by showing the percentage of saved energy with respect to the standard approach (only PID controllers with no

Table 12.1 Performance indices

Topology	MSE	Settling time (s)	Energy saving (%)
1	0.94	250	10
2	0.53	150	11
3	0.46	70	11.5

Table 12.2 Qualitative features

Topology	Null steady-state error	Actuator constraints	
		Pump speed N	Heater power P
1	✓		
2		✓	✓
3	✓		✓

MPC) in which no energy saving is considered (i.e., temperature T_2 is kept at 75°C all the time). Although all topologies perform similarly, Topology 3 shows better results.

Quantitatively speaking, it would be quite risky to determine the *best* control topology from those proposed here. From the point of view of energy savings, tuning procedures for cost functions in (12.9a), (12.14a) and (12.15a) might improve the performance of the controllers but the design and application of tuning strategies are out of the scope of this study. On the other hand, MSE index could also be improved by different and accurate tuning criteria, which in turn would allow to reduce or even eliminate some overshoots in temperature dynamics.

Finally, Table 12.2 summarizes the main features presented by the three topologies. It is worth to highlight that Topology 2 presents some small steady-state error as a consequence of model inaccuracies. However, it is the only one that can handle both actuator constraints. From this perspective, Topology 3 can be chosen as an intermediate solution, since null steady-state error is achieved and constraints on the electrical resistor can be handled.

12.5 Conclusions

Although the pasteurization process can be perfectly accomplished by standard PID controllers, in this chapter energy saving is also claimed as a control objective, which makes the MPC approach suitable for operating the plant. Therefore, three different control topologies based on MPC have been proposed in order to study which the role of the MPC should be, i.e., MPC as a supervisor or as a regulatory controller. The study has been carried out from a practical perspective, where conclusions have been drawn from experimental data. Therefore, typical problems in real implementations have been encountered, e.g., noisy signals, model inaccuracies, hardware limitations.

Topology 1 shows a proper performance against model uncertainties because the PID controllers locally compensate the model mismatches. However, the MPC is merely used to compute the set-point for the local controllers wherein some useful dynamics for the optimality of the control objective are not used. In addition, actuator constraints cannot be handled. Notice here that MPC might be suitably replaced by other optimal control techniques such as LQR. On the opposite side, an MPC commanding the actuators directly is tested in Topology 2. In this case, the model of the plant without controllers is used in the optimization despite the inaccuracies. This leads to a loss of optimality due to the optimal for the model differs from the optimal for the plant which, in turn, results in a loss of control performance. Finally, Topology 3 is intended so that the advantages of the two previous topologies are preserved. Therefore, Topology 3 is presented as an intermediate solution between the other two topologies, wherein a local PID controller is used together with the MPC.

According to the results discussion in Sect. 12.4, Topology 3 results to be the most convenient one. This means that the choice of a control topology is not obvious since many intermediate solutions could be possible in a real plant, where a certain number of control loops need to be considered. Therefore, the results obtained in this work motivate and justify the need of further studies in order to determine the best control topology for a given plant.

Acknowledgments This work has been partially supported by the project ECOCIS (Ref. DPI2013-48243-C2-1-R).

References

1. P. Aadaleesan, N. Miglan, R. Sharma, P. Saha, Nonlinear system identification using Wiener type Laguerre-Wavelet network model. Chem. Eng. Sci. **63**(15), 3932–3941 (2008)
2. Ch. Anil, R. Padma-Sree, Tuning of PID controllers for integrating systems using direct synthesis method, in *ISA Transactions* (2015). (in press)
3. Armfield, *Process Plant trainer PTC23-MKII, Instruction Manual* (2015)
4. K.J. Åström, B. Wittenmark, *Computer Controlled Systems: Theory and Design*, 3 edn. Prentice Hall (1996)
5. M. García-Sanz, J.C. Guillén, J.J. Ibarrola, Robust controller design for uncertain systems with variable time delay. Control Eng. Pract. **9**(9), 961–972 (2001)
6. J.M. Grosso, C. Ocampo-Martinez, V. Puig, Adaptive multilevel neuro-fuzzy model predictive control for drinking water networks, in *Proceedings of the 20th Mediterranean Conference on Control Automation (MED)* (Barcelona, Spain, July 2012)
7. K. Holmström, M. Edvall, A. Göran, Tomlab—for large-scale robust optimization, in *Proceedings of the Nordic MATLAB Conference* (Copenhagen, Denmark, 2003)
8. D. Hrovat, S. Di Cairano, H.E. Tseng, I.V. Kolmanovsky, The development of model predictive control in automotive industry: A survey, in *Proceedings of the IEEE International Conference on Control Applications (CCA)* (Dubrovnik, Croatia, 2012), pp. 295–302
9. J.J. Ibarrola, J.C. Guillén, J.M. Sandoval, M. García-Sanz, Modelling of a high temperature short time pasteurization process. Food Control **9**(5), 267–277 (1998)
10. J.J. Ibarrolà, J.M. Sandoval, M. García-Sanz, M. Pinzolas, Predictive control of a high temperature-short time pasteurisation process. Control Eng. Pract. **10**(7), 713–725 (2002)
11. M. Kaufman, *Principles of Thermodynamics* (CRC Press, 2002)
12. M.T. Khadir, J.V. Ringwood, Linear and nonlinear model predictive control design for a milk pasteurization plant. Control Intell. Syst. **31**(1), 1–10 (2003)
13. L. Ljung, *System Identification: Theory for the User*, (Prentice Hall, 1999)
14. L. Ljung, *System Identification Toolbox* (The MathWorks Inc, Users Guide, 2012)
15. W.M.F. Wan Mokhtar, F.S. Taip, N. Abdul Aziz, S.B. Mohd Noor, Process control of pink guava puree pasteurization process: Simulation and validation by experiment. Int. J. Adv. Sci. Eng. Inf. Technol. **2**(4) 31–34 (2012)

16. S. Niamsuwan, P. Kittisupakorn, I.M. Mujtaba, Control of milk pasteurization process using model predictive approach. Comput. Chem. Eng. **66**, 2–11 (2014)
17. C. Ocampo-Martinez, V. Puig, J.M. Grosso, S. Montes de Oca, Chapter multi-layer decentralized model predictive control of large-scale networked systems, in *DMPC Made Easy* (Springer, 2013)
18. T. O'Mahony, L. Torres, A virtual control engineering laboratory based on an industrial pasteurisation process, in *Proceedings of the 7th IFAC Symposium on Advances in Control Education* (Sheffield, United Kingdom, 2006)
19. K. Pirabakaran, V.M. Becerra, Automatic tuning of PID controllers using model reference adaptive control techniques, in *Proceedings of the 27th Annual Conference of the IEEE Industrial Electronics Society (IECON)* (2001), pp.736–740
20. S. Joe Qin, A. Badgwell, A survey of industrial model predictive control technology. Control Eng. Pract. **11**(7), 733–764 (2003)
21. R. Sánchez-Pena, F. Bianchi, Model selection: From LTI to switched-LPV, in *Proceedings of the American Control Conference* (Montreal, Canada 2012)
22. L. Torres, T. O'Mahony, P. Kellyand B. O'Connor, Controller design and implementation on a pilot-scale pasteurisation plant, in *Proceedings of the IEE Irish Signals and Systems Conference* (Dublin, Ireland, 2005)
23. Watson-Marlow, *Peristaltic Pump Drive Units* (2012)

Part V
Optimization-Based Analysis and Design for Particular Classes of Dynamical Systems

Chapter 13
An Optimization-Based Framework for Impulsive Control Systems

Fernando Lobo Pereira, Fernando A.C.C. Fontes, António Pedro Aguiar and João Borges de Sousa

Abstract This chapter concerns a discrete-time sampling state feedback control optimizing framework for dynamic impulsive systems. This class of control systems differs from the conventional ones in that the control space is enlarged to contain measures and, thus, the associated trajectories are merely of bounded variation. In other words, it may well exhibit jumps. We adopt the most recent impulsive control solution concept that pertains to important classes of engineering systems and, in this context, present impulsive control theory results on invariance, stability, and sampled data trajectories having in mind the optimization-based framework that relies on an MPC-like scheme. The stability of the proposed MPC scheme is addressed.

Keywords Impulsive control · Model predictive control · Stability · Measure-driven differential equations · Invariance · Stability

F. Lobo Pereira (✉) · A. Pedro Aguiar
LSTS—Laboratory of Underwater Systems and Technologies,
SYSTEC—Research Center for Systems and Technologies,
ISR—Institute for Systems and Robotics,
Faculty of Engineering, Porto University (FEUP),
4200-465 Porto, Portugal
e-mail: flp@fe.up.pt

A. Pedro Aguiar
e-mail: pedro.aguiar@fe.up.pt

F.A.C.C. Fontes
SYSTEC—Research Center for Systems and Technologies,
ISR—Institute for Systems and Robotics,
Faculty of Engineering, Porto University (FEUP),
4200-465 Porto, Portugal
e-mail: faf@fe.up.pt

J. Borges de Sousa
LSTS—Laboratory of Underwater Systems and Technologies,
Faculty of Engineering, Porto University (FEUP), 4200-465 Porto, Portugal
e-mail: jtasso@fe.up.pt

S. Olaru et al. (eds.), *Developments in Model-Based Optimization and Control*,
Lecture Notes in Control and Information Sciences 464,
DOI 10.1007/978-3-319-26687-9_13

13.1 Introduction

This chapter concerns the control of a large class of impulsive systems by using a model predictive control (MPC)-like scheme that combines optimization and state feedback at discrete points in time. This is a practical approach that makes the most of optimal control theory in order to take into account perturbations that affect the behavior of real-life systems, and, at the same time, mitigates the huge computational burden underlying the online computation associated with optimal feedback control, which, in general, requires solving a certain partial differential equations of the Hamilton–Jacobi–Bellman type.

From the application perspective, the motivation for considering the control of impulsive systems has been growing over the years with the increasing number of engineering applications involving systems exhibiting jumps in their trajectories. These include mechanical systems subject to collisions in which the velocity has a discontinuity at the collision times—for example, robot manipulators, and walking robots—, spacecraft navigation often associated with minimum fuel problems whose solutions involve impulses, multiphase systems—for example, car engines with a discrete number of gears, for which engine revolutions have a discontinuity during a gear change—, reposition of a stock of a product in inventory control, investment policies in economic systems, among many others (see [8, 13, 14, 16, 20, 38, 46]). Along this vein, Aubin, [12], noticed that the impulsive framework is well suited to model hybrid systems in which discrete events and continuum dynamics drive the evolution of the state variable together in such a way that the effect of events in the latter can be regarded as jumps.

Intuitively, the need to adopt an impulsive control framework arises when the control systems exhibits very fast and very slow dynamics abstraction and the optimal control problem of interest is such that these two components of the dynamics cannot be dealt with separately. In this context, the velocity set associated with the very fast dynamics can become unbounded, and the existence of solution can be guaranteed only if the space of conventional controls in $\mathbb{L}_1$ can be extended to a larger space, [10, 34], which, in the simpler context in which the unbounded control enters linearly in the dynamics, could be the space of regular Borel measures, [35].

This chapter follows along the line of the work in [30], by extending in a significantly more sophisticated and nontrivial way the class of impulsive control systems considered previously, so that more realistic challenges can be addressed. In this sense, the work presented here improves strongly upon the one in the above-mentioned article.

Given time interval $[0, T]$, the impulsive dynamic control system is specified by

$$\begin{aligned} &dx = f(t, x, u)dt + G(t, x, u)d\vartheta \\ &(x(0), x(T)) \in C_0 \times C_T \\ &u \in \mathcal{U}, \quad \vartheta \in \mathcal{I}, \end{aligned}$$

where $t \in [0, T]$, x, u, and ϑ are, respectively, the time, state, conventional control, and the impulsive control variables whose values are restricted to certain sets and are related by given mappings. This class of dynamic impulsive control systems was first introduced in [8], being the necessary conditions of optimality in the form of a Maximum Principle proved in [9] for an even more complete formulation of the optimal control problem. All the items involved in the equation describing the impulsive dynamics, and a proper solution concept associated with it are explained in detail in the next Sect. 13.2.

For now, we remark that this formulation is strongly justified by substantive practical engineering reasons. Consider this simple abstract, yet representative, example of driving a disk in $\mathbb{R}^2$ between a given pair of distinct points while keeping its angular velocity equal to zero by controlling it with four jet thrusters (see [8]). Assume that the center of gravity of disk changes as the fuel is consumed when the jets are fired. Clearly, the thrusters have to be restricted to yield a resultant force vector whose line intersects the center of gravity, see Fig. 13.1. Thus, the inclination angles, θ_i, $i = 1, \ldots, 4$, of the thrusters have to be controlled to satisfy the angular velocity constraint during the firing of the jets. It is well known that in space navigation, as well as, in many other applications that the optimal solution to the minimum fuel problem is, in general, of an impulsive nature. Thus, this example clearly shows that a control of conventional type is required to act "while" control impulses are active. It is important to remark that impulses are, in fact, mathematical abstractions, whose compact representation facilitates not only the description of these systems, but also the derivation of the various types of conditions, such as invariance, stability, and optimality, supporting control design, thus providing guidance for practical or implementable approximations.

The current state of the art in MPC is extremely vast. This is not surprising due to the extremely wide variety of successful applications. Thus, stabilizing properties have been investigated in a wide number of publications in various contexts that

Fig. 13.1 The angle of the thrusters are restricted by the center of gravity location

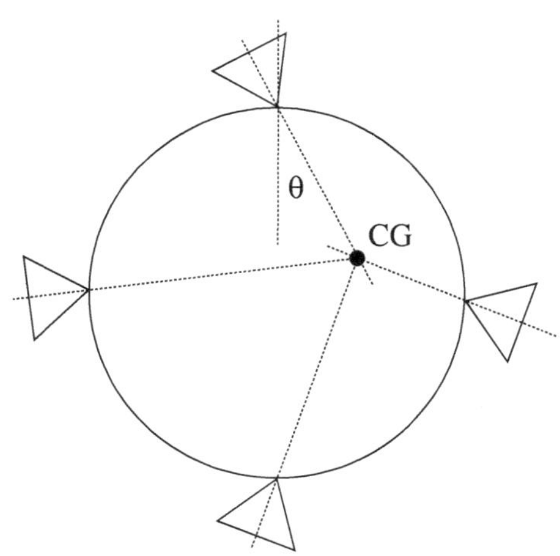

concern the wide variety of requirements exhibited by the various classes of applications. For example, and, to name just a few general contexts, absolutely continuous trajectories in continuum-time are considered in [2, 19, 24, 26, 29, 33], discontinuous controls, and discontinuous feedbacks are dealt with in [1, 25, 27], and switched systems are addressed in [37, 45].

In what concerns impulsive control systems, there are several results for optimal control, both on the maximum principle, [4–7, 9, 11, 40, 43, 50] and Hamilton–Jacobi–Bellman equation, [32] that extend those pertaining the conventional control systems, [3, 44, 49]. There are also results for sampled-data and feedback solution concept, [31] invariance, [32, 42], and stability, [41, 47]. In [30], these results were readily used, and adapted to a sampled-data MPC framework in the context of the measure driven dynamics given by

$$dx = f(t, x, u)dt + g(t, x)d\mu$$

where $(u, \mu) \in \mathcal{U} \times \mathcal{K}$ is the control variable formed by a pair of a conventional and an impulsive controls, and x is the state variable. In this work, sufficient conditions for the stability of the controlled system are given, being a key role played by the specification of the sampling times in the context of a trajectory that exhibits discontinuities generated by the control impulses. Clearly, the set of sampling times of the sampled-data MPC framework have to include the points at which the trajectory jumps. This is a critical issue due to the fact that the sampled-data MPC law is a function of the state at the last sampling instant rather than of the state at the current time. Therefore, the measurement error at the times of discontinuity may become unbounded, and, thus, jeopardize the stability and inherent robustness of the MPC. This issue is addressed by considering the sampling times as an additional variable provided by the impulsive optimal control optimization which determines the times at which the optimal impulses will occur in the optimization time horizon. For the details, check [30]. The simpler case of linear impulsive control systems is addressed in [48].

As it follows from above, the practical implementation of impulsive control results necessary involves the construction of some sort of approximation to the impulsive controls. In the context of the sampled-data MPC, the approximation to the optimal control solution encompasses also a sampling strategy whose frequency is proportional to the velocity of state variable variation on the support of the measure considered as the approximation to the impulsive control. This is an extremely welcome feature of the results discussed in this chapter as resources for sampling (which in many applications might be extremely significant) are recruited at the stages of the trajectory evolution in which they are most needed in order to capture the wealth of the system's dynamics.

This chapter is organized as follows. In the next Sect. 13.2, we discuss the impulsive dynamical system, which includes the main assumptions on the data of the problem as well as the solution concept and some related work. Then, in Sect. 13.3, we briefly introduce some key results on impulsive control that are relevant for the impulsive MPC (IMPC) framework which is introduced and discussed in ensuing

Sect. 13.4. In Sect. 13.5, the main stability results for this framework are given, and discussed. The ensuing Sect. 13.6, includes an extended outline of the proof. In the final Sect. 13.7, we summarize the conclusions.

13.2 The Impulsive Control Framework

Let us present and discuss the class of impulsive control systems that we are going to consider in this work. This includes the assumptions on the optimal control data, as well as, the adopted solution concept, first introduced in [8].

Let us state the optimal impulsive control problem that will play a central role in this work.

$$(P_I)\ \text{Minimize}\quad h(x(T)) \tag{13.1}$$
$$\text{subject to}\quad dx = f(t,x,u)dt + G(t,x,u)d\vartheta \tag{13.2}$$
$$(x(0), x(T)) \in C_0 \times C_T \tag{13.3}$$
$$u \in \mathcal{U} \tag{13.4}$$
$$\vartheta \in \mathcal{I}, \tag{13.5}$$

where

- $h : \mathbb{R}^n \mapsto \mathbb{R}$ specifies the cost functional;
- $f : [0,T] \times \mathbb{R}^n \times \mathbb{R}^m \to \mathbb{R}^n$, and $G : [0,T] \times \mathbb{R}^n \times \mathbb{R}^m \to \mathbb{R}^{n\times k}$ are given mappings defining, respectively the absolutely continuous and the singular dynamics;
- C_0 and C_T are compact subsets of $\mathbb{R}^n$;
- $\mathcal{U} = \{u \in L_\infty([0,T];\mathbb{R}^m) : u(t) \in \Omega\}$, being Ω a compact subset of $\mathbb{R}^m$;
- $\vartheta = (\mu, \{u_\tau, v_\tau\})$ is the impulsive control, being the first component $\mu \in \mathcal{K}$ a Borel measure μ with range in K, i.e., $\forall A \subset [0,T]$, $\mu(A) \in K$, being K a convex, closed and pointed cone in $\mathbb{R}^k$, and $\{u_\tau, v_\tau\}$ is a pair of functions associated with μ at time τ in the support of its atomic component that are specified by:

  $u_\tau \in \mathcal{U}_\tau$, i.e., $u_\tau \in L_\infty([0,\bar{\mu}_\tau]];\mathbb{R}^m)$ s. t. $u_\tau(s) \in \Omega$ $\mathcal{L}$-a.e. in $[0,\bar{\mu}_\tau]$, and
  $v_\tau \in \mathcal{V}_\tau$, i.e., $v_\tau \in L_\infty([0,\bar{\mu}_\tau];\mathbb{R}^k)$ s. t. $v_\tau(s) \in K$ with $\sum_{j=1}^k |v_\tau^j(s)| = 1$ $\mathcal{L}$-a.e. in $[0,\bar{\mu}_\tau]$ and $\int_0^{\bar{\mu}_\tau} v_\tau^j(s)ds = \mu^j(\{\tau\})$, $j = 1,\ldots,k$;

  and finally,
- $\mathcal{I} = \mathcal{K} \times (\mathcal{U}_\tau \times \mathcal{V}_\tau)$ is the impulsive control constraint set.

In the above and from now on, for any atom τ of the measure μ, $\bar{\mu}_\tau := |\mu|(\{\tau\})$, being $d|\mu|$ the total variation measure associated with $d\mu$.

We remark that this is by no means the most general formulation. For example, in [35], a formulation for nonlinear dependence of the dynamics on the impulsive control is addressed. Thus, it covers the following famous minimal surface problem provided by Euler.

Example 13.1 (*Euler minimal surface problem*)

$$\begin{aligned} &\text{Minimize } y(1) \\ &\text{subject to } \dot{y}(t) = x(t)\sqrt{1+u^2(t)}, \quad \dot{x}(t) = u(t), \quad u(t) \in \mathbb{R}, \\ &\qquad y(0) = 0, \quad (x(0), x(1)) = (x_0, x_1) \text{ with } x_0, x_1 \geq 0. \end{aligned} \tag{13.6}$$

This dynamic system clearly exhibits nonlinear dependence on the control u, and the solution to this problem only exists if the control u is in the space of Borel measures.

An example is the following extended impulsive version of the well-known Goddard rocket problem (see, e.g., [15, 17]) allowing mass drops.

Example 13.2 (*Extended Goddard problem with mass drops*)

$$\begin{aligned} \dot{h} &= v, \\ \dot{v} &= \frac{1}{m}[u - D(v,h)] - g_1(h), \\ dm &= -\frac{1}{c}udt - g_2(m,u)d\mu \end{aligned} \tag{13.7}$$

where the $col(h, v, m)$ is the state variable, whose components are, respectively, altitude, vertical velocity, and the mass of the vehicle, $col(u, \mu)$ are the controls, whose components are, respectively, the thrust (conventional control) and the mass drop (impulsive control), D represents drag, g_1 gravity acceleration, and g_2 a function reflecting the way the mass is dropped.

Let us define the trajectory solution concept adopted for (13.2). Consider, the impulsive control $\vartheta = (\mu, \{u_\tau, v_\tau\})$, the number $\tau \in [0, T]$ and the arbitrary vector $x \in \mathbb{R}^n$. Denote by $\chi_\tau(\cdot) = \chi_\tau(\cdot, x)$ the solution to the following dynamical system

$$\begin{cases} \dot{\chi}_\tau(s) = G(\tau, \chi_\tau(s), u_\tau(s))v_\tau(s), \ s \in [0, \bar{\mu}_\tau], \\ \chi_\tau(0) = x. \end{cases} \tag{13.8}$$

The function of bounded variation $x(t)$ on the interval $[0, T]$ is a solution to the differential equation (13.2), associated with the control (u, ϑ) and the initial value x_0, if $x(0) = x_0$ and, for every $t \in (0, T]$,

$$\begin{aligned} x(t) = x_0 &+ \int_0^t [f(\tau, x(\tau), u(\tau)) + G(\tau, x(\tau), u(\tau))w_{ac}(\tau)]\, d\tau \\ &+ \int_{[0,t]} G(\tau, x(\tau), u(\tau))d\mu_{sc} + \sum_{\tau \leq t} \big[\chi_\tau(\bar{\mu}_\tau, x(\tau^-)) - x(\tau^-)\big]. \end{aligned} \tag{13.9}$$

Here, $w_{ac}(\tau)dt$ and $d\mu_{sc}$ are, respectively, the absolutely continuous and the singular continuous components of the canonical decomposition of the measure μ, i.e., $d\mu = w_{ac}dt + d\mu_{sc} + d\mu_a$. The last term concerns the atomic component. Note also that in spite of $\mu_{sc}(\{a\}) = 0 \ \forall a \in \mathbb{R}$, the representation of the second integral in

(13.9) is more natural since, in general, μ_{sc} might be supported on sets that are not defined by a countable number of set operations.

We remark that, by solving the Eq. (13.8) at each atom τ of the control measure, we obtain a well-defined arc joining the jump endpoints. This is a key issue for the well posedness of this solution concept as it endows it with robustness in the sense that any trajectory, solution to the system, can be arbitrarily approximated by a certain sequence of conventional trajectories, i.e., whose dynamics involve only controls of conventional type.

The triple (x, u, ϑ) is called a control process, if it satisfies (13.9). A control process is said to be admissible, if it satisfies all the constraints of problem (P_I), i.e., (13.3), (13.4) and (13.5). An admissible process (x^*, u^*, ϑ^*) is said to be optimal if, for any admissible process (x, u, ϑ), the inequality $h(x^*(T)) \leq h(x(T))$ holds.

This trajectory solution concept is well defined under the following standing assumptions on the data of the problem (P_I):

S1 The mappings f and G are Lipschitz continuous with respect to x, $\forall (t, u) \in [0, T] \times \Omega$;
S2 The mapping f is $\mathcal{B} \times \mathcal{L}$-measurable with respect to (t, u), $\forall x \in \mathbb{R}^n$;
S3 The mapping G is continuous with respect to all its arguments and such that the set $G(t, x, \Omega)$ is compact and convex $\forall (t, x) \in [0, T] \times \mathbb{R}^n$;
S4 The set $\Omega \subset \mathbb{R}^m$ is compact.

These assumptions are not the weakest under which this trajectory solution concept holds but are reasonable and simple to state. These conditions do not suffice to ensure the existence of solution to problem (P_I) and, moreover, they will have to be supplemented with additional ones in order to enable the proof of some results—to be addressed in the next section—required to show the stability of our MPC scheme.

13.3 Some Results on Impulsive Systems

In this section, we provide some notation, concepts, and restate for the first time some previous results for impulsive control systems, such as invariance, [22, 42], and stability, [41, 47], to the class of systems addressed in this chapter. Moreover, since we are dealing with a MPC scheme, we also introduce a solution concept for our system—strongly related to the one above—that is relevant for sampled-data feedback control, as well as the associated results on invariance of sets and stability that are pertinent to establish the stability of the proposed MPC scheme.

13.3.1 Some Notation and Preliminary Definitions

Let us recall some notation and preliminary definitions, mostly extracted from [21, 39]. Consider, a closed set S and a point y not in S. If x is the point in S that is the nearest one to y, then the direction $y - x$ is called a proximal normal direction to S at x.

The set of all proximal normals is the proximal normal cone defined by

$$N_S^P(x) = \{\zeta \in \mathbb{R}^n : \exists \sigma \geq 0 \text{ s.t. } \zeta \cdot (\bar{x} - x) \leq \sigma \|\bar{x} - x\|^2, \ \forall \bar{x} \in S\}.$$

Given, a lower semi-continuous function $f : \mathbb{R}^n \to \mathbb{R} \cup \{+\infty\}$, the proximal subdifferential of f at a point $x \in$ dom can be defined by

$$\partial^P f(x) := \left\{ \zeta \in \mathbb{R}^n : (\zeta, -1) \in N_{\text{epi}\, f}^P((x, f(x))) \right\}.$$

For a given $f : D \mapsto \mathbb{R}$, with $D \subset \mathbb{R}^n$, its epigraph is the set defined by epi $f := \{(x, y) : x \in D, \ y \geq f(x)\}$. Conditions for invariance and stability using these objects for conventional nonlinear dynamic control systems can be found in [21].

A technique that will play a role in impulsive control consists in re-parameterizing the time variable in order to obtain a detailed characterization of the arc joining the endpoints of any of the trajectory jumps. The original time variable t is replaced by the new time variable $s \in \bar{\eta}(t) := [\eta(t^-), \eta(t)]$ where $\eta \in BV^+([0, T]; \mathbb{R}^+)$ defined by $\eta(0) = 0$ and, for $t > 0$, by

$$\eta(t) := t + \int_{[0,t]} d|\mu|(\tau). \tag{13.10}$$

where, as above, $d|\mu|$ denotes the total variation measure associated with $d\mu$ and by $BV^+([0, T]; \mathbb{R}^+)$ we mean the space of scalar functions of bounded variation on $[0, T]$ which are monotonically increasing and take on positive values. The set-valued map $\bar{\eta}(\cdot)$ is monotone, and thus has an inverse that we denote by θ. Thus, $\theta \circ \bar{\eta}(t) = t$. Let us denote by S_μ^a and S_μ^{sc}, the support of, respectively, the atomic and singular continuous components of the measure $d\mu$. Thus, $\eta(T) = T + |\tilde{S}_\mu^{sc}| + \sum_{\tau \in S_\mu^a} |I_\tau^\mu| + \int_0^T |w_{ac}(s)| ds$ where $|A|$ is a short notation for the Lebesgue measure of the set A, $I_\tau^\mu = [\eta(\tau^-), \eta(\tau)] \, \forall \tau \in S_\mu^a$, and $\tilde{S}_\mu^{sc}$ is the union of all sets $B \subset [0, \eta(T)]$ is such that $d|\mu_{sc}|$ is supported on $\theta(B) \subset [0, T]$.

13.3.2 Weak Invariance and Stability

In this subsection, we state invariance and stability results for the class of impulsive control systems currently considered. Under some additional assumptions, it adapts the corresponding results for different classes of measure-driven differential systems, notably, differential inclusions, which have been developed over the years in [32, 41, 42, 47], by migrating some results derived for the conventional context, [21].

From now on, let (f, G) denote the measure-driven differential system given by (13.2). Consider a closed set $S \subset \mathbb{R}^n$.

We say that the system $((f, G), S)$ is weakly invariant if $\forall x_0 \in S$, $\exists (x, (u, \vartheta))$ with u and ϑ satisfying (13.4) and (13.5) such that (13.2) holds with $x(0) = x_0$, then $x(t) \in S\ \forall t$. If this inclusion holds for all feasible controls, then, we say that the strong invariance property is satisfied.

The lower Hamiltonian of system (13.2) is defined as the set-valued map

$$H^l_{f,G}(t, x, \xi) := \min\{H^{l,a}_{f,G}(t, x, \xi), H^{l,c}_{f,G}(t, x, \xi)\},$$

where

$$\begin{aligned} H^{l,a}_{f,G}(t, x, \xi) &= \min\{\langle \xi, G(t, x, \tilde{u}(s))\tilde{v}(s)\rangle : \tilde{u}(s) \in \Omega, \tilde{v}(s) \in V_t\}, \\ &\qquad \forall s \in [\eta(t^-), \eta(t)] \text{ if } t \in S^a_\mu, \\ H^{l,c}_{f,G}(t, x, \xi) &= \min\{\langle \xi, E(t, x, u, v, w)\rangle : u \in \Omega, v \in \bar{V}_t(w), \\ &\qquad w \in [0, 1]\}, \text{ if } t \in [0, T] \setminus S^a_\mu, \end{aligned}$$

where $E(t, x, u, v, w) := f(t, x, u)w + G(t, x, u)v(1 - w)$, $\bar{V}_t(w) = K$ if $w > 0$ and $\bar{V}_t(0) = \tilde{V}(t)$, being the graph of $\tilde{V}(t)$, $\forall t \in S^{sc}_\mu$, defined by the set of pairs $(t, v) = (\theta(s), \tilde{v}(s))$, such that $\tilde{v}(s) \in K \cap B^k_1(0)\ \forall s \in \tilde{S}^{sc}_\mu$, and $\int_A \tilde{v}^j(s)ds = \mu^j_{sc}(\theta(A))$, $j = 1, \ldots, k$, $\forall A \subset \tilde{S}^{sc}_\mu$. Note that, the function $\theta(\cdot)$ is injective in $\tilde{S}^{sc}_\mu$. Since $w \in [0, 1]$ asserts whether, at a given point in time the measure is absolutely continuous—if $w > 0$—or singular—if $w = 0$—, the graph of $\bar{V}_t$ is given by $(0, \tilde{V}(t))$ for $w = 0$, and (w, K), where K is the cone defining the range of the control measure, for $w \in (0, 1]$. Above and in what follows, $B^k_1(0)$ denotes the unit ball in $\mathbb{R}^k$ centered at the origin.

The fact that we consider controlled measure-driven differential equations—where the measure μ might exhibit the three components of its canonical decomposition, absolutely continuous, singular continuous, and atomic—makes the expressions of the "Hamiltonian" rather complex. In order to connect with the concepts for conventional systems with which the reader might be more familiar, it should be noted that Hamiltonian in Mechanics corresponds to the so-called Pontryagin function when evaluated along the optimal control process that reflects the maximal (w.r.t. to the defined performance criterion) total energy of system. The term "lower Hamiltonian" used by the authors in the chapter reflects the total energy of the control system which becomes minimal by selecting an admissible control function.

In the control context, weak invariance is a property of a pair (dynamic control system specified by a certain velocity vector field f, set of points of the state space denoted by S) that holds when there is at least one control function for which the state trajectory remains within the given set once it was initiated there. Thus, if the lower Hamiltonian, $H^l(t, x, \xi)$, is always nonnegative at any point (t, x)—where $x \in S$ is the value of the state variable at time t—whenever, the adjoint variable takes on some value ξ in the proximal normal cone of the set S at x, then this means that there is at least one admissible control function whose value is such that does not point the state trajectory outwards the set S.

Now, we are ready to state the main invariance result which extends previous results, [32, 41, 42], to the class of impulsive systems considered here.

Proposition 13.1 *Consider the system* $((f, G), S)$ *as defined above.*

If $\forall x \in S$, *and* $\xi \in N_S^P(x)$, *we have that* $H^l_{f,G}(t, x, \xi) \leq 0$, *then the system* $((f, G), S)$ *is weakly invariant.*

The proof of this result is inspired in the one of [42] and requires additional hypotheses, notably,

H1 $\forall (t, x) \in [0, T] \times \mathbb{R}^n$, $\bar{E}(t, x)$ and $\bar{G}(t, x)$ are nonempty, convex, and compact sets.

H2 The set-valued maps $\bar{E}(\cdot, \cdot)$ and $\bar{G}(\cdot, \cdot)$ are upper semi-continuous.

H3 There are constants a and b such that, $\forall (x, t) \in [0, T] \times \mathbb{R}^n$, $\forall v_E \in \bar{E}(t, x)$ and $\forall v_G \in \bar{G}(t, x)$

$$\max\{\|v_E\|, \|v_G\|\} \leq a\|x\| + b.$$

where $\bar{G}(t, x) = \{G(t, x, u)v : u \in \Omega, v \in K \cap MB_1(0)\}$ and $\bar{E}(t, x) = \{E(t, x, \Omega, \bar{V}_t(w), w) : w \in [0, 1]\}$, being M some sufficiently large in the sense that $M \geq \|d\mu\|_{TV}$ for any $((f, G), S)$ invariant control process. Here, $\|d\mu\|_{TV}$ denotes the total variation norm of the measure μ. The outline of a simple proof of this result, in a context that requires the Lipschitz continuity of the mappings f and G w.r.t. also to the time variable, consists of using the above-mentioned re-parameterization procedure in order to construct an equivalent conventional dynamical control system or which invariance results are already available. Then, the above proposition results from expressing these conditions in terms of the data of the original system by writing the conditions in the original parametrization.

Moreover, the above proposition can be extended from finite horizon to $[0, \infty)$ if we impose the additional assumption

H4 $\|\mu\|_{TV} < \infty$, and $\forall \tau > 0$, $\lim_{t \to \infty} |\mu|([t, t + \tau]) = 0$.

Remark that the invariance conditions provided in Proposition 13.1 are given in the form of a decreasing function condition, which, for a certain choices of functions, notably, a pair of Lyapunov functions, yield stability conditions.

Therefore, we use this result in order to provide stability conditions for the class of impulsive control systems considered in this chapter, and thus, extending the work in [41, 47].

Proposition 13.2 *Let us assume all the hypotheses considered above for the data of our impulsive dynamic control system in the infinite horizon context.*

Assume also that there is a Lyapunov pair (V, W) *that satisfies the following conditions:*

- *Positive definiteness:* $V(t,x) \geq 0$ *and* $W(t,x) \geq 0$ *for all* $(t,x) \in [0,\infty) \times \mathbb{R}^n$, *and* $W(t,x) = 0$ *if and only if* $x = 0$;
- *Growth: the set* $\{x \in \mathbb{R}^n : V(t,x) \leq r\}$ *is compact for all* $\forall r \geq 0$ *and* $\forall t$;
- *Infinitesimal decay:* $\forall (t,x) \in [0,\infty) \times \mathbb{R}^n$, $\exists (u,\vartheta) \in \mathcal{U} \times \mathcal{I}$ *s.t.*, $\forall (\nu, \zeta^e) \in \partial^P V(t, x^e)$
$$H^l_{f,G}(t, x^e, \nu, \zeta^e) \leq -W(t,x).$$

Then, $x = 0$ *is an asymptotically stable equilibrium.*

The lower Hamiltonian is now adapted to accommodate both the extended trajectory and the time dependence, i.e.,

$$H^l_{f,G}(t, x^e, \nu, \zeta^e) = \min\{H^{l,a}_{f,G}(t, x^e, \zeta^e), H^{l,c}_{f,G}(t, x, \nu, \zeta)\},$$

where $H^{l,c}_{f,G}(t,x,\nu,\zeta) = \min\limits_{u\in\Omega, v\in\bar{V}_t(w), w\in[0,1]} \{\bar{\nu}(w) + \langle \zeta, E(t,x,u,v,w)\rangle\}$, *with* $\bar{\nu}(w) = \nu$ *if* $w > 0$, *and* $\bar{\nu}(0) = 0$, *and* ζ^e *and* x^e *denotes that the evolution along the jump is considered, i.e.,* $\tilde{\zeta}_\tau(s) \in \partial^P_x V(t, \tilde{x}_\tau(s))$, $\forall s \in \bar{\eta}(s) = \theta^{-1}(\tau)$.

Regarding the literature on optimal control for impulsive systems, we refer to [4–6, 9, 10, 40, 50] for necessary conditions of optimality, to [32] for Hamilton–Jacobi–Bellman results, and to [34] for existence results.

13.3.3 Euler Solution

In this subsection, we introduce the notion of Euler solution for the considered class of impulsive control systems which is an appropriate concept when we have to deal with sampled-data trajectories and feedback control as it is the case of the implementation of MPC schemes.

Under the standing assumptions and hypothesis H1-H4, the Euler solution is well defined. However, the MPC scheme requires additional assumptions in order to ensure existence of solution for the associated family of optimal control problems, [34], and, in line with [30], conditions on its design parameters.

Let us fix a feasible control strategy formed by a conventional control $u \in \mathcal{U}$, and an impulsive control $\vartheta \in \mathcal{I}$, whose associated measure $d\mu \in \mathcal{K}$ defined on a given general time interval $[a,b]$ has a finite total variation measure $d|\mu|$. Then, consider the following initial-value problem:

$$dx(t) = f(t, x(t), u(t))dt + G(t, x(t), u(t))d\vartheta(t), \ t \in [a,b], \tag{13.11}$$
$$x(a) = x(a^-) = x_0.$$

Along the lines of [32], we consider, the impulsive Euler solution by first recalling the re-parameterization function (13.10), and the set-valued map $\bar{\eta} : [a,b] \to \mathcal{B}([a,b])$ defined above to be

$$\bar{\eta}(t) = \begin{cases} \{\eta(t)\} & \text{if } \mu(\{t\}) = 0 \\ [\eta(t^-), \eta(t)] & \text{if } \mu(\{t\}) \neq 0 \end{cases}$$

Then, we define a partition on the range of $\bar{\eta}_N$ as follows: $\pi = \{s_0, s_1, \dots, s_N\}$, with $s_0 = a$ and $s_N = \eta(b^+)$. The diameter of the partition π is defined by $\alpha_\pi := \max\{s_i - s_{i-1} : 1 \leq i \leq N\}$ as depicted in Fig. 13.2.

For $i = 1, \dots N$, we associate to each node point s_i of the partition π_N, a node point, t_i, in the t-domain by $t_i = \theta^{-1}(s_i)$, being $t_0 = a$. Remark that, for each partition point on the range of $\bar{\eta}$ there corresponds only one point in π_N if is in a subset of the original time on which the measure μ is continuous or exhibits sufficiently small atoms. However, if there is a significant (relatively to the distance between two samples in the re-parameterized domain) atom at some time τ, then the interval $\bar{\eta}(\tau)$ might contain multiple consecutive points of π_N. Let $s_{i_\tau^-} = \min\{s_i \in \pi_N : s_i \in \bar{\eta}(\tau)\}$ and $s_{i_\tau^+} = \max\{s_i \in \pi_N : s_i \in \bar{\eta}(\tau)\}$, then we may define the approximating measure $d\mu_N$ as follows. For $i = 0, \dots, N-1$:

$$d\mu_N(t) := \begin{cases} d\mu(t), & \text{if } t \in [t_i, t_{i+1}) \text{ with } t_{i+1} > t_i \\ \displaystyle\sum_{i=i_\tau^-}^{i_\tau^+} (s_{i+1} - s_i) v_i \delta_{t_i}, & \text{if } t \text{ is such that } \eta(t) > \eta(t^-) \end{cases} \tag{13.12}$$

where the vectors v_i, for $i = i_\tau^- \dots i_\tau^+$, are such that $|v_i| = 1$, $\displaystyle\sum_{i=i_\tau^-}^{j} (s_{i+1} - s_i) v_i \in K$ and $\displaystyle\sum_{i=i_\tau^-}^{i_\tau^+} (s_{i+1} - s_i) v_i = \mu(\{t\})$. Notice that $d\mu_N$ coincides with $d\mu$ whenever the singular atomic component is absent.

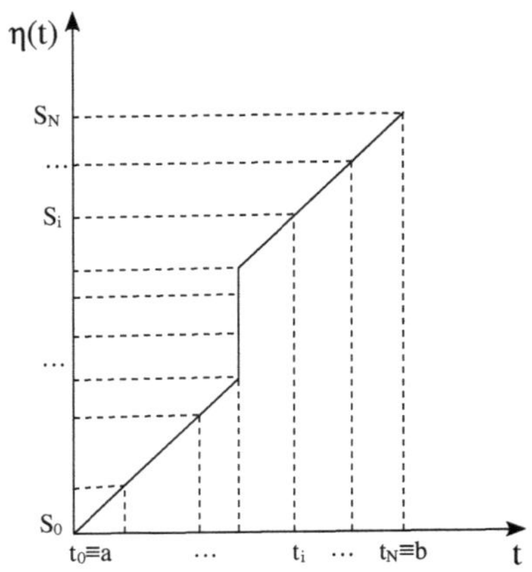

Fig. 13.2 Sampling scheme (courtesy from [31])

The impulsive Euler polygonal multifunction $x^e_{\pi_N}$—meaning the superscript e that the solution is to be interpreted in the extended sense, and the subscript π_N that a partition with π_N is being considered—is defined recursively as follows:

We start with the subinterval $[t_0, t_1]$, and the dynamical system

$$dx(t) = f(t, x_0, u(t))dt + G(t, x_0, u(t))d\vartheta_0(t),$$

with $x(t_0) = x_0$. By integration, the node point $x_1 := x(t_1)$ is uniquely defined. If the measure $d\mu_0$ associated with the impulsive control $d\vartheta_0$ is singular atomic, then $[t_0, t_1]$ reduces to one point, i.e., $t_1 = t_0$ and the node x_1 is computed by integrating the singular dynamics with the sampled measure, given by (13.12) with total variation $s_1 - s_0$, and along the extended trajectory and control. For a general x_i, we obtain the next node point $x_{i+1} := x(t_i)$ by integrating $dx(t) = f(t, x_i, u(t))dt + G(t, x_i, u(t))d\vartheta_i(t)$ on $[t_i, t_{i+1}]$ with $x(t_i) = x_i$.

This procedure is carried out until all elements of the partition π_N are covered, and an impulsive Euler polygonal multifunction $x^e_{\pi_N}(t)$ is obtained on $[a, b]$. We recall that, on the support of the atomic component of the measure $d\mu$, the trajectory $x^e_{\pi_N}$ exhibits jumps with well defined arcs joining their endpoints.

By Impulsive Euler Solution (IES) of the initial-value problem (13.11), we mean any set-valued map $x^e(\cdot)$ which is the uniform limit of impulsive Euler polygonal set-valued map $x^e_{\pi_N}(\cdot)$, corresponding to some sequence π_N as $N \to \infty$ and, α_{π_N}, the largest subinterval of the partition, is such that $\alpha_{\pi_N} \downarrow 0$. Of course, this uniform limit also holds for the sequence of arcs joining the endpoints of the jumps.

Similar arguments to those in [32] that, under all the assumptions considered in this chapter, a IES exists and have a number of desirable properties, namely:

(a) At least an IES $x^e(t)$ exists for system (13.11);
(b) Any IES solution to (13.11), interpreted in the extended sense that is by considering the variation along the arc joining the jump endpoints whenever t is an atom of the control measure, has linear growth;
(c) Any IES solution to (13.11) is consistent with other solution concepts, namely the one of robust solution presented in Sect. 13.2.

13.4 Model Predictive Control of Impulsive Systems

Now, we introduce the sampled-data MPC scheme for the class of impulsive control systems (IMPC), (13.2), considered in this chapter.

The construction of the feedback control law is achieved by using a sampled-data IMPC strategy.

Consider a sequence of sampling instants $\pi := \{t_i\}_{i \geq 0}$ in $[0, +\infty)$ with inter-sampling times $\delta_i > 0$ such that $t_{i+1} = t_i + \delta_i$ for all $i \geq 0$. The feedback control is obtained by solving iteratively online open-loop optimal control problems $\mathcal{P}(t_i, x_i, T)$ at each sampling instant $t_i \in \pi$, every time using the current measure of the state of the plant x_i.

$$\mathcal{P}(t_i, x_i, T) \text{ Minimize } W(t_i + T, x(t_i + T)) + \int_{t_i}^{t_i+T} L_{ac}(s, x(s), u(s))ds$$
$$+ \int_{[t_i, t_i+T]} L_s(s, x(s), u(s))d\vartheta(s) \quad (13.13)$$

$$\text{subject to } dx(t) = f(t, x(t), u(t))dt + G(t, x(t), u(t))d\vartheta(t)\, \forall t \in [t_i, t_i + T], \quad (13.14)$$

$$x(t_i) = x_i, \quad (13.15)$$
$$u \in \mathcal{U}_{|[t_i, t_i+T]},$$
$$\vartheta \in \mathcal{I}_{|[t_i, t_i+T]},$$
$$x(t_i + T) \in S, \quad (13.16)$$

where $S \subset \mathbb{R}^n$ is a given closed set, and the mappings $W : [t_i, t_i + T] \times \mathbb{R}^n \to \mathbb{R}$, $L_{ac} : [t_i, t_i + T] \times \mathbb{R}^n \times \mathbb{R}^m \to \mathbb{R}$ and $L_s : [t_i, t_i + T] \times \mathbb{R}^n \times \mathbb{R}^m \to \mathbb{R}^k$ are chosen in order to ensure the purposes of the IMPC scheme and, in particular, its stability. Here, $\mathcal{V}_{|A}$ denotes the set of functions $\mathcal{V}$ restricted to the set A. The remaining ingredients were introduced in Sect. 13.2.

The IMPC algorithm is defined according to a receding horizon scheme that takes into account the specificities of the impulsive control considered in this work.

1. Initialization. Set parameters, specify initial data, and iteration counter $i = 0$.
2. Sample the current state of the plant $x(t_i) = x_i$.
3. Solve problem $\mathcal{P}(t_i, x_i, T)$ to obtain the open-loop optimal conventional control $\bar{u}^i \in \mathcal{U}_{|[t_i, t_i+T]}$ and impulsive control $\bar{\vartheta}^i \in \mathcal{I}_{|[t_i, t_i+T]}$.[1]
   Obviously that whenever $\bar{\mu}^i(\{t\}) \neq 0$ (i.e., if the optimal control measure has an atom, including the time endpoints t_i and $t_i + T$), then, the optimal arc joining the associated trajectory endpoints has to be defined by computing the optimal pair of functions $(\bar{u}_t^i(\cdot), \bar{v}_t^i(\cdot))$ defined on the associated emerging interval $[t, t + |\bar{\mu}^i(\{t\})|]$.
4. Determination of the next sampling instant. This is given by $t_i + \delta_i$ where $\delta_i := \min\{\delta, \tau\}$ and $\tau = \min\{t \in [0, \delta] : \bar{\mu}^i(\{t_i + t\}) \neq 0\}$. That is, the next sampling instant is the earliest time in which either a time interval of duration δ elapses, or an atom of $\bar{\mu}^i$ occurs.
   It is important to remark that the case of $\tau = 0$ makes sense when the perturbations affecting the system are extremely fast, and it is "physically" feasible to "increase the sampling speed"[2] so that the state can be sampled at several midway points of the jump at t_i.

[1] This problem can be solved by using direct methods or indirect methods which take advantage of necessary conditions of optimality, possibly, in the form of a maximum principle such as the ones proved in [9, 10].

[2] Here, we consider a mathematical abstraction of a scheme whose physical realization may involve sampling "during" the "atomic activities" at a frequency several orders of magnitude higher than the one when the trajectory is evolving continuously.

5. Apply to the plant the control pair $\bar{u}^i$ and $\bar{\vartheta}^i$ during the interval $[t_i, t_i + \delta_i]$, being the control strategy values computed for $t \geq t_i + \delta_i$ discarded.
6. Now the optimization time horizon slides, i.e., we consider $t_{i+1} = t_i + \delta_i$, we let $i = i + 1$ and repeat the procedure from step 2.

One might wonder as a remark to step 4, whether an infinite number of jumps might occur leading to, what in the context o hybrid control, is designated by Zeno behavior. Under the assumptions of existence of solutions that we impose in our framework which lead to the existence of feasible trajectories whose associated control measures have finite total variation in any finite time interval, the answer is no. Thus, there are no mathematical difficulties in dealing with the accumulation of a countable number of jumps, or even with an uncountable number of singular continuous evolutions if the finiteness of their total variation is observed. Obviously, in our context, there is no need of imposing the assumption/restriction of a threshold time interval separating consecutive jumps. It is important to have in mind that the control synthesis leads to control processes which, in the engineering practice, are approximated by "sufficiently close implementable/computable" controls.

For a given partition π, the control law $(u^*_\pi, \vartheta^*_\pi)$ resulting from the concatenation of the sequentially computed $(\bar{u}^i, \bar{\vartheta}^i)$ is a "sampling-feedback" control law, l, since during each sampling interval, the control $(u^*_\pi, \vartheta^*_\pi)$ depends on the state $x^*(t_i)$ in the extended sense as defined in Sect. 13.2. More precisely, the resulting trajectory is given by

$$\begin{aligned} &dx^*_\pi(t) = f(t, x^*_\pi(t), u^*_\pi(t))dt + G(t, x^*_\pi(t), u^*_\pi(t))d\vartheta^*_\pi(t), \quad t \geq t_0, \\ &x^*_\pi(t_0) = x_0 \end{aligned}$$

where

$$\begin{aligned} (u^*_\pi, d\vartheta^*_\pi)(t) &= (l_u, dl_\vartheta)(t, x^*_\pi(\lfloor t \rfloor_\pi)), \quad \forall t \geq t_0 \\ &:= (\bar{u}^i, d\bar{\vartheta}^i)(t; \lfloor t \rfloor_\pi, x^*_\pi(\lfloor t \rfloor_\pi)), \; t \in [t_i, t_i + \delta_i], \; i = 1, \ldots. \end{aligned}$$

Here, the function $t \mapsto \lfloor t \rfloor_\pi$ gives the last sampling instant just before t that is

$$\lfloor t \rfloor_\pi := \max_i \{t_i \in \pi : t_i \leq t\}.$$

Remark that this does not preclude the possibility of several consecutive t_i's taking on the same value.

Similar sampled-data frameworks, using continuous-time models and sampling the state of the plant at discrete instants of time, were adopted in [18, 23, 26, 27, 29, 36]. The main difference here concerns the new steps in the receding horizon strategy that had to be introduced to deal with the discontinuities of the trajectory.

A key idea in the above IMPC scheme consists in the fact that at least one sampling point—and, in general, more than one—at every time the trajectory jumps, has to be ensured (see Fig. 13.3).

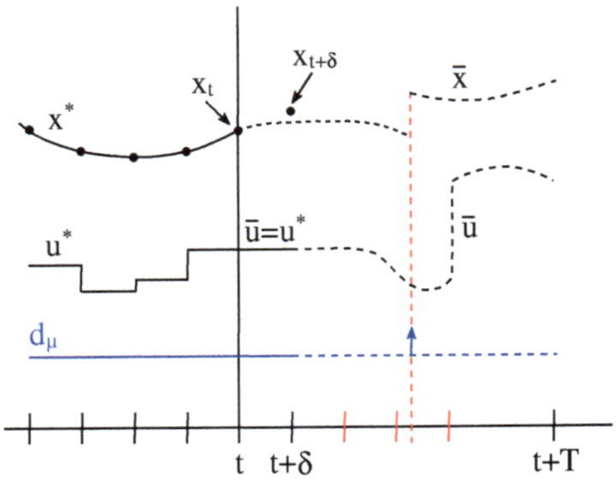

Fig. 13.3 Sample times include the support of atoms of the measure μ associated with the impulsive control

Suppose that this were not the case, and that $\tau \in (t_i, t_i + \delta)$ is a point in time in which the trajectory has a discontinuity. Then, for $t \in (\tau, t_i + \delta)$, neither $\|x^*_\pi(\lfloor t \rfloor) - x^*_\pi(t)\|$ nor $\|l(x^*_\pi(\lfloor t \rfloor)) - l(t, x^*_\pi(t))\|$ can be ensured to be smaller than an arbitrary quantity by reducing the size of $\delta \geq 0$. Thus, a bound on the error between the sampled-data feedback and the "ideal" continuous feedback required to establish stability can not be guaranteed.

However, in the impulsive sampled-data framework adopted for the IMPC scheme proposed here, the solution to optimal impulsive control problem provides not only the currently conventional control function but also the sequence of future sampling instants through its impulsive component. Note, that the fact that the trajectory in an extended sense can be sampled along the arc joining the jump endpoints is critical to ensure the continuity property relating the error bound and sampling frequency required to guarantee the stability of the IMPC scheme.

13.5 Stability of the Impulsive MPC Scheme

In this section, stability results for this framework are given. Without any loss of generality, we assume that the origin of the state space is the equilibrium point of interest and the goal of the MPC scheme is to drive the system to this equilibrium point. We show that stability is guaranteed if the design parameters of the optimal control problem, notably the optimization time horizon T, the mappings L_{ac}, L_s and W, and terminal constraint set S satisfy certain conditions. While, the ingredients of the objective function $((L_{ac}, L_s), W)$ have to satisfy the properties of a control Lyapunov pair within the set S, the system $((f, G), S)$ must be weakly invariant. Obviously, the set feasible control processes of $(\mathcal{P}(t_i, x_i, T))$ must be nonempty, and a solution to $(\mathcal{P}(t_i, x_i, T))$ must exist.

We start by addressing the feasibility and existence of solution to the optimal control problems to be solved recursively in the IMPC scheme, and, then, we provide sufficient stability conditions of the Lyapunov type.

In order to show the sufficient conditions for the nonemptiness of the set of feasible control processes as well as the existence of solution of each one of the $(\mathcal{P}(t_i, x_i, T))$'s we need additional hypotheses that also encompass the IMPC scheme design ingredients.

Consider the following assumptions, collectively designated by feasibility condition (FC), on the design parameters time horizon T, functions $((L_{ac}, L_s), W)$, and terminal constraint set S.

F1 The set S is closed and contains the origin.
F2 The functions L_{ac} and L_s satisfy the following conditions: continuity in their arguments, for $L(t, 0, 0) = 0$, and there is a continuous positive definite and radially unbounded function $M : \mathbb{R}^n \to \mathbb{R}_+$ such that $\|L(t, x, u)\| \geq M(x)$ for all $u \in \Omega$, where $L = col(L_{ac}, L_s)$.
F3 The extended "velocity" set V^e, defined by

$$V^e = \{(r, (f(t, x, u), 0, 1)w + (G(t, x, u)v, 1, 0)(1 - w), w,)) : \\ r \geq L_{ac}(t, x, u)w + L_s(1, x, u)v(1 - w), \\ u \in \Omega,\ v \in V_t(w),\ w \in [0, 1]\},$$

where (as before) $V_t(w) = K$ if $w > 0$ and $V_t(0) = K \cap B_1(0)$, is convex.
F4 The function W is positive semi-definite and continuously differentiable.
F5 The set S is reachable within a time interval of duration T from any initial state. That is, for every x_0 there exists a control pair $u \in \mathcal{U}_{|[t_0, t_0+T]}$ and $\vartheta \in \mathcal{I}_{|[t_0, t_0+T]}$ satisfying

$$x(t_0 + T; t_0, x_0, u, \vartheta) \in S.$$

Now, we are in position to state the following result guaranteeing the feasibility of the IMPC scheme.

Theorem 13.1 *Assume that the standing hypothesis and that H1–H4 hold.*
If the design parameters satisfy (FC), then a sequence of control processes which are optimal solutions to the problems $\mathcal{P}(t_i, x_{t_i}, T)$, $i = 1, \ldots,$ exists and the trajectory generated by the IMPC strategy has no finite escape times.

The proof is a straightforward extension and combination of previous results on (i) existence of solution to impulsive control problems from [11, 34], and (ii) the overall trajectory associated with the control strategy generated by the proposed IMPC scheme has no finite escape times, [26, Sect. 6.2]. For the latter conclusion, just note that, by re-parameterizing the time variable as previously, a sequence of equivalent conventional optimal control problems is obtained (this is possible under our assumptions) to which the arguments in [30] can be applied.

Now, in order to state our result on sufficient conditions for the stability of control strategies generated by the IMPC scheme expressed in terms of a pair of control Lyapunov functions, let us define the following Lyapunov-like stability condition (SC) to be satisfied by (13.2).

For a given time interval of duration T, there are mappings $((L_{ac}, L_s), W)$ and an endpoint constraint S as defined in the statement of $\mathcal{P}(t_i, x_{t_i}, T)$, such that the set of controls $(u, \vartheta) \in \mathcal{U} \times \mathcal{I}$ restricted to the considered optimization horizon simultaneously satisfying the conditions (S1) and (S2) below is nonempty. These conditions are:

$$(S1) \quad \forall \xi \in N_S(x),$$
$$0 \geq \begin{cases} \langle \xi, G(t, x, \tilde{u}(s))\tilde{v}(s)\rangle, & \forall s \in \bar{\eta}(t), \ \forall t \in S^a_\mu \\ \langle \xi, E(t, x, u, v, w)\rangle, \text{where } v \in \bar{V}_t(w), \ w \in [0, 1], & \forall t \in \left(S^a_\mu\right)^c, \end{cases}$$

and

$$(S2) \quad \forall (\nu, \zeta^e) \in \partial_P W(t, x^e),$$
$$0 \geq \begin{cases} \langle \tilde{\zeta}_t(s), G(t, \tilde{x}_t(s), \tilde{u}(s))\tilde{v}(s)\rangle + L_s(t, \tilde{x}_t(s), \tilde{u}_t(s))\tilde{v}_t(s), \\ \qquad \forall s \in \bar{\eta}(t), \ \forall t \in S^a_\mu \\ \bar{\nu}(w) + \langle \zeta, E(t, x, u, v, w)\rangle + e(t, x, u, v, w), \\ \qquad \text{where } v \in \bar{V}_t(w), \ w \in [0, 1], \ \forall t \in \left(S^a_\mu\right)^c. \end{cases}$$

Here, $e(t, x, u, v, w) = L_{ac}(t, x, u)w + L_s(t, x, u)v(1 - w)$ and all other ingredients—functions, sets, and constraints on control values—are as defined in Sect. 13.3.2.

Our main result is as follows:

Theorem 13.2 *Assume that hypotheses H1–H4 and (FC) hold. Moreover, consider that the design parameters in $\mathcal{P}(t_i, x_{t_i}, T)$ satisfy (SC), then the generated IMPC control strategy is stabilizing, i.e., for a sufficiently small inter-sample time δ, we have $\|x^*(t)\| \to 0$ as $t \to +\infty$.*

Let us consider, the following impulsive dynamic control system and let us construct a feedback impulsive control strategy (u, ϑ) that can be generated by the IMPC scheme to drive the system from its initial time to the origin as $t \to \infty$. The data is as follows:

Example 13.3

$$\begin{cases} dx = Axdt + g(x, u)d\mu, & x(0) = col(1, 0), \\ u(t) \in [-1, 1], & \mu \in C^*([0, \infty), \mathbb{R}^+), \end{cases} \tag{13.17}$$

where $A = \begin{bmatrix} 0 & 1 \\ -1 & 0 \end{bmatrix}$, $g(x, u) = h(x)\bar{g}(x, u)$, $h(x) = \max\{sgn_0^+(|x_2| - |x_1|), 0\}$, and $\bar{g}(x, u) = (A - sgn_0^+(x_1)uI)x$.

Here, I is the identity matrix in $\mathbb{R}^2$, $sgn_0^+(a) = 1$ if $a \geq 0$ and $sgn_0^+(a) = -1$, otherwise, and $C^*([0, \infty); \mathbb{R}^+)$ is the space of positive measures on $[0, \infty)$.

From a simple inspection, it is easy to conclude that there are regions of the state space in which the system is not controllable. Let R_i, $i = 1, 2, 3$, and 4, are defined by the set of pairs (x_1, x_2), respectively, satisfying $-x_1 \leq x_2 \leq x_1$ for $x_1 \geq 0$, $-x_2 \leq x_1 \leq x_2$ for $x_2 \geq 0$, $x_1 \leq x_2 \leq -x_1$ for $x_1 \leq 0$, and $-x_2 \leq x_21 \leq x_2$ for $x_2 \leq 0$. Notice that $\mathbb{R}^2 = \bigcup_{i=1}^{4} R_i$. It is easy to see that it is not possible to control the system in $R_1 \bigcup R_3$, where the trajectory follows arcs of circumference with a radius defined by the entry point at either R_1 or R_3. On the other hand, there are many control strategies acting in the region $R_2 \bigcup R_4$ that drive the system to the origin, some of which even do not require impulses at all. So, in order to show our point, let us consider that the absolutely component of the measure μ is absent.

The system dynamics clearly suggests that the Lyapunov function $V(x_1, x_2) = \frac{1}{2}(x_1^2 + x_2^2)$ is a suitable one for the control synthesis. Clearly, under the assumptions on $d\mu$, we have:

- If $x \in R_1 \bigcup R_3$, then $\dot{V}(x) = x_1\dot{x}_1 + x_2\dot{x}_2 = 0$.
- If $x \in R_2 \bigcup R_4$, then $dV(x) = x_1 dx_1 + x_2 dx_2$.

Thus, one may choose $u(t) = sgn_0^+(x_1(t))$ and $d\mu = \frac{\pi}{2}\sum_{k=1}^{\infty} \delta_{(k+\frac{1}{4})\pi}(t)$, where $\delta_\tau(t)$ represents the unit Dirac impulse at $t = \tau$. The choice of the impulses is natural since, otherwise, the state variable would not move at the points were $x_1 = x_2$. It is important to remark that the sign of x_1 changes during the jump, and the ordinary control u also has to change as the jump occurs. It is clear that the chosen control function implies that V will decrease in the region $R_2 \bigcup R_4$ when our extended solution concept—the ordinary control may change as a jump occurs—is adopted. After re-parameterizing the dynamics as defined above, we have, for the parameter s varying from 0 to $\frac{\pi}{2}$, that

$$\begin{bmatrix} \dot{\tilde{x}}_1(s) \\ \dot{\tilde{x}}_2(s) \end{bmatrix} = \begin{bmatrix} -1 & 1 \\ -1 & -1 \end{bmatrix} \begin{bmatrix} \tilde{x}_1(s) \\ \tilde{x}_2(s) \end{bmatrix}$$

and, thus, $\dot{\tilde{V}}(\tilde{x}(s)) = -2\tilde{x}_1^2(s) - \tilde{x}_2^2(s)$, which is clearly negative for $x \neq 0$. Notice that the "magnitude" of the jump, $\frac{\pi}{2}$, is needed in order to fully transverse either the region R_2 or R_4 and that the jump is not just specified by the pair of its endpoints but rather an arc that satisfies the singular dynamics. As shown in Fig. 13.4, in the absence of perturbations, the trajectory will have jumps at the times $t_k = \frac{2k+1}{4}\pi$, satisfying $\|x_{t_k^+}\| = e^{-\frac{\pi}{2}}\|x_{t_k^+}\|$. The jumps will be from the line $x_1 = x_2$ to the line $x_1 = -x_2$, alternating in the $x_2 > 0$ and the $x_2 < 0$ half spaces.

The data of this example satisfies strong assumptions which makes the asymptotic stability to be established with the usual Lyapunov method, thus dispensing with the sophistication of the above results. However, it is not difficult to check the conditions derived in this section are sufficient to ensure the stability of the proposed control

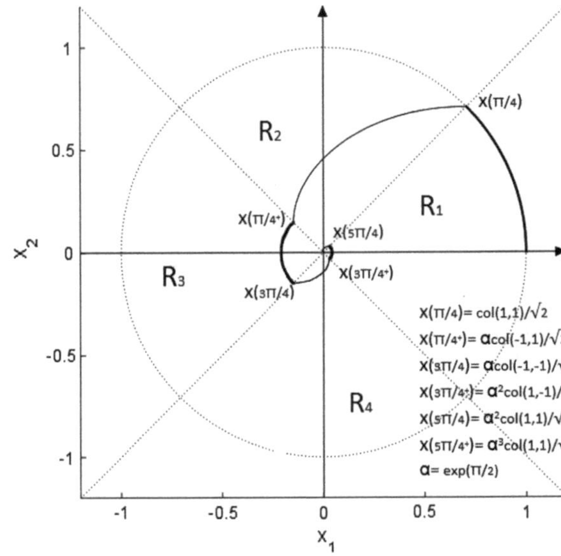

Fig. 13.4 The *thicker* and the *thinner lines* represents, respectively, the absolutely continuous and the jump evolutions

strategy, being the sets S chosen as $S_k = e^{-(\frac{\pi}{2})^k} B_1(0)$ where $B_1(0)$ is the unit ball centered at the origin.

In what concerns the sampling strategy, the sparsest one should include the points t_k (just before the jump) and other points short while after the jump. As mentioned above, ideally, one can consider taking several samples in between t_k and t_k^+ in order to check the variation of the sign of x_1. Of course, in the engineering practice, this abstraction is considered by approximating the impulsive controls by appropriate absolutely continuous controls, and, then, practical high-frequency sampling could take place in the support of the chosen $d\mu$.

13.6 Outline of the Proof for Stability Result

The assumption that the set S will be reached by the proposed IMPC scheme allows us to assert that there exists some time t such that $x(t) = x_t \in S$. Remark that this "crude" assumption could have been replaced by a more natural one involving the system's controllability that would allow us to show that the set S would be attained at some finite time.

Hence, it is natural to assume (S1) component of the stability condition (SC). Thus, a straightforward application of Proposition 13.1 in Section 13.3.2, which can be regarded as an extension of the invariance result in [42], implies that there exists a control pair $(\bar{u}_t, \bar{\vartheta}_t)$ such that, for some $\delta_1 > 0$, and $\forall \tau \in [t, t + \delta_1)$, we have

$$x(\tau; t, x_t, (\bar{u}_t, \bar{\vartheta}_t)) \in S \tag{13.18}$$

Here and in what follows, $x(b; a, x_a, (u_a, \vartheta_a))$ is a short notation for the value of the state variable at time b when $x(a) = x_a$ and the control (u_a, ϑ_a) is applied in the interval $[a, b]$.

From the fact that there is a set of feasible control processes (u, ϑ) for which (S1) and (S2) hold simultaneously, we may conclude from Proposition 13.2 in Section 13.3.2, which can be regarded as an extension of the Lyapunov stability result in [47] that the control $(\bar{u}_t, \bar{\vartheta}_t)$ may also be such that, for some $\delta_2 > 0$, we may have that, $\forall \tau \in [t, t + \delta_2)$,

$$W(\tau, x(\tau; t, x_t, (\bar{u}_t, \bar{\vartheta}_t))) - W(t, x_t) \\ \le - \int_t^{\tau} L_{ac}(s, x(s), \bar{u}(s))ds - \int_{[t,\tau]} L_s(s, x(s), \bar{u}(s))d\vartheta(s). \quad (13.19)$$

Clearly, both conditions hold for $\bar{\delta} = \min\{\delta_1, \delta_2\}$. A straightforward application of the arguments from [26] (recall that, under our assumptions, the impulsive control problem considered here can be re-parameterized in the time variable so that a conventional optimal control problem is obtained) are used to show that if both (13.18) and (13.19) are satisfied, then a certain MPC value function V is shown to be monotone decreasing. More precisely, for some $\delta > 0$, $\delta < \bar{\delta}$, small enough and for any $t'' > t' > 0$,

$$V(t'', x^*(t'')) - V(t', x^*(t')) \le - \int_{t'}^{t''} M(x^*(s))ds. \quad (13.20)$$

where M is the continuous, radially unbounded, positive definite function considered in the assumption F2.

Now, let us consider a time partition π defining the sampling strategy, and denote the associated MPC value function by V_π. In a sufficiently small neighborhood of t is defined as

$$V_\pi(t, x) := V_{\lfloor t \rfloor_\pi}(t, x)$$

where $V_{\lfloor t \rfloor_\pi}(t, x_t)$ is the value function for the optimal control problem $\mathcal{P}(t, x_t, T - (t - \lfloor t \rfloor_\pi))$ (the optimal control problem defined where the horizon is shrank in its initial part by $t - \lfloor t \rfloor_\pi$).

Now, by patching together a sufficiently large number of adjoint small intervals $[t', t'')$ so that (13.20) holds in each one of them, we can then write that for any $t \ge t_0$

$$0 \le V_\pi(t, x^*(t)) \le V_\pi(t_0, x^*(t_0)) - \int_{t_0}^{t} M(x^*(\tau))d\tau.$$

Since $V_\pi(t_0, x^*(t_0))$ is finite, we conclude that the function $t \mapsto V_\pi(t, x^*(t))$ is bounded and, thus that $t \mapsto \int_{t_0}^t M(x^*(\tau))d\tau$ is also bounded. Therefore, $t \mapsto x^*(t)$ is bounded and, since both f and G are continuous and takes values on bounded sets of (t, x, u), it is clear that, after the re-parameterization considered in Sect. 13.2

that the map $s \mapsto \dot{\tilde{x}}^*(s)$, for all $s \in \bar{\eta}(t)$ is also bounded. All the conditions required to apply Barbalat's Lemma [28] are met, it follows from its application that the trajectory $\tilde{x}^*(\cdot)$ converges asymptotically to the origin and, since $x^*(t) = \tilde{x}^*(\eta(t))$ $\mathcal{L} - a.e.$ so does the $x^*(\cdot)$.

We just proved the attractiveness property, i.e., for a sufficiently small inter-sample time δ we have $\|x^*(t)\| \to 0$ as $t \to +\infty$. The stability in the Lyapunov sense follows from the continuity of V.

13.7 Conclusion

In this chapter, a novel impulsive model predictive control scheme was introduced. The novelty resides in the sampling scheme as well as in the fact that the "singular" dynamics, i.e., the ones responsible for the trajectory jumps depends also on the conventional control, besides the usual dependence on the time and state variables. The motivation for this is clear from the wide classes of problems that can be considered as it was illustrated in the introduction.

A key feature of the proposed IMPC scheme consists on the fact that the optimal impulsive control strategy determines an appropriately adapted sampling strategy, in the sense that the state variable is sampled, at least once, at the points of discontinuity. This feature allows to ensure the stability of the overall IMPC scheme under assumptions which are standard for the sufficiency of the Lyapunov type of stability. This is the main result discussed and whose proof is outlined in this article.

It is important to note that the impulsive paradigm can be regarded as a very convenient mathematical abstraction to deal with systems with slow and fast dynamics. An important feature of the proposed IMPC scheme in the context of a practical implementation approximating the impulsive control is that it yields a feedback control strategy that includes information to adapt the sampling frequency to the "speed" of the dynamics.

Acknowledgments All the authors gratefully acknowledge the support of FCT under the projects "Advanced control of a fleet of heterogeneous autonomous vehicles" (PESSOA program), PTDC/EEI-AUT/1450/2012 and "Incentivo/EEI/UI0147/2014." The first three authors also acknowledge the support of FCT under the projects PEst-OE/EEI/UI0147/2014, and the R&D unit SYSTEC—UID/EEA/00147/2013.

References

1. M. Alamir, *Stabilization of Nonlinear Systems Using Receding-Horizon Control Schemes: A Parameterized Approach for Fast System*, Lecture Notes in Control and Information Sciences (Springer, London, 2006)
2. M.R. Almassalkhi, I. Hiskens, Model-predictive cascade mitigation in electric power systems with storage and renewables—part I: theory and implementation. IEEE Trans. Power Syst. **30**(1), 67–77 (2015)
3. A.V. Arutyunov, *Optimality Conditions: Abnormal and Degenerate Problems* (Kluwer Academic Publishers, London, 2000)

4. A.V. Arutyunov, V. Dykhta, F.L. Pereira, Necessary conditions for impulsive nonlinear optimal control problems without a priori normality assumptions. J. Optim. Theory Appl. **124**(1), 55–77 (2005)
5. A.V. Arutyunov, D. Yu. Karamzin, F.L. Pereira, A nondegenerate maximum principle for the impulse control problem with state constraints. SIAM J. Control Optim. **43**(5), 1812–1843 (2005)
6. A.V. Arutyunov, D. Yu. Karamzin, F.L. Pereira, On constrained impulsive control problems. J. Math. Sci. **165**(6), 654–688 (2010)
7. A.V. Arutyunov, D. Yu. Karamzin, F.L. Pereira, Pontryagin's maximum principle for optimal impulsive control problems. Dokl. Math. **81**(3), 418–421 (2010)
8. A.V. Arutyunov, D. Yu. Karamzin, F.L. Pereira, On a generalization of the impulsive control concept: controlling system jumps. Discret. Contin. Dyn. Syst **29**(2), 403–415 (2011)
9. A.V. Arutyunov, D. Yu. Karamzin, F.L. Pereira, Pontryagin's maximum principle for constrained impulsive control problems. Nonlinear Anal., Theory, Meth. Appl. **75**(3), 1045–1057 (2012)
10. A.V. Arutyunov, D. Yu. Karamzin, F.L. Pereira, State constraints in impulsive control problems: Gamkrelidze-like conditions of optimality. J. Optim. Theory Appl. **166**(2), 440–459 (2015)
11. A.V. Arutyunov, D. Yu. Karamzin, F.L. Pereira, G.N. Silva, Investigation of regularity conditions in optimal control problems with geometric mixed constraints. Optim.: A J. Math. Progr. Oper. Res. **48**(7), 1–22 (2015). doi:10.1080/02331934.2015.1014478
12. J.-P. Aubin, *Impulse Differential Equations and Hybrid Systems: A Viability Approach*, Lecture Notes (University of California, Berkeley, 2000)
13. J.-P. Aubin, J. Lygeros, M. Quincampoix, S. Sastry, N. Seube, Impulse differential inclusions: a viability approach to hybrid systems. IEEE Trans. Autom. Control **47**(1), 2–20 (2002)
14. J. Baumeister: On optimal control of a fishery, in *Proceedings of the NOLCOS'01—IFAC Symposium on Nonlinear Control System* (2001)
15. J.T. Betts, *Practical Methods for Optimal Control Using Nonlinear Programming* (SIAM Publication, Philadelphia, 2001)
16. B. Brogliato, *Nonsmooth Impact Mechanics: Models, Dynamics and Control*, vol. 220, LNCIS (Springer, Berlin, 1996)
17. A.E. Bryson, Optimal control—1950 to 1985. IEEE Control Syst. **16**(3), 26–33 (1996)
18. H. Chen, F. Allgöwer. Nonlinear model predictive control schemes with guaranteed stability, in *Nonlinear Model Based Process Control* eds. by R. Berber, C. Kravaris (Kluwer, London, 1998)
19. H. Chen, F. Allgöwer, A quasi-infinite horizon nonlinear model predictive control scheme with guaranteed stability. Automatica **34**(10), 1205–1217 (1998)
20. C. Clark, F.H. Clarke, G. Munro, The optimal exploitation of renewable stocks. J. Evol. Econom. **47**, 25–47 (1979)
21. F.H. Clarke, Y.S. Ledyaev, R.J. Stern, P.R. Wolenski, *Nonsmooth Analysis and Control Theory* (Springer, New York, 1998)
22. V.A. Oliveira, F. Lobo Pereira, G.N. Silva, Invariance for impulsive control systems, in *IEEE International Conference on Decision and Control*. (Maui, EUA, 9–12 Dec 2003)
23. R. Findeisen, L. Imsland, F. Allgöwer, B. Foss, State and output feedback nonlinear model predictive control: an overview. Eur. J. Control **9**, 190–206 (2003)
24. R. Findeisen, L. Imsland, F. Allgöwer, B. Foss, Towards a sampled-data theory for nonlinear model predictive control, in *New Trends in Nonlinear Dynamics and Control and their Applications*, Lecture Notes in Control and Information Sciences, ed. by W. Kang, C. Borges, M. Xiao (Springer, Heidelberg, 2003), pp. 295–311. doi:10.1007/978-3-540-45056-6-19
25. F.A.C.C. Fontes, Discontinuous feedback stabilization using nonlinear model predictive controllers, in *Proceedings of CDC 2000–39th IEEE Conference on Decision and Control* (Sydney, Australia, December 2000)
26. F.A.C.C. Fontes, A general framework to design stabilizing nonlinear model predictive controllers. Syst. Control Lett. **42**, 127–143 (2001)

27. F.A.C.C. Fontes, Discontinuous feedbacks, discontinuous optimal controls, and continuous-time model predictive control. Int. J. Robust Nonlinear Control **13**(3–4), 191–209 (2003)
28. F.A.C.C. Fontes, L. Magni, A generalization of Barbalat's lemma with applications to robust model predictive control, in *Proceedings of Sixteenth International Symposium on Mathematical Theory of Networks and Systems (MTNS2004)*, Leuven, Belgium, 5–9 July 2004
29. F.A.C.C. Fontes, L. Magni, E. Gyurkovics, Sampled-data model predictive control for nonlinear time-varying systems: Stability and robustness, in Assessment and Future Directions of Nonlinear Model Predictive Control, ed. by F. Allgower, R. Findeisen, L. Biegler, Lecture Notes in Control and Information Systems, vol. 358 (Springer, 2007), pp. 115–129
30. F.A.C.C. Fontes, F.L. Pereira, Model predictive control of impulsive dynamical systems. Nonlinear Model Predict. Control **4**, 305–310 (2012)
31. S.L. Fraga, F.L. Pereira, On the feedback control of impulsive dynamic systems, in *Proceedings of the IEEE Conference on Decision and Control* (2008), pp. 2135–2140
32. S.L. Fraga, F.L. Pereira, Hamilton–Jacobi–Bellman equation and feedback synthesis for impulsive control. IEEE Trans. Autom. Control **57**(1), 244–249 (2012)
33. L. Grüne, D. Nesic, J. Pannek, Model predictive control for nonlinear sampled-data systems, in *Proceedings of the Assessment and Future Directions of Nonlinear Model Predictive Control (NMPC05)*, eds. by R. Findeisen, F. Allgöwer, L. Biegler, Lecture Notes in Control and Information Sciences, vol. 358 (Springer, Heidelberg, 2007), pp. 105–113
34. D.Yu. Karamzin, V.A. Oliveira, F. Lobo Pereira, G.N. Silva, On the properness of the extension of dynamic optimization problems to allow impulsive controls. ESAIM: Control Optim. Calc. Var. **21**(3), 857–875 (2015)
35. D. Yu. Karamzin, V.A. Oliveira, F.L. Pereira, G.N. Silva, On some extension of optimal control theory. Eur. J. Control **20**(6), 284–291 (2014)
36. L. Magni, R. Scattolini, Model predictive control of continuous-time nonlinear systems with piecewise constant control. IEEE Trans. Autom. Control **49**, 900–906 (2004)
37. L. Magni, R. Scattolini, M. Tanelli, Switched model predictive control for performance enhancement. Int. J. Control **81**(12), 1859–1869 (2008)
38. J.-P. Marec, *Optimal Space Trajectories* (Elsevier, Amsterdam, 1979)
39. B.S. Mordukhovich, *Variational Analysis and Generalized Differentiation, I: Basic Theory, II: Applications* (Springer, Berlin, 2006)
40. F.L. Pereira, G.N. Silva, Necessary conditions of optimality for vector-valued impulsive control problems. Syst. Control Lett. **40**(3), 205–215 (2000)
41. F.L. Pereira, G.N. Silva, Stability for impulsive control systems. Dyn. Syst. **17**(4), 421–434 (2002)
42. F.L. Pereira, G.N. Silva, V.A. Oliveira, Invariance for impulsive control systems. Autom. Remote Control **69**(5), 788–800 (2008)
43. F. Lobo Pereira, G.N. Silva, A maximum principle for infinite time asymptotically stable impulsive dynamic control systems. *NOLCOS 2010, 8th IFAC Symposium on Nonlinear Control Systems*. Bologna, Italy, 13–16 Sept 2010
44. L.S. Pontragin, V.G. Boltyanskii, R.V. Gamkrelidze, E.F. Mishchenko, *Mathematical Theory of Optimal Processes* (Interscience Publishers, Wiley, New York, 1962). (English translation)
45. P. Riedinger, I.-C. Morarescu, A numerical framework for optimal control of switched input affine nonlinear systems subject to path constraints. Math. Comput. Simul. **95**, 63–77 (2013)
46. R. Rishel, An extended pontryagin principle for control systems whose control laws contain measures. SIAM J. Control **3**, 191–205 (1965)
47. G.N. Silva, F.L. Pereira, Lyapounov stability for impulsive dynamical systems. Proc. IEEE Conf. Decis. Control **2**, 2304–2309 (2002)
48. P. Sopasakis, P. Patrinos, H. Sarimveis, A. Bemporad. Model predictive control for linear impulsive systems, in *Proceedings of the 2012 IEEE 51st Annual Conference on Decision and Control (CDC)* (IEEE, 2012) pp. 5164–5169
49. R.B. Vinter, *Optimal Control* (Birkhuser, Boston, 2000)
50. R.B. Vinter, F.L. Pereira, Maximum principle for optimal processes with discontinuous trajectories. SIAM Journal on Control and Optimization **26**(1), 205–229 (1988)

Chapter 14
Robustness Issues in Control of Bilinear Discrete-Time Systems—Applied to the Control of Power Converters

Mohsen Vatani, Morten Hovd and Sorin Olaru

Abstract Recently, a controller design method based on Sum of Squares programming has been developed for the control of discrete-time bilinear systems, and applications to power converters have been studied. In the present work, robustness issues arising in these designs are studied. First, the issue of change of operating point is addressed, and relevant stability analysis is developed. For linear systems, one can simply "shift the origin" of the deviation variables to obtain the same behavior for a new operating point. For nonlinear systems, in contrast, one will experience changed dynamics when applying the same controller at a new operating point (even after "shifting the origin"). New criteria are introduced to verify the stability of designed controller for other desired operating points. A related topic that is covered is the introduction of integral action in the bilinear controller design, giving offset-free control for persistent disturbances. The effectiveness of the proposed methods are evaluated based on time-domain simulations of a boost converter.

Keywords Bilinear systems · Sum of squares · Lyapunov techniques · Control of power convertors

M. Vatani (✉) · M. Hovd
Department of Engineering Cybernetics, Norwegian University of Science and Technology, 7491 Trondheim, Norway
e-mail: mohsen.vatani@itk.ntnu.no

M. Hovd
e-mail: morten.hovd@itk.ntnu.no

S. Olaru
Laboratory of Signals and Systems, CentraleSupelec-CNRS-Univ. Paris-Sud, Université Paris-Saclay, 3 Rue Joliot Curie, 91192 Gif-sur-Yvette Cedex, France
e-mail: sorin.olaru@supelec.fr

S. Olaru et al. (eds.), *Developments in Model-Based Optimization and Control*,
Lecture Notes in Control and Information Sciences 464,
DOI 10.1007/978-3-319-26687-9_14

14.1 Introduction

The Sum of Squares (SOS) technique has received a lot of attention in control theory, a reference point being Parrilo's thesis, [15], in the year 2000, which showed SOS effectiveness as a method for proving the non-negativity of polynomial functions in control analysis and design. Ever since, SOS programming has been used in implementation of different numerical analysis tools, [3–6, 9, 14, 16].

The main idea of using SOS in the stability analysis of discrete-time systems is to prove that the decrease in the candidate Lyapunov function in each sampling time is a SOS, and thus, positive. Using the SOS method for stability investigations in the literature upon these principles is well understood and represents a systematic methodology for constructing a Lyapunov function. Furthermore, the design of stabilizing controller can benefit from SOS programming as long as the control degrees of freedom can be enhanced to ensure that the decrease in the Lyapunov function is a SOS, see for example [19] as an exemplification on this line.

It is a well-known fact that the state space averaged model of a class of dc/dc power converters, including boost, buck–boost, Cuk, flyback, and so on, are described by bilinear models which include the product of the duty cycle and states. In addition, it is recently shown that the dynamics of modular multilevel converters are also described by a discrete-time bilinear model, [20]. In most of the previous works, the bilinear terms have usually been neglected and the controller design process is performed for the linear approximated model by the small signal assumption. This strategy cannot guarantee the stability of the equilibrium point especially for large disturbances.

In recent years, several works have been devoted to the problem of stabilizing the bilinear state space model of power converters. In [1], the stabilization and controller design process of a continuous-time bilinear system using LMI feasibility problems is discussed and applied on a Cuk converter. In [7], a piecewise-affine approximation of the bilinear dynamic is presented and an estimation of the stability region is found by using a piecewise-quadratic Lyapunov function. Afterward, the stability analysis problem is converted into an optimization problem with LMI constraints and the proposed method is applied on a buck–boost converter. In [12, 13], the robust control design of bilinear dc/dc converters by LMI approaches is considered. The stability conditions for affine state-feedback control laws by considering input and state constraints for bilinear discrete-time systems are discussed in [17] and are applied to a noninverting buck–boost converter.

The application of SOS programming methods in power engineering applications have been recently reported in the literature, [2, 20]. In [18], the SOS programming method is used to design a controller for a class of power converters with a discrete-time bilinear averaged dynamic model and the superiority in the performance of the proposed SOS-based control strategy are evaluated and compared with the available control strategies in the literature for various dc/dc converters.

In this work, robustness issues arising in the design of SOS-based controllers for dc/dc power converters are discussed. First, the problem of shifting the origin of the deviation variables is discussed. For linear systems, the same behavior is expected after shifting the origin to the new equilibrium point. However, for nonlinear systems, changes in the dynamical behavior may be observed when the system moves to a new operating point. As a result, the designed controller for nominal operating point is not necessarily stable for all other equilibrium points of the system. A new criterion is introduced here to check the stability of the designed controller for other desired operating points. Afterward, the robustness of the SOS controller to persistent disturbances is improved by augmenting the bilinear model of the system by an integration state which provides offset free control for the desired states. The effectiveness of the proposed methods are evaluated based on time-domain simulations of a boost converter.

In the following, first, the general averaged discrete-time bilinear model of dc/dc converters is developed in Sect. 14.2 and a coordinate transformation is performed to transform the equilibrium point of the model to the origin. Then, the controller design process for the discrete-time bilinear system based on SOS programming is described in Sect. 14.3. In Sect. 14.4, the issues related to change of operating point is discussed. The stability analysis for a set of desired operating points is investigated while the system is controlled by using the SOS controller designed for the nominal operating point. Section 14.5 is devoted to offset the free control of desired states by introducing an integral action in the bilinear model of the system. Finally, Sect. 14.6 concludes this work.

14.2 Averaged Model of the Power Converters

A wide variety of power converters are modeled as switched systems with a specific model for each switching status as

$$\begin{aligned} &\text{switch status } on: \quad \dot{\mathbf{x}}(t) = \mathbf{A}_1\mathbf{x}(t) + \mathbf{B}_1\mathbf{v}, \\ &\text{switch status } off: \quad \dot{\mathbf{x}}(t) = \mathbf{A}_2\mathbf{x}(t) + \mathbf{B}_2\mathbf{v}, \end{aligned} \tag{14.1}$$

where state $\mathbf{x}$ represents capacitor voltages and inductor currents and vector $\mathbf{v}$ represents source voltages and diode voltages. A pulse width modulation (PWM) signal with switching frequency $f_s = 1/T_s$ controls the on/off status of the converter switches. The sum of t_{on} (time period in which the switch is in *on* state) and t_{off} (the time period in which the switch is in *off* state) for each switch is equal to the switching period T_s. The duty cycle d is defined as the ratio of t_{on}/T_s and consequently $t_{off} = (1-d)T_s$. Assuming that the inductor current is not saturated and by considering the duty cycle definition, the average model of the converter is formulated as

$$\dot{\mathbf{x}}(t) = (d(t)\mathbf{A}_1 + (1-d(t))\mathbf{A}_2)\mathbf{x}(t) + (d(t)\mathbf{B}_1 + (1-d(t))\mathbf{B}_2)\mathbf{v}$$

which can be simplified and reformulated as

$$\dot{\mathbf{x}}(t) = \underbrace{\mathbf{A}_2}_{\mathbf{A}_c} \mathbf{x}(t) + \underbrace{(\mathbf{A}_1 - \mathbf{A}_2)}_{\mathbf{B}_{cb}} \mathbf{x}(t)d(t) + \underbrace{(\mathbf{B}_1\mathbf{v} - \mathbf{B}_2\mathbf{v})}_{\mathbf{B}_c} d(t) + \underbrace{\mathbf{B}_2\mathbf{v}}_{\mathbf{d}_c}. \quad (14.2)$$

Equation (14.2) is in the form of a standard bilinear continuous-time system:

$$\dot{\mathbf{x}}(t) = \mathbf{A}_c\mathbf{x}(t) + \mathbf{B}_{cb}\mathbf{x}(t)u(t) + \mathbf{B}_c u(t) + \mathbf{d}_c,$$

where $u(t) = d(t)$ is the input of the system. In general, for converters with more than one switch, the averaged continuous-time bilinear model of the converter is represented by

$$\dot{\mathbf{x}}(t) = \mathbf{A}_c\mathbf{x}(t) + \sum_{i=1}^{m}(\mathbf{B}_{cb,i}\mathbf{x}(t) + \mathbf{B}_{c,i})\mathbf{u}_i(t) + \mathbf{d}_c, \quad (14.3)$$

where $u_i = d_i$ is the duty cycle of the ith switch and m is the number of switches.

The desired equilibrium operating point of the bilinear model of the converter is nonzero. By defining the desired equilibrium state vector and input as $\mathbf{x}^{ss}$ and $\mathbf{d}^{ss}$, respectively, the coordinate transformation is defined by

$$\tilde{\mathbf{x}} = \mathbf{x} - \mathbf{x}^{ss}, \quad \tilde{\mathbf{d}} = \mathbf{d} - \mathbf{d}^{ss}. \quad (14.4)$$

Substituting for the state variables and input from (14.4) in (14.3) yields

$$\dot{\tilde{\mathbf{x}}}(t) = \mathbf{A}_c(\tilde{\mathbf{x}}(t) + \mathbf{x}^{ss}) + \sum_{i=1}^{m}(\mathbf{B}_{cb,i}(\tilde{\mathbf{x}}(t) + \mathbf{x}^{ss}) + \mathbf{B}_{c,i})(\tilde{d}_i(t) + d_i^{ss}) + \mathbf{d}_c. \quad (14.5)$$

Equation (14.5) can be decomposed into two equations. The first equation represents the relation between the desired equilibrium operating point and equilibrium input as

$$\mathbf{A}_c\mathbf{x}^{ss} + \sum_{i=1}^{m}(\mathbf{B}_{cb,i}\mathbf{x}^{ss} + \mathbf{B}_{c,i})d_i^{ss} + \mathbf{d}_c = 0, \quad (14.6)$$

while the second equation represents the dynamic of the converter in the form of a standard bilinear system with its equilibrium point at the origin.

$$\dot{\tilde{\mathbf{x}}}(t) = \left(\mathbf{A}_c + \sum_{i=1}^{m}\mathbf{B}_{cb,i}d_i^{ss}\right)\tilde{\mathbf{x}}(t) + \sum_{i=1}^{m}(\mathbf{B}_{cb,i}\tilde{\mathbf{x}}(t) + \mathbf{B}_{cb,i}\mathbf{x}^{ss} + \mathbf{B}_{c,i})\tilde{d}_i(t). \quad (14.7)$$

With respect to the static operation, Eq. (14.6) describes a hyperbolic curve in the extended $(\mathbf{x}^{ss}, \mathbf{d}^{ss})$ space. For a given constant input vector $\mathbf{d}^{ss}$, one has:

$$\left(\mathbf{A}_c + \sum_{i=1}^{m} d_i^{ss}\mathbf{B}_{cb,i}\right)\mathbf{x}^{ss} = -\sum_{i=1}^{m} \mathbf{B}_{c,i}d_i^{ss} - \mathbf{d}_c \tag{14.8}$$

and the existence/uniqueness of an equilibrium point is related to the singularity of the matrix $\mathbf{A}_c + \sum_{i=1}^{m} d_i^{ss}\mathbf{B}_{cb,i}$. Conversely, the viability of the state $\mathbf{x}^{ss}$ as an equilibrium point is related to the feasibility of the set of linear equations:

$$\sum_{i=1}^{m}(\mathbf{B}_{cb,i}\mathbf{x}^{ss} + \mathbf{B}_{c,i})d_i^{ss} = -\mathbf{A}_c\mathbf{x}^{ss} - \mathbf{d}_c, \tag{14.9}$$

which depends on the number of inputs (m) with respect to the number of states (n) and ultimately on the full row rank condition of the matrix:

$$\left[(\mathbf{B}_{cb,1}\mathbf{x}^{ss} + \mathbf{B}_{c,1}) \quad \ldots \quad (\mathbf{B}_{cb,m}\mathbf{x}^{ss} + \mathbf{B}_{c,m})\right]. \tag{14.10}$$

Based on (14.7) and assuming a sampling period of T_s, the discrete-time bilinear model of the converter, based on a forward Euler approximation, becomes

$$\tilde{\mathbf{x}}_{k+1} = \underbrace{(T_s\mathbf{A}_c + T_s\sum_{i=1}^{m}\mathbf{B}_{cb,i}d_i^{ss} + \mathbf{I})}_{\mathbf{A}}\tilde{\mathbf{x}}_k + \sum_{i=1}^{m}(\underbrace{T_s\mathbf{B}_{cb,i}}_{\mathbf{B}_{b,i}}\tilde{\mathbf{x}}_k + \underbrace{T_s\mathbf{B}_{cb,i}\mathbf{x}^{ss} + T_s\mathbf{B}_{c,i}}_{\mathbf{B}_i})\tilde{d}_{i,k}.$$

which is in the form of a standard discrete-time bilinear system as

$$\tilde{\mathbf{x}}_{k+1} = \mathbf{A}\tilde{\mathbf{x}}_k + \sum_{i=1}^{m}(\mathbf{B}_{b,i}\tilde{\mathbf{x}}_k + \mathbf{B}_i)\tilde{u}_{i,k}. \tag{14.11}$$

Throughout this chapter, the subscript k will refer to a discrete-time instant while a subscript i will identify a specific matrix (as in (14.11) above) or a particular element of a vector.

Example 14.1 The circuit diagram of a dc-dc boost converter is shown in Fig. 14.1 [12]. The states of the system are considered as the inductor current $x_1 = i_L$ and the capacitor voltage $x_2 = v_C$. The system parameters are $R_L = 0\,\Omega$, $L = 100\,\mu\text{H}$, $C = 200\,\mu\text{F}$, $R = 10\,\Omega$, $v_g = 12\,\text{V}$, and $T_s = 5\,\mu\text{s}$. During the *on* state, the status of the switch is $S = 1$ for the period of t_{on} and the system matrices with respect to

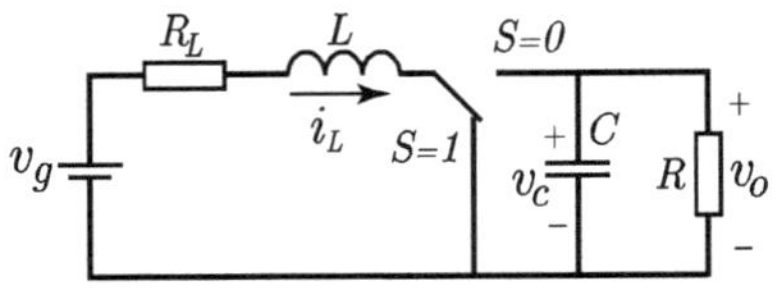

Fig. 14.1 Circuit diagram of boost converter [12]

(14.1) is as follows

$$\mathbf{A}_1 = \begin{bmatrix} -\frac{R_L}{L} & 0 \\ 0 & -\frac{1}{RC} \end{bmatrix}, \quad \mathbf{B}_1 = \begin{bmatrix} \frac{1}{L} \\ 0 \end{bmatrix}, \quad v = v_g,$$

while during the *off* state, the status of the switch is $S = 0$ for the period of $t_{off} = T_s - t_{on}$ and the system matrices with respect to (14.1) is as follows

$$\mathbf{A}_2 = \begin{bmatrix} -\frac{R_L}{L} & -\frac{1}{L} \\ \frac{1}{C} & -\frac{1}{RC} \end{bmatrix}, \quad \mathbf{B}_2 = \begin{bmatrix} \frac{1}{L} \\ 0 \end{bmatrix}, \quad v = v_g.$$

The load voltage $v_o = v_C = x_2$ is considered as the output which should be kept at the desired voltage $x_2^{ss} = v_{ref} = 24\ V$. By solving (14.6), the desired equilibrium operating point and the equilibrium input are calculated as

$$x_1^{ss} = \frac{v_{ref}^2}{Rv_g} = 4.8, \quad d^{ss} = \frac{v_{ref} - v_g}{v_{ref}} = 0.5\,. \tag{14.12}$$

The discrete-time bilinear model of the converter, with the origin as the equilibrium point, is calculated based on (14.11) as

$$\begin{aligned} \tilde{\mathbf{x}}_{k+1} &= \begin{bmatrix} 1 & \frac{T_s(u^{ss}-1)}{L} \\ -\frac{T_s(u^{ss}-1)}{L} & 1 - \frac{T_s}{RC} \end{bmatrix} \tilde{\mathbf{x}}_k + \left(\begin{bmatrix} 0 & \frac{T_s}{L} \\ -\frac{T_s}{C} & 0 \end{bmatrix} \tilde{\mathbf{x}}_k + \begin{bmatrix} \frac{T_s x_2^{ss}}{L} \\ -\frac{T_s x_1^{ss}}{C} \end{bmatrix} \right) \tilde{u}_k \\ &= \begin{bmatrix} 1 & -0.025 \\ 0.0125 & 0.9975 \end{bmatrix} \tilde{\mathbf{x}}_k + \left(\begin{bmatrix} 0 & 0.05 \\ -0.025 & 0 \end{bmatrix} \tilde{\mathbf{x}}_k + \begin{bmatrix} 1.2 \\ -0.12 \end{bmatrix} \right) \tilde{u}_k \end{aligned} \tag{14.13}$$

14.3 Controller Design Based on the Sum of Squares

This section presents a brief introduction to the SOS-based controller design procedure for the stabilization of the discrete-time bilinear system (14.11) to the origin, proposed in [19].

The main tool for analyzing the stability of nonlinear systems is the Lyapunov's direct method:

Theorem 14.1 *Lyapunov stability for discrete-time systems: If in a positive invariant neighborhood* $D \subseteq \mathbb{R}^n$ *of the equilibrium state* $\tilde{\boldsymbol{x}}_e = \boldsymbol{0}$ *of a discrete-time system represented by* $\tilde{\boldsymbol{x}}_{k+1} = f(\tilde{\boldsymbol{x}}_k)$*, there exists a function* $V(.) : D \rightarrow \mathbb{R}$ *such that*

$$W_1(\|\tilde{\boldsymbol{x}}\|) \leq V(\tilde{\boldsymbol{x}}) \leq W_2(\|\tilde{\boldsymbol{x}}\|), \quad \forall x \in D$$

where W_1 *and* W_2 *are* K_∞ *functions,*[1] *and the rate of change* $\Delta V(\tilde{\boldsymbol{x}}) = V(f(\tilde{\boldsymbol{x}})) - V(\tilde{\boldsymbol{x}})$ *is negative definite in* $D \backslash \{0\}$*, then the equilibrium state is asymptotically stable in* D.

A quadratic Lyapunov function $V(\tilde{\mathbf{x}}) = \tilde{\mathbf{x}}^T \mathbf{P} \tilde{\mathbf{x}}$ is used in this paper for some given weighting matrix $\mathbf{P} > 0$. The first assumption in Theorem 14.1 is satisfied for this function by considering:

$$\lambda_{min}(\mathbf{P}) \|\tilde{\mathbf{x}}\|_2^2 \leq \tilde{\mathbf{x}}^T \mathbf{P} \tilde{\mathbf{x}} \leq \lambda_{max}(\mathbf{P}) \|\tilde{\mathbf{x}}\|_2^2$$

where λ_{min} and λ_{max} are the minimum and maximum eigenvalues of $\mathbf{P}$. Consequently, the closed-loop stability is guaranteed by ensuring that the candidate quadratic Lyapunov function is decreasing in each time step:

$$V(\tilde{\mathbf{x}}_k) - V(\tilde{\mathbf{x}}_{k+1}) = \tilde{\mathbf{x}}_k^T \mathbf{P} \tilde{\mathbf{x}}_k - \tilde{\mathbf{x}}_{k+1}^T \mathbf{P} \tilde{\mathbf{x}}_{k+1} > 0. \tag{14.14}$$

To fulfill (14.14) within all time steps, a stabilizing controller should be designed. The stabilizing controller is considered in the form of ratio of two polynomials as

$$\tilde{u}_i(\tilde{\mathbf{x}}) = \frac{c_i(\tilde{\mathbf{x}})}{c_0(\tilde{\mathbf{x}})}, \tag{14.15}$$

where $c_i(\tilde{\mathbf{x}})$ are polynomials of orders within $[1, n_n]$, $c_0(\tilde{\mathbf{x}})$ is a polynomial of order within $[0, n_d]$, and $\tilde{u}_i$ is the number i input. Although the example described here has a single input, the controller design is presented in a general way, allowing for multiple-input systems. All of the inputs have the same denominator polynomial $c_0(\tilde{\mathbf{x}})$. The controller is obliged to satisfy the control constraints of the form

$$|\tilde{u}_i(\tilde{\mathbf{x}})| \leq \tilde{u}_{i,max}. \tag{14.16}$$

To design the controller (14.15) such that it satisfies (14.14) and (14.16), the SOS programming method is exploited. The basic idea behind the SOS programming method for checking the nonnegativity of a polynomial $p(\mathbf{y})$ is to replace the nonnegativity with the condition that the polynomial can be transformed in the SOS form [15].

---

[1] See [8] for the definition of K_∞ functions.

Definition 14.1 For $\mathbf{y} \in \mathbb{R}^n$, a polynomial $p(\mathbf{y})$ is a SOS if there exist some polynomials $f_j(\mathbf{y}),\ j = \{1, 2, \ldots m\}$, such that

$$p(\mathbf{y}) = \sum_{j=1}^{m} f_j^2(\mathbf{y}).$$

The SOS decomposition can be computed by semi-definite programming by using the available software. In this paper, the software package YALMIP [10, 11] is used for solving the SOS decomposition.

In the following, the controller design procedure, using SOS programming method, to stabilize the system to the origin is discussed. The denominator polynomial $c_0(\tilde{\mathbf{x}})$ is assumed to be an SOS polynomial. However, to guard against excessively large inputs, the denominator polynomial is specified as $c_0(\tilde{\mathbf{x}}) = \acute{c_0}(\tilde{\mathbf{x}}) + 1$, with $\acute{c_0}(\tilde{\mathbf{x}})$ being an SOS polynomial, thus ensuring that the denominator polynomial cannot be very small anywhere in $\mathbb{R}^n$. Furthermore, the controller is reformulated as:

$$\tilde{u}_i(\tilde{\mathbf{x}}) = \frac{\mathbf{C}(\tilde{\mathbf{x}})\tilde{\mathbf{x}}}{\acute{c_0}(\tilde{\mathbf{x}}) + 1}, \tag{14.17}$$

where $\mathbf{C}(\tilde{\mathbf{x}})$ is a polynomial matrix.

For the sake of simplicity, the bilinear system dynamics is expressed as

$$\tilde{\mathbf{x}}_{k+1} = \mathbf{A}\tilde{\mathbf{x}}_k + (\mathbf{B}_{\tilde{\mathbf{x}}} + \mathbf{B})\tilde{\mathbf{u}}_k, \tag{14.18}$$

where $\mathbf{B}_{\tilde{\mathbf{x}}} = \left[\ \mathbf{B}_{b,1}\tilde{\mathbf{x}}_k\ \mathbf{B}_{b,2}\tilde{\mathbf{x}}_k\ \ldots\ \mathbf{B}_{b,m}\tilde{\mathbf{x}}_k\ \right]$, $\mathbf{B} = [\ \mathbf{B}_1\ \mathbf{B}_2\ \ldots\ \mathbf{B}_m\ ]$, and m is the number of inputs.

Theorem 14.2 *Region of convergence: Given a quadratic function $V(\tilde{\boldsymbol{x}}) = \tilde{\boldsymbol{x}}^T \boldsymbol{P}\tilde{\boldsymbol{x}}$, a scalar $\gamma > 0$, polynomials $c_i(\tilde{\boldsymbol{x}}), i \in [1, \ldots, m]$, and SOS polynomials $\acute{c_0}(\tilde{\boldsymbol{x}})$ and $s_1(\tilde{\boldsymbol{x}}, z)$, a bilinear discrete-time system (14.18) in closed loop with the control law (14.17) is stable $\forall \tilde{\boldsymbol{x}}$ with $\tilde{\boldsymbol{x}}^T \boldsymbol{P}\tilde{\boldsymbol{x}} < \gamma$, provided that*

$$\begin{bmatrix} \tilde{\boldsymbol{x}} \\ z \end{bmatrix}^T \boldsymbol{M}(\tilde{\boldsymbol{x}}) \begin{bmatrix} \tilde{\boldsymbol{x}} \\ z \end{bmatrix} - s_1(\tilde{\boldsymbol{x}}, z)(\gamma - \tilde{\boldsymbol{x}}^T \boldsymbol{P}\tilde{\boldsymbol{x}}) > 0, \tag{14.19}$$

where $\boldsymbol{M}(\tilde{\boldsymbol{x}})$ is defined as

$$\begin{bmatrix} (\acute{c_0}(\tilde{\boldsymbol{x}}) + 1)\boldsymbol{P} & ((\acute{c_0}(\tilde{\boldsymbol{x}}) + 1)\boldsymbol{A} + (\boldsymbol{B}_{\tilde{x}} + \boldsymbol{B})\boldsymbol{C}(\tilde{\boldsymbol{x}}))^T \boldsymbol{P} \\ \boldsymbol{P}((\acute{c_0}(\tilde{\boldsymbol{x}}) + 1)\boldsymbol{A} + (\boldsymbol{B}_{\tilde{x}} + \boldsymbol{B})\boldsymbol{C}(\tilde{\boldsymbol{x}})) & (\acute{c_0}(\tilde{\boldsymbol{x}}) + 1)\boldsymbol{P} \end{bmatrix}.$$

Theorem 14.3 *Given the polynomial $c_i(\tilde{\boldsymbol{x}})$, SOS polynomials $\acute{c_0}(\tilde{\boldsymbol{x}})$ and $q_i(\tilde{\boldsymbol{x}})$, the input constraint in (14.16) is satisfied $\forall\ \tilde{\boldsymbol{x}}$ with $\tilde{\boldsymbol{x}}^T \boldsymbol{P}\tilde{\boldsymbol{x}} < \gamma$ provided*

$$\begin{bmatrix} (\acute{c_0}(\tilde{\boldsymbol{x}}) + 1)\tilde{u}_{max,i}^2 - q_i(\tilde{\boldsymbol{x}})(\gamma - \tilde{\boldsymbol{x}}^T \boldsymbol{P}\tilde{\boldsymbol{x}}) & c_i(\tilde{\boldsymbol{x}}) \\ c_i(\tilde{\boldsymbol{x}}) & \acute{c_0}(\tilde{\boldsymbol{x}}) + 1 \end{bmatrix} > 0. \tag{14.20}$$

For a given Lyapunov function weighting matrix **P**, Theorems 14.2 and 14.3 allow for controller design according to

$$\begin{aligned} \max_{\acute{c}_0(\tilde{\mathbf{x}}), c_i(\tilde{\mathbf{x}}), s_1(\tilde{\mathbf{x}}, z), q_i(\tilde{\mathbf{x}})} \quad & \gamma \\ \text{such that} \quad & (14.18) \text{ and } (14.19) \text{ hold} \\ & \acute{c}_0(\tilde{\mathbf{x}}), s_1(\tilde{\mathbf{x}}, \mathbf{z}), q_i(\tilde{\mathbf{x}}) \text{ SOS.} \end{aligned} \tag{14.21}$$

The coefficients in the polynomials $\acute{c}_0(\tilde{\mathbf{x}})$, $\mathbf{C}(\tilde{\mathbf{x}})$, and $s_1(\tilde{\mathbf{x}}, \mathbf{z})$ enter linearly in (14.19). Thus, for given γ and **P**, (14.19) can be verified with the polynomial coefficients as free variables. This corresponds to finding a feasible point for a constrained semi-definite programming problem, and can be easily formulated and solved using readily available software such as YALMIP (with an appropriate semi-definite programming solver). As a result of Theorem 14.2, the calculated control input (14.17) stabilizes the system within the region defined by $\tilde{\mathbf{x}}^T \mathbf{P} \tilde{\mathbf{x}} < \gamma$.

Example 14.2 The problem to be solved here is the determination of the controller which stabilizes the discrete-time bilinear average model of the boost converter, presented in Example 14.1, in the region determined by $\tilde{\mathbf{x}}^T \mathbf{P} \tilde{\mathbf{x}} < \gamma$. The matrix **P** is selected as:

$$\mathbf{P} = \begin{bmatrix} 0.4 & 0.6 \\ 0.6 & 1.75 \end{bmatrix},$$

and $\gamma = 4$. The control effort constraint is set to $\tilde{u}_{max} = 0.4$. Based on the controller design method by SOS programming, using the YALMIP software, the designed controller based on (14.15) is calculated as follows:

$$\tilde{u}(\tilde{\mathbf{x}}) = \frac{-9.1\tilde{x}_1 - 7.21\tilde{x}_2 - 0.14\tilde{x}_1^2 - 0.003\tilde{x}_1\tilde{x}_2 + 0.16\tilde{x}_2^2}{38.29 - 0.46\tilde{x}_1 + 0.23\tilde{x}_2 + 14.71\tilde{x}_1^2 - 2.13\tilde{x}_1\tilde{x}_2 + 11.56\tilde{x}_2^2}.$$

The simulation of the boost converter with the designed SOS controller is performed in PLECS/MATLAB software and the results are shown in Fig. 14.2. The initial state is set at $\mathbf{x}_0 = [0.3, 25.4]^T$. The output voltage of the boost converter, $x_2 = v_o$, is shown in Fig. 14.2a and the inductor current, $x_1 = i_L$, is shown in Fig. 14.2b where the oscillations are the consequence of simulating the detailed switching model. The response of the system is fast without overshoot for the SOS controllers. Figure 14.2c presents the duty cycle of switch S_1 which satisfies the input constraints for designed controller. The cost function $\tilde{\mathbf{x}}_k^T \mathbf{P} \tilde{\mathbf{x}}_k$ is shown in Fig. 14.2d which verifies the decrease in the cost function within each time step. Figure 14.2e represents the state trajectories of the boost converter controlled by the SOS controller. The initial condition is selected on the boundary of the mentioned region of convergence. It is shown that all trajectories converge to the origin ($\tilde{\mathbf{x}}_e = 0$).

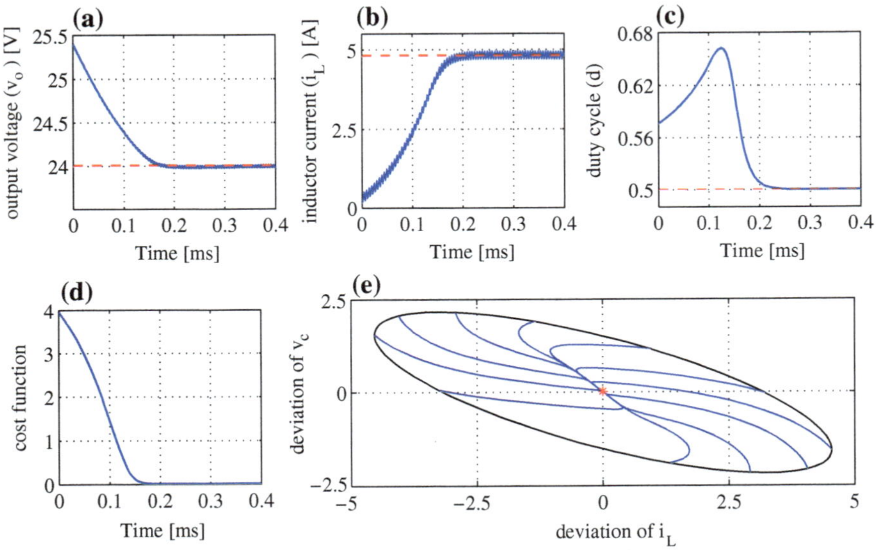

Fig. 14.2 Simulation results of the boost converter with SOS controller: **a** Output voltage, **b** inductor current, **c** duty cycle of the switch, **d** Lyapunov cost function $\tilde{\mathbf{x}}_k^T \mathbf{P} \tilde{\mathbf{x}}_k$, and **e** state trajectories started from boundary of region of convergence

14.4 Change of Operating Point

In the last section, the stability of the discrete-time bilinear system is investigated by designing a rational controller guaranteeing convergence of the state trajectories to the nominal operating point of the system. For linear systems, the same behavior is observed by shifting the origin of the deviation variables when a new operating point is desired. However, for nonlinear systems, shifting the origin of the deviation variables results in new dynamics which is not necessarily stabilized by the designed controller for nominal operating point. In this section, the stability of the discrete-time bilinear system with the stabilizing controller designed for nominal operating point is investigated for the cases when the operating point of the system is changed.

By defining the new operating point of the system as $\mathbf{x}_{new}^{ss}$ and $\mathbf{u}_{new}^{ss}$, the new deviation variables are defined as

$$\hat{\mathbf{x}} = \mathbf{x} - \mathbf{x}_{new}^{ss}, \quad \hat{\mathbf{u}} = \mathbf{u} - \mathbf{u}_{new}^{ss}, \tag{14.22}$$

and the distance of the new operating point from the nominal operating point is defined as

$$\tilde{\mathbf{x}}^{ss} = \mathbf{x}_{new}^{ss} - \mathbf{x}^{ss}, \quad \tilde{\mathbf{u}}^{ss} = \mathbf{u}_{new}^{ss} - \mathbf{u}^{ss}, \tag{14.23}$$

By substituting for $\mathbf{x}$ and $\mathbf{u}$ from (14.4) in (14.22), the following equations are deducted:

$$\hat{\mathbf{x}} = \tilde{\mathbf{x}} + \mathbf{x}^{ss} - \mathbf{x}^{ss}_{new} = \tilde{\mathbf{x}} - \tilde{\mathbf{x}}^{ss}, \tag{14.24}$$

$$\hat{\mathbf{u}} = \tilde{\mathbf{u}} + \mathbf{u}^{ss} - \mathbf{u}^{ss}_{new} = \tilde{\mathbf{u}} - \tilde{\mathbf{u}}^{ss}, \tag{14.25}$$

Substituting for $\tilde{\mathbf{x}}$ and $\tilde{\mathbf{u}}$ from (14.24) and (14.25) in (14.11), the system dynamics in the new operating point are calculated as follows

$$\hat{\mathbf{x}}_{k+1} + \tilde{\mathbf{x}}^{ss} = \mathbf{A}(\hat{\mathbf{x}}_k + \tilde{\mathbf{x}}^{ss}) + \sum_{i=1}^{m} (\mathbf{B}_{b,i}(\hat{\mathbf{x}}_k + \tilde{\mathbf{x}}^{ss}) + \mathbf{B}_i)(\hat{u}_{i,k} + \tilde{u}_i^{ss}). \tag{14.26}$$

Equation (14.26) consists of two separate equations. The first equation represents the dynamics of the system for the new deviation variables as

$$\hat{\mathbf{x}}_{k+1} = \left(\underbrace{\mathbf{A} + \sum_{i=1}^{m} \mathbf{B}_{b,i} \tilde{u}_i^{ss}}_{\Delta \mathbf{A}} \right) \hat{\mathbf{x}}_k + \sum_{i=1}^{m} (\mathbf{B}_{b,i} \hat{\mathbf{x}}_k + \mathbf{B}_i + \underbrace{\mathbf{B}_{b,i} \tilde{\mathbf{x}}^{ss}}_{\Delta \mathbf{B}_i}) \hat{u}_{i,k}. \tag{14.27}$$

where the new dynamics are represented as matrices $\Delta\mathbf{A}$ and $\Delta\mathbf{B}_i$. For the sake of simplicity, we define $\Delta\mathbf{B} = \left[\Delta\mathbf{B}_1\tilde{\mathbf{x}}^{ss}\ \Delta\mathbf{B}_2\tilde{\mathbf{x}}^{ss}\ \ldots\ \Delta\mathbf{B}_m\tilde{\mathbf{x}}^{ss}\right]$, so (14.27) is simplified to

$$\hat{\mathbf{x}}_{k+1} = (\mathbf{A} + \Delta\mathbf{A})\hat{\mathbf{x}}_k + (\mathbf{B}_{\hat{\mathbf{x}}} + \mathbf{B} + \Delta\mathbf{B})\hat{\mathbf{u}}_k. \tag{14.28}$$

The second equation represents the trivial relation between the new operating point and the nominal one as

$$\tilde{\mathbf{x}}^{ss} = \mathbf{A}\tilde{\mathbf{x}}^{ss} + \sum_{i=1}^{m} (\mathbf{B}_{b,i}\tilde{\mathbf{x}}^{ss} + \mathbf{B}_i)\tilde{u}^{ss}. \tag{14.29}$$

Note that (14.29) is also an immediate result of subtracting (14.6) for new and nominal operating point.

Corollary 14.1 *Given $\boldsymbol{P}$, γ, c_i, and $\acute{c}_0$ from (14.21) (design of the controller for the nominal operating point), and the SOS polynomial $s_2(\hat{\boldsymbol{x}}, z)$, the bilinear discrete-time system with the control law (14.17) is closed-loop stable for the new operating point $\boldsymbol{x}^{ss}_{new}$ and $\boldsymbol{u}^{ss}_{new}$ in the region $\forall\hat{\boldsymbol{x}}$ with $\hat{\boldsymbol{x}}^T\boldsymbol{P}\hat{\boldsymbol{x}} < \gamma$, provided that*

$$\begin{bmatrix} \hat{\boldsymbol{x}} \\ z \end{bmatrix}^T \boldsymbol{M}(\hat{\boldsymbol{x}}) \begin{bmatrix} \hat{\boldsymbol{x}} \\ z \end{bmatrix} - s_2(\hat{\boldsymbol{x}}, z)(\gamma - \hat{\boldsymbol{x}}^T\boldsymbol{P}\hat{\boldsymbol{x}}) > 0, \tag{14.30}$$

where

$$M(\hat{x}) = \begin{bmatrix} (\acute{c_0}(\hat{x})+1)P & \left((\acute{c_0}(\hat{x})+1)(A+\Delta A)+(B_{\hat{x}}+B+\Delta B)C(\hat{x})\right)^T P \\ P\left((\acute{c_0}(\hat{x})+1)(A+\Delta A)+(B_{\hat{x}}+B+\Delta B)C(\hat{x})\right) & (\acute{c_0}(\hat{x})+1)P \end{bmatrix} \quad (14.31)$$

Based on Corollary 14.1, considering the controller and region of convergence calculated for nominal operating point, it is enough to find the SOS polynomial $s_2(\hat{\mathbf{x}}, \mathbf{z})$ in (14.30) to prove that the system is stable in the new operating point $\mathbf{x}^{ss}_{new}$. However, if it is required to prove the stability of the system for a set of equilibrium points, usage of Corollary 14.1 is cumbersome.

Definition 14.2 The convex hull of a set of points $\mathbf{Q} = \{q_i\}_{i=1}^{n_q}$, with $q_i \in \mathbb{R}^n$, is a polytope defined as

$$\mathrm{conv}(\mathbf{Q}) = \left\{ x \in \mathbb{R}^n : x = \sum_{i=1}^{n_q} \alpha_i q_i,\ \alpha_i \geq 0, \sum_{i=1}^{n_q} \alpha_i = 1 \right\} \quad (14.32)$$

Each point $q_i \in \mathbf{Q}$ that is not in the convex hull of the other points, i.e., $q_i \notin \mathrm{conv}(\mathbf{Q}\backslash\{q_i\})$ is called a vertex of $\mathrm{conv}(\mathbf{Q})$.

It follows directly from the definition of the convex hull above that any point in a polytope can be expressed as an interpolation between the vertices of the polytope.

Lemma 14.1 *Consider the set of points $V = \{\mathbf{v}_i\}_{i=1}^{n_v}$ as the set of vertices of a convex polytope. If there is a linear function f such that $f(\mathbf{v}_1) \geq 0$, $f(\mathbf{v}_2) \geq 0, \ldots,$ and $f(\mathbf{v}_{n_v}) \geq 0$ then $f(\mathbf{v}) \geq 0$ where $\mathbf{v}$ is any point inside the convex polytope.*

Proof Every point in a convex polytope is the convex hull of its vertices. As a result:

$$f(\mathbf{v}) = f(\alpha_1\mathbf{v}_1 + \alpha_2\mathbf{v}_2 + \cdots + \alpha_{n_v}\mathbf{v}_{n_v}), \quad \alpha_i \geq 0, \quad \sum_{i=1}^{n_v} \alpha_i = 1.$$

The function f is assumed to be linear. As a result

$$f(\mathbf{v}) = \alpha_1 f(\mathbf{v}_1) + \alpha_2 f(\mathbf{v}_2) + \cdots + \alpha_{n_v} f(\mathbf{v}_{n_v}) \geq 0.$$

Equation (14.29) can be solved for $\tilde{\mathbf{x}}^{ss}$ and $\tilde{\mathbf{u}}^{ss}$ to calculate the set containing the desired equilibrium points of the system in which the stability analysis of Corollary 14.1 should be performed. Several methods are suggested in the literature to calculate an outer approximating polytope that covers a set of points. The discussion of these methods are out of the scope of this work. Here, we assume that a polytope V with vertices v_i, covering all the desired equilibrium points of the discrete-time bilinear system, is available, such that any desired equilibrium point

$$\mathbf{v} = \begin{bmatrix} \tilde{\mathbf{x}}^{ss} \\ \tilde{\mathbf{u}}^{ss} \end{bmatrix}$$

can be expressed as an interpolation between vertices v_i of the polytope V.

Theorem 14.4 *If an SOS polynomial $s_2(\hat{x}, z)$ is found which satisfies (14.30) for all vertices of the polytope V, then all steady state operating points in V are stable equilibrium points with the same controller and region of convergence calculated for the nominal operating point.*

Proof Observe that $\Delta\mathbf{A}$ is linear in $\tilde{\mathbf{x}}^{ss}$ and $\Delta\mathbf{B}$ is linear in $\tilde{\mathbf{u}}^{ss}$, respectively, and that $\Delta\mathbf{A}$ and $\Delta\mathbf{B}$ enter linearly in (14.31). For any given $(\hat{\mathbf{x}}, \mathbf{z})$, the left-hand side of (14.30) may therefore be expressed as a function $m(\mathbf{v})$, i.e., (14.30) may be expressed as

$$m(\mathbf{v}) > 0.$$

We note that the function m is linear in the argument $\mathbf{v}$. The theorem then follows directly from Lemma 14.1.

Example 14.3 Here, the boost converter with the SOS controller designed in Example 14.2 for nominal operating point is considered and the stability of the system for other desired operating points is analyzed. The reference for the output voltage is assumed to be in the range of $v_{ref} = 20\,\text{V}–v_{ref} = 30\,\text{V}$. The set of new operation points is calculated by (14.12) and (14.23) and is shown in Fig. 14.3, together with a polytope covering the equilibrium points.

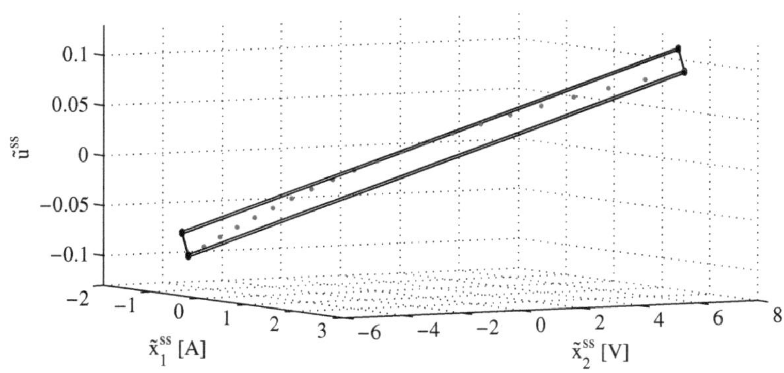

Fig. 14.3 The set of new operating points and the outer approximated polytope which covers the points

Theorem 14.4 is then applied to the vertices of the convex polytope and the following common SOS polynomial is found

$$s_2(\hat{\mathbf{x}}, \mathbf{z}) = 1.65\hat{x}_1^2 + 5.52\hat{x}_1\hat{x}_2 + 7.94\hat{x}_2^2 + 2\hat{x}_1 z_1 + 4.44\hat{x}_2 z_1 + 1.7z_1^2 + 4.07\hat{x}_1 z_2 + 14.63\hat{x}_2 z_2 + 5.75 z_1 z_2 + 8.16 z_2^2,$$

which shows that the boost converter is stable in the mentioned range of operation with the same controller and region of convergence as calculated in Example 14.2 for nominal operating point.

The simulation results of the boost converter with SOS controller for two new operating points are shown in Fig. 14.4. The initial state is set at $\mathbf{x}_0 = [12,\ 28.6]^T$ and the first operating point of the converter is selected as $\mathbf{x}^{ss}_{new,1} = [7.5, 30]^T$ and $\mathbf{u}^{ss}_{new,1} = 0.6$. Then, at $t = 0.1$ s, the operating point is changed to $\mathbf{x}^{ss}_{new,2} = [6.5325, 28]^T$ and $\mathbf{u}^{ss}_{new,2} = 0.5714$. Figure 14.4a, b show the output voltage and inductor current of the converter, respectively. It is shown that both states of the system converge to their references. Figure 14.4c represents the duty cycle of the semiconductor switch which remains in the predefined bound of input. Finally, Fig. 14.4d shows the state trajectories of the boost converter controlled by the SOS controller to the operating point with $v_{ref} = 30$ V. The initial states are selected on the boundary of region of convergence. It is shown that all trajectories converge to the new operating point by the same controller designed for nominal operating point.

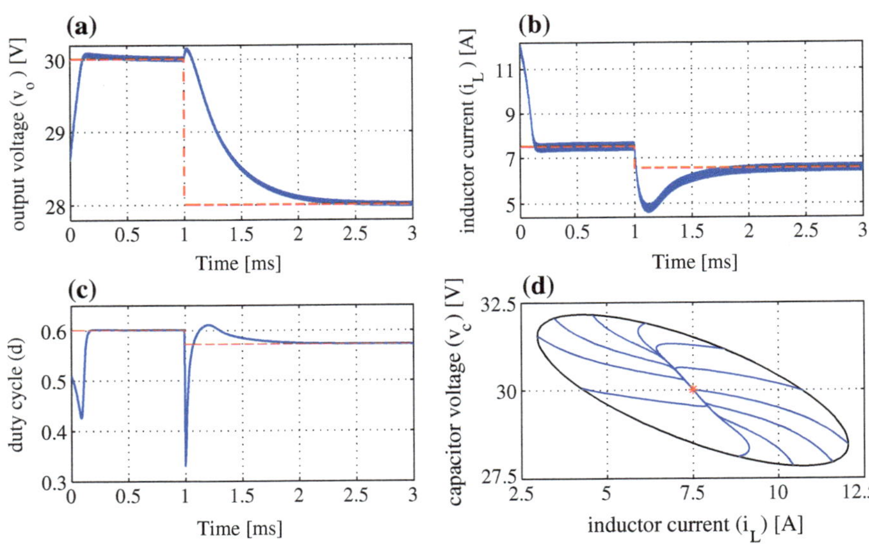

Fig. 14.4 Simulation results of the boost converter with SOS controller for various operating points: **a** Output voltage, **b** inductor current, **c** duty cycle of the switch, and **d** state trajectories started from boundary of region of convergence for the operating point at $v_{ref} = 24$ V

14.5 Introducing Integral Action in the SOS Controller Design

It is a common practice for dc/dc power converters to control their output voltage close to its reference even in the presence of persistent disturbances. In many practical applications, the accurate values of load parameters are unknown and only the time variant and the range of parameters are known. The resistor loads are sensitive to heat and their values change gradually due to heating when conducting electricity. In addition, the dc source voltage may experience fluctuations due to aging or improper design. In this section, the SOS controller is improved to provide robustness to persistent disturbances.

In this regard, a new state is added to the system which represents an integration action over the error between the state and desired reference value as

$$\mathbf{x}_{II,k+1} = \mathbf{x}_{II,k} + \tilde{\mathbf{x}}_{I,k} \tag{14.33}$$

where $\mathbf{x}_{II,k}$ is the integration state and $\tilde{\mathbf{x}}_I$ is the state which is desired to follow its reference in the case of disturbances. If the reference values are nonzero, the deviation variables should be calculated as described in (14.4). By introducing the integration state as a new state, the discrete-time bilinear model of the system in (14.18) is rewritten as follows:

$$\begin{bmatrix} \tilde{\mathbf{x}}_o \\ \tilde{\mathbf{x}}_I \\ \mathbf{x}_{II} \end{bmatrix}_{k+1} = \underbrace{\begin{bmatrix} \mathbf{A} & \mathbf{0} \\ [\mathbf{0}\ \mathbf{I}] & \mathbf{I} \end{bmatrix}}_{\mathbf{A}_I} \begin{bmatrix} \tilde{\mathbf{x}}_o \\ \tilde{\mathbf{x}}_I \\ \mathbf{x}_{II} \end{bmatrix}_k + \left(\underbrace{\begin{bmatrix} \mathbf{B}_{\tilde{\mathbf{x}}} \\ \mathbf{0} \end{bmatrix}}_{\mathbf{B}_{\tilde{\mathbf{x}},I}} + \underbrace{\begin{bmatrix} \mathbf{B} \\ \mathbf{0} \end{bmatrix}}_{\mathbf{B}_I} \right) \tilde{\mathbf{u}}_k. \tag{14.34}$$

where $\tilde{\mathbf{x}} = [\tilde{\mathbf{x}}_o, \tilde{\mathbf{x}}_I]^T$. Equation (14.34) is in the form a standard discrete-time bilinear systems in (14.18) as

$$\tilde{\mathbf{y}}_{k+1} = \mathbf{A}_I \tilde{\mathbf{y}}_k + (\mathbf{B}_{\tilde{\mathbf{y}},I} + \mathbf{B}_I)\tilde{\mathbf{u}}_k.$$

The control design method presented in (14.21) can be used for the new system model in (14.34) to find a controller in the form of (14.15) which stabilizes the converter.

Example 14.4 The robustness of the SOS controller with an augmented integrating state is investigated here for the boost converter presented in Example 14.1 with disturbances in the parameter values of load resistor (R) and source voltage (v_g). First, the new system matrices are calculated using (14.13) and (14.34) and then the SOS controller is found using (14.21) as

$$\begin{aligned} c_1(\tilde{\mathbf{x}}) = & -43.4\tilde{x}_1 - 170.8\tilde{x}_2 - 5.4\tilde{x}_3 + 5.3\tilde{x}_1^2 + 28.8\tilde{x}_1\tilde{x}_2 - 32.4\tilde{x}_2^2 \\ & +0.9\tilde{x}_1\tilde{x}_3 - 2.0\tilde{x}_2\tilde{x}_3 - 0.1\tilde{x}_3, \end{aligned}$$

$$c_0(\tilde{\mathbf{x}}) = 39.0 - 7.0\tilde{x}_1 + 0.9\tilde{x}_2 + 1.3\tilde{x}_3 + 4.6\tilde{x}_1^2 - 2.2\tilde{x}_1\tilde{x}_2 + 3.8\tilde{x}_2^2 \\ +0.7\tilde{x}_1\tilde{x}_3 + 0.2\tilde{x}_2\tilde{x}_3 + 1.4\tilde{x}_3.$$

which is in the form of ratio of two polynomials as presented in (14.15). The simulation results of the mentioned controller are shown in Figs. 14.5 and 14.6. The initial state is set at $\mathbf{x}_0 = [0.5; 24.7]$.

Figure 14.5 shows the simulation results of the boost converter with disturbances in the load resistor value. The load resistor is set to its nominal value $R = 10\,\Omega$ at the beginning of the simulation and is then changed to $R = 8\,\Omega$ at $t = 1.5$ ms and to $R = 12\,\Omega$ at $t = 3$ ms. The output voltage (v_o) of the converter is shown in Fig. 14.5a where the waveform follows its reference value even after step changes in the load resistor value. The inductor current (i_L) is depicted in Fig. 14.5b and the duty cycle of the switch is depicted in Fig. 14.5c where it is shown that the duty cycle satisfies the input constraint.

The simulation results of the boost converter with disturbances in the source voltage (v_g) is shown in Fig. 14.6. The source voltage is set to its nominal value $v_g = 12\,V$ at the beginning of the simulation and is then changed to $v_g = 13$ V at $t = 1.5$ ms and $v_g = 10.5$ V at $t = 3$ ms. Figures 14.6a–c show the corresponding changes in the output voltage, inductor current, and duty cycle, respectively. The results of Fig. 14.6 shows that SOS controller with augmented integral action is able to handle persistent disturbances in the source voltage of the boost converter.

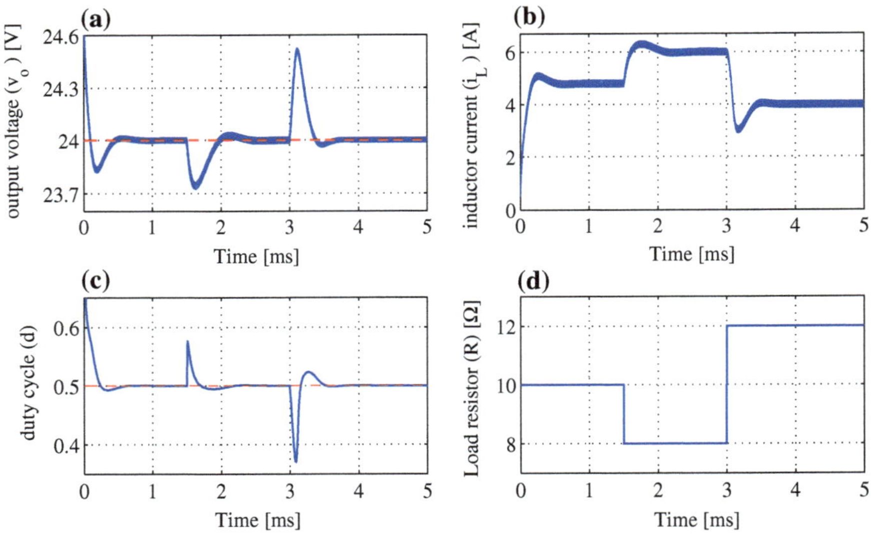

Fig. 14.5 Simulation results of the boost converter for SOS controller with integral action: **a** Output voltage, **b** inductor current, **c** duty cycle of the switch, and **d** load resistor

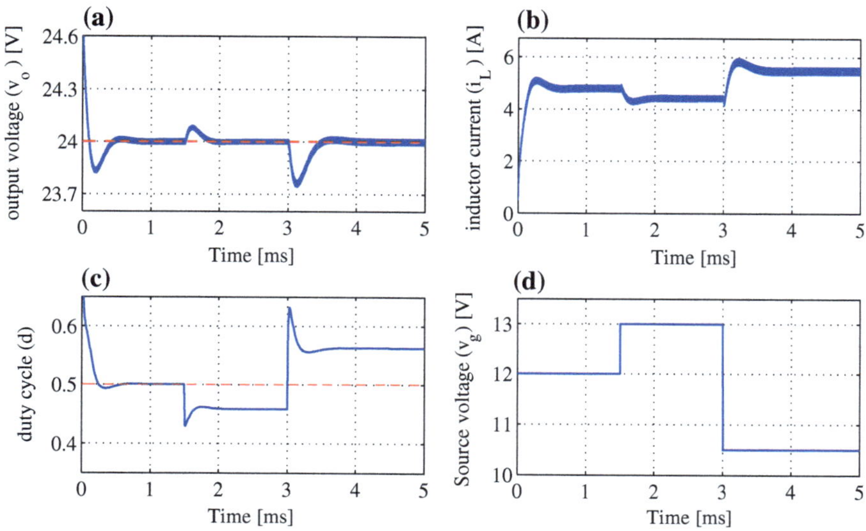

Fig. 14.6 Simulation results of the boost converter for SOS controller with integral action: **a** Output voltage, **b** inductor current, **c** duty cycle of the switch, and **d** source voltage

14.6 Conclusion

In this work, robustness issues concerning the control of discrete-time bilinear systems using SOS programming methods are investigated. First, it is shown that a wide variety of power converters are represented by bilinear averaged models. Then, a controller design method is described which guarantees the Lyapunov stability of the system. This controller is designed for the nominal operating point of the system. It is shown that the change of operating point for bilinear systems results in new dynamic in the system model which is not trivially stable. A new theorem is proposed to prove the stability of the SOS controller designed for nominal operating point for a range of desired operating points. In addition, an integration state is included in the model to improve the robustness of the controller to persistent disturbances. The effectiveness of the proposed methods are verified through time-domain simulations of a boost converter.

References

1. F. Amato, C. Cosentino, A.S. Fiorillo, A. Merola, Stabilization of bilinear systems via linear state-feedback control. IEEE Trans. Circuits Syst. II: Express briefs **56**, 76–80 (2009)
2. M. Anghel, F. Milano, A. Papachristodoulou, Algorithmic construction of lyapunov functions for power system stability analysis. IEEE Trans. Circuits Syst. I: Regul. Pap. **60**, 2533–2546 (2013)

3. G. Chesi, LMI techniques for optimization over polynomials in control: a survey. IEEE Trans. Autom. Control **55**, 2500–2510 (2010)
4. G. Chesi, *Domain of Attraction, Analysis and Control via SOS Programming* (Springer, London, 2011)
5. E.J. Hancock, A. Papachristodoulou, Structured sum of squares for networked systems analysis, in *Proceedings of the 50th IEEE Conference on Decision and Control and European Control Conference (CDC-ECC)* (2011), pp. 7236–7241
6. E.J. Hancock, A. Papachristodoulou, Generalised absolute stability and sum of squares. Automatica **49**, 960–967 (2013)
7. T. Hu, A nonlinear-system approach to analysis and design of power-electronic converters with saturation and bilinear terms. IEEE Trans. Power Electron. **26**, 399–410 (2011)
8. H.K. Khalil, *Nonlinear Systems*, 3rd edn. (Prentice Hall, Upper Saddle River, 2002)
9. J. Lavaei, A.G. Aghdam, A necessary and sufficient condition for robust stability of LTI discrete-time systems using sum-of-squares matrix polynomials, in *Proceedings of the 45th IEEE Conference on Decision and Control* (2006), pp. 2924–2930
10. J. Löfberg, YALMIP: a toolbox for modeling and optimization in MATLAB, in *Proceedings of the CACSD Conference*, Taipei, Taiwan (2004)
11. J. Löfberg, Pre- and post-processing sum-of-squares programs in practice. IEEE Trans. Autom. Control **54**, 1007–1011 (2009)
12. C. Olalla, I. Queinnec, R. Leyva, A. El Aroudi, Robust control design of bilinear DC-DC converters with guaranteed region of stability, in *IEEE International Symposium on Industrial Electronics (ISIE)* (2010), pp. 3005–3010
13. C. Olalla, I. Queinnec, R. Leyva, A. El Aroudi, Optimal state-feedback control of bilinear DCDC converters with guaranteed regions of stability. IEEE Trans. Ind. Electron. **59**, 3868–3880 (2012)
14. A. Papachristodoulou, S. Prajna, A tutorial on sum of squares techniques for systems analysis, in *Proceedings of the American Control Conference* (2005), pp. 2686–2700
15. P.A. Parrilo, Structured semidefinite programs and semialgebraic geometry methods in robustness and optimization, Ph.D. dissertation, California Institute of Technology, Pasadena, CA (2000)
16. S. Prajna, A. Papachristodoulou, W. Fen, Nonlinear control synthesis by sum of squares optimization: a Lyapunov-based approach, in *Proceedings of the 5th Asian Control Conference* (2004), pp. 157–165
17. V. Spinu, N. Athanasopoulos, M. Lazar, G. Bitsoris, Stabilization of bilinear power converters by affine state feedback under input and state constraints. IEEE Trans. Circuits Syst. II: Express Briefs **59**, 520–524 (2012)
18. M. Vatani, M. Hovd, Control of bilinear power converters using sum of squares programming, in *Proceedings of the 14th European Control Conference* (2015)
19. M. Vatani, M. Hovd, S. Olaru, Control design and analysis for discrete time bilinear systems using sum of squares methods, in *Proceedings of the 53rd IEEE Conference on Decision and Control* (2014), pp. 3143–3148
20. M. Vatani, M. Hovd, M. Saeedifard, Control of the modular multilevel converter based on a discrete-time bilinear model using the sum of squares decomposition method. IEEE Trans. Power Deliv. **30**, 2179–2188 (2015)

Chapter 15
On the LPV Control Design and Its Applications to Some Classes of Dynamical Systems

Franco Blanchini, Daniele Casagrande, Giulia Giordano and Stefano Miani

Abstract In this chapter, a control design approach based on linear parameter-varying (LPV) systems, which can be exploited to solve several problems typically encountered in control engineering, is presented. By means of recent techniques based on Youla–Kucera parametrization, it is shown how it is possible not only to design and optimize stabilizing controllers, but also to exploit the structure of the Youla–Kucera parametrized controller to face and solve side problems, including: (a) dealing with nonlinearities; (b) taking into account control input constraints; (c) performing controller commutation or online adaptation, e.g., in the presence of faults; and (d) dealing with delays in the system. The control scheme is observer-based, namely a prestabilizing observer-based precompensator is applied. Consequently, a Youla–Kucera parameter is applied to produce a supplementary input ignition, which is a function of the residual value (the difference between the output and the estimated output). Based on the fact that any stable operator which maps the residual to the supplementary input preserves stability, several additional features can be added to the compensator, without compromising the loop stability.

Keywords Linear parameter-varying (LPV) systems · Youla–Kucera parametrization · Actuator and sensor faults · Time-delay systems · Control of saturated systems

---

F. Blanchini · G. Giordano
DIMI, University of Udine, Via Delle Scienze 206, Udine, Italy
e-mail: blanchini@uniud.it

G. Giordano
e-mail: giulia.giordano@uniud.it

D. Casagrande · S. Miani (✉)
DIEGM, University of Udine, Via Delle Scienze 206, Udine, Italy
e-mail: casagrande@uniud.it

S. Miani
e-mail: miani.stefano@uniud.it

S. Olaru et al. (eds.), *Developments in Model-Based Optimization and Control*,
Lecture Notes in Control and Information Sciences 464,
DOI 10.1007/978-3-319-26687-9_15

15.1 Introduction

Linear parameter-varying (LPV) systems constitute a class of linear time-varying systems which lay in between uncertain systems and linear systems and allow for an elegant and effective description of many dynamic systems for which the knowledge of the current configuration is known pointwise in time, [29]. Such systems have been studied since the 1990s (see [13, 15, 20, 22, 24, 26, 28, 33]) and can be thought of as linear time-invariant plants with time-varying, uncertain but pointwise-in-time known, parameters. The parametric structure may be intrinsic in the physical system or may appear in the model, resulting, e.g., from the linearization of nonlinear systems in different operating points, [1, 30], or from the adoption of different controllers, each acting according to the designer-specified switching role, [5, 7, 14]. This last point of view, say the analysis of LPV systems as the result of the combination of a linear (possibly time-varying) system along with a scheduled controller, has provided (i) a full comprehension of phenomena occurring during the system commutation, and (ii) a full and exhaustive characterization of the stabilizing controllers that can be adopted for this class of systems.

In this chapter, we will review the newly proposed results in this area and we will provide several examples of systems for which the provided theory guarantees an effective solution. These examples, aimed at bridging different fields under the common denominator of LPV systems, span from the case of control in the presence of actuator and sensor faults, for which effective results have been provided in [32], to the case of control of time-delay systems, [6, 18, 21], and to the case of control of saturated systems, [12].

15.2 LPV Systems: Definition and Main Results

The class of LPV systems is described by the n-dimensional system with m inputs and p outputs

$$\begin{aligned} \sigma(x(t)) &= A(w(t))x(t) + B(w(t))u(t), \\ y(t) &= C(w(t))x(t), \end{aligned} \tag{15.1}$$

where $w(\cdot)$ is a function taking values in an assigned compact set $\mathcal{W}$ and $\sigma(x(t))$ represents the differential operator in the continuous-time case and the single step shift in the discrete-time case. The current value of $w(t) \in \mathcal{W}$ is known and available for control purposes, whereas its future evolution is not.

Example 15.1 Consider a simple pendulum of length l and mass M. Its dynamics are

$$\begin{aligned} \dot{x}_1(t) &= x_2(t), \\ \dot{x}_2(t) &= -\frac{g}{l}\sin(x_1(t)) - \frac{b}{Ml^2}x_2(t) + \frac{1}{Ml^2}u(t), \\ y(t) &= x_1(t), \end{aligned}$$

where g is the gravity acceleration, b is the constant viscous coefficient, x_1 represents the angle, x_2 the angular velocity and u the control input. Since the current value of $x_1(t) = y(t)$ is known, the system can be "embedded" in an LPV system (precisely, this particular case of LPV is also known in the literature as *quasi-LPV*, see [23] and the references therein), e.g., by rewriting the nonlinear term $\sin(x_1(t))$ as $\frac{\sin(x_1(t))}{x_1(t)} x_1(t)$ and by defining $w(t) = \frac{\sin(x_1(t))}{x_1(t)}$, with $\mathcal{W} = [-0.2172\ 1]$. By means of this parameter, the dynamics of the systems can be written in the form (15.1) with

$$A(w(t)) = \begin{bmatrix} 0 & 1 \\ -\frac{gw(t)}{l} & -\frac{b}{Ml^2} \end{bmatrix}, \quad B(w(t)) = \begin{bmatrix} 0 \\ \frac{1}{Ml^2} \end{bmatrix}, \quad C(w(t)) = [1 \quad 0].$$

In this particular case, matrices B and C do not actually depend on w. ◊

Other significant examples of LPV systems will be reported along the chapter.

Definition 15.1 System (15.1) is *LPV stable* if the zero equilibrium is asymptotically stable for any function $w : [0, +\infty) \to \mathcal{W}$.

It is a rather established fact that stability of $A(w)$ for every constant value of $w \in \mathcal{W}$ is just a necessary condition for the stability of (15.1) in the sense of Definition 15.1.

The asymptotic stability of the zero equilibrium is equivalent to the existence of a Lyapunov function.

The peculiarity of LPV systems, as far as control design is concerned, lies in the fact that the future evolution of the time-varying parameter $w(t)$ is unknown, but its current value is known. This characteristic allows us to derive nonconservative conditions for the existence of an LPV stabilizing regulator based on linear matrix inequalities (LMIs), [2, 3, 9, 20, 25], as per the following result proved in [3, 5].

Theorem 15.1 *The LPV system (15.1) is (quadratically) stabilizable via a n-dimensional observer—based LPV regulator if and only if there exist two symmetric positive—definite matrices P and Q, both in $\mathbb{R}^{n\times n}$, and two matrices $U(w) \in \mathbb{R}^{m\times n}$ and $Y(w) \in \mathbb{R}^{n\times p}$ such that the following set of LMIs (in the continuous-time and in the discrete-time case, respectively) is satisfied for every $w \in \mathcal{W}$.*

- *Continuous-time:*

$$PA(w)^\top + A(w)P + B(w)U(w) + U(w)^\top B(w)^\top \prec 0, \tag{15.2}$$

$$A(w)^\top Q + QA(w) + Y(w)C(w) + C(w)^\top Y(w)^\top \prec 0. \tag{15.3}$$

- *Discrete-time:*

$$\begin{bmatrix} P & (A(w)P + B(w)U(w))^\top \\ A(w)P + B(w)U(w) & P \end{bmatrix} \succ 0, \tag{15.4}$$

$$\begin{bmatrix} Q & (QA(w) + Y(w)C(w))^\top \\ QA(w) + Y(w)C(w) & Q \end{bmatrix} \succ 0. \tag{15.5}$$

If the above conditions are satisfied for every $w \in \mathcal{W}$, then the observer—based control law

$$\begin{aligned}\sigma(\hat{x}(t)) &= [A(w(t)) + L(w(t))C(w(t)) + B(w(t))J(w(t))]\,\hat{x}(t)\\ &\quad -L(w(t))y(t) + B(w(t))v(t),\\ u(t) &= J(w(t))\hat{x}(t) + v(t),\end{aligned} \tag{15.6}$$

with $v(t) = 0$,[1] *is stabilizing. The observer and estimated state gains are*

$$J(w(t)) = U(w(t))P^{-1} \tag{15.7}$$

and

$$L(w(t)) = Q^{-1}Y(w(t)). \tag{15.8}$$

Remark 15.1 The conditions reported in the previous theorem guarantee the existence of a Luenberger observer-based controller of the same dimension of the plant. The interested reader is referred to [5] for necessary and sufficient conditions for the existence of a **linear** extended observer when the LMI conditions just reported fail.

It is worth stressing that the stabilizability conditions might correspond to an infinite number of LMIs. However, there are important cases in which such a set of LMIs reduces to a finite number, thus the solution is easily computable by means of standard software tools. Two interesting cases are the following.

1. Systems where the input and output matrices are constant, while $A(w(t))$ is polytopic [1]: $A(w(t)) = \sum_{i=1}^{r} A_i w_i(t)$, with $w_i(t) \geq 0\ \forall i$ and $\sum_{i=1}^{r} w_i(t) = 1$.
2. Systems belonging to the class of switching systems, [7, 16], described by

$$\begin{aligned}\sigma(x(t)) &= A_{i(t)}x(t) + B_{i(t)}u(t),\\ y(t) &= C_{i(t)}x(t),\end{aligned}$$

where $i(t) \in \{1, \ldots, r\}$ is a switching signal.

In both cases, the set of $r + r$ LMIs to be solved is the following (the continuous-time case LMIs only are reported):

$$PA_i^\top + A_iP + B_iU_i + U_i^\top B_i^\top \prec 0, \qquad i = 1, \ldots, r \tag{15.9}$$

$$A_i^\top Q + QA_i + Y_iC_i + C_i^\top Y_i^\top \prec 0, \qquad i = 1, \ldots, r \tag{15.10}$$

with the understanding that $B_i = B$ for all i and $C_i = C$ for all i if the input and output matrices are constant.

---

[1] The signal $v(t)$ is introduced in (15.6) to avoid duplicating the observer-based stabilizing regulator equations and will be used later, when the Youla–Kucera parameter will come into play.

In case 1, the stabilizing observer—based controller is

$$\begin{aligned}\sigma(\hat{x}(t)) &= \sum_{i=1}^{r} [A_i + L_i C + BJ_i]\, w_i(t)\hat{x}(t) - \sum_{i=1}^{r} L_i w_i(t) y(t) + Bv(t),\\ u(t) &= \sum_{i=1}^{r} J_i w_i(t)\hat{x}(t) + v(t),\end{aligned} \tag{15.11}$$

whereas in case 2 it is

$$\begin{aligned}\sigma(\hat{x}(t)) &= \left(A_{i(t)} + L_{i(t)}C_{i(t)} + B_{i(t)}J_{i(t)}\right)\hat{x}(t) - L_{i(t)}y(t) + B_{i(t)}v(t),\\ u(t) &= J_{i(t)}\hat{x}(t) + v(t)\end{aligned} \tag{15.12}$$

and, again, $J_i = U_i P^{-1}$, $L_i = Q^{-1} Y_i$, while $v(t) = 0$ (see Footnote 1).

The signal $v(t)$ appearing in Theorem 15.1, and so far set to 0, can be successfully employed to parametrize all the linear stabilizing compensators via Youla–Kucera parametrization. Moreover, if it is generated as the output of an LPV stable operator whose input is the estimation error, then the overall system remains stable, as per the next result.

Theorem 15.2 *Assume the stabilizability conditions in Theorem 15.1 are satisfied. Let*

$$o(t) = C(w(t))\hat{x}(t) - y(t) \tag{15.13}$$

and consider any globally asymptotically stable operator mapping $o(t)$ into $v(t)$, i.e.,

$$\begin{aligned}\sigma(z(t)) &= g(z(t), o(t), t),\\ v(t) &= h(z(t), o(t), t).\end{aligned} \tag{15.14}$$

Then the observer—based regulator (15.6), with $v(\cdot)$ defined by (15.14) and (15.13), is stabilizing.

Under the additional assumption of the existence of a single quadratic Lyapunov function for the closed-loop system, the converse is also true. Consider the closed-loop system obtained from (15.1) when the stabilizing controller

$$\begin{aligned}\sigma(q(t)) &= f(q(t), y(t), t),\\ u(t) &= k(q(t), y(t), t)\end{aligned} \tag{15.15}$$

is adopted and assume that such system admits a single quadratic Lyapunov function. Then the stabilizing controller (15.15) can be parametrized as in (15.6) for a proper stable operator (15.13), (15.14), which is known as the Youla–Kucera *parameter.*

The constructive proof of this theorem is reported in the Appendix, along with the procedure to compute the Youla–Kucera parameter. Here it is worth stressing that the parameterization of all the stabilizing controllers, depicted in Fig. 15.1, is exactly the classical Youla–Kucera parameterization.

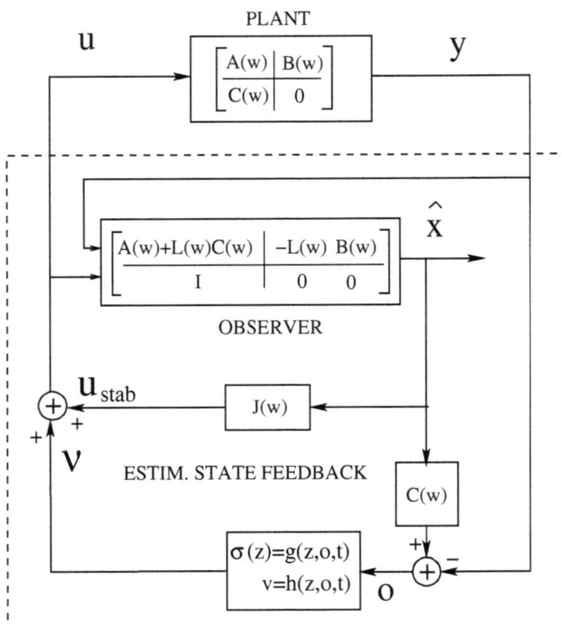

Fig. 15.1 Youla–Kucera parameterization

Apart from the case in which the stabilizing operator is itself LPV, the determination and realization of the Youla–Kucera parameter are far from being simple and will not be investigated here (the interested reader is referred to [4] and the references therein). We rather stress, once again, that the freedom in the choice of the stable operator can be successfully exploited to cope with different problems, as will be seen in the next section.

15.3 Exploiting the Youla–Kucera Parameter Choice

In this section, some applications of the proposed results will be presented, to show how it is possible to take advantage of the freedom in the choice of the Youla–Kucera parameter, so as to deal with several side problems.

15.3.1 Online Controller Adaptation Induced by Faults

In the recent literature, several LPV fault-tolerant control schemes have been analyzed (see, for instance, [27, 31]). Here, the case of real over-actuated control systems with multiple sensing channels is considered. The input and output channel redundancy is supposed to be introduced for security reasons (think, as an example, of the flap control of an airplane).

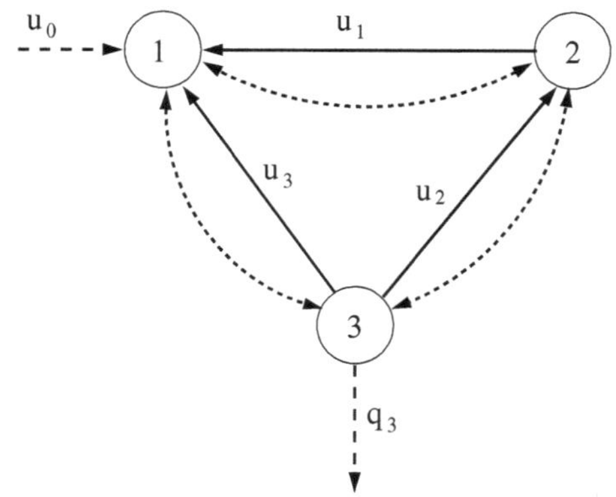

Fig. 15.2 The system of three tanks considered in Sect. 15.3.1. *Dashed lines* represent normal flows while *solid lines* represent additional flows

In such systems, rather than designing a single **robust** controller to cope with every fault scenario, it is natural to design different controllers for every possible fault configuration, so as to fully exploit the available sensors and actuators in each configuration and then commute among the different controllers when a fault is sensed. Unfortunately, since the commutations can be assumed to be random, the switching between stabilizing controllers can result in an unstable behavior. To overcome this limit, it is sufficient to verify the LMI conditions in Theorem 15.1, design the observer-based controller and then realize the obtained controllers via the Youla–Kucera parametrization.

In this example, to keep the exposition simple, only the determination of the observer—based controllers will be dealt with and the Youla–Kucera parameters will be set to zero.

The dynamics of the system we consider are

$$\dot{x}(t) = A_{w(t)}x(t) + B_{w(t)}u(t),$$
$$y(t) = C_{w(t)}x(t),$$

where the integer parameter $w(t)$ is associated with the wth fault scenario. To illustrate the idea, consider the simple system formed by three connected tanks as in Fig. 15.2, in which a natural flow occurs between the different tanks, proportional to their relative levels.[2] The flow from tank k to tank h is $q_{kh}(t) = \alpha(x_k(t) - x_h(t))$, with $\alpha > 0$. Moreover, tank 3 has a discharge channel, whose flow is proportional to its level, $q_3(t) = \beta x_3(t)$, and tank 1 is fed by u_0. An additional flow between the tanks can be forced by three connecting valves (one for each pair of tanks). In the

[2]Levels and flows have to be intended as the deviation from the steady–state values.

nominal condition, the three sensors measuring the levels work, as well as the three valves controlling the additional flow. In the present example, it is assumed that the system is always working in some faulty condition corresponding to one sensing channel and one valve working. Thus the parameter $w(t)$ can be represented as a pair (i, j) belonging to the set $\mathcal{W} = \{1, 2, 3\}^2$; the corresponding LPV system is

$$\begin{aligned}\dot{x} &= Ax + B_{i(t)}u,\\ y &= C_{j(t)}x,\end{aligned}$$

where

$$A = \begin{bmatrix} -2\alpha & \alpha & \alpha \\ \alpha & -2\alpha & \alpha \\ \alpha & \alpha & -2\alpha-\beta \end{bmatrix},$$

$$B_1 = \begin{bmatrix} 1 & 1 & 0 & 0 \\ 0 & -1 & 0 & 0 \\ 0 & 0 & 0 & 0 \end{bmatrix}, \quad B_2 = \begin{bmatrix} 1 & 0 & 0 & 0 \\ 0 & 0 & 1 & 0 \\ 0 & 0 & -1 & 0 \end{bmatrix}, \quad B_3 = \begin{bmatrix} 1 & 0 & 0 & 1 \\ 0 & 0 & 0 & 0 \\ 0 & 0 & 0 & -1 \end{bmatrix}$$

and

$$C_1 = \begin{bmatrix} 1 & 0 & 0 \\ 0 & 0 & 0 \\ 0 & 0 & 0 \end{bmatrix}, \quad C_2 = \begin{bmatrix} 0 & 0 & 0 \\ 0 & 1 & 0 \\ 0 & 0 & 0 \end{bmatrix}, \quad C_3 = \begin{bmatrix} 0 & 0 & 0 \\ 0 & 0 & 0 \\ 0 & 0 & 1 \end{bmatrix}.$$

with $\alpha = 1$ and $\beta = 0.5$.

There are nine possible configurations, but given the input and output matrices combinations are independent, it is sufficient to solve six LMIs (15.2) and (15.3) so as to obtain the following gains:

$$J_1 = \begin{bmatrix} 0.8018 & 1.0654 & -4.4610 \\ 0.8977 & -1.1775 & 1.4267 \\ 0 & 0 & 0 \\ 0 & 0 & 0 \end{bmatrix}, \quad J_2 = \begin{bmatrix} 1.9252 & -5.4114 & 1.2044 \\ 0 & 0 & 0 \\ 1.1919 & 0.4867 & -1.3340 \\ 0 & 0 & 0 \end{bmatrix},$$

$$J_3 = \begin{bmatrix} 0.9773 & -4.0721 & 1.0690 \\ 0 & 0 & 0 \\ 0 & 0 & 0 \\ 0.7539 & 1.3342 & -1.4642 \end{bmatrix}, \quad L_1 = \begin{bmatrix} 0.8372 & 0 & 0 \\ -0.9527 & 0 & 0 \\ -0.8679 & 0 & 0 \end{bmatrix},$$

$$L_2 = \begin{bmatrix} 0 & -0.9527 & 0 \\ 0 & 0.8372 & 0 \\ 0 & -0.8679 & 0 \end{bmatrix}, \quad L_3 = \begin{bmatrix} 0 & 0 & -0.7796 \\ 0 & 0 & -0.7796 \\ 0 & 0 & 0.9233 \end{bmatrix}.$$

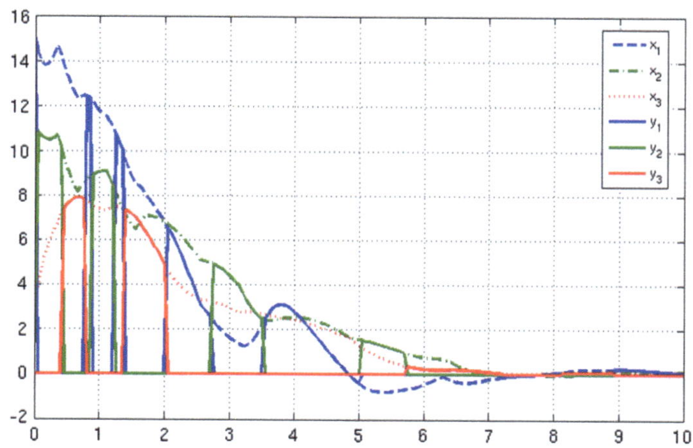

Fig. 15.3 Time histories of the tank level deviations from steady–state values

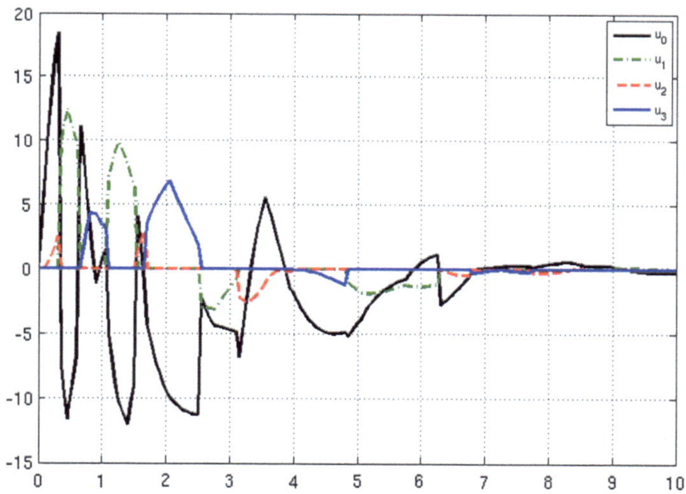

Fig. 15.4 Time histories of the flow deviations from steady–state values

During the system evolution, the ith sensor and the jth actuator faults are sensed and the corresponding input and output gains are used in the observer-based regulator (15.12).

Figures 15.3 and 15.4 depict the system signals evolution during the transient when the initial condition is $x(0) = [15\ 11\ 3]^\top$, an arbitrary sequence of faults occurs, and each of the Youla–Kucera parameters is set to zero (i.e., the standard Luenberger observer-based controllers are used).

To be more precise, Fig. 15.3 depicts the evolution of the state and actual output deviations from steady-state values. The representation in Fig. 15.3 has to be interpreted as follows: the solid line is the active output and the dotted lines are the state values. To clarify the representation, it can be noted that during the first instants, the second sensing channel is working, then the third, then the first just before $t = 1s$, then the second again, etc.

Figure 15.4 depicts the evolution of the system input deviation from the steady-state value, and the pictorial representation is as follows: the active output (corresponding to the working actuator) is working whenever it is different from zero. To make things clear, the second valve is active during the first instants, then the first, then the third, etc.

15.3.2 Discrete-Time Delays in Network Controlled Systems

Consider a network controlled n-dimensional discrete-time plant with m inputs and p outputs in the presence of unknown—but—bounded integer observation delay[3] $\tau(t) \in \{0, 1, \dots, \tau_{\max}\}$ (and without delays in the actuator channel). This system can be modeled by the equations

$$\begin{aligned} x(t+1) &= Ax(t) + Bu(t), \\ y(t) &= Cx(t - \tau(t)), \end{aligned}$$

and can be alternatively represented by the following switching system, by adding delayed copies of the output to the state:

$$\begin{aligned} x_e(t+1) &= A^e x_e(t) + B^e u(t) \\ y(t) &= C^e_{\tau(t)} x_e(t) \end{aligned}$$

with $\tau(t) \in \{0, 1, \dots, \tau_{\max}\}$ and

$$A^e = \begin{bmatrix} A & 0^{n\times(\tau_{\max}-1)p} & 0^{n\times p} \\ C & 0^{p\times(\tau_{\max}-1)p} & 0^{p\times p} \\ 0^{(\tau_{\max}-1)p\times n} & I^{(\tau_{\max}-1)p} & 0^{(\tau_{\max}-1)\times p)} \end{bmatrix},$$

$$B^e = \begin{bmatrix} B \\ 0^{p\times m} \\ 0^{(\tau_{\max}-1)p\times m} \end{bmatrix},$$

$$C^e_0 = \begin{bmatrix} C \; 0^{p\times(\tau_{\max}-1)p} \; 0^{p\times p} \end{bmatrix},$$

$$C^e_i = \begin{bmatrix} 0^{p\times(n+(i-1)p)} \; I^p \; 0^{p\times(\tau_{\max}-i)p} \end{bmatrix},$$

[3]The delay is assumed to be a multiple of the sampling period if the discrete-time system is obtained as the discretization of a continuous-time system.

for $i = 1, \ldots, \tau_{\max}$. The above system falls in the class of LPV systems described in the previous sections. Hence, assuming the satisfaction of the LMIs (15.4) and (15.5), it is possible to parametrize every stabilizing observer by means of the proposed Youla–Kucera structure. This in turn allows the designer to choose $\tau_{\max} + 1$ compensators, each stabilizing the system for a constant value of the delay τ, and realize such compensators by means of a proper Youla–Kucera parameter, thus guaranteeing stability of the system in the presence of arbitrary (bounded) sequences of delays. Note that the number of variables of the augmented system may be large if $\tau_{\max}$ is large, which may result in a heavy computational load. In addition, when the variation of the delay is very fast, the system may not be robust. We do not consider these questions here, since an analysis of the robustness of the method and of its efficiency when applied to a real problem is beyond the scopes of the chapter. For numerical examples of network controlled systems, we refer the reader to [18], or to [17].

15.3.3 Smith Predictor for LPV Stable Plants

The combination between time-delay and parameter-dependency may lead to several different linear parameter-varying time-delay systems (see, for instance, [10]). Here we focus on the problem of controlling an LPV stable continuous-time system in the presence of a known, time-varying but bounded, delay. We show that this problem can be solved with the technique described in Sect. 15.2. The considered dynamics are

$$\begin{aligned} \dot{x}(t) &= A(w)x(t) + B(w)u(t), \\ y(t) &= C(w)x(t - \tau(w)), \end{aligned}$$

with $\tau(w) \in [0, \bar{\tau}]$, for some $\bar{\tau} > 0$, for all $w \in \mathcal{W}$. We denote by $\Pi(w) = \{A(w), B(w), C(w)\}$ the state-space representation of the delay—free part of the plant, by $W_P(s, w)$ its transfer function, by $W_C(s, w)$ the (parametric) transfer function of the controller and by $\Delta(s, w)$ the block corresponding to the delay, namely $\Delta(s, w) = e^{-\tau(w)s}$. These notations are used in the block diagram of Fig. 15.5 that represents the Smith predictor scheme.

Apart from the block associated with the delay, this scheme is analogous to the scheme depicted in Fig. 15.1, with the negative feedback loop inside the smaller dashed rectangle playing the role of the Youla–Kucera parameter; however, the block corresponding to $J(w)$ is absent, since the plant is assumed to be stable. The (parametric) transfer function of the Youla–Kucera parameter is

$$W_{YK}(s, w) = [I + W_C(s, w)W_P(w, s)]^{-1} W_C(s, w). \tag{15.16}$$

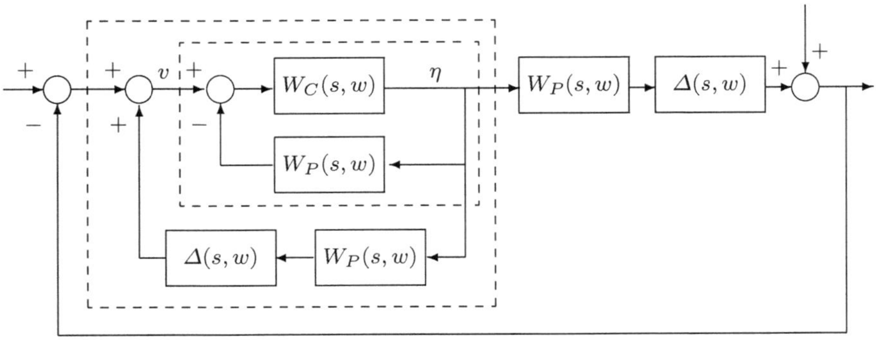

Fig. 15.5 Smith predictor control scheme

Consider, now, a state-space realization $\Theta(w) = \{F(w), G(w), H(w), K(w)\}$ of the transfer function (15.16) associated with the equations

$$\begin{aligned} \dot{z}(t) &= F(w(t))z(t) + G(w(t))v(t), \\ \eta(t) &= H(w(t))z(t) + K(w(t))v(t). \end{aligned}$$

The overall system equations, in the absence of an external input, are

$$\dot{x}(t) = A(w(t))x(t) + B(w(t))u(t), \tag{15.17}$$
$$y_0(t) = C(w(t))x(t), \tag{15.18}$$
$$\dot{x}_c(t) = A(w(t))x_c(t) + B(w(t))u(t), \tag{15.19}$$
$$y_{c0}(t) = C(w(t))x_c(t), \tag{15.20}$$
$$\dot{z}(t) = F(w(t))z(t) + G(w(t))v(t), \tag{15.21}$$
$$\eta(t) = H(w(t))z(t) + K(w(t))v(t), \tag{15.22}$$
$$v(t) = y_{c0}(t-\tau) - y_0(t-\tau), \tag{15.23}$$

where x denotes the plant state, x_c denotes the state of the copy of the plant in the feedback loop and z denotes the state of the Youla–Kucera block Θ.

The main result concerning the stability of LPV control systems with delay is the following [6].

Theorem 15.3 *The control system (15.17)–(15.23), as in Fig. 15.5, is LPV stable if the state-space realization $\Theta(w)$ is LPV stable.*

In order to apply Theorem 15.3, one needs an LPV stable realization of the controller that can be obtained by steps 5 and 6 of the procedure reported in the Appendix.

Remark 15.2 The scheme based on the Smith predictor can be fragile with respect to delay uncertainties (see [19]). However, we stress that in the present setting the time-varying parameters (one of which is the delay) are assumed to be exactly known at each time instant. A discussion on the robustness of the proposed method with respect to uncertainties in the plant only can be found in [6].

15.3.3.1 Example

As an example, we consider the dam–gallery system analyzed in [8] and we focus on the transfer function between the upstream flow Q_U and the downstream flow Q_D. By using simplified Saint-Venant's equations, and choosing the downstream flow as scheduling parameter, this transfer function turns out to be, [8],

$$W_P(w, s) = \frac{e^{-s\tau(w)}}{1 + k_1(w)s + k_2(w)s^2}, \tag{15.24}$$

where $w = Q_D$ and where $k_1(w)$, $k_2(w)$ and $\tau(w)$ are polynomials in w of degree three. A stability analysis, omitted for brevity, shows that, with the polynomial coefficients reported in [8], system (15.24) is LPV stable. Therefore, Theorem 15.3 can be applied. The LPV controller

$$R(s, w) = K\frac{1 + k_1(w)s + k_2(w)s^2}{s(1 + sT)},$$

for instance, which cancels the system dynamics and introduces the new poles 0 and $-1/T$, can be adopted. If $R(s, w)$ is realized according to steps 4, 5, and 6 of the procedure reported in the Appendix, then the closed-loop system is stable for any time-varying w.

15.3.4 Youla–Kucera Parameter as an Input Limiter for Constrained Systems

Another interesting problem is the control of LPV systems with input saturation (see, for instance, [11, 34]). Here we propose a solution based on the idea that, since the Youla–Kucera parameter can be any stable operator that maps the signal $o(t)$ to the signal $v(t)$, it can be exploited to achieve override control, [12]. When the absolute value of the control input is constrained to remain below a threshold $\bar{u}$, the principle of the override control is to consider the actual control input $u(t)$ as the sum of the ideal stabilizing control $u_{stab}(t)$ and of an additional signal $v(t)$ defined by

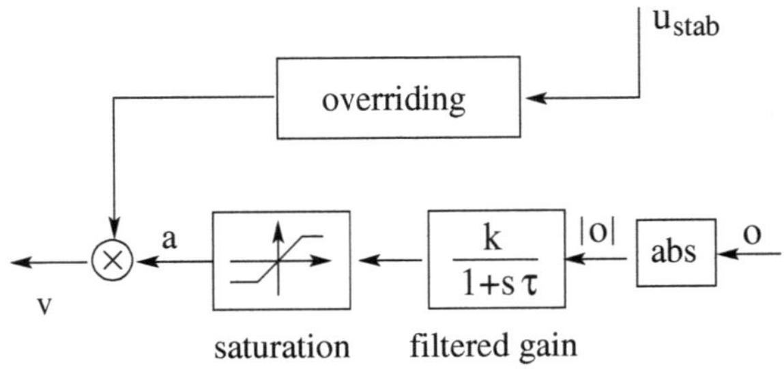

Fig. 15.6 The overriding Youla–Kucera block

$$v(t) = \begin{cases} \bar{u} - u_{stab}(t) & \text{if} \quad u_{stab}(t) > \bar{u} \\ 0 & \text{if} \quad |u_{stab}(t)| \leq \bar{u} \\ -\bar{u} - u_{stab}(t) & \text{if} \quad u_{stab}(t) < -\bar{u} \end{cases}$$

thus guaranteeing $|u(t)| \leq \bar{u}$. Since the actual control is different from the ideal one, the problem is how to ensure stability of this type of scheme. A possible solution, justified by the fact that the override control is useful at the beginning of the transient, is to activate it only when the absolute value of the observer error is above a given threshold $\bar{o}$. When $|o(t)| < \bar{o}$, the state is suitably reconstructed and the override signal is inactivated. A realization of this idea is depicted in Fig. 15.6. The absolute value of $o(t)$ is filtered by a filter with transfer function $\frac{k}{1+s\tau}$ and then saturated to 1 to obtain the activation signal $a(t)$. The aim of the filter is twofold. First, it avoids the action to be disabled if the signal $|o(t)|$ is less than the threshold for a too short interval (for instance when it "passes through zero"). Second, the constant gain k can be chosen so as to scale the estimation error to a magnitude suitable for the saturation function. For large values of $o(t)$, $a(t) = 1$ and the override control is active; after the transient, $o(t)$ goes to zero and so does $a(t)$.
Figs. 15.7 and 15.8 report the transient for the system with

$$A = \begin{bmatrix} 0 & 1 \\ 0 & 0 \end{bmatrix}, \quad B = \begin{bmatrix} 1 \\ 0 \end{bmatrix}, \quad C = \begin{bmatrix} 1 & 0 \end{bmatrix}, \tag{15.25}$$

and $\bar{u} = 1$. The matrices provided by the LMIs (15.2) and (15.3) are

$$J = \begin{bmatrix} -1 & -2 \end{bmatrix}, \quad L = \begin{bmatrix} -1 \\ -1 \end{bmatrix}.$$

The parameters of the filter are $k = 10$ (corresponding to $\bar{o} = 0.1$) and $\tau = 3$. Figure 15.7 shows the transient from the initial condition $[2 \;\; 2]^\top$ without the overriding scheme: it can be seen that the control bound is deeply violated. Conversely, Fig. 15.8 shows the transient from the same initial condition $[2 \;\; 2]^\top$ with the overriding scheme: control bounds are not violated. The scheme can be applied in the same way also when observer and feedback gain are gain–scheduled.

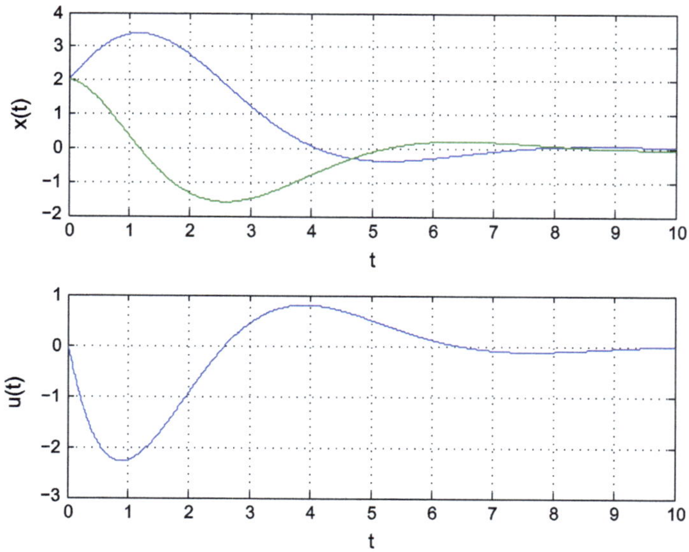

Fig. 15.7 The transient of the states (*top*) and of the control input (*bottom*) for the system (15.25) without the overriding scheme

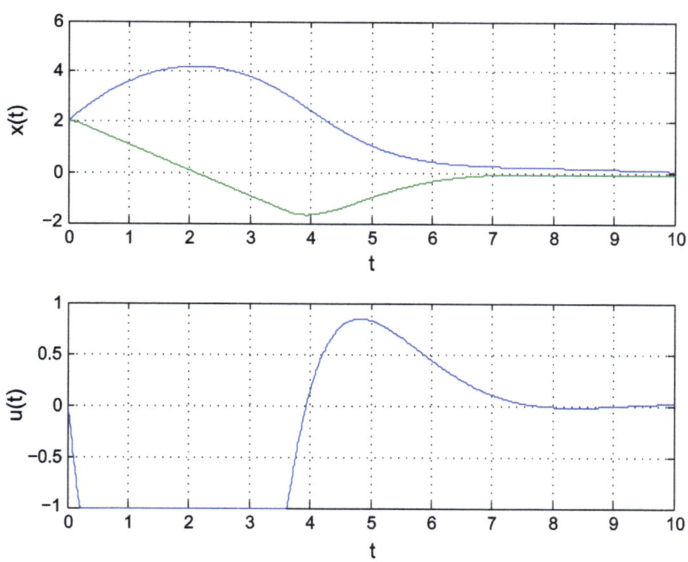

Fig. 15.8 The transient of the states (*top*) and of the control input (*bottom*) for the system (15.25) with the overriding scheme

15.4 Conclusions

In the present work, the class of LPV systems and some recent fundamental results, concerning the possibility of resorting to the classical Youla–Kucera parametrization of stabilizing controllers, have been introduced. The presented results have been accompanied by several case studies, to illustrate the potential benefits of the proposed approach in different areas. Future research directions include a full characterization of override control schemes in an LPV framework, an investigation of the benefits of the LPV approach in fault-tolerant multisensor control schemes, [21], and the exploitation of recent set-theoretic results in the time-delay framework, [32], to provide a solution to the control of LPV continuous-time open-loop unstable plants in the presence of time-varying bounded delays.

Appendix

Proof of Theorem 15.2

Proof We sketch the proof of the first part, and of the second part for the case in which the operator is linear. The interested reader is referred to [4] for the general case.

Consider the variables $e(t) = \hat{x}(t) - x(t)$, $x(t)$ and $z(t)$, so that the resulting dynamic system is

$$\begin{aligned}
\sigma(e) &= [A(w) + L(w)C(w)]e \\
\sigma(x) &= [A(w) + B(w)J(w)]x + B(w)J(w)e + B(w)v \\
\sigma(z) &= g(z, o, t) \\
o &= Ce
\end{aligned}$$

In view of the quadratic stabilizability conditions in Theorem 15.1, both matrices $A(w) + L(w)C(w)$ and $A(w) + B(w)J(w)$ are asymptotically stable (since each of them admits a single quadratic Lyapunov function). Hence, the variable $e(t) \to 0$ as $t \to \infty$. This in turn implies that also $o(t) \to 0$ and, since the operator that maps the output error o into v is asymptotically stable, also $v \to 0$. Now, going back to the state evolution, this is governed by an asymptotically stable system fed by two signals that vanish as $t \to \infty$, which is enough to conclude that $x \to 0$.

As far as the converse part is concerned, assume there exists an LPV stabilizing regulator

$$\begin{aligned}
\sigma(q(t)) &= F(w)q(t) + G(w)y(t) \\
u(t) &= H(w)q(t) + K(w)y(t)
\end{aligned} \tag{15.26}$$

so that the closed-loop system

$$\begin{bmatrix} \sigma(x) \\ \sigma(q) \end{bmatrix} = \begin{bmatrix} A(w) + B(w)K(w)C(w) & B(w)H(w) \\ G(w)C(w) & F(w) \end{bmatrix} \begin{bmatrix} x \\ q \end{bmatrix} = A_{cl}(w) \begin{bmatrix} x \\ q \end{bmatrix}$$

is LPV stable and admits a quadratic Lyapunov function. This means that there exists

$$P = \begin{bmatrix} P_{11} & P_{12} \\ P_{12}^\top & P_{22} \end{bmatrix} \succ 0$$

such that (in the continuous-time case), the following Lyapunov inequality is satisfied

$$A_{cl}(w)P + PA_{cl}^\top(w) \prec 0. \tag{15.27}$$

Denoting by $U(w) \doteq K(w)C(w)P_{11} + H(w)P_{12}^\top$, the upper–left block of (15.27) gives

$$\begin{aligned} &(A(w) + B(w)K(w)C(w))P_{11} + (B(w)H(w))P_{12}^\top \\ &\quad + P_{11}(A(w) + B(w)K(w)C(w))^\top + P_{12}(B(w)H(w))^\top \\ &\quad = A(w)P_{11} + P_{11}A(w)^\top + B(w)U(w) + U(w)^\top B(w) \prec 0, \end{aligned}$$

which corresponds to (15.2) with $P = P_{11}$. Similarly, if we pre- and post-multiply (15.27) by $P^{-1} = Q = \begin{bmatrix} Q_{11} & Q_{12} \\ Q_{12}^\top & Q_{22} \end{bmatrix}$, we obtain $QA_{cl}(w) + A_{cl}^\top(w)Q \prec 0$ and, by considering the upper-left block of this expression, condition (15.3) with $Q = Q_{11}$ is obtained for $Y(w) \doteq Q_{11}B(w)K(w) + Q_{12}G(w)$. To conclude, it must be shown that the stabilizing LPV regulator can be realized as (15.6) and (15.13) for a proper stable operator (15.14). To this aim, set

$$\begin{aligned} J(w) &= U(w)P_{11}^{-1} \\ L(w) &= P_{22}^{-1}Y(w) \end{aligned}$$

and consider the observer-based stabilizing regulator (15.6). By resorting to the standard Youla–Kucera parameterization, since for every fixed value of w the resulting closed-loop system is stable, it is known that for each value of w the stabilizing regulator (15.26) can be realized as (15.6) and (15.13) where the stable operator (15.14) is linear, say

$$\begin{aligned} \sigma(z(t)) &= F_o(w)z + G_o(w)o(t) \\ v(t) &= H_o(w)z + K_o(w)o(t) \end{aligned} \tag{15.28}$$

Since the matrices $F_o(w)$ are Hurwitz stable (Schur stable, in the discrete-time case), each of them satisfies the Lyapunov equation

$$F_o(w)^\top P(w) + P(w)F_o(w) = -I \tag{15.29}$$

for $P(w) \succ 0$. Now, let $\Omega(w)$ be the square root of $P(w) = \Omega^\top(w)\Omega(w)$ and apply, for each w, the following similarity transformation:

$$\begin{aligned} \tilde{F}_o(w) &= \Omega(w)F_o(w)\Omega(w)^{-1}, & \tilde{G}_o(w) &= \Omega(w)G_o(w), \\ \tilde{H}_o(w) &= H_o(w)\Omega^{-1}(w), & \tilde{K}_o(w) &= K_o(w) \end{aligned} \tag{15.30}$$

Notice that, in view of the applied transformation, all of the matrices $\tilde{F}_o(w)$ share the same quadratic Lyapunov function with $\tilde{P} = I$, since

$$\tilde{F}_o(w)^\top + \tilde{F}_o(w) = -\Omega^{-\top}(w)\Omega^{-1}(w) \prec 0$$

for every w. This amounts to saying that the Youla–Kucera parameter

$$\begin{aligned} \sigma(z(t)) &= \tilde{F}_o(w)z + \tilde{G}_o(w)o(t) \\ v(t) &= \tilde{H}_o(w)z + \tilde{K}_o(w)o(t) \end{aligned} \tag{15.31}$$

has the same input–output behavior as the "original" one and is LPV stable. □

The constructive proof described above for the determination of the Youla–Kucera parameter is summarized in the next procedure.

Given the LPV plant (15.1) and a family of LPV regulators of the form (15.6)–(15.14), each stabilizing the LPV plant for a constant value of w:

1. solve the LMIs (15.2) and (15.3) (or (15.4) and (15.5) in the discrete-time case);
2. compute the gains (15.7) and (15.8);
3. compute the Luenberger observer-based controller (15.6);
4. for every stabilizing regulator (15.6)–(15.14), compute the Youla–Kucera parameter (15.28);
5. solve the Lyapunov equation (15.29) (or the discrete-time Lyapunov equation, in the discrete-time case) and, for every w, determine the corresponding square root $\Omega(w)$ (such that $P(w) = \Omega^\top(w)\Omega(w)$);
6. apply the suggested transformation to derive the LPV stabilizing Youla–Kucera parameter (15.31).

Note that the described procedure has to be repeated for all $w \in \mathcal{W}$.

References

1. P. Apkarian, P. Gahinet, G. Becker, Self-scheduled $\mathcal{H}_\infty$ control of linear parameter-varying systems: a design example. Automatica **31**(9), 1251–1261 (1995)
2. P. Apkarian, P. Gahinet, A convex characterization of gain-scheduled $\mathcal{H}_\infty$ controllers. IEEE Trans. Automat. Control **40**(5), 853–864 (1995)
3. G. Becker, A. Packard, Robust performance of linear parametrically varying systems using parametrically-dependent linear feedback. Syst. Control Lett. **23**(3), 205–215 (1994)

4. F. Blanchini, D. Casagrande, S. Miani, Parametrization of all stabilizing compensators for absorbable nonlinear systems, in *Proceedings of the 49th IEEE Conference on Decision and Control* (2010), pp. 5943–5948
5. F. Blanchini, D. Casagrande, S. Miani, U. Viaro, Stable LPV realization of parametric transfer functions and its application to gain-scheduling control design. IEEE Trans. Autom. Control **55**(10), 2271–2281 (2010)
6. F. Blanchini, D. Casagrande, S. Miani, U. Viaro, Stable LPV realization of the Smith predictor. Int. J. Syst. Sci. to appear, (2015)
7. F. Blanchini, S. Miani, F. Mesquine, A separation principle for linear switching systems and parametrization of all stabilizing controllers. IEEE Trans. Autom. Control **54**(2), 279–292 (2009)
8. Y. Bolea, V. Puig, J. Blesa, Gain-scheduled Smith predictor PID-based LPV controller for open-flow canal control. IEEE Trans. Control Syst. Technol. **22**(2), 468–477 (2014)
9. S. Boyd, L. El Ghaoui, E. Feron, V. Balakrishnan, *Linear Matrix Inequalities in System and Control Theory* (SIAM, Philadelphia, 2004)
10. C. Briat, Linear Parameter-Varying and Time-Delay Systems. Analysis, Observation, Filtering and Control, Advanced in Delays and Dynamics, vol. 3, Springer, Hidelberg (2015)
11. A. L. Do, J. M. G. Da Silva, O. Sename, L. Dugard, Control design for LPV systems with input saturation and state constraints: an application to a semi-active suspension, in *Proceedings of the 50th IEEE Conference on Decision and Control and European Control Conference* (2011), pp. 3416–3421
12. A.H. Glattfelder, W. Schaufelberger, Stability of discrete override and cascade-limiter single-loop control systems. IEEE Trans. Autom. Control **33**(6), 532–540 (1988)
13. A. Helmerson, μ synthesis and LFT gain scheduling with real uncertainties. Int. J. Robust Nonlinear Control **8**(7), 631–642 (1998)
14. J. Hespanha, A.S. Morse, Switching between stabilizing controllers. Automatica **38**(11), 1905–1917 (2002)
15. D.A. Lawrence, W.J. Rugh, Gain scheduling dynamic controllers for a nonlinear plant. Automatica **31**(3), 381–390 (1995)
16. D. Liberzon, Switching in systems and control, in *Systems and Control: Foundations and Applications*, Birkhäuser (2003)
17. http://www.diegm.uniud.it/smiani/Ongoing/NCS/NCS.html
18. S. Miani, A.C. Morassutti, Switching controllers for networked control systems with packet dropouts and delays in the sensor channel, in *Proceedings of the 1st IFAC Workshop on Estimation and Control of Networked Systems* (2009), pp. 334–339
19. W. Michielis, S.-I. Niculescu, On the delay sensitivity of Smith predictors. Int. J. Syst. Sci. **34**(8–9), 543–551 (2003)
20. J. Mohammadpour, C.W. Scherer (eds.), *Control of Linear Parameter Varying Systems with Applications* (Springer Science and Business Media, LLC, 2012)
21. S. Olaru, J.A. De Donà, M.M. Seron, F. Stoican, Positive invariant sets for fault tolerant multisensor control schemes. Int. J. Control **83**(12), 2622–2640 (2010)
22. A. Packard, G. Becker, Quadratic stabilization of parametrically-dependent linear systems using parametrically-dependent linear, dynamic feedback, in *Advances in Robust and Nonlinear Control Systems* (1992 ASME Winter Annual Meeting, Anaheim, CA, USA, 8–13 November 1992), DCS-vol. 43 (1992), pp. 29–36
23. D. Rotondo, F. Nejjari, V. Puig, Quasi-LPV modeling, identification and control of a twin rotor MIMO system. Control Eng. Pract. **21**(6), 829–846 (2013)
24. W.J. Rugh, J.S. Shamma, Research on gain scheduling. Automatica **36**(10), 1401–1425 (2000)
25. C.W. Scherer, LPV control and full block multipliers. Automatica **37**(3), 361–375 (2001)
26. O. Sename, P. Gaspar, J. Bokor eds. Robust Control and Linear Parameter Varying Approaches. Applications to Vehicle Dynamics, Lecture notes in control and information science, vol. 437, Springer (2013)
27. O. Sename, J. C. Tudon-Martinez, S. Fergani, LPV methods for fault-tolerant vehicle dynamic control, in *Proceedings of the Conference on Control and Fault-Tolerant Systems (SysTol)* (2013), pp. 116–130

28. S.M. Shahruz, S. Behtash, Designing controllers for linear parameter-varying systems by the gain scheduling technique. J. Math. Anal. Appl. **168**(1), 195–217 (1992)
29. J.S. Shamma, An overview of LPV systems, in *Control of Linear Parameter Varying Systems with Applications*, eds. by J. Mohammadpour and C.W. Scherer, Part I, Chapter 1 (Springer, New York, 2012), pp. 63–26
30. J.S. Shamma, M. Athans, Guaranteed properties of gain scheduled control for linear parameter-varying plants. Automatica **27**(3), 559–564 (1991)
31. C. Sloth, T. Esbensen, J. Stoustrup, Robust and fault-tolerant linear parameter-varying control of wind turbines. Mechatronics **21**(4), 645–659 (2011)
32. F. Stoican, S. Olaru, M.M. Seron, J.A. De Donà, A fault tolerant control scheme based on sensor-actuation channel switching and dwell time. Int. J. Robust Nonlinear Control **24**(4), 775–792 (2014)
33. R. Tóth, Modeling and Identification of Linear Parameter-Varying Systems, Lecture notes in control and information science, vol. 403 (Springer 2010)
34. F. Wu, K.M. Grigoriadis, A. Packard, Anti-windup controller design using linear parameter-varying control methods. Int. J Control **73**(12), 1104–1114 (2000)

Chapter 16
Ultimate Bounds and Robust Invariant Sets for Linear Systems with State-Dependent Disturbances

Sorin Olaru and Vasso Reppa

Abstract The objective of this chapter is to present a methodology for computing robust positively invariant sets for linear, discrete time-invariant systems that are affected by additive disturbances, with the particularity that these disturbances are subject to state-dependent bounds. The proposed methodology requires less restrictive assumptions compared to similar established techniques, while it provides the framework for determining the state-dependent (parameterized) ultimate bounds for several classes of disturbances. The added value of the proposed approach is illustrated by an optimization-based problem for detecting the mode of functioning of a switching system.

Keywords Invariant sets · Ultimate bounds · State dependent disturbances

16.1 Introduction

The analysis and control design for linear dynamics with *set constrained disturbances* is a mature subject in control theory, [5, 12]. The analysis of dynamical systems affected by *state- and control-dependent disturbances* is also a well-established subject which can be traced back to the early 1970s, [20, 29], mainly in the stochastic control system framework and latter in works on the mismatched uncertainties, [3, 19]. We recall for example some historical observations made in [20] stating that "the general conclusion is that control-dependent noise calls for conservative control (small gains) while state-dependent noise calls for vigorous control (large gains)."

Results dealing with both *set-theoretic* notions and *state-dependent disturbances* can be found in various studies dedicated to viability theory, [1]; propagation of

S. Olaru (✉) · V. Reppa
Laboratory of Signals and Systems (UMR CNRS 8506),
CentraleSupélec-CNRS-Univ. Paris-Sud, Université Paris-Saclay , 91192 Gif-sur-Yvette, France
e-mail: sorin.olaru@centralesupelec.fr

V. Reppa
e-mail: vasiliki.reppa@centralesupelec.fr

S. Olaru et al. (eds.), *Developments in Model-Based Optimization and Control*,
Lecture Notes in Control and Information Sciences 464,
DOI 10.1007/978-3-319-26687-9_16

parametric uncertainties, [4]; control of system with uncertainties in the parameters as well as in the input, [18]; ultimate boudedness control via set-induced Lyapunov functions, [7]; reachability analysis, [22]; and minimal invariant sets, [9, 25]. In the present paper, we are interested in the characterization of the ultimate boundedness of linear dynamical systems affected by additive disturbances with the particularity that these disturbances are subject to state-dependent bounds.

From the theoretical point of view, there are different connections with the mature literature on minimal invariant sets of dynamic systems with bounded disturbances, [13, 15, 17, 23], and in a broader sense to the set-theoretic methods in control, [6, 8, 10]. From a practical standpoint, the characterization of positive invariant sets can be used for diagnosis. Recently, the set-theoretic methods have been used in model-based fault diagnosis (FD) and the design of fault tolerant control (FTC) laws, [28]. The positive invariance enables the analysis and offers guarantees for FD/FTC performance, that is, robustness, fault detectability and isolability, and fault tolerance under strict set-inclusion or set-separation conditions, [11, 21, 24, 26, 27].

The goal and the main contribution of this chapter is to establish a methodology for computing ultimate bounds and robust positively invariant (RPI) sets for a class of discrete linear systems, affected by disturbances bounded by a state-dependent function. In order to highlight the contribution, in Sect. 16.2 we present some established methodologies for computing state-independent RPI sets and extensions for state-dependent RPI sets and ultimate bounds, which however require more restrictive assumptions compared to those used in the present work. Then, in Sect. 16.3 we describe a new approach for the computation of the state-dependent (parameterized) RPI sets, providing some examples for illustrating both the applicability and limitations of the proposed methodology. In Sect. 16.4, we initially discuss the design of the parameterized RPI sets for some special classes of systems and the state-dependent function that bounds the additive system disturbances. Then, the computation of parameterized sets will be formulated as an optimization problem that can be applied for detecting the system mode switching (e.g., due to a fault). Section 16.5 offers some concluding remarks and future directions.

Notation: $\mathbb{Z}$ is the set of all integers. $\mathbb{Z}_+$ is the set of all nonnegative integers and $\mathbb{Z}_{[k_0,k_1]}$ the set of nonnegative integers in the interval $[k_0, k_1]$. $\mathbb{R}^n$ is the n-dimensional Euclidean space with $\|.\|$ denoting the prescribed norm (Euclidean norm for simplicity). The closed convex hull of a set S will be denoted $Conv\{S\}$. The Minkowski sum of two sets $S_1, S_2 \subset \mathbb{R}^n$ will be denoted by $S_1 \oplus S_2 = \{x_1 + x_2 : x_1 \in S_1, x_2 \in S_2$. The image of $S \subset \mathbb{R}^n$ via $g : \mathbb{R}^n \to \mathbb{R}^m$ is described by the set $g(S) = \{g(x) : x \in S\}$.

Given a function $f : \mathbb{R}^n \to \mathbb{R}$, its *level set* (contour) for $c \in \mathbb{R}$ is defined as

$$\mathcal{L}_f(c) = \left\{x \in \mathbb{R}^n : f(x) = c\right\}, \tag{16.1}$$

while its *sublevel set* for given $c \in \mathbb{R}$ is described as

$$\mathcal{L}_f^-(c) = \left\{x \in \mathbb{R}^n : f(x) \le c\right\}. \tag{16.2}$$

Definition 16.1 A set $S \subset \mathbb{R}^n$ is *star-shaped* in $\bar{x} \in S$ if for any point $x \in S$ and $0 \leq \beta \leq 1$ it holds that $\beta x + (1-\beta)\bar{x} \in S$.

Definition 16.2 A function $f : \mathbb{R}^n \to \mathbb{R}_+$ is called *increasing* from $\bar{x} \in \mathbb{R}^n$ if any sublevel set $\mathcal{L}_f^-(c)$, $c \in \mathbb{R}_+$ is a star-shaped set in $\bar{x}$.

16.2 Background

16.2.1 Problem Formulation

Consider a discrete-time linear system affected by additive uncertainty:

$$x_{k+1} = Ax_k + Bw_k, \tag{16.3}$$

where $x_k \in \mathbb{R}^n$ is the state vector at the time $k \in \mathbb{Z}_+$ and $A \in \mathbb{R}^{n \times n}$, $B \in \mathbb{R}^n$. It is considered that the dynamics in (16.3) correspond to a closed-loop system, for which the exponential stability is guaranteed in the disturbance-free case according to the following assumption:

Assumption 16.1 The matrix A is Schur (all the eigenvalues are inside the unit circle).

In the present chapter we concentrate on the case of a matrix B represented by a column vector, which is related to a signal $w_k \in \mathbb{R}$ representing the additive disturbance. The main characteristic of the additive disturbance will be its boundedness by a state-dependent function $f : \mathbb{R}^n \to \mathbb{R}_+$ such that

$$|w_k| \leq f(x_k). \tag{16.4}$$

Definition 16.3 Let us consider the solution $x_k : \mathbb{Z}_+ \to \mathbb{R}^n$ of (16.3) denoted as $x_k = x(x_0, \mathbf{w}^{0:k-1})$ for a given initial condition x_0 and disturbance sequence $\mathbf{w}^{0:k-1} = \left[w_0, \ldots, w_{k-1}\right] \in \mathbb{R}^k$ satisfying (16.4). We say that

- x_k is *bounded* if there exists a positive constant $d(x_0) < \infty$ such that the inequality $\|x(x_0, \mathbf{w}^{0:k-1})\| \leq d(x_0)$ holds for all $k \in \mathbb{Z}_+$. If the initial condition x_0 can be chosen arbitrarily large then the solutions of (16.3) are *globally bounded.*[1]
- x_k is *ultimately bounded* if there exists a bounded set $S \subset \mathbb{R}^n$, possibly dependent on x_0, and a nonnegative integer $T(x_0, S) < \infty$, such that $x_k \in S$ for all $k \geq T(x_0, S)$.

[1]Often the notion of *global boundedness* is complemented by the attribute *uniform* to emphasize the possible dependence of the bound on the initial condition x_0 but not on the initial moment in time. This addition is superfluous for time-invariant dynamics and will be abandoned here.

- The bounded set S represents an *ultimate bound* for the trajectories initiated in x_0 if the sequence of sets $X_0 = \{x_0\},\ X_{k+1} = AX_k \oplus Bf(X_k)$ satisfies[2]

$$\limsup_{k\to\infty} X_k \subset S.$$

The set S is a *global ultimate bound* if it is an ultimate bound for any $x_0 \in \mathbb{R}^n$.

The problem to be considered in this paper can be outlined as follows:

Objective 1: Find sufficient conditions for ultimate boundedness of the state trajectories satisfying (16.3) taking into account (16.4).

Objective 2: Describe the parameterization of ultimate bounds with respect to the initial conditions, when these ultimate bounds are not global. In this respect, we seek to express ultimate bounds in terms of compact and convex sets.

From the analysis point of view, the robust positive invariance of a set with respect to the system dynamics is another important notion in the present work.

Definition 16.4 (*RPI set*) A set $\Omega \subset \mathbb{R}^n$ is a *robust positively invariant (RPI) set* with respect to (16.3) if $Ax + Bw \in \Omega$ for all $x \in \Omega$ and for all w satisfying $|w| \leq f(x)$.

Even if an ultimate bound set is not guaranteed to be RPI, we will be interested in finding sets which are both RPI and ultimate bounds for the state trajectories.

16.2.2 A Detour on the State-Independent Disturbance Bounds Case

Strong results are available with respect to the RPI sets description in the case of state-independent bounds on disturbance. Let us consider the class of dynamics presented in (16.3) with

$$w_k \in F, \tag{16.5}$$

with a closed, convex bounded set $F \subset \mathbb{R}$. The state-independent bounds can be interpreted as a particular case of (16.4), with a constant limiting function $f(x) = \bar{w}$ which leads to $F = \{w : |w| \leq \bar{w}\}$.

In order to differentiate the robust positive invariance in the case of state-independent bounds on disturbances (16.5) from the state-dependent counterpart in (16.4), the following definitions are introduced.

Definition 16.5 (*F-invariance*) A set $\Omega \subset \mathbb{R}^n$ is *F-invariant* with respect to (16.5) if $Ax + Bw \in \Omega$ for all $x \in \Omega$ and all $w \in F$.

---

[2]The set S is a *proper superset* of $\limsup_{k\to\infty} X_k$. The meaning of the outer limit (lim sup) is particular in this context as it is understood in a set-theoretic framework [2] as the set of cluster points of sequences $x(x_0, \mathbf{w}^{0:k-1}) \in X_k$.

Definition 16.6 *The minimal F-invariant set*, denoted by $M(F)$, is defined as the F-invariant set contained in any closed F-invariant set.

The minimal F-invariant set is unique, compact and contains the origin if F contains the origin. Its ϵ-neighborhood (i.e., $M(F) \oplus \mathbb{B}^n_\epsilon$, where $\mathbb{B}^n_\epsilon$ is the ϵ ball in $\mathbb{R}^n$) represents an ultimate bound for (16.5). It is well known [8, 15, 17] that a Schur matrix A leads to a minimal F-invariant set described as

$$M(F) = \lim_{k\to\infty} \bigoplus_{i=0}^{k} A^i B F. \tag{16.6}$$

16.2.3 F-invariant Sets for the Case of State-Dependent Disturbances

The work in of Kuntsevich and Pshenichnyi, [17], concentrates principally on dynamic systems affected by bounded disturbances as in Sect. 16.2.2, but presents an extension to systems with bounded nonlinearity which gives a basic idea on the possible analysis of the state-dependent disturbances in a set-theoretic framework. We reformulate here the main developments by adapting the construction in [17] to the present framework in order to analyze its mechanism and stress the working hypothesis.

Assumption 16.2 The state-dependent bound on the disturbances for system (16.3) is described by a function $f(.)$ for which $f(\mathbb{R}^n) = \{f(x) : x \in \mathbb{R}^n\}$ is bounded.

Under the Assumption 16.2 on the global bounds of the disturbance, one can initialize a set sequence by choosing $M_0 = \mathbb{R}^n$ and $F_0 = Conv\{f(M_0)\}$. By assuming M_j to be F_j-invariant for $F_j = Conv\left\{f(M_j)\right\}$ (which trivially holds for M_0) the following set sequence can be defined:

$$M_{j+1} = M(F_j), \quad j = 0, 1, \ldots, \infty \tag{16.7}$$

as the minimal F_j-invariant for

$$x_{k+1} = Ax_k + Bw_k, \quad w_k \in F_j. \tag{16.8}$$

Exploiting the inclusion $M_{j+1} \subset M_j$, one can conclude on the existence of convex sets defined as

$$M_\infty = \bigcap_{j=0}^{\infty} M_j \quad \text{and } F_\infty = \bigcap_{j=0}^{\infty} F_j \tag{16.9}$$

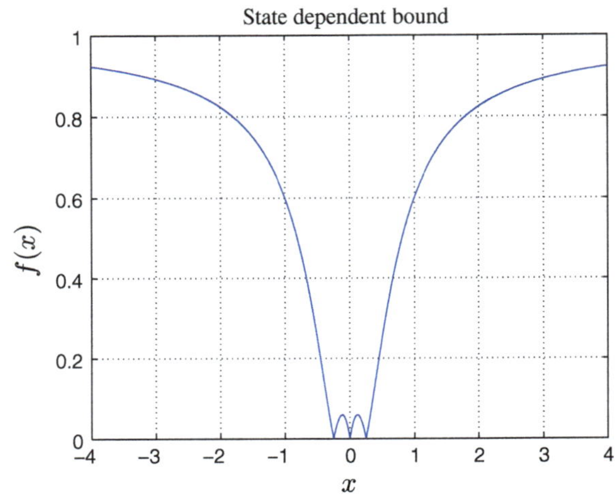

Fig. 16.1 Graphical description of the state-dependent bound for the dynamics in (16.10)

Theorem 16.1 ([17]) *Under the Assumptions 16.1 and 16.2, each of the sets M_j, $j = 0, 1, \dots, \infty$ is a robust positively invariant set for (16.3). If the initial state x_0 belongs to M_j for some $j \in \mathbb{Z}_+$, then the trajectory of the system (16.3) reaches in finite-time the ϵ-neighborhood of the set M_{j+1} and remains in that neighborhood. Thus the system is ultimately bounded and the ϵ-neighborhood of M_∞ represents an ultimate bound.* ■

The following remarks motivate our study.

Remark 16.1 The boundedness restrictions in Assumption 16.2 are relatively conservative. The procedure proposed in [17] cannot be initialized in the presence of radially unbounded functions[3] describing the state-dependent limitations on the uncertainties (16.4) as long as M_0 will be unbounded.

Remark 16.2 Even in the case that Assumption 16.2 holds, Theorem 16.1 proposes a sequence of invariant sets but the asymptotic construction of M_∞ leads in practice to an approximation of the limit set (in fact its ϵ-neighborhood) which represents a *state-independent ultimate bound.* This raises a question about the existence of a state-dependence (or parameterization) of the ultimate bounds with respect to the initial conditions. Note that a parametrization can contribute to the reduction of conservativeness in the description of the ultimate bounds.

There are simple examples to show that the above procedure can be refined by computing *state-dependent ultimate bounds*. Consider the dynamics

$$x_{k+1} = 0.5x_k + w_k, \quad |w_k| \le f(x_k) = \left| \frac{4x_k^2 - |x_k|}{4x_k^2 - 1} \right|, \tag{16.10}$$

[3]A function $f : \mathbb{R}^n \to \mathbb{R}$ is radially unbounded if $\|x\| \to \infty \implies f(x) \to \infty$.

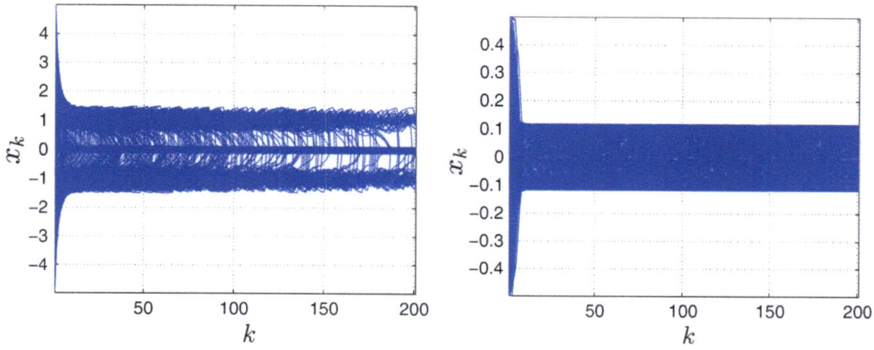

Fig. 16.2 Simulations for the dynamics in (16.10) with random initial conditions in $[-5; 5]$—*left*; $[-0.5; 0.5]$—*right*

where $f(x)$ is shown in Fig. 16.1. Using the methodology in [17] that is described through (16.7)–(16.9), the nested set construction results in $M_\infty = [-1.5; 1.5]$ and fails to identify all the intervals $[-c; c]$ with $\frac{\sqrt{5}-2}{2} \leq c \leq 0.5$ which are robust positively invariant and represent tighter ultimate bounds if $x_0 \in [-c; c]$.

Figure 16.2 presents the time-domain simulations for different initial conditions illustrating the existence of invariant sets $[-1.5; 1.5]$ and $[-\frac{\sqrt{5}-2}{2}; \frac{\sqrt{5}-2}{2}]$.

16.2.4 Jordan Decomposition Approach for the Design of Ultimate Bounds in the Presence of State-Dependent Disturbances

The work of Kofman et al in [14] expresses or maybe deals with the computation of the ultimate bounds for linear systems with state-dependent perturbation bounds. The systematic method for their characterization under monotonicity conditions is provided in the next theorem.

Theorem 16.2 ([14]) *Consider the system (16.3), satisfying Assumption 16.1 with a Jordan canonical form of the transition matrix* $A = V^{-1}\Lambda V$. *Suppose that*

$$|Bw| \leq g(|x|),\ \textit{for all}\ x \in \mathbb{R}^n \tag{16.11}$$

with a continuous map $g : \mathbb{R}^n_+ \to \mathbb{R}^n_+$ *satisfying*

$$|y_1| \leq |y_2| \implies g(|y_1|) \leq g(|y_2|). \tag{16.12}$$

Consider the map:

$$T(y) \triangleq |\Lambda|y + |V^{-1}|g(|V|y) \tag{16.13}$$

and suppose that a point b satisfying $b = T(b)$ *exists. Let* $y_m \in \mathbb{R}^n$ *denote any point satisfying* $\lim_{k\to\infty} T^k(|V^{-1}y_m|) = b$. *If the initial condition* x_0 *satisfies* $|V^{-1}x_0| \leq |V^{-1}y_m|$, *then, for any* $\epsilon \in \mathbb{R}^n_+$, *there exists* $l = l(\epsilon, y_m)$ *such that for all* $k \geq l$,

$$|V^{-1}x_k| \leq b + \epsilon \tag{16.14}$$

The result is remarkable in several respects. By building the ultimate bounds on the existence of fixed points for (16.13), the theorem opens the door for the characterization of different ultimate bounds (16.14) for the same dynamics. Such ultimate bound has an associated domain of attraction identified via the collection of points y_m in the statement. Moreover, the ultimate bound (16.14) is robust positively invariant. All these characteristics provide suitable solutions with respect to the objectives of the present work. The assumptions are however relatively restrictive as detailed in the following remarks. This fact motivates the main results presented in the next section.

Remark 16.3 The monotonicity condition (16.12) is not fulfilled by the *increasing functions* from a particular point $\bar{x}$ (see Definition 16.2), which are related to the star-shape property of the level set of the boundary function, and not to classical convexity of the level set of the bounding function. The monotonic functions are only a particular subclass of these increasing functions.

Remark 16.4 Theorem 16.2 builds on a fixed-point type of condition. This condition involves the map (16.13) which in turn builds on a series of over-approximations of the nonlinear state-dependent bounding function as for example those related to the component-wise absolute values of the matrices V and V^{-1}. Moreover, a qualitative analysis of the number of such fixed points and the correspondent bassins of attraction are only implicitly embedded in the result. The set-theoretic methods can offer alternative condition and describe parameters for the existence of ultimate bounds.

16.3 Main Result

For a Schur matrix $A \in \mathbb{R}^{n\times n}$ and vector $B \in \mathbb{R}^n$ let us define the limit set

$$\mathbb{M} = M(\mathbb{B}_1) = \lim_{k\to\infty} \bigoplus_{i=0}^{k} A^i B\mathbb{B}_1, \tag{16.15}$$

where $\mathbb{B}_1$ stands for the unit ball.[4] The set $\mathbb{M}$ is *minimal robust positively invariant (RPI)* set with respect to $x_{k+1} = Ax_k + Bw_k \; forall w_k \in \mathbb{B}_1$, or equivalently for all $|w_k| \leq 1$.

In order to establish the main results we consider the Minkowski function $g_{\mathbb{M}} : \mathbb{R}^n \to \mathbb{R}_+$ associated to the set $\mathbb{M}$:

$$g_{\mathbb{M}}(x) = inf \{\lambda \in \mathbb{R}_+ : x \in \lambda\mathbb{M}\}. \tag{16.16}$$

Theorem 16.3 *Let us consider the LTI system (16.3) with the state-dependent bound $f(x)$ presented in (16.4), an increasing function from $\bar{x} = 0 \in \mathbb{R}^n$ in the sense of Definition 16.2. Consider the Minkowski function $g_{\mathbb{M}}(x)$ of the set $\mathbb{M}$ defined through (16.15)–(16.16), where the matrices (A, B) characterize system (16.3). The parameterized set*

$$\Omega_{\mathbb{M}}(\alpha) = \alpha\mathbb{M} \tag{16.17}$$

is RPI with respect to (16.3) for any scalar positive parameters α in the set

$$S_{\mathbb{M}} = \left\{\alpha \in \mathbb{R}_+ | f(x) \leq g_{\mathbb{M}}(x) \; \forall x \in \mathcal{L}_{g_{\mathbb{M}}}(\alpha)\right\} \tag{16.18}$$

where $\mathcal{L}_{g_{\mathbb{M}}}(.)$ is the level set defined according to (16.1).

Proof First note that $\Omega_{\mathbb{M}}(\alpha)$ is invariant with respect to $x_{k+1} = Ax_k + Bw_k$, with $w_k \in \alpha\mathbb{B}_1$. Indeed using the definition (16.15) we have

$$\Omega_{\mathbb{M}}(\alpha) = \alpha\mathbb{M} = \alpha \lim_{k\to\infty} \bigoplus_{i=0}^{k} A^i B\mathbb{B}_1 = \lim_{k\to\infty} \bigoplus_{i=0}^{k} A^i B \{\alpha\mathbb{B}_1\}, \tag{16.19}$$

which shows that $\Omega_{\mathbb{M}}(\alpha)$ is RPI with respect to (16.3) when $w_k \in \alpha\mathbb{B}_1$.

Consider now a scalar $\alpha \in S_{\mathbb{M}}$. From the definition of the set $S_{\mathbb{M}}$ in (16.18),

$$f(x_1) \leq g_{\mathbb{M}}(x_1) = \alpha, \; \forall x_1 \in \mathcal{L}_{g_{\mathbb{M}}}(\alpha). \tag{16.20}$$

The fact that $f(x)$ is an increasing function from $0 \in \mathbb{R}^n$ ensures on one hand that $\mathcal{L}_f^-(c_1) \supseteq \mathcal{L}_f^-(c_2) \supset \{0\}$ if $c_1 \geq c_2 > 0$, and on the other hand the star-shape property of the sublevel set of $f(.)$. Exploiting this last property, for any $x_2 \in \Omega_{\mathbb{M}}(\alpha)$ there exists a scalar $0 \leq \beta \leq 1$ such that $x_2 = \beta x_1$ and $x_1 \in \mathcal{L}_{g_{\mathbb{M}}}(\alpha)$. From the definition of the sublevel set (16.1) we have

$$f(x_2) \leq f(x_1) \tag{16.21}$$

---

[4]The unit ball is defined with respect to a predefined norm $|.|_p$. In the present case the matrix $B \in \mathbb{R}^{n\times 1}$ and thus the corresponding unit ball is defined in $\mathbb{R}$ where the $|.|_p$ are equivalent for $p \in [1, \infty)$.

From (16.20) and (16.21) it yields $f(x_2) \leq \alpha$ and subsequently

$$|w_k| \leq \alpha \; \forall x_k \in \alpha\mathbb{M} = \Omega_{\mathbb{M}}(\alpha). \tag{16.22}$$

Thus, the proof of invariance of $\Omega_{\mathbb{M}}(\alpha)$ with respect to (16.3)–(16.4) is completed. ∎

The problems formulated in Sect. 16.2.1 can be addressed in light of Theorem 16.3. The next corollaries resume these basic sufficient conditions for the existence of ultimate bounds represented by convex sets and their parametrization with respect to the initial conditions.

Corollary 16.1 *The solution $x_k = x(x_0, \mathbf{w}^{0:k-1})$ of (16.3) is globally ultimately bounded if the set $S_{\mathbb{M}}$ in (16.18) exists and is unbounded. Additionally, if $S_{\mathbb{M}} = \mathbb{R}_+$ then the origin is a robustly asymptotically stable equilibrium point.*

Proof If the set $S_{\mathbb{M}}$ is unbounded, then there exists a subset $[\bar{c}, \infty) \subseteq S$ such that $\Omega_{\mathbb{M}}(\alpha)$ is RPI for all $\alpha \in [\bar{c}, \infty)$. But these sets are also attractive (in the virtue of the properties of the minimal RPI sets with constant bounds) and by consequence $\Omega_{\mathbb{M}}(\bar{c})$ will represent a global ultimate bound for the state trajectories. For the second part of the corollary, it is easy to observe that $\bar{c} = 0$ and thus any neighborhood of the origin $[-\epsilon, \epsilon]$ can be reached in a finite number of iterations independently of the initial conditions. It follows that $x_k \to 0$ as $k \to \infty$. ∎

Corollary 16.1 offers a sufficient condition for global ultimate boundedness and robust asymptotic stability in the presence of state-dependent disturbances. It is worth to mention that this condition admits state-dependent bounds described by radially unbounded functions $f(.)$ (which is not the case of Theorem 16.1). Indeed, the only condition to be satisfied in Theorem 16.3 is $f(x) < g_{\mathbb{M}}(x)$, $\forall x \in \mathbb{R}^n$ and, by definition, the function $g_{\mathbb{M}}(.)$ is radially unbounded.

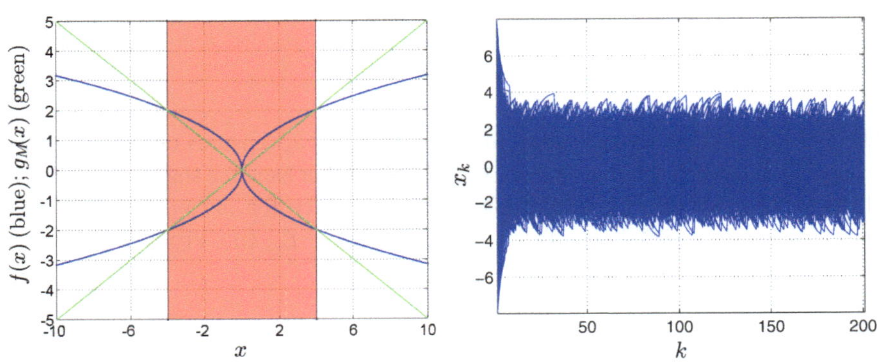

Fig. 16.3 *Left* the graph of $f(x)$ (*blue*) and $g_{\mathbb{M}}(x)$ (*green*)—describing the envelope (positive and negative) bounds of the disturbances for Example 16.1. The *red* interval represents the region for which $f(x) > g_{\mathbb{M}}(x)$ and thus correspond to scaling factors $\alpha \notin S$. *Right* time simulations with random initial conditions in $[-8, 8]$

Example 16.1 Consider the dynamical system:

$$x_{k+1} = 0.5x_k + w_k, |w_k| \leq f(x_k) = |x_k|^{\frac{1}{2}}. \tag{16.23}$$

Figure 16.3 shows the relationship between $f(x)$ and $g_{\mathbb{M}}(x)$. The last one represents the Minkowski function that corresponds to the minimal invariant set $\mathbb{M} = [-2, 2]$ in (16.15). Note that the function $f(x)$ is radially unbounded and thus Theorem 16.1 cannot be applied. Using Theorem 16.3 one can describe the set of admissible parameters $S_{\mathbb{M}} = [2, \infty)$ leading to admissible invariant sets $\Omega_{\mathbb{M}}(\alpha) = \alpha[-2; 2]$, $\forall \alpha \in S_{\mathbb{M}}$ and the global ultimate bound $\Omega_{\mathbb{M}}((1+\epsilon)2) = 2(1+\epsilon)[-2; 2] = (1+\epsilon)[-4; 4]$, $\epsilon > 0$. Note also, in the virtue of the Corollary 16.1, that the origin is not a robust asymptotically stable equilibrium point for this example.

Corollary 16.2 *Consider the system (16.3) under the assumptions of Theorem 16.3 and the set $S_{\mathbb{M}} \subseteq \mathbb{R}_+$ in (16.18). If $S_{\mathbb{M}}$ is unbounded and described by a (possibly infinite) union of disjoint intervals*

$$S_{\mathbb{M}} = [\underline{c}_1, \bar{c}_1) \cup [\underline{c}_2, \bar{c}_2) \cup \cdots \cup [\underline{c}_i, \bar{c}_i) \ldots \tag{16.24}$$

with $0 = \bar{c}_0 \leq \underline{c}_1 < \bar{c}_1 < \underline{c}_2 < \bar{c}_2 < \ldots \underline{c}_i < \bar{c}_i < \ldots$, then $\Omega_{\mathbb{M}}((1+\epsilon)\underline{c}_{i+1})$ is an ultimate bound for $\epsilon \in \left(0, \frac{\bar{c}_{i+1} - \underline{c}_{i+1}}{\underline{c}_{i+1}}\right)$ and any $x_0 \in \Omega_{\mathbb{M}}(c)$ with $c \in [\bar{c}_i, \bar{c}_{i+1})$.

Proof We split the interval $[\bar{c}_i, \bar{c}_{i+1}) = [\bar{c}_i, \underline{c}_{i+1}) \cup [\underline{c}_{i+1}, \bar{c}_{i+1})$. On one hand, for any initial condition in $\Omega_{\mathbb{M}}(\underline{c})$ with $\underline{c} \in [\underline{c}_{i+1}, \bar{c}_{i+1})$, an ultimate bound can be obtained using the ϵ-outer approximation of $\Omega_{\mathbb{M}}(\underline{c}_{i+1})$ with a similar argument used in Corollary 16.1. On the other hand, it can be observed that $x_0 \in \Omega(c) \subset \Omega(\underline{c}_{i+1})$ for any $c \in [\bar{c}_i, \underline{c}_{i+1})$ and the ultimate boundedness of the trajectories follows from the robust positive invariance of of $\Omega(\underline{c}_{i+1})$. ■

The construction of the set $S_{\mathbb{M}}$ in Theorem 16.3 might be a difficult task as long as it involves the Minkowski function of minimal robust positive invariant set $\mathbb{M}$ in (16.6). It is known that $\mathbb{M}$, being the limit set of an infinite Minkowski sum, has a finite (explicit) representation in terms of generators only for restricted classes of LTI dynamics (in the case $A^k B = \alpha B$ for some $k \in \mathbb{N}_+$ and $0 \leq \alpha \leq 1$). As such, for practical reasons, the use of approximations is enabled along the lines of the next results.

Theorem 16.4 *Let $\mathbb{U}$ be a* polyhedral *RPI set with respect to $x_{k+1} = Ax_k + Bw_k$ with $w_k \in \mathbb{B}_1$. The parameterized set $\Omega_{\mathbb{U}}(\alpha) = \alpha\mathbb{U}$ is RPI with respect to (16.3) for all α in*

$$S_{\mathbb{U}} = \left\{\alpha \in \mathbb{R}_+ | f(x) \leq g_{\mathbb{U}}(x)\ \forall x \in \mathcal{L}_{g_{\mathbb{U}}}(\alpha)\right\} \tag{16.25}$$

The proof is similar to the one in Theorem 16.3 and is omitted.

Example 16.2 Consider the second-order dynamical system:

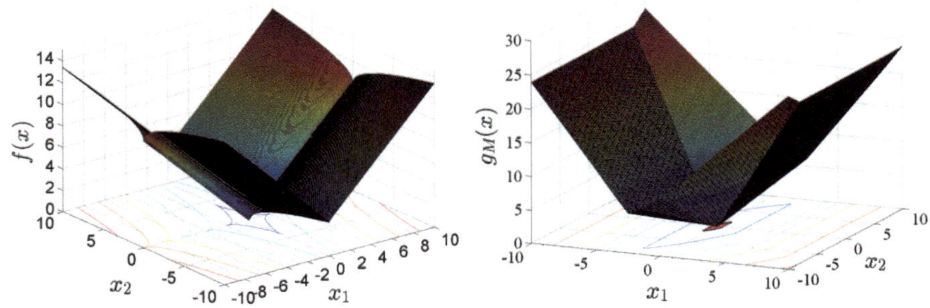

Fig. 16.4 *Left* the graph of $f(x)$. *Right* the graph of $g_{\mathbb{U}}(x)$ in Example 16.2

$$\begin{bmatrix} x_{1,k+1} \\ x_{2,k+1} \end{bmatrix} = \begin{bmatrix} 0.2 & 0.2 \\ 0 & 0.4 \end{bmatrix} \begin{bmatrix} x_{1,k} \\ x_{2,k} \end{bmatrix} + \begin{bmatrix} 0 \\ 1 \end{bmatrix} w_k, \tag{16.26}$$

where $|w_k| \leq f(x_k) = |x_{1,k}| + 0.1|x_{1,k}|^{0.2}|x_{2,k}|^{0.2} + |x_{2,k}|^{0.5}$ and $f(.)$ fulfills the increasing from 0 assumption as illustrated in Fig. 16.4. The minimal invariant set $\mathbb{M}$ in this case will be replaced by a tight outer invariant approximation $\mathbb{U}$, illustrated in Fig. 16.4 with its level sets.[5] The superposition of the functions $f(x)$ and $g_{\mathbb{U}}(x)$ is given in Fig. 16.5 where it can be seen that their intersection is done along non-convex curves. At the bottom of the same figure, we can see the regions for which $f(x) \geq g_{\mathbb{U}}(x)$ together with a value of the scalar α such that the parameterized set $\Omega_{\mathbb{U}}(\alpha)$ is guaranteed to be RPI. This provides an exemplification of the analysis tools available via Theorem 16.3.

Remark 16.5 For constructing the parameterized set $\Omega_{\mathbb{U}}(\alpha)$, one can use low complexity invariant approximation, [14, 21, 23], of the set $\mathbb{M}$ in (16.6), as for example,

$$\mathbb{U} = \left\{ x : |V^{-1}x| \leq (I - |\Lambda|)^{-1}|V^{-1}||B| \right\} \tag{16.27}$$

with $\Lambda = V^{-1}AV$, corresponding to the Jordan canonical form of the transition matrix in (16.3). The function $g_{\mathbb{U}}(.)$ corresponds to a polyhedral cone and is piecewise linear over a cone partition of the state space.

Proposition 16.1 *Let $\mathbb{M}$ be the minimal RPI set with respect to $x_{k+1} = Ax_k + Bw_k$ with $w_k \in \mathbb{B}_1$. If $\mathbb{U}$ is a polyhedral RPI approximation of $\mathbb{M}$, then $S_{\mathbb{M}} \supseteq S_{\mathbb{U}}$, where $S_{\mathbb{M}}$ and $S_{\mathbb{U}}$ are constructed based on (16.18) and (16.25) for a given function $f(.)$ increasing from* 0.

Proof Note that $\mathbb{U} \supseteq \mathbb{M}$ based on the properties of the minimal RPI set. This fact implies $g_{\mathbb{U}}(x) \leq g_{\mathbb{M}}(x)$ $\forall x \in \mathbb{R}^n$ and this relationship can be related to the inequalities involved in (16.18)–(16.25) where $f(x) \leq g_{\mathbb{U}}(x)$ only if $f(x) \leq g_{\mathbb{M}}(x)$. Under

[5]This particular function is increasing from $\bar{x} = 0$ in the sense of Definition 16.2 but not monotonic according to (16.12) and thus the hypothesis of Theorem 16.2 is not satisfied in this case.

the star-shaped assumption for $f(x)$ it follows that $S_{\mathbb{M}} \supseteq S_{\mathbb{U}}$ and thus the approximation will be inherited by the parameterized set of RPI sets for which the sufficient conditions hold. ■

An illustration of the impact of the invariant set approximation on the function entering in the comparison with the state-dependent bound in Theorem 16.3 is given in Fig. 16.6.

Remark 16.6 Theorem 16.3 builds on the assumption of a bounding function increasing from 0 which is satisfied for Examples 16.1 and 16.2. However, the system in (16.10) violates this assumption which is based on the star-shape property of the sublevel set. Indeed, a simple check shows that

$$\mathcal{L}_f^-(0.05) = [-0.071; 0.071] \cup [0.167; 0.306] \cup [-0.306; -0.167].$$

Another example of bounding function which does not satisfy the increasing assumption will be the Himmelblau's function for the bound of the disturbance with respect the system dynamics in (16.26). This function is presented in Fig. 16.7 with the corresponding contour (level sets) which are non-connected and cannot lead to RPI sets centered in the origin.

In order to apply Theorem 16.3 for any state-dependent bound on the disturbance (see Remark 16.6), embedding via a star-shaped envelope can be used as follows:

Proposition 16.2 *Let $f : \mathbb{R}^n \to \mathbb{R}_+$ and a point $\bar{x} \in \mathbb{R}^n$. The function $h : \mathbb{R}^n \to \mathbb{R}_+$ defined as*

$$h(x) = \max_{0 \le \gamma \le 1} f(\gamma x + (1 - \gamma)\bar{x}) \tag{16.28}$$

is increasing from $\bar{x}$ and $f(x) \le h(x)$ $\forall x \in \mathbb{R}^n$.

Proof Proposition 16.2 can be proved via the direct application of the star-shape properties described in Definitions 16.1–16.2. ■

For the system in (16.10), the use of the star-shape embedding (Fig. 16.8) leads to the identification of the set of admissible parameters $S_{\mathbb{M}} = [\frac{\sqrt{5}-2}{4}, 0.25) \cup [0.75, \infty)$ for the parameterized invariant sets $\Omega_{\mathbb{M}} = \alpha[-2, 2]$, $\alpha \in S_{\mathbb{M}}$ illustrated by the comparison between $h(x)$ and $g_{\mathbb{M}}(x)$ in Fig. 16.9.

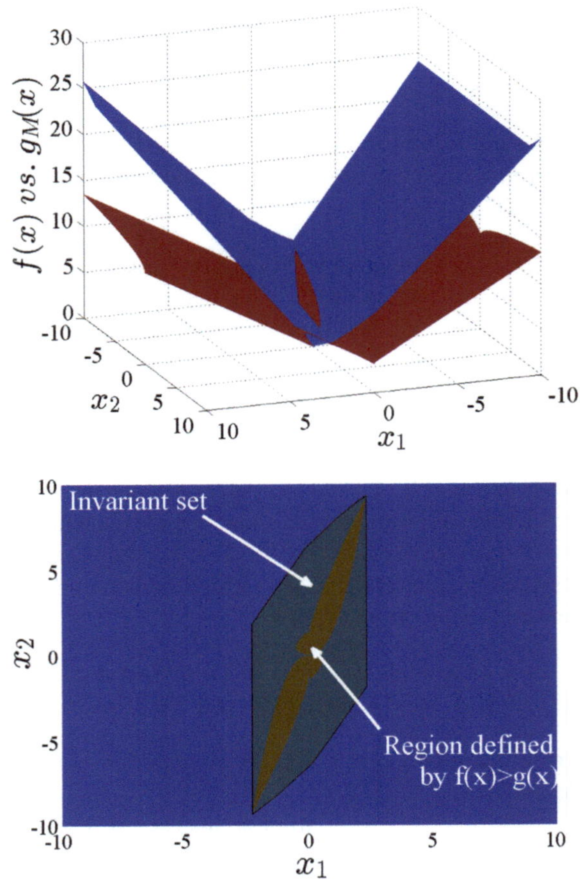

Fig. 16.5 *Top* the graph of $f(x)$ (*red*) and $g_{\mathbb{M}}(x)$ (*blue*) in Example 16.2. *Bottom* 2D illustration of the shape of the RPI set in comparison with the region for which $f(x) \geq g_{\mathbb{U}}(x)$

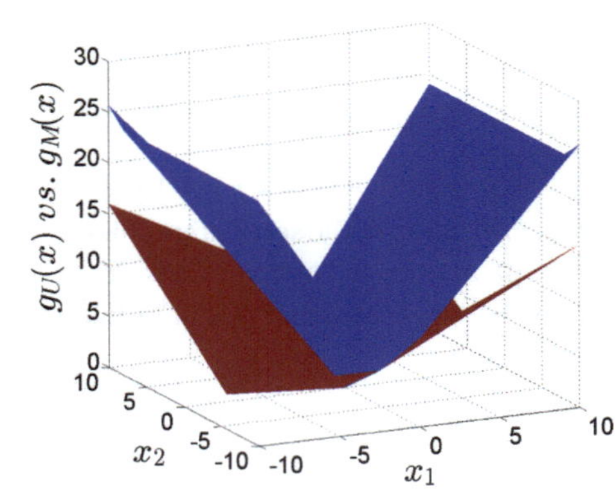

Fig. 16.6 The comparision of a tight approximation of the graph of $g_{\mathbb{M}}(x)$ (*blue*) based on the tight approximation of the minimal RPI set in (16.6) and $g_{\mathbb{U}}(x)$ (*red*) based on (16.27) for the LTI system in Example 16.2

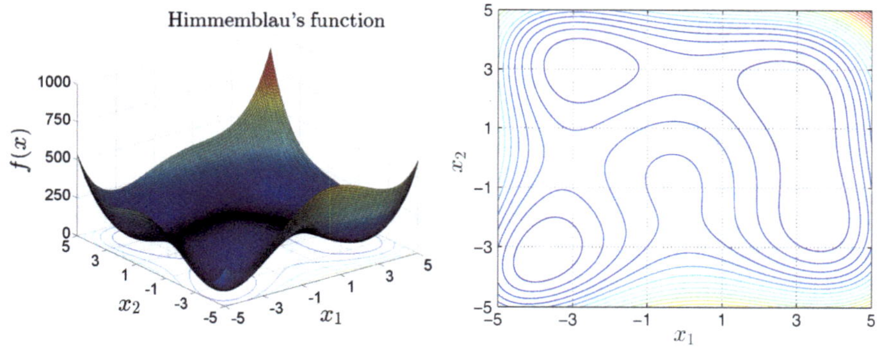

Fig. 16.7 *Left* the graph of Himmelblau's function $f([x_1 x_2]^T) = (x_1^2 + x_2 - 11)^2 + (x_1 + x_2^2 - 7)^2$. *Right* the corresponding (non-connected) level sets

Fig. 16.8 $f(x)$ (*blue*) in (16.10) and its star-shaped embedding $h(x)$ (*red*)

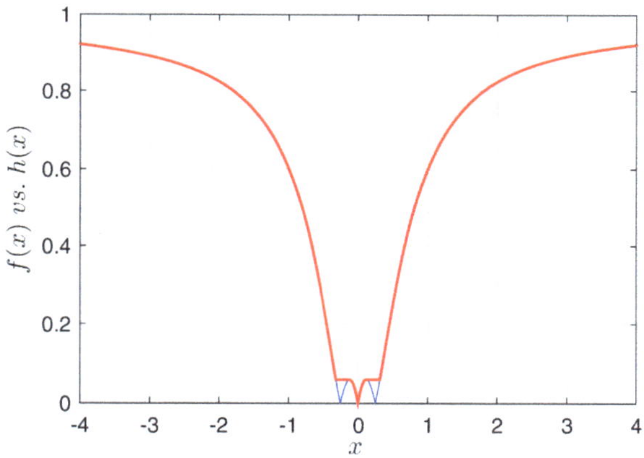

Fig. 16.9 $h(x)$ (*red*) compared to the Minkowski function $g_{\mathbb{M}}(x)$ (*green*)

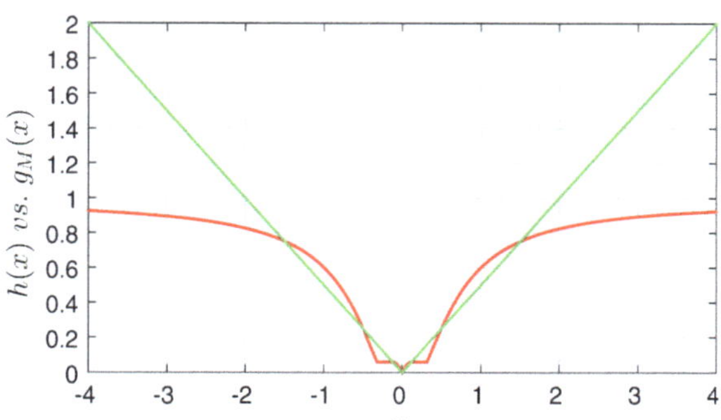

16.4 Extensions and Connections with Optimization-Based Design

16.4.1 Extensions

Based on the main results of this study presented in Sect. 16.3, it is worth to determine the parameterized sets $\Omega_{\mathbb{M}}(\alpha)$ for special cases of the bound-function $f(x)$ summarized in the following corollaries.

Corollary 16.3 *Consider system (16.3). If $f(x)$ is convex, then $\Omega_{\mathbb{M}}(\alpha) = \alpha\mathbb{M}$ is an invariant set for all $\alpha_m \leq \alpha \leq \alpha_M$ with*

$$\alpha_m = \inf\left\{\alpha \in \mathbb{R}_+ | f(x) \leq g_{\mathbb{M}}(x)\ \forall x \in \mathcal{L}_{g_{\mathbb{M}}}(\alpha)\right\} \quad (16.29)$$

$$\alpha_M = \sup\left\{\alpha \in \mathbb{R}_+ | f(x) \leq g_{\mathbb{M}}(x)\ \forall x \in \mathcal{L}_{g_{\mathbb{M}}}(\alpha)\right\} \quad (16.30)$$

Definition 16.7 The nonlinear function $f : \mathbb{R}^n \to \mathbb{R}_+$ is continuous and cone-bounded over $\mathbb{R}^n$, if there exist nonnegative constants λ_0 and λ_1 such that

$$\|f(x)\| \leq \lambda_0 + \lambda_1\|x\|\ \forall x \in \mathbb{R}^n. \quad (16.31)$$

Corollary 16.4 *Consider the system (16.3) where the additive uncertainties satisfy (16.4) with a cone-bounded function $f(.)$ as described in Definition 16.7. The parameterized set $\Omega_{\mathbb{M}}(\alpha) = \alpha\mathbb{M}$ is RPI for all $\alpha \geq \alpha_m$ with α_m given by (16.29).*

Remark 16.7 In [16], several classes of uncertain nonlinear dynamics have been mentioned in the context of state-dependent uncertainties, as for example:

$$x_{k+1} = Ax_k + c_k g(x_k) + z_k, \quad (16.32)$$

with $c_k, z_k \in \mathbb{R}^n$ satisfying elementwise the inequality $\|c_k\| \leq C_{\max}$, $\|z_k\| \leq Z_{\max}$ with $C_{\max}, Z_{\max} \in \mathbb{R}^n_+$, and a scalar function of a vector argument $g : \mathbb{R}^n \to \mathbb{R}_+$. The system (16.32) is an example of dynamics which can be regarded as linear subject to a cone-bounded uncertainty and parameterized sets $\Omega_{\mathbb{M}}(\alpha)$ can be determined according to Corollary 16.4.

Corollary 16.5 *Consider the system*

$$x_{k+1} = Ax_k + Bw_k + B_u\bar{u},\ |w_k| \leq f(x_k), \quad (16.33)$$

satisfying Assumption 16.1, with $f(.)$ an increasing function from $(I - A)^{-1}B_u\bar{u}$, $B_u \in \mathbb{R}^n$ and a constant signal $\bar{u}$. The parameterized set $\Omega_{\mathbb{M}}(\alpha, \bar{u}) = \alpha\mathbb{M} \oplus (I - A)^{-1}B_u\bar{u}$ is RPI for all $\alpha \in \mathcal{S}_{\mathbb{M}}(\bar{u})$ with $\mathcal{S}_{\mathbb{M}}(\bar{u})$ computed based on Theorem 16.3 for the system $\xi_{k+1} = A\xi_k + Bw_k$ subject to constraints $|w_k| \leq f(\xi_k + (I - A)^{-1}B_u\bar{u})$.

Proof We observe that in the absence of disturbances, the trajectories converge to $\bar{x} = (I - A)^{-1}B_u\bar{u}$. Then the analysis can be done with respect to the shifted dynamics $x_k = \bar{x} + \xi_k$ with the particularity that the function $g_{\mathbb{M}}(\xi_k)$ is star-shaped in 0 while the original bounding function $f(.)$ is described in the original state space and thus it is computed according to the change of variable $f(\xi_k + \bar{x})$. ■

16.4.2 Detection of Mode Switching via Set Invariance

The positive invariance of a set with respect to the nominal dynamics is a strong notion and can be exploited for the detection of a switch in the dynamics, [28]. The basic idea is to construct off-line the family of invariant sets and monitor in real-time the inclusion of the state in the respective set. In case that the invariance is violated, a change of mode is detected. Subsequently, the convergence to a different invariant (limit set) can lead to the identification of the current mode of functioning. This mechanism has been documented and is well understood for linear dynamical systems in the presence of bounded disturbances. We will show in the next paragraphs the that way the theoretical developments on the state-dependent ultimate bounds can be used in practice.

The off-line construction of the family of invariant sets is realized based on the following proposition (the proof is omitted for brevity).

Proposition 16.3 *Let us consider a dynamical system described by*

$$x_{k+1} = A_i x_k + B_i w_k + B_{iu}\bar{u},\ |w_k| \leq f_i(x_k) \tag{16.34}$$

with $i \in \{1, 2\}$ *a switching signal in between two modes. It is considered that for each mode we can construct independently the parameterized invariant sets* $\Omega_{\mathbb{M}}^{(i)}(\alpha, \bar{u})$, $\alpha \in \mathcal{S}_{\mathbb{M}}^{(i)}(\bar{u})$ *according to Corollary 16.5.*

- *If* $x_k \in \Omega_{\mathbb{M}}^{(i)}(\alpha, \bar{u})$, *where* $\Omega_{\mathbb{M}}^{(i)}$ *is RPI with respect to the* i*-th mode of the dynamics of system (29) with* $i \in \{1, 2\}$, $\alpha \in \mathcal{S}_{\mathbb{M}}^{(i)}$ *and* $x_{k+1} \notin \Omega_{\mathbb{M}}^{(i)}(\alpha, \bar{u})$, *then a switch took place.*
- *Consider additionally that*

$$S_{\mathbb{M}}^{(i)}(\bar{u}) = [\underline{c}_1^{(i)}, \bar{c}_1^{(i)}) \cup [\underline{c}_2^{(i)}, \bar{c}_2^{(i)}) \cup \cdots \cup [\underline{c}_j^{(i)}, \bar{c}_j^{(i)}) \ldots;\ i \in \{1, 2\}, \tag{16.35}$$

and $x_0 \in \Omega_{\mathbb{M}}^{(1)}(\alpha^{(1)}, \bar{u}) \cap \Omega_{\mathbb{M}}^{(2)}(\alpha^{(2)}, \bar{u})$ *for* $\alpha^{(1)} \in [\bar{c}_j^{(1)}, \bar{c}_{j+1}^{(1)}) \subseteq \mathcal{S}_{\mathbb{M}}^{(1)}(\bar{u})$, $\alpha^{(2)} \in [\bar{c}_l^{(2)}, \bar{c}_{l+1}^{(2)}) \subseteq \mathcal{S}_{\mathbb{M}}^{(2)}(\bar{u})$. *The time-invariant mode of functioning can be identified if* $\Omega_{\mathbb{M}}^{(1)}(\underline{c}_{j+1}^{(1)}, \bar{u}) \cap \Omega_{\mathbb{M}}^{(2)}(\underline{c}_{l+1}^{(2)}, \bar{u}) = \emptyset$.

A simple way to exploit the result is to consider the auxiliary signal $\bar{u}$ as a degree of freedom for the separation of the ultimate bounds for the modes of functioning in a switching dynamical system. The basic idea is to find the auxiliary signal $\bar{u}$, minimum in norm such that the ultimate bounds, which corresponds to a given state, are separated. Taking into account Proposition 16.3, we formulate for $x \in \mathbb{R}^n$ the following optimization problem in a compact form:

$$\min_{\bar{u}} \ |\bar{u}| \tag{16.36}$$

$$\text{subject to:} \ \ \Omega_{\mathbb{M}}^{(1)}(\underline{c}_{j+1}^{(1)}, \bar{u}) \cap \Omega_{\mathbb{M}}^{(2)}(\underline{c}_{l+1}^{(2)}, \bar{u}) = \emptyset \tag{16.37}$$

where j and l are such that $x \in \Omega_{\mathbb{M}}^{(1)}(\alpha^{(1)}, \bar{u}) \cap \Omega_{\mathbb{M}}^{(2)}(\alpha^{(2)}, \bar{u})$. The optimization is nonlinear and highly correlated with the state-dependent bounds.

If $S_{\mathbb{M}}^{(i)}$ are non-connected sets, then each subinterval should be treated independently. Note however that the resultant ultimate bounds for each interval of parame-

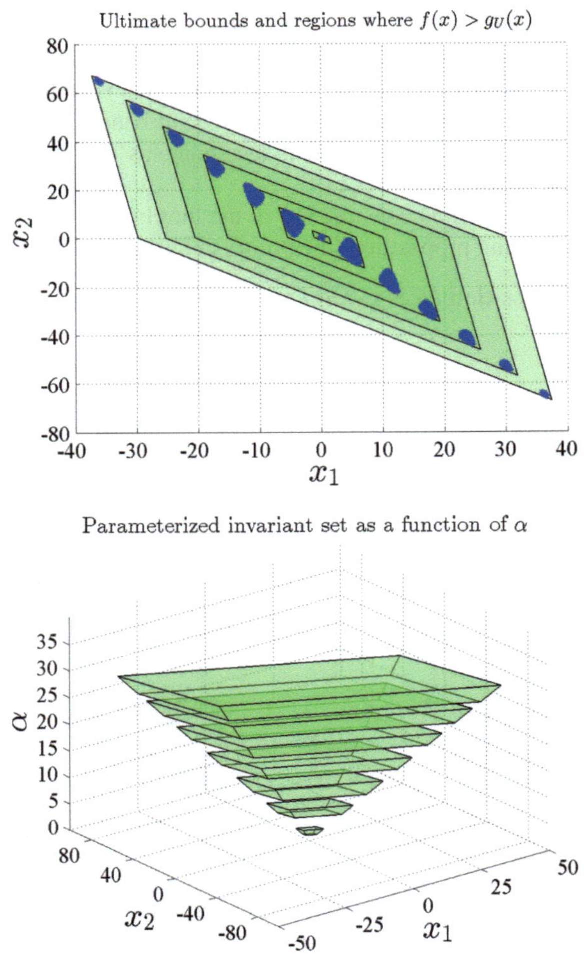

Fig. 16.10 The ultimate bounds $\Omega_{\mathbb{M}}^{(1)}(\alpha, \bar{u})$ obtained by exploiting the relationship between the state-dependent noise-bounding function $f(x)$ and the function $g_{\mathbb{U}}(x)$ on the *top*. The interval of scaling factors α corresponding to invariant sets $\Omega_{\mathbb{M}}^{(1)}(\alpha)$—*bottom* figure

ters are convex sets and that for specific classes of state-dependent bounds (convex, cone-bounded) $S_{\mathbb{M}}^{(i)}$ is a connected set, see Corollary 16.3–16.4.

Example 16.3 Consider the dynamical system (16.34) with

$$A_1 = \begin{bmatrix} 1 & 0.1 \\ -0.9 & 0 \end{bmatrix}; B_1 = \begin{bmatrix} 0 \\ 1 \end{bmatrix}; B_{1u} = \begin{bmatrix} 0 \\ 0 \end{bmatrix}; \tag{16.38}$$

$$A_2 = \frac{1}{3}\begin{bmatrix} 1 & -0.2 \\ -0.2 & 0.5 \end{bmatrix}; B_2 = \begin{bmatrix} 1 \\ 1 \end{bmatrix}; B_{2u} = \begin{bmatrix} 1 \\ 0 \end{bmatrix};$$

$$f_1(x_k) = 0.1 + |0.7 * sin(x_{k,1}) - x_{k,1}|; \; f_2(x_k) = 1.$$

The dynamics of the first mode is not affected by the exogenous signal $\bar{u}$ and as such the parameterized family of ultimate bounds (Fig. 16.10) will be described by the union of intervals:

$$S_{\mathbb{U}}^{(1)}(\bar{u}) = [1.1; 2.2) \cup [5; 6.8) \cup [8.9; 11.5) \cup [13.8; 16.1) \cup [18.5; 20.8) \\ \cup [22.8; 25.5) \cup [26.8, \ldots \tag{16.39}$$

The second mode of functioning is linear and the parameterized invariant set is given by

$$\Omega_{\mathbb{M}}^{(2)}(\alpha, \bar{u}) = \alpha\mathbb{M} \oplus (I - A)^{-1} B_{2u}\bar{u} \tag{16.40}$$

with $\alpha \in [1, \infty)$ and $\bar{u} \in \mathbb{R}$.

Solving four linear programming problems for each of the intervals in (16.39), one can find the level of $\bar{u}$ which ensures asymptotic mode detection—Fig. 16.11.

16.5 Conclusion and Further Research

The chapter revisited the ultimate bounds for linear systems in the presence of additive disturbances. Their characterization was extended from the classical case of fixed bounds to the state-dependent bounds. It is shown that a particular set-induced function can be defined in the state space and serve as a comparison for the state-dependent bounds.

In the case of multiple sources of additive disturbance affecting linear dynamics, the present study can be extended to account for elementwise state-dependent bounds by analyzing independently each column of the input matrix $B \in \mathbb{R}^{n \times m}$ and aggregating their effects.

Acknowledgments The results presented in this chapter were initiated in the framework of the French-Italian collaborative research project Galileo 2014. The authors would like to thank Prof.

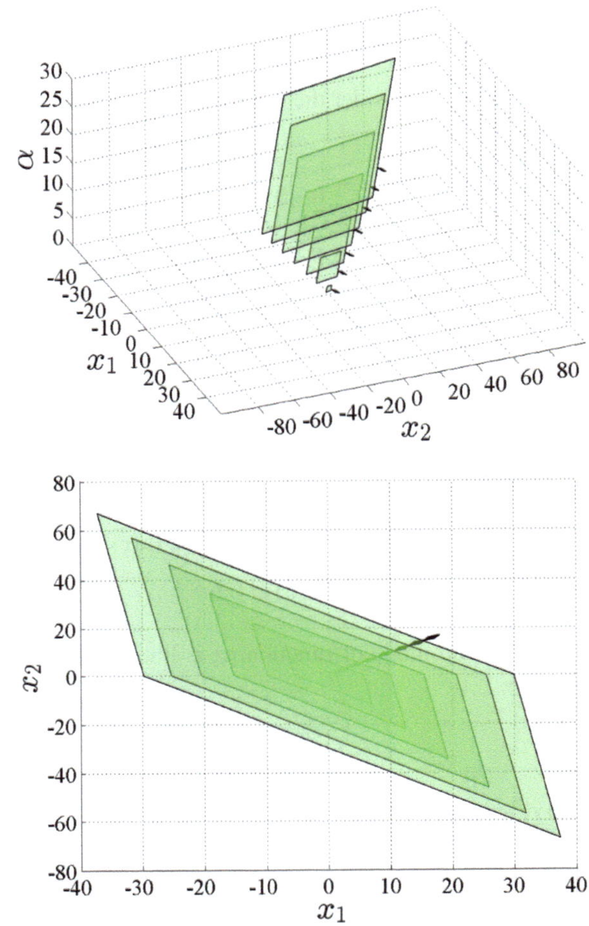

Fig. 16.11 The separation of ultimate bounds for different intervals of scaling coefficients $\Omega_{\mathbb{M}}^{(1)}(\underline{c}_{j+1}^{(1)}) \cap \Omega_{\mathbb{M}}^{(2)}(1, \bar{u}) = \emptyset$ *Bottom* Projection of the ultimate bound sets

F. Blanchini, Prof. D. Casagrane, Prof. S. Miani, and G. Giordano for the fruitful discussion on the topic. This work was supported by the People Programme (Marie Curie Actions) of the European FP7 (2007–2013) under REA grant agreement n° 626891 (FUTuRISM).

References

1. J.-P. Aubin, *Viability Theory* (Springer Science & Business Media, Heidelberg, 2009)
2. J.-P. Aubin, H. Frankowska, *Set-Valued Analysis*. (Springer, Heidelberg, 2009)
3. B.R. Barmish, G. Leitmann, On ultimate boundedness control of uncertain systems in the absence of matching assumptions. IEEE Trans. Autom. Control **27**(1), 153–158 (1982)
4. B.R. Barmish, J. Sankaran, The propagation of parametric uncertainty via polytopes. IEEE Trans. Autom. Control **24**(2), 346–349 (1979)
5. D.P. Bertsekas, I.B. Rhodes, Recursive state estimation for a set-membership description of uncertainty. IEEE Trans. Autom. Control **16**(2), 117–128 (1971)

6. G. Bitsoris, Positively invariant polyhedral sets of discrete-time linear systems. Int. J. Control **47**(6), 1713–1726 (1988)
7. F. Blanchini, Ultimate boundedness control for uncertain discrete-time systems via set-induced lyapunov functions, in *Proceedings of the 30th IEEE Conference on Decision and Control* (1991), pp. 1755–1760
8. F. Blanchini, S. Miani, *Set-Theoretic Methods in Control* (Springer, Heidelberg, 2007)
9. E. De Santis, Invariant sets: a generalization to constrained systems with state dependent disturbances, in *Proceedings of the 37th IEEE Conference on Decision and Control*, vol. 1 (1998), pp. 622–623
10. E. De Santis, On positively invariant sets for discrete-time linear systems with disturbance: an application of maximal disturbance sets. IEEE Trans. Autom. Control **39**(1), 245–249 (1994)
11. X. Feng, V. Puig, C. Ocampo-Martinez, S. Olaru, F. Stoican, Set-theoretic methods in robust detection and isolation of sensor faults. Int. J. Syst. Sci. **46**(13), 2317–2334 (2015)
12. J.D. Glover, F.C. Schweppe, Control of linear dynamic systems with set constrained disturbances. IEEE Trans. Autom. Control **16**(5), 411–423 (1971)
13. E.C. Kerrigan, Robust constraint satisfaction: invariant sets and predictive control, Ph.D. thesis. University of Cambridge, 2001
14. E. Kofman, H. Haimovich, M.M. Seron, A systematic method to obtain ultimate bounds for perturbed systems. Int. J. Control **80**(2), 167–178 (2007)
15. I. Kolmanovsky, E.G. Gilbert, Theory and computation of disturbance invariant sets for discrete-time linear systems. Math. Probl. Eng. **4**(4), 317–367 (1998)
16. A.V. Kuntsevich, V.M. Kuntsevich, Invariant sets for families of linear and nonlinear discrete systems with bounded disturbances. Autom. Remote Control **73**(1), 83–96 (2012)
17. V.M. Kuntsevich, B.N. Pshenichnyi, Minimal invariant sets of dynamic systems with bounded disturbances. Cybern. Syst. Anal. **32**(1), 58–64 (1996)
18. G. Leitmann, Guaranteed asymptotic stability for a class of uncertain linear dynamical systems. J. Optim. Theory Appl. **27**(1), 99–106 (1979)
19. G. Leitmann, On the efficacy of nonlinear control in uncertain linear systems. J. Dyn. Syst. Meas. Control **103**(2), 95–102 (1981)
20. P. McLane, Optimal stochastic control of linear systems with state- and control-dependent disturbances. IEEE Trans. Autom. Control **16**(6), 793–798 (1971)
21. S. Olaru, J.A. De Doná, M.M. Seron, F. Stoican, Positive invariant sets for fault tolerant multisensor control schemes. Int. J. Control **83**(12), 2622–2640 (2010)
22. S.V. Raković, E.C. Kerrigan, D.Q. Mayne, Reachability computations for constrained discrete-time systems with state-and input-dependent disturbances, in *Proceedings of the 42nd IEEE Conference on Decision and Control*, vol. 4 (2003), pp. 3905–3910
23. S.V. Raković, E.C. Kerrigan, K.I. Kouramas, D.Q. Mayne, Invariant approximations of the minimal robust positively invariant set. IEEE Trans. Autom. Control **50**(3), 406–410 (2005)
24. V. Reppa, S. Olaru, M.M. Polycarpou, Structural detectability analysis of a distributed sensor fault diagnosis scheme for a class of nonlinear systems, in *Proceedings of the 9th IFAC SAFEPROCESS* (Paris, France, 2015), pp. 1485–1490
25. R.M. Schaich, M. Cannon, Robust positively invariant sets for state dependent and scaled disturbances, in *Proceedings of the 54th IEEE Conference on Decision and Control* (2015)
26. M.M. Seron, X.W. Zhuo, J.A. De Doná, J.J. Martínez, Multisensor switching control strategy with fault tolerance guarantees. Automatica **44**(1), 88–97 (2008)
27. M.M. Seron, J.A. De Doná, Robust fault estimation and compensation for LPV systems under actuator and sensor faults. Automatica **52**, 294–301 (2015)
28. F. Stoican, S. Olaru, *Set-theoretic Fault-tolerant Control in Multisensor Systems* Wiley-ISTE, (2013)
29. J.L. Willems, J.C. Willems, Feedback stabilizability for stochastic systems with state and control dependent noise. Automatica **12**(3), 277–283 (1976)

Chapter 17
RPI Approximations of the mRPI Set Characterizing Linear Dynamics with Zonotopic Disturbances

Florin Stoican, Cristian Oară and Morten Hovd

Abstract In this chapter we provide a robust positive invariance (RPI) outer-approximation of the minimal RPI (mRPI) set associated to linear dynamics with zonotopic disturbances. We prove that the candidate sets considered are either RPI or become so with a scaling factor. The results base on the concomitant computation of extremal points and their extremal hyperplanes. Further, we consider the equivalence with ultimate bounds constructions and show that successive RPI representations become monotonically "tighter" as their complexity increases. The results are tested in illustrative examples.

Keywords Minimal robust positive invariant (mrpi) sets · Extremal trajectories · Zonotopic disturbances

17.1 Introduction

The notions of *positive invariance (PI)* and its robust counterpart, *robust positive invariance (RPI)* are fundamental in a large number of control topics. We may mention reference governor synthesis, [8], predictive controllers with terminal constraints [13], robust time-optimal control, [2], safe collision avoidance, [18, 19], or fault tolerant control, [22], as areas which make use of these notions.

F. Stoican (✉) · C. Oară
Department of Automation Control and Systems Engineering,
Politehnica University of Bucharest, Bucharest, Romania
e-mail: florin.stoican@acse.pub.ro

C. Oară
e-mail: cristian.oara@acse.pub.ro

M. Hovd
NTNU, Trondheim, Norway
e-mail: morten.hovd@itk.ntnu.no

S. Olaru et al. (eds.), *Developments in Model-Based Optimization and Control*,
Lecture Notes in Control and Information Sciences 464,
DOI 10.1007/978-3-319-26687-9_17

The common thread in all these applications is that the (robust) invariance permits to 'push' the difficult computations (the set construction) into an offline phase and then, at runtime, only simple computations remain to be made.

As the domain is quite large we concentrate on the rest of the chapter on the *minimal RPI (mRPI) set* which represents the smallest set which is still RPI for a given dynamic (its counterpart is the maximal RPI (MRPI) set, which is the largest invariant set respecting given constraints). Further, we will consider linear dynamics with zonotopic disturbances (a symmetric subset of the polyhedral sets). These assumptions not only allow easier computations but actually provide a theoretical framework for the results shown hereafter.

The mRPI set, barring some particular cases, has no finite description of its boundary. Consequently, a great deal of research was directed toward finding RPI approximations of the mRPI set. The existing results can be classified mainly into: (i) iterative and (ii) explicit methods. In the former class we consider all the methods which take an initial set and through a recursive iteration improve the approximation by exploiting the contractive properties of the linear mapping, this includes [1, 20] or [15]. The latter gives an explicit formulation of the boundary of the set, a classic example being the sublevels of the quadratic Lyapunov functions or ultimate bounds formulation, [11, 16]. The main parameters that characterize these methods are fidelity of the representation and numerical complexity. With respect to these requirements, the aforementioned procedures have a complementary behavior. The iterative procedures can approximate arbitrarily well ("ϵ"-outer-approximations) but have an exponential increase in the computational effort, whereas the explicit formulations are simple to deduce but are conservative.

It is then worthwhile to seek a method which combines the best aspects of both classes: fast computation and accurate representation of the mRPI set. To this end, we consider [9, 10] where an explicit representation of the boundary of the null-controllable set was provided (apparently a well-known technique in the state of the art, [17]). We have studied these constructions from the point of view of extremal representations in a previous work, [21]. We will point in the present chapter that the notions are readily adapted to the mRPI set and use the techniques described in the respective context for our own ends: the construction of an RPI approximation of the mRPI set with a minimum of computations.

We expand by exploiting additional structure of the boundary. That is, we consider both the extremal points defining the boundary and their associated extremal hyperplanes. This allows to obtain outer-approximations which may be directly RPI or, if not, become so after scaling with a finite scalar. This is in contrast with the work in [21] where all the constructions start with inner-approximations and therefore require a scaling factor regardless of the set. Further, we show that the construction is strongly related to the notion of ultimate bounds, [11], and that the RPI approximations which we obtain are increasingly 'tighter' around the mRPI set proportionally with the complexity of the representation.

Notation

Hereafter, e_i denotes the i-th standard basis vector. $[X]_i$ denotes the i-th row of matrix X. The Minkowski sum of two sets, A and B, is denoted as $A \oplus B = \{x : x = a + b,\ a \in A, b \in B\}$. $\mathrm{conv}(S)$ denotes the convex hull of set S. $\#S$ denotes the cardinality of set S.

17.2 Preliminaries

In order to introduce the main ideas of the chapter we recapitulate some basic invariance results (see, e.g., [5]). Let us consider the following LTI system in $\mathbb{R}^n$:

$$x^+ = Ax + \delta \tag{17.1}$$

where $A \in \mathbb{R}^{n\times n}$ is a Schur matrix; x and x^+ represent the current and successor states of the system, respectively and δ is a disturbance bounded by $\Delta \subset \mathbb{R}^n$ which is convex, compact, contains the origin in its nonempty interior and is bounded.

System (17.1) is used next for defining basic invariance notions [4].

Definition 17.1 Under dynamics (17.1), a set $\Omega \subset \mathbb{R}^n$ is called positive invariant if $A\Omega \subseteq \Omega$ and robust positive invariant set if $A\Omega \oplus \Delta \subseteq \Omega$. ♦

The mRPI set associated to (17.1), denoted for further use as $\Omega_\infty(A, \Delta)$, is defined as the RPI set contained in any closed RPI set. This is known to be unique, compact and—in the case when Δ contains the origin—to contain the origin, [12].

An alternative representation is to define it as the limit set to the set recurrence relation[1] $\Omega_k(A, \Delta) = A\Omega_{k-1}(A, \Delta) \oplus \Delta$ (with $\Omega_0 = \{0\}$):

$$\Omega_\infty(A, \Delta) \triangleq \lim_{k\to\infty} \Omega_k(A, \Delta) = \bigoplus_{i=1}^{\infty} A^i \Delta. \tag{17.2}$$

The set $\Omega_k(A, \Delta)$ is the k-reachable set under dynamics (17.1) starting from $\{0\}$.

As stated earlier, it is not possible to compute an exact representation of (17.2), except under restrictive assumptions such as when matrix A is nilpotent, [14]. Hence, approximations have to be used and various algorithms for the construction of RPI approximations exist in the literature. We point to the necessity that the approximation is itself an RPI set (otherwise the guarantee that a certain signal remains in a bounded area is no longer valid).

The above invariance notions hold for any kind of bounded disturbance set. Hereafter, in order to apply the techniques from [9] (and obtain meaningful results), we restrict our analysis to the case of zonotopic disturbances.

[1] The convergence to a finite limit is guaranteed by the stability of the system and by the boundedness of the perturbation.

Zonotopes represent a particular class of polytopes characterized by the following definition:

Definition 17.2 The subset of R^n with center c and set of generators $\mathcal{B} \triangleq \{b_1, \ldots, b_m\} \subset \mathbb{R}^n$, such that

$$Z(c, \mathcal{B}) = \left\{ x \in \mathbb{R}^n : x = c + \sum_{i=1}^{m} \lambda_i b_i \,, |\lambda_i| \leq 1, \; b_i \in \mathcal{B} \right\}$$

with $i = 1, \ldots, m$ is called a zonotope. ♦

A zonotope has the following properties, [7]:

(i) is closed under linear transformation:

$$LZ(c, \mathcal{B}) = Z(Lc, L\mathcal{B}); \tag{17.3}$$

(ii) is closed under Minkowski sum:

$$Z(c_1, \mathcal{B}_1) \oplus Z(c_2, \mathcal{B}_2) = Z(c_1 + c_2, \mathcal{B}_1 \cup \mathcal{B}_2). \tag{17.4}$$

These properties together with the associative property of the Minkowski addition lead to some interesting results.

Proposition 17.1 (Proposition 1, [21]) *Consider the dynamics (17.1) with a zonotopic disturbance set Δ. Then, its associated mRPI set, $\Omega_\infty(A, \Delta)$, verifies the following relations:*

(i) for any $k \geq 1$, $\Omega_\infty(A, \Delta) = \Omega_\infty(A^k, A^{k-1}\Delta \oplus \cdots \oplus \Delta)$;
(ii) given $\Delta = Z(c, \mathcal{B})$, the mRPI set can be decomposed as $\Omega_\infty(A, \Delta) = \left\{(I - A)^{-1} c\right\} \oplus \bigoplus_{i=1}^{m} \Omega_\infty(A, \Delta_i)$, where $\Omega_\infty(A, \Delta_i)$ is the mRPI set associated with dynamics $x^+ = Ax + \delta_i$ and where $\delta_i \in \Delta_i \triangleq \{\lambda_i b_i, \; |\lambda_i| \leq 1\}$.

Proof See the proof of Proposition 1, [21]. □

Note that the relation between the mRPI set described in (ii) of Proposition 17.1 also holds for the more general case where the disturbance set Δ is a Minkowski sum of sets: $\Delta = \Delta_1 \oplus \cdots \oplus \Delta_m$.

In the rest of the paper we consider a particular case of dynamics of form (17.1).

Assumption 17.1 Consider the dynamics

$$x^+ = Ax + b\lambda \tag{17.5}$$

with the following particularities:

(i) the eigenvalues of matrix $A \in \mathbb{R}^{n \times n}$ are real and positive;
(ii) the disturbance varies along a segment defined by the fixed vector $b \in \mathbb{R}^n$. With respect to (17.1) this is equivalent with $\delta \in \Delta = \{\lambda b, \ |\lambda| \leq 1\}$. ♦

Taking into account Proposition 17.1 we observe that the assumptions affecting (17.5) are manageable, that is, the dynamics can be relaxed up to a system which has: 1) a state matrix with real eigenvalues and 2) zonotopic disturbances. First, notice that since A^2 has only positive eigenvalues regardless of the sign of the eigenvalues of A (as long as they are real) it follows that we can compute the mRPI set for dynamics $A^2 \to A$ and $A\Delta \oplus \Delta \to \Delta$ and use it for the original dynamics (17.5) (see (i) for $k = 2$) without any loss in the mRPI representation. Second, using (ii) we have that it suffices to compute the mRPI sets for each of the segments composing the zonotopic disturbance and to add them (in the Minkowski sense) at the end of the procedure to recover the mRPI corresponding to dynamics (17.1).

Remark 17.1 The above discussion is important since it makes the link to more general dynamics, in the sense of a state matrix which is Schur, regardless of the sign of the matrix eigenvalues. Coupled with the cases covered earlier we cover a reasonably large number of dynamics (keep in mind that usually the perturbations are given as magnitude conditions which can be then modeled as zonotopic sets). Note also that complex eigenvalues and the continuous time case can be discussed [9]. ♦

17.3 Explicit Representation of the mRPI Set

The results described here are based on [9] where the shape of the null-controllable set of an anti-stable dynamic is studied. With some minor changes, the same reasoning can be applied to the computation of minimal RPI sets for stable dynamics. Recalling (17.2), the mRPI set for dynamics (17.5) can be seen as the collection of trajectories starting from an initial state $x_0 = \{0\}$ and taking all possible noise realizations:

$$\Omega_\infty(A, \Delta) = \lim_{k \to \infty} \bigcup_{|\lambda_l| \leq 1} \left\{ \sum_{l=1}^{k} A^{k-l} b \lambda_l \right\} \tag{17.6}$$

Not all these trajectories will result in extremal points (points which lie on the boundary of $\Omega_\infty(A, \Delta)$). A sequence of disturbances $(\lambda_1, \lambda_2, \dots)$ is called extremal if it steers the trajectory starting from $\{0\}$ into an extremal point (a vertex) of $\Omega_\infty(A, \Delta)$. The collection of these extremal sequences is denoted as Σ and is given by all "bang-bang" disturbance realizations which have at most $n - 1$ switches (see Lemma 3 of [10]):

$$\Sigma = \left\{ \pm(\lambda_1, \lambda_2, \dots) : \lambda_l = \begin{cases} 1, & 1 \le l < i_1, \\ (-1)^j, & i_j \le l < i_{j+1}, \\ (-1)^{n-1}, & i_{n-1} \le l \le \infty \end{cases} (i_1, \dots, i_{n-1}) \in \mathbf{I}^n_\infty \right\}, \tag{17.7}$$

where

$$\mathbf{I}^n_k = \{(i_1, \dots, i_{n-1}) : 1 \le i_1 \le \dots \le i_{n-1} \le k\} \tag{17.8}$$

denotes the collection of all switching sequences from $\mathbb{R}^n$ where we have at most '$n-1$' switches (to which correspond 'n' intervals along the time domain) and where the last switch happens no later than the k instant.

Lemma 17.1 *To an extremal sequence from (17.7) characterized by switching times* $\mathbf{i} = \{i_1, \dots i_{n-1}\}$ *as in (17.8) corresponds a pair of normal vector and extremal points* $\left(c_\mathbf{i}^\top, \pm x_\mathbf{i}^*\right)$ *defined as:*

(i) the normal vector $c_\mathbf{i} \in \mathbb{R}^n$ *which respects*

$$c_\mathbf{i}^\top A^{i_j} b = 0, \quad \forall i_j \in \mathbf{i}; \tag{17.9}$$

(ii) extremal point $x_\mathbf{i}^* \in \mathbb{R}^n$

$$x_\mathbf{i}^* = \left(2 \sum_{j=1}^{n-1} (-1)^j A^{i_j} + (-1)^n I \right) (I - A)^{-1} b. \tag{17.10}$$

Proof These results are derived in [9] for the continuous domain where it is noted that $c^\top e^{At} b$ has at most $n-1$ sign changes ('switches'). This holds for a matrix A with positive eigenvalues. After various computations it is shown that the disturbance sequence $\lambda(l) = \operatorname{sgn}(c^\top A^l b)$ is extremal. Based on this, in [10] it is shown that for a given switching sequence correspond both the extremal hyperplane characterized by the normal vector verifying (17.9) and the extremal point (17.10). □

Using Lemma 17.1 we provide a dual representation of the mRPI set (17.6).

Proposition 17.2 *For dynamics (17.5), the associated mRPI set is given in*

(i) half-space representation[2]*:*

$$\Omega_\infty(A, \Delta) = \{x : |c_\mathbf{i}^\top x| \le c_\mathbf{i}^\top x_\mathbf{i}^*, \quad \forall \mathbf{i} \in \mathbf{I}^n_\infty\} \tag{17.11}$$

(ii) generator representation (i.e., as convex hull of its extremal points):

$$\Omega_\infty(A, \Delta) = \operatorname*{conv}_{\mathbf{i} \in \mathbf{I}^n_\infty} \left(\pm x_\mathbf{i}^*\right) \tag{17.12}$$

[2]Hereinafter, without any loss of generality, we make the convention that $c_\mathbf{i}^\top x_i^* \ge 0$.

Proof (i) The vector $c_\mathbf{i}$ represents the direction along which we find the maximal and minimal values of the mRPI, the extremal points $\pm x_\mathbf{i}^*$. Therefore we obtain two extremal hyperplanes which bound the mRPI: $c_\mathbf{i}x = c_\mathbf{i}x_\mathbf{i}^*$ and $c_\mathbf{i}x = -c_\mathbf{i}x_\mathbf{i}^*$. Assuming, as in footnote 2 that the product $c_\mathbf{i}x_\mathbf{i}^*$ is positive, it follows that the mRPI set lies in $|c_\mathbf{i}^\top x| \leq c_\mathbf{i}^\top x_\mathbf{i}^*$ which leads to (17.11).
(ii) Since by definition the points $\pm x_\mathbf{i}^*$ are extremes of the mRPI, the relation (17.12) follows immediately. □

Several remarks are in order.

Remark 17.2 Note that there are three elements which interlock to give the boundary description: the switching sequence $\mathbf{i}$, the normal vector $c_\mathbf{i}^\top$ and the extremal point $x_\mathbf{i}^*$. Depending on the desired approach, we may: (i) provide a switching sequence $\mathbf{i}$ and introduce it in (17.9) and (17.10) to obtain $c_\mathbf{i}^\top$ and $x_\mathbf{i}^*$ respectively; or (ii) provide a vector $c_\mathbf{i}^\top$, obtain the sequence $\mathbf{i}$ for which (17.9) is verified and use it in (17.10) to obtain $x_\mathbf{i}^*$. ♦

Lastly, we recall the dynamical interpretation given to the extreme points (17.10).

Remark 17.3 Let $x_{e,\pm} = \pm(I - A)^{-1}b$, the equilibrium point of (17.5) for constant disturbance $\lambda(k) = \pm 1$. It can be noted that half of the points in (17.10) are formed by trajectories of (17.5) starting from $x_{e,+}$ under any bang-bang sequence with $n - 2$ or less switches. The other half is symmetric with the first part and consists of the trajectories starting from $x_{e,-}$ under any bang-bang sequence with $n - 2$ or less switches, see [10] for further details. In particular, this means that no point (except $x_{e,\pm}$) remains fixed on the boundary, that is, the boundary itself is invariant but it *flows* in-between the two equilibrium points, see [6] for an analysis of the mRPI boundary behavior for nonlinear dynamics. ♦

Illustrative Example

For the purpose of illustration let us consider the two-dimensional case. Keeping the notation of Lemma 17.1, the switching sequence reduces to a single switch: $\mathbf{i} = \{i_1\}$, as implied by $n - 1 = 1$. Consequently, the sequence of extremal noise realizations becomes

$$\Sigma = \left\{ \pm(\lambda_1, \lambda_2, \dots) : \lambda_l = \left\{ \begin{array}{c} 1,\ 1 \leq l < i_1, \\ -1,\ i_1 \leq l \leq \infty \end{array} \right.,\ i_1 \in \mathbf{I}_\infty^2 \right\} \tag{17.13}$$

which simplifies the formulations of both (17.9) and (17.10) into:

$$c_\mathbf{i}^\top A^{i_1} b = 0, \tag{17.14a}$$

$$x_\mathbf{i}^* = \pm \left(2A^i - I\right)(I - A)^{-1}b. \tag{17.14b}$$

Further, let us consider the LTI dynamics

$$x^+ = \begin{bmatrix} 0.91 & -0.07 \\ 0.01 & 0.79 \end{bmatrix} x + \begin{bmatrix} 0.03 \\ 0.31 \end{bmatrix} \lambda, \quad |\lambda| \leq 1. \tag{17.15}$$

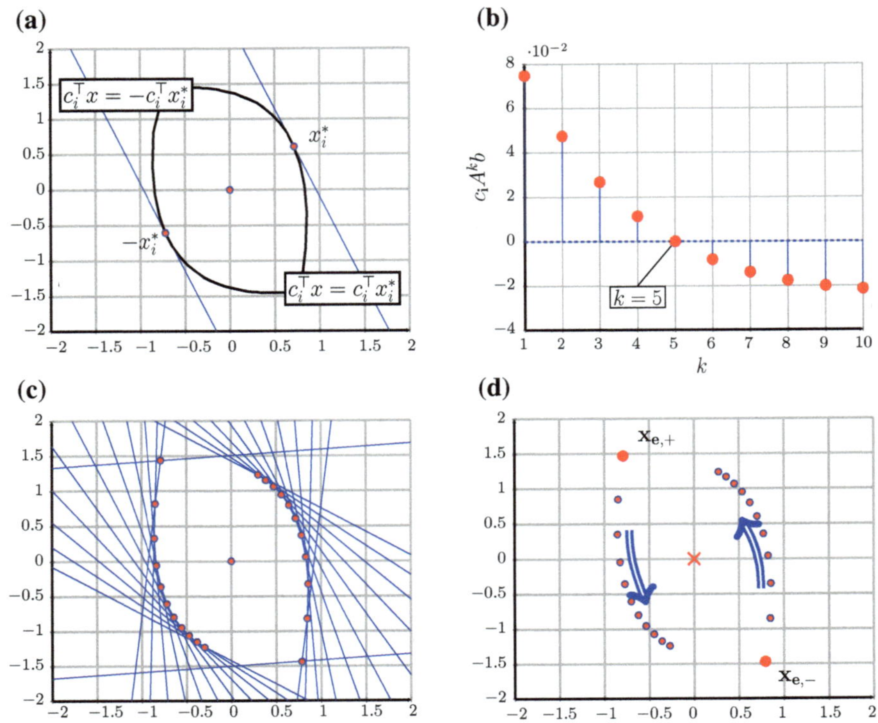

Fig. 17.1 Illustrations of the mRPI set, **a** extremal hyperplanes and extremal points, **b** mapping $cA^k b$, **c** dual representation, **d** equilibrium points and extremal trajectories

Over these dynamics we highlight the theoretical results from above. First, we take an arbitrary switching time $i_1 = 5$ and depict in Fig. 17.1a the resulting extremal hyperplanes and corresponding extremal points, as resulting from Lemma 17.1:

$$c_\mathrm{i} = \begin{bmatrix} 0.7114 \\ 0.2886 \end{bmatrix}, \quad x_\mathrm{i}^* = \pm \begin{bmatrix} -0.7176 \\ -0.6062 \end{bmatrix}.$$

Note that (17.14a) does not uniquely define $c_\mathrm{i} \in \mathbb{R}^2$—it only defines the normal vector direction but not its length. To take a unique c_i we add an arbitrary constraint, $c_\mathrm{i}^\top \cdot \begin{bmatrix} 1 & 1 \end{bmatrix}^\top = 1$. Further, as in Proposition 17.2, we depict the resulting extremal hyperplanes. We illustrate these against the 'real' mRPI set (obtained as in (17.12) with a large number of extremal points) and it can be seen that indeed (17.14a) and (17.14b) define extremal hyperplanes for the mRPI set. Moreover, we check in Fig. 17.1b that $c_\mathrm{i} A^{i_1} b$ switches its sign at the desired value $i_1 = 5$ which means that the product $c_\mathrm{i}^\top x$ is maximized for the sequence of disturbances $\{1, 1, 1, 1, -1, -1, \ldots\}$.

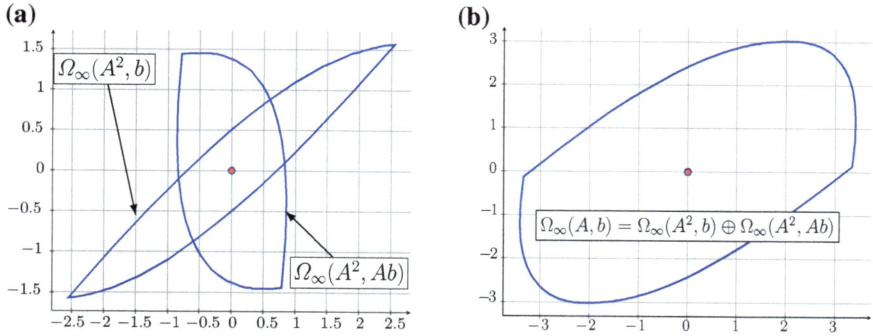

Fig. 17.2 Illustrations of the mRPI set for state matrix with negative eigenvalues, **a** the sets $\Omega_\infty(A^2, b)$, $\Omega_\infty(A^2, Ab)$, **b** the set $\Omega_\infty(A, b)$

Further we show in Fig. 17.1c the half-space construction (17.11) by enumerating pairs (17.14a)–(17.14b) for a sequence of switching instants $\mathbf{I}_{10}^2 = \{0, 1, \ldots, 10\}$ and $\{\infty\}$.

As discussed in Remark 17.3, it is possible to describe the extreme points of the mRPI as two trajectories starting from $x_{e,+}$ and $x_{e,-}$, respectively. E.g., for the two-dimensional case, it means that starting from $x_{e,-}$ and considering a constant ($n - 2 = 0$ switches) disturbance, $\lambda(l) = 1$ for any $l \geq 0$ we pass through half of the extreme points (17.14b). The converse holds when starting from $x_{e,+}$ and considering a constant disturbance $\lambda(l) = -1$ for any $l \geq 0$. We depict this in Fig. 17.1d.

As stated earlier these mRPI constructions hold for the restricted dynamics (17.5). In particular, they are not applicable for state matrices with negative eigenvalues. For illustration we take dynamics

$$x^+ = \begin{bmatrix} -1.0559 & 1.1978 \\ -0.1711 & 0.9975 \end{bmatrix} x + \begin{bmatrix} 0.03 \\ 0.31 \end{bmatrix} \lambda, \quad |\lambda| \leq 1. \tag{17.16}$$

The state matrix has a negative eigenvalue and therefore we apply Proposition 17.1 and consider two auxiliary dynamics where the state matrices are positive '$x^+ = A^2x + Ab\lambda$' and '$x^+ = A^2x + b\lambda$' (case (i) of the proposition) and then combine the resulting mRPI sets to retrieve the mRPI set for dynamics (17.16), case (ii) of the proposition. These operations are depicted in Fig. 17.2a and b, respectively.

17.4 RPI Constructions Using $\boldsymbol{\Omega}_\infty(A, \boldsymbol{\Delta})$

The main idea of this chapter is to use the constructions stated in Sect. 17.3 to provide RPI approximations of the mRPI set. With respect to previous work in [21], we consider here both the extremal points and their corresponding extremal hyperplanes.

This allows to provide half-space outer-approximations of the mRPI set (in contrast with the inner-approximations from [21]).

Having an explicit description of the mRPI's boundary does not suffice since in practice we can consider only a finite complexity (finite number of extremal hyperplanes or extremal points[3]). In [21], finite sequences of extremal points $\{\pm x_{\mathbf{i}}^*\}$ have been considered together with scaling factors which make the resulting inner-approximation into an RPI set.

Here we pursue the dual approach, that is, we consider the half-space representation and inquire about the positive robust invariance of the resulting sets. To this end, let us consider a sequence of $N \geq n$ pairs of extremal hyperplanes and extremal points $\{c_{\mathbf{i}}^\top, \pm x_{\mathbf{i}}^*\}_{\mathbf{i}\in\mathbf{I}}$, where $\mathbf{I} \subset I_\infty^n$ gathers a collection of switching sequences $\mathbf{i} = \{i_1, \ldots, i_{n-1}\}$.

We can now define the set

$$S(\mathbf{I}) = \{z : \ |c_{\mathbf{i}}^\top x| \leq d_{\mathbf{i}}, \forall \mathbf{i} \in \mathbf{I}\}, \tag{17.17}$$

where[4] $d_{\mathbf{i}} \triangleq c_{\mathbf{i}}^\top x_{\mathbf{i}}^*$ and for compactness we denote $C_{\mathbf{I}}^\top \triangleq \left[\ldots c_{\mathbf{i}} \ldots\right]^\top$ and $d_{\mathbf{I}} = \left[\ldots d_{\mathbf{i}} \ldots\right]^\top$.

Proposition 17.3 *Assuming the set* $S(\mathbf{I})$ *defined as in* (17.17) *and a scalar* $\mu \in \mathbb{R}^+$, *the set* $\mu S(\mathbf{I})$ *is RPI under dynamics* (17.5) *iff*

$$|c_{\mathbf{i}}^\top b| \leq \mu[I - |H|]_i \cdot \left[c_{\mathbf{i}}^\top C_{\mathbf{I}}^+ (I - (C_{\mathbf{I}}^\top C_{\mathbf{I}}^+ - H))^{-1} C_{\mathbf{I}}^\top \cdot \left(2\sum_{j=1}^{n-1}(-i)^j A^{i_j} + (-1)^n I\right) b\right]_{\mathbf{i}\in\mathbf{I}} \quad \forall \mathbf{i} \in \mathbf{I}. \tag{17.18}$$

Proof As per [3], it is known that the robust invariance of $\mu S(\mathbf{I})$ under dynamics (17.5) is validated iff $\exists H \in \mathbb{R}^{N\times N}$ s.t. $C_{\mathbf{I}}^\top A = H C_{\mathbf{I}}^\top$ and $|C_{\mathbf{I}}^\top b| \leq \mu \cdot (I - |H|) d_{\mathbf{I}}$. The later inequality comes from forcing $|C_{\mathbf{I}}^\top x^+| \leq d_{\mathbf{I}}$ and using the equality given a priori.

Since we assumed that $N \geq n$ it follows that $C_{\mathbf{I}}^\top$ is 'tall' and full rank which means that it accepts a left pseudoinverse defined as $C_{\mathbf{I}}^+ = (C_{\mathbf{I}} C_{\mathbf{I}}^\top)^{-1} C_{\mathbf{I}}$. This guarantees the existence of H. Note that the i-th element of $d_{\mathbf{I}}$ is given by $c_{\mathbf{i}}^\top x_{\mathbf{i}}^* = c_{\mathbf{i}}^\top (I - A)^{-1}\left(2\sum_{j=1}^{n-1}(-i)^j A^{i_j} + (-1)^n I\right) b$ where term $(I - A)^{-1} = (I - C_{\mathbf{I}}^+ H C_{\mathbf{I}}^\top)^{-1}$, which, via the 'Woodbury matrix identity' transforms into $C_{\mathbf{I}}^+ (I - (C_{\mathbf{I}}^\top C_{\mathbf{I}}^+ - H))^{-1} C_{\mathbf{I}}^\top$ which means that $|c_{\mathbf{i}}^\top b| \leq [I - |H|]_i \, [d_{\mathbf{i}}]_{\mathbf{i}\in\mathbf{I}}$ becomes (17.18). The minimal

[3] This dual approach comes from the polyhedral sets definition which allows the equivalence between generator representation and half-space representation.

[4] We assume without loss of generality that $c_{\mathbf{i}}^\top$ is taken such that $d_{\mathbf{i}} \geq 0$.

scaling factor is then found by searching iteratively for the minimal μ which validates the constraint for all $\mathbf{i} \in \mathbf{I}$. □

While μ can be easily obtained as the result of an LP optimization problem, it is less obvious what is the link between μ and the collection of switching sequences $\mathbf{I}$ or if the LP problem is feasible at all. Still, it has been shown in [21] that two particular sequences led to monotonically decreasing scaling factors, so we can expect that similar inferences can be drawn as well here. In addition, we note that any formulation of form (17.17) is an outer-approximation (i.e., it contains the mRPI set).

From Lemma 17.1 and Remark 17.2 we recall that we may choose the normals of the extremal hyperplanes as we desire. Taking into account the structure of matrix A we have the following lemma:

Lemma 17.2 *Let there be $V, \Lambda \in \mathbb{R}^{n\times n}$ such that they describe the eigendecomposition of matrix A (i.e., $A = V\Lambda V^{-1}$ where $\Lambda = \text{diag}(\ldots\lambda_i\ldots)$ and λ_i are the eigenvalues of matrix A). We take the normal vector $c_{\mathbf{i}}^\top = [V^{-1}]_i$ as the i-th row of the inverse of matrix V and we have that*

(i) the switching sequence which define $c_{\mathbf{i}}$ is

$$\mathbf{I}_e = \{i_1 = \cdots = i_{n-1} = \infty\}, \tag{17.19}$$

(ii) to which corresponds the extremal point

$$x_{\mathbf{i}} = (I - A)^{-1}b. \tag{17.20}$$

Proof Note that $A^k = V\Lambda^k V^{-1}$ which means that $c_{\mathbf{i}}^\top A^k b$ becomes $c_{\mathbf{i}}^\top V\Lambda^k V^{-1}b$. Coupled with the choice of $c_{\mathbf{i}} = [V^{-1}]_i$ we have that $c_{\mathbf{i}}^\top A^k b = \lambda^k \cdot e_i^\top V^{-1}b$ which means that the mapping never changes sign. From this it follows that the switching happens at 'infinity' and thus we reach (17.19) and (17.20). □

Corollary 17.1 *The set $S(\mathbf{I}_e)$ is RPI.*

Proof We revisit (17.18) with the normal vectors being the rows of V^{-1} (taken as in Lemma 17.2) which means that $N = n$ and therefore $C_{\mathbf{I}}^\top$ is an invertible matrix, i.e., $C_{\mathbf{I}}^+ = (C_{\mathbf{I}}^\top)^{-1} = V$ and $H = \Lambda$. This simplifies the constraint (17.18) into a more manageable form: the i-th element of the right-side term becomes $e_i^\top \mu \cdot (I - H)d_{\mathbf{I}} = [\mu \cdot (I - H)d_{\mathbf{I}}]_i$. This reduces to $\mu(1 - \lambda_i)d_i = \mu(1 - \lambda_i)c_{\mathbf{i}}^\top x_{\mathbf{i}}^*$. Noting that $(I - A)^{-1} = V(I - \Lambda)^{-1}V^{-1}$, the fact that $A^{i_j}(I - A)^{-1} = (I - A)^{-1}A^{i_j}$ and using these in the definition of $x_{\mathbf{i}}^*$ we reach

$$|c_{\mathbf{i}}^\top b| \leq \mu(1 - \lambda_i)c_{\mathbf{i}}^\top V(I - \Lambda)^{-1}V^{-1} \cdot \left(2\sum_{j=1}^{n-1}(-i)^j A^{i_j} + (-1)^n I\right) b \tag{17.21}$$

which becomes (by noting that $c_{\mathrm{i}}^{\top} V = e_i^{\top}$)

$$|c_{\mathrm{i}}^{\top} b| \leq \mu(1-\lambda_i) e_i^{\top} (I-\Lambda)^{-1} V^{-1} \cdot \left(2\sum_{j=1}^{n-1} (-i)^j A^{i_j} + (-1)^n I\right) b \tag{17.22}$$

Then, we have that (by noting that $e_i^{\top}(I-\Lambda)^{-1}V^{-1} = \dfrac{1}{1-\lambda_i} e_i^{\top} V^{-1} = \dfrac{1}{1-\lambda_i} c_{\mathrm{i}}^{\top}$)

$$|c_{\mathrm{i}}^{\top} b| \leq \mu \frac{1-\lambda_i}{1-\lambda_i} c_{\mathrm{i}}^{\top} \cdot \left(2\sum_{j=1}^{n-1} (-i)^j A^{i_j} + (-1)^n I\right) b \tag{17.23}$$

where, recalling that $c_{\mathrm{i}}^{\top} A^{i_j} b = 0$ by construction, we simplify to

$$|c_{\mathrm{i}}^{\top} b| \leq \mu \cdot (-1)^n c_{\mathrm{i}}^{\top} b. \tag{17.24}$$

With $c_{\mathrm{i}}^{\top}$ chosen such that the right side of the equation is positive, we have that the right side is in fact equal to the left side for $\mu = 1$. In other words, $S(\mathbf{I}_e)$ is RPI. □

Remark 17.4 In [11] and related papers, the 'ultimate bounds' construction is employed. For dynamics (17.5) the set is described as

$$S_{UB}(A, \Delta) = \{x : |V^{-1}x| \leq (I - |\Lambda|)^{-1} |V^{-1} b|\}. \tag{17.25}$$

With the additional assumptions made in this chapter, $\Lambda > 0$ and that $c_i^{\top} b > 0$ (17.25) reduces to

$$S_{UB}(A, \Delta) = \{x : |V^{-1}x| \leq (I - \Lambda)^{-1} V^{-1} b\} \tag{17.26}$$

which is in fact equivalent to the construction from Corollary 17.1. Further, this matches with a result from [23] where it has been shown that under certain assumptions the set (17.25) is tight (i.e., it touches the mRPI set). This is also the case here since (17.20) are extremal points of the mRPI set. ♦

As stated earlier, (17.10) provides an explicit description of the mRPI set boundary. This description involves an infinity of terms and thus cannot be used in practice. We observe that the extremal points given in (17.10) agglomerate toward either of the fixed points $x_{e,-}$ and $x_{e,+}$. Further, the points closer to the fixed points are generated with *bang-bang* subsequences appearing at a latter index in the construction (17.10). Using these two facts we have that:

(i) we can keep a finite subset of points by discarding the ones closer to $x_{e,+}$ and $x_{e,-}$;
(ii) this subset of points is defined by bang-bang sequences happening in a finite time (that is, all the $n-1$ switches are done in a finite time).

Lemma 17.3 *Under dynamics (17.5), for any $k \in \mathbb{N}$, the set*

$$R_k^e(A, \Delta) \triangleq S(\mathbf{I}_k^n \cup \mathbf{I}_e) \tag{17.27}$$

respects relation

$$R_{k+1}^e(A, \Delta) = AR_k^e(A, \Delta) \oplus \Delta. \tag{17.28}$$

Proof The proof follows from the definition of the mRPI's boundary and Remark 17.3. □

Lemma 17.3 allows to describe the following RPI constructions.

Proposition 17.4 *Under dynamics (17.5), for any $k \in \mathbb{N}$,*

(i) $R_k^e(A, \Delta)$ is RPI;
(ii) for a set

$$\Omega(\mathbf{I}) = S(\mathbf{I} \cup \mathbf{I}_e) \tag{17.29}$$

there exists $\gamma \geq 1$ s.t. $\gamma\Omega(\mathbf{I})$ is RPI.

Proof (i) We have that $\mathbf{I}_k^n \subset \mathbf{I}_{k+1}^n$ since all switching sequences where the last switch happens not later than k are automatically happening not later than $k+1$. This means that $R_{k+1}^e(A, \Delta) \subset R_k^e(A, \Delta)$ since the former contains all the extremal hyperplanes of the later. Considering Lemma 17.3 as well, it follows that $AR_k^e(A, \Delta) \oplus \Delta \subset R_k^e(A, \Delta)$ which means that $R_k^e(A, \Delta)$ is RPI.
(ii) Let there be $k = \min_l l$ s.t. $(i_1, \ldots, i_{n-1}) \in \mathbf{I}$, $i_{n-1} \leq l$, that is, we identify the latest switch from any of the switching sequences of $\mathbf{I}$. It follows then that $R_k^e(A, \Delta) \subseteq \Omega(\mathbf{I})$ since $R_k^e(A, \Delta)$ contains all the extremal hyperplanes which appear in $\Omega(\mathbf{I})$. Further, let there be a scalar $\gamma \geq 1$ s.t. $\Omega(\mathbf{I}) \subset R_k^e(A, \Delta) \oplus \frac{\gamma-1}{\gamma} A^{-1}\Delta$. It follows then that $A\Omega(\mathbf{I}) \oplus \frac{1}{\gamma}\Delta \subset AR_k^e(A, \Delta) \oplus \frac{\gamma-1}{\gamma}\Delta \oplus \frac{1}{\gamma}\Delta = AR_k^e(A, \Delta) \oplus \Delta$. Combined with the initial inclusion and case (i) it follows that $A\Omega(\mathbf{I}) \oplus \frac{1}{\gamma}\Delta \subset \Omega(\mathbf{I})$ which means there exists $\gamma \geq 1$ s.t. $\gamma\Omega(\mathbf{I})$ is RPI. □

The constructions discussed in Sect. 17.4 are obtained directly through the enumeration of extremal hyperplanes (and eventually a scaling). Thus, the computationally difficult task of calculating recursive Minkowski sums is avoided (as would be the case in the iterative procedures which assure arbitrarily close outer-approximations). To provide a measure of the storage requirements we give the next result.

Proposition 17.5 *The number of extremal hyperplanes defining the boundary of R_k^e (A, Δ) is given by:*

$$\# R_k^e(A, \Delta) = 2 \cdot \left(1 + \sum_{i=0}^{n-1} \binom{k}{i}\right). \tag{17.30}$$

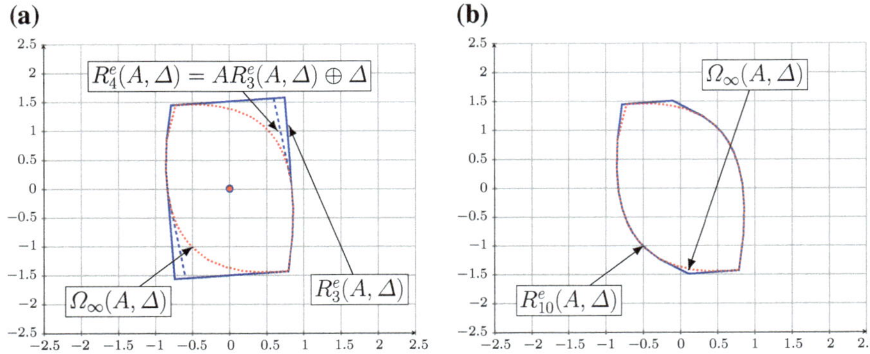

Fig. 17.3 RPI approximation of the mRPI set. **a** RPI approximation of the mRPI set using $R^e_3(A, \Delta)$. **b** RPI approximation of the mRPI set using $R^e_{10}(A, \Delta)$

Proof The number of normal vectors is actually the number of sequences of at most '$n - 1$' ordered elements taken from the first 'k' natural numbers. Thus, we choose first 0, then 1 and so forth until $n - 1$ from the first k integers and obtain the terms in (17.30), with the addition of the two extremal hyperplanes 'at infinity.' □

Illustrative Example
Let us consider again the dynamics (17.15) and apply the RPI approximations presented in Proposition 17.4. We start with the constructions (17.27) and depict in Fig. 17.3a the set $R^e_3(A, \Delta) = S(\mathbf{I}^2_3 \cup \mathbf{I}_e)$—solid blue contour. For comparison we add the set $R^e_4(A, \Delta)$—dashed blue contour and the 'real' mRPI set—dotted red contour. Note that, as per Lemma 17.3, we have that $R^e_4(A, \Delta) = AR^e_3(A, \Delta) \oplus \Delta$.

A larger number of switching sequences will result in a closer approximation of the mRPI set. In Fig. 17.3b we depict the set $R^e_{10}(A, \Delta)$—solid blue contour, against the mRPI set—dashed red contour. As it can be seen the difference between the two sets is almost invisible. To illustrate the point we consider the sequence of sets $R^e_k(A, \Delta)$ with $k \in \{0, 1, \ldots, 10\}$ and analyze the differences between consecutive elements. For this purpose we measure the set volume (in $\mathbb{R}^2$ it is actually its area) and see how it varies. In Table 17.1 the first row denotes the volume of the current set; the second row denotes the variation of the area between two consecutive sets and the third row enumerates the decrease in volume as a percentage. As it can be seen the values decrease (second to third row) monotonically. Moreover the variation is

Table 17.1 Volume variation for the $R^e_k(A, \Delta)$ sets

k	0	1	2	3	4	5	6	...	10
$\text{vol}(R^e_k(A, \Delta))$	5.62	5.19	4.88	4.66	4.50	4.38	4.30	...	4.14
$\Delta\text{vol}(R^e_k(A, \Delta))$	*	0.43	0.30	0.22	0.16	0.11	0.08	...	0.02
$\dfrac{\Delta\text{vol}(R^e_k(A, \Delta))}{\text{vol}(R^e_k(A, \Delta))}[\%]$	*	7.65	5.96	4.56	3.44	2.56	1.89	...	0.53

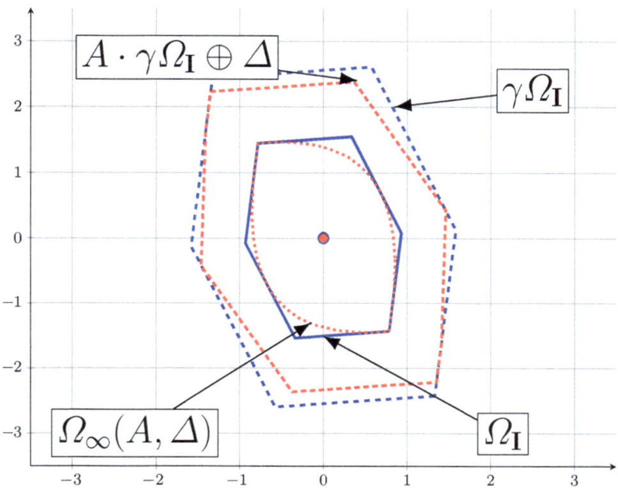

Fig. 17.4 RPI approximation of the mRPI set using $\Omega_{\mathbf{I}} = S(\{5\} \cup \mathbf{I}_e)$

quite fast which means that after a relatively small number of steps the approximation becomes 'good enough.'

Secondly, we consider the construction (17.29) and test that indeed there exists a scaling factor which makes it invariant. We take $\mathbf{I} = \{5\}$ and compute $\Omega_{\mathbf{I}} = S(\mathbf{I} \cup \mathbf{I}_e)$ and illustrate it in Fig. 17.4—solid blue contour, against the mRPI set—dotted red contour. To obtain the RPI set, we solve an LP optimization to find $\gamma = 1.6921$. The set $\gamma\Omega_{\mathbf{I}}$ and its iteration under dynamics (17.15) are depicted as well, dashed blue and densely dashed red contour, respectively.

17.5 Conclusions

In this chapter we have provided an explicit description for the boundary of the mRPI set characterizing dynamics with zonotopic disturbances. Further we have constructed RPI outer-approximations from pairs of extremal hyperplanes and extremal points and discussed their properties. We have studied the RPI property of the candidate sets considered, and, where was the case, computed the necessary scaling factors. We have also analyzed the scaling factors associated to each of them.

Further advances are possible. For example, there are results treating the null-controllable set for complex eigenvalues and the continuous time case, [9]. These should be relatively easy to adapt for the present mRPI approximations. Another direction to be explored is the analytic computation of the scaling factors as a function of the selected switching sequences.

Acknowledgments The work has been partially funded by the Sectorial Operational Programme Human Resources Development 2007–2013 of the Ministry of European Funds through the Financial Agreement [grant number POSDRU/159/1.5/S/132395] and by a grant of the Romanian National Authority for Scientific Research, CNCS - UEFISCDI, project identification number PN-II-ID-PCE-2011-3-0235.

References

1. Z. Artstein, S.V. Raković, Feedback and invariance under uncertainty via set-iterates. Automatica **44**(2), 520–525 (2008)
2. D.P. Bertsekas, I.B. Rhodes, On the minimax reachability of target sets and target tubes. Automatica **7**(2), 233–247 (1971)
3. G. Bitsoris, On the positive invariance of polyhedral sets for discrete-time systems. Syst. Control Lett. **11**(3), 243–248 (1988)
4. F. Blanchini, Set invariance in control—a survey. Automatica **35**(11), 1747–1767 (1999)
5. F. Blanchini, S. Miani, *Set-Theoretic Methods in Control* (Birkhauser, Boston, 2007)
6. J.A. De Dona, J. Lévine, On barriers in state and input constrained nonlinear systems. SIAM J. Control Optim. **51**(4), 3208–3234 (2013)
7. K. Fukuda, From the zonotope construction to the Minkowski addition of convex polytopes. J. Symb. Comput. **38**(4), 1261–1272 (2004)
8. E.G. Gilbert, I.V. Kolmanovsky, Fast reference governors for systems with state and control constraints and disturbance inputs. Int. J. Robust Nonlinear Control **9**(15), 1117–1141 (1999)
9. T. Hu, Z. Lin, L. Qiu, An explicit description of null controllable regions of linear systems with saturating actuators. Syst. Control Lett. **47**(1), 65–78 (2002)
10. T. Hu, D.E. Miller, L. Qiu, Null controllable region of lti discrete-time systems with input saturation. Automatica **38**(11), 2009–2013 (2002)
11. E. Kofman, H. Haimovich, M.M. Seron, A systematic method to obtain ultimate bounds for perturbed systems. Int. J. Control **80**(2), 167–178 (2007)
12. I. Kolmanovsky, E.G. Gilbert, Theory and computation of disturbance invariant sets for discrete-time linear systems. Math. Probl. Eng. **4**, 317–367 (1998)
13. D.Q. Mayne, Control of constrained dynamic systems. Eur. J. Control **7**(2–3), 87–99 (2001)
14. D.Q. Mayne, W.R. Schroeder, Robust time-optimal control of constrained linear systems. Automatica **33**, 2103–2118 (1997)
15. S. Olaru, J.A. De Doná, M.M. Seron, F. Stoican, Positive invariant sets for fault tolerant multisensor control schemes. Int. J. Control **83**(12), 2622–2640 (2010)
16. S. Olaru, V. Reppa, Ultimate bounds for linear discrete-time systems with state dependent disturbances. to be pusblished, in *Developments in Model-Based Optimization and Control*, ed. by S. Olaru, A. Grancharova, F.L. Pereira (Springer, Berlin, 2015)
17. C. Olech, Extremal solutions of a control system. J. Differ. Equ. **2**(1), 74–101 (1966)
18. I. Prodan, S. Olaru, C. Stoica, S.-I. Niculescu, On the tight formation for multi-agent dynamical systems, *Agents and Multi-agent Systems Technologies and Applications*, vol. LNAI 7372 (Springer, Berlin, 2012), pp. 554–565
19. I. Prodan, S. Olaru, C. Stoica, S.I. Niculescu, Predictive control for trajectory tracking and decentralized navigation of multi-agent formations. Int. J. Appl. Math. Comput. Sci. **23**(1), 91–102 (2013)
20. S.V. Raković, E.C. Kerrigan, K.I. Kouramas, D.Q. Mayne, Invariant approximations of the minimal robust positively invariant set. IEEE Trans. Autom. Control **50**(3), 406–410 (2005)
21. F. Stoican, M. Hovd, S. Olaru, Explicit invariant approximation of the mRPI set for LTI dynamics with zonotopic disturbances, in *52nd IEEE Conference on Decision and Control* (Florence, Italy, 2013), pp. 3237–3242

22. F. Stoican, S. Olaru, *Set-Theoretic Fault Tolerant Control in Multisensor Systems*, Engineering and Materials Science edition (Wiley - ISTE, London, 2013)
23. F. Stoican, S. Olaru, J.A. De Doná, M.M. Seron, Zonotopic ultimate bounds for linear systems with bounded disturbances, in *Proceedings of the 18th IFAC World Congress* (Milano, Italy, 2011), pp. 9224–9229, 28 August–2 September 2011

Index

S. Olaru et al. (eds.), *Developments in Model-Based Optimization and Control*,
Lecture Notes in Control and Information Sciences 464,
DOI 10.1007/978-3-319-26687-9